ETHER SPACE-TIME
&
COSMOLOGY

Volume 3

PHYSICAL VACUUM RELATIVITY AND QUANTUM PHYSICS

Michael C. Duffy and Joseph Levy Editors

A book dealing with experimental and theoretical studies
devoted to the exploration of the modern ether concept, evidence
of its reality and implications for modern physics.

Apeiron
Montreal

Published by C. Roy Keys Inc
4405 Rue St Dominique
Montreal Quebec H2W 2B2 Canada
http://reshift.vif.com

First published 2009

Library and Archives Canada, Cataloguing in Publication

    Ether space-time & cosmology / Michael C. Duffy and Joseph Lévy, editors.

Includes bibliographical references.
Contents: v. 1. Modern ether concepts, relativity and geometry. -- v. 2. New insights into a key physical medium. -- v. 3. Physical vacuum, relativity and quantum physics.
ISBN 978-0-9732911-8-6.--ISBN 978-0-9864926-0-0

    1. Ether (Space). 2. Cosmology. 3. Relativity (Physics). 4. Space and time. I. Duffy, Michael Ciaran II. Lévy, Joseph, 1936- III. Title: Ether space-time and cosmology.

QC177.E84 2009   530.11   C2009-900610-3

Illustration, front: Andromeda Galaxy by Robert Gendler
http://www.robgendlerastropics.com

ETHER SPACE-TIME
&
COSMOLOGY

Volume 3

Physical vacuum, relativity and quantum physics

Web Site of the Program

http://www.physicsfoundations.org

Contents of Volume 3

Foreword

The program "Ether Space-Time & Cosmology," comprises several volumes designed to inform the physics community of the resurgence of the ether in modern science. The reality of the concept and its importance, which were denied at the beginning of the 20th century, aroused renewed interest by the end of the 20th century and the beginning of the 21st century.

Even though it is not yet officially recognized, and if names such as physical vacuum, plenum or cosmic substratum are not identified with the aether, its role is now expected by many, and there is no doubt that its importance will gradually appear evident. But this has not yet happened, and, still today, it rarely receives all the attention it deserves; this is certainly why there is no total consensus today about the nature and properties of the aether.

It is therefore not surprising that the articles presented in these volumes may put forward different points of view, allowing the reader to compare the different arguments presented and make an informed choice. One of the objectives of this series of books is to progressively disclose its properties.

The introduction of the aether as a key actor in physical processes will resolve a number of paradoxes in 20th century physics which arose because it had been dismissed. There is no doubt that all fields of physics are concerned.

This third volume deals in particular with topics relating to quantum theory, dark energy and dark matter, and with critical studies of the basic assumptions of relativity.

The main approaches presented in the different articles are based on the neo Lorentzian ether concept, which assumes the existence of a preferred ether frame, and the Einstein ether concept based on the complete equivalence of all inertial frames. Of course these concepts evolved because of the developments in quantum mechanics concerning the vacuum.

For much of the experimental data, and at least before further analysis, it is commonly admitted that these approaches anticipate the same results; this is why they may seem difficult to differentiate. And another objective pursued by many authors is to find criteria capable of discriminating between them. In the subset of the Lorentzian approach, there is a current which aims at demonstrating that the conventional space-time transformations conceal hidden variables which need to be disclosed for a full understanding of physics.

"Ether Space-Time & Cosmology" is a development of the Physical Interpretations of Relativity Theory conferences, which began in 1988, in London, and which now take place in London, Moscow, Calcutta and Budapest. Details of these conferences, including names and addresses of contacts and sponsors, are given on the PIRT web site <www.physicsfoundations.org>

Michael C. Duffy & Joseph Levy

Introduction

This third volume of Ether Space-Time & Cosmology, like the preceding ones, presents works by physicists recognized for their creativity and experience. The subjects covered by them deal with various aspects of the aether concept and its relationship with different fields of fundamental physics, such as quantum theory, dark energy and dark matter, and critical study of the basic assumptions of relativity, among others. This latter study, which was initiated and partly discussed in the first volumes, is of the utmost importance for the development of physics. It is necessary to set the record straight about this point, over which there is still no consensus among physicists.

The rejection of the aether by Einstein in 1905 has had a tremendous influence on the teaching and research in physics from the beginning of the century to the present time. The fact that Einstein reversed his position in 1916 has been almost completely ignored by the physics community. Yet numerous quotations from Einstein show no reservations about his new conviction, for example:

> According to the general theory of relativity, space without ether is unthinkable.

There is no doubt that the existence of physical variables, such as permeability and permittivity, would not have any rational explanation if the vacuum was deprived of aether. This is also the case for physical processes such as the Casimir effect and the ability to transmit electromagnetic waves. It is commonly admitted today that the vacuum contains a large amount of energy; how could this be possible without the existence of a medium supporting this energy? This simple fact makes it incomprehensible the denial of the aether by a significant part of the physics community, an entirely ambiguous attitude, since, although the properties assigned to the vacuum require a substratum, this substratum is negated, such that the role and the investigations it deserves are not granted to it. Ignorance of this physical reality, which pervades the entire universe, can only have serious consequences for the development of science. It certainly explains a large part of the difficulties encountered by physics in recent decades, despite its successes.

There are numerous unresolved problems in contemporary physics. The failure to recognize the role of the aether is ignoring a key parameter involved in physical interactions and is source of error. It is comparable to ignoring the action of a magnetic field on iron. Its influence may be relatively weak on *certain* processes occurring at low absolute speeds where the gamma factor is near unity, but it becomes substantial as speeds reach a significant fraction of the speed of light.

Several experimental and theoretical studies, today, lend support to the existence of a preferred aether frame. In this case, there is no doubt that many physical laws will be revised once the aether is officially recognized, because, due to the fact that meter sticks contract and clocks do not tick at the same rate as a function of their absolute speed, the information they provide varies and they do not allow a reliable assessment of physical data, a fact to be considered especially for processes occurring at high absolute velocities.

Among the many unsolved problems, modern physics is faced with enigmas that conventional theories cannot explain and which bring their contradictions into relief.

Although quantum theory does not officially recognize the ether and therefore does not provide a thorough analysis of all its aspects, it nevertheless deduces from its equations, that the vacuum should contain an enormous amount of energy per cm^3, a fact which is untenable and inconsistent with general relativity. No satisfactory way to make these two theories mutually compatible has been found.

The big bang theory faces a number of unresolved difficulties. Explanations for the dark energy, the physical entity which seems to accelerate the expansion of the universe by its antigravitational effect, have given rise to various theories; most have recourse to the old but revived ad hoc hypothesis known as cosmological constant, whose meaning is not elucidated. Moreover, all attempts made to relate dark energy to the energy of the vacuum have failed, because the calculations based on quantum field theory attribute to this energy a value incommensurable with any value that might explain the accelerating expansion (10^{120} times greater). And even though it has eminent defenders, no explanation of dark mass has been accepted by all as yet.

We do not know with any certainty today what dark matter consists of, and we have even less idea what dark energy is. Their very existence is disputed by several authors.

Although their common goal was to unify relativity and quantum theory, string theories differ in the number of dimensions they attribute to space-time, which, in any case, is much higher than the number 4 assumed by relativity. In one of them, the number is 26 dimensions, an assumption that is difficult to justify without any experimental basis to support it.

String theory and superstring theory cannot be made compatible with a positive cosmological constant and the accelerated expansion of the universe without extreme complications. These theories are challenged today by loop quantum theory, which assumes a quantified space-time in contrast to the space-time continuum of general relativity

The "Ether space-time and cosmology" program is intended to seek solutions, based mainly on aether theory, to solve the difficulties encountered by physics today. Contrary to what is often believed, this project is not in disagreement with the ideas of eminent founders of modern physics, such as Dirac or Bell.

Of course the nature of the aether is difficult to describe, and today there are a number of hypotheses about its composition and properties; but we are convinced that research into the aether will be one of the main objectives of 21st century physics, which will enable us to solve a number of paradoxes that now obstruct its progress. Actually some have already been resolved.

I wish to express my gratitude to Michael Duffy for the decisive role he has played in promoting the development of new theories to address the problems facing modern physics.

Joseph Levy

Quotations about the aether by Maxwell, Dirac and Einstein.

Introductory Lecture on Experimental Physics
James Clerk Maxwell. (1871)

On the plenum
Is it true then that our scientific speculations have really penetrated beneath the visible appearance of things, which seem to be subject to generation and corruption, and reached the entrance of that world of order and perfection, which continues this day as it was created, perfect in number and measure and weight?
We may be mistaken. No one has as yet seen or handled an individual molecule, and our molecular hypothesis may, in its turn, be supplanted by some new theory of the constitution of matter; but the idea of the existence of unnumbered individual things, all alike and all unchangeable, is one which cannot enter the human mind and remain without fruit.
But what if these molecules, indestructible as they are, turn out to be not substances themselves, but mere affections of some other substance?
According to Sir W. Thomson's theory of Vortex Atoms, the substance of which the molecule consists is a uniformly dense plenum, the properties of which are those of a perfect fluid, the molecule itself being nothing but a certain motion impressed on a portion of this fluid, and this motion is shown, by a theorem due to Helmholtz, to be as indestructible as we believe a portion of matter to be.
If a theory of this kind is true, or even if it is conceivable, our idea of matter may have been introduced into our minds through our experience of those systems of vortices which we call bodies, but which are not substances, but motions of substance; and yet the idea which we have thus acquired of matter, as a substance possessing inertia, may be truly applicable to that fluid of which the vortices are the motion, but of whose existence, apart from the vortical motion of some of its parts, our experience gives us no evidence whatever.
It has been asserted that metaphysical speculation is a thing of the past, and that physical science has extirpated it. The discussion of the categories of existence, however, does not appear to be in danger of coming to an end in our time, and the exercise of speculation continues as fascinating to every fresh mind as it was in the days of Thales.

Extracts from an article by Paul Dirac
"Physical knowledge has advanced much since 1905, notably by the arrival of quantum mechanics, and the situation has again changed. If one examines the question in the light of present-day knowledge, one finds that the aether is no

longer ruled out by relativity, and good reasons can now be advanced for postulating an aether. . . .

We can now see that we may very well have an aether, subject to quantum mechanics and conformable to relativity, provided we are willing to consider a perfect vacuum as an idealized state, not attainable in practice.

From the experimental point of view there does not seem to be any objection to this. We must make some profound alterations to the theoretical idea of the vacuum. . . . <u>Thus, with the new theory of electrodynamics we are rather forced to have an aether</u>" P.A.M. Dirac, Nature, 1951, vol. 168, pp. 906-907

Others quotations from Dirac
When, in 1952, Leopold Infeld pointed out that one could accept all of the conclusions of Dirac's new electrodynamics without postulating an aether, Dirac responded as follows:

"Infeld has shown how the field equations of my new electrodynamics can be written so as not to require an aether. This is not sufficient to make a complete dynamical theory. It is necessary to set up an action principle and to get a Hamiltonian formulation of the equations suitable for quantization purposes, and <u>for this the aether velocity is required</u>" (Dirac 1952).

Quotations from Einstein
"According to the general theory of relativity, <u>space without ether is unthinkable</u> for in such space there not only would be no propagation of light, but also no possibility of existence for standards of space and time (measuring rods and clocks)".
Albert Einstein 1920

and

There is a weighty argument to be adduced in favour of the ether hypothesis. To deny the ether is ultimately to assume that empty space has no physical qualities whatever. The fundamental facts of mechanics do not harmonize with this view.
A. Einstein 1920

"In theoretical physics we cannot do without aether, that is, a continuum endowed with physical properties, because general relativity ... excludes immediate action at a distance. Any theory of action at a distance implies the existence of continuous fields, hence of the aether".
A. Einstein 1924.

Ether, nonlocal Quantum effects, and new tests of Fundamental Physics

G. Spavieri, J. Erazo, and A. Sanchez

Centro de Física Fundamental, Facultad de Ciencias,
Universidad de Los Andes, Mérida, 5101-Venezuela
spavieri@ula.ve

G. T. Gillies

Department of Physics, University of Virginia
Charlottesville, VA 22901-4714 USA

Abstract

The momentum of the electromagnetic (em) fields P_e appears in several areas of modern physics. In both the equations for matter and light wave propagation P_e represents the relevant em interaction. According to some modern aether models, the speed of light in moving rarefied media differs from that foreseen by special relativity. As an application of wave propagation properties, a first order optical experiment which tests the speed of light in moving rarefied gases is presented. We recall that P_e is also the link to the unitary vision of the quantum effects of the Aharonov-Bohm (AB) type and that, besides the traditional classical approaches to the limit of the photon mass m_{ph}, effects of the AB type provide a powerful quantum approach for the limit of m_{ph}. Table-top experiments based on a new effect of the AB type, together with the scalar AB effect, seem capable of yielding the limit $m_{ph} = 9,4 \times 10^{-52}$g, a value that would improve upon the results achieved with other approaches.

PACS: 03.30.+p, 03.65.Ta, 01.55.+b, 42.15.-i

1 Introduction

The *interaction* em momentum $\mathbf{P}_e$ has attracted the attention of physicists as it arises in different scenarios of modern physics involving em interactions. One of these scenarios is that of light propagation in slowly moving media [1], [2]. Another is that of a unitary view of quantum nonlocal effects of the Aharonov-Bohm (AB) type [3], [4]. More commonly, the interaction em momentum $\mathbf{P}_e$ appears as a nonvanishing quantity in em experiments involving "open" or convection currents, while $\mathbf{P}_e$ vanishes in the usual em experiments or interactions with closed currents or circuits [2], [5].

The main purpose of this article is to review the recent advances of physics involving the em momentum $\mathbf{P}_e$ and its role in the proposal of new tests or in making other advances, such as setting a new limit on the photon mass.

In the field of electromagnetism, a growing number of articles questioning the standard interpretation of special relativity have appeared [6]-[8]. Some of the authors of Refs. [6] and [7] adhere to a point of view close to the historical works of Lorentz and Poincaré, who maintained the existence of a preferred frame referred to as the aether or, archaically, as the luminiferous aether. It has been argued that these different formulations of Special Relativity are truly compatible only in vacuum, as differences may appear when light propagates in transparent moving media. Thus, Consoli and Costanzo [8], Cahill and Kitto [9], and Guerra and de Abreu [7], point out that, for experiments of the Michelson–Morley type, which are often said to have given a null-result, this is not the case and cite the famous work by Miller [10]. The claim of these authors is that the available data point towards a consistency of non-null results when the interferometer is operated in the "gas-mode", i.e., corresponding to light propagating through a gas [8] such as air or helium, for instance, even in modern maser versions of optical tests.

Moreover, tests that involve em interactions in open currents or circuits have been reconsidered lately by Indorato and Masotto [11] who point out that these experiments are not completely reliable and may be inconclusive [2]. Because of all this, physicists have recently proposed experiments about those predictions of the theory that have not been fully tested, or they have formulated untested assumptions that differ from the standard interpretation of Special Relativity [2], [5], [7], [8].

The concept of nonlocality is linked also to quantum effects of the Aharonov-Bohm type and, in this case, $\mathbf{P}_e$ is also the link to the unitary vision of these quantum effects. It turns out that, besides the traditional classical approaches to the limit of the photon mass m_{ph}, effects of the AB type represent a powerful tool for placing limits on m_{ph}.

The interesting point is that all the above-mentioned scenarios and polemical hypotheses are linked to the interaction em momentum. Therefore, throughout this article we highlight the role of $\mathbf{P}_e$ in each one of these scenarios.

2   Wave equations for matter and light waves

To elucidate the role of em momentum in modern physics, we start by considering the wave equations for matter and light waves and show how the interaction term $\mathbf{Q}$ of these equations is related to $\mathbf{P}_e$ [12]. In general, with T_{ik}^M the Maxwell stress-tensor, the covariant description of the em momentum leads to the four-vector em momentum P_e^α expressed as

$$P_e^i c = \gamma \int (c\mathbf{g} + T_{ik}^M \beta^i) d^3\sigma \qquad cP_e^0 = \gamma \int (u_{em} - \mathbf{v} \cdot \mathbf{g}) d^3\sigma \qquad (1)$$

where $\beta = v/c$, and the em energy and momentum are evaluated in a special frame $K^{(0)}$ moving with velocity $\mathbf{v}$ with respect to the laboratory frame. Here, u_{em} is the energy density and $\mathbf{S} = \mathbf{g}c$ is the energy flux or flow.

The analogy between the wave equation for light in moving media and that for charged matter waves has been pointed out by Hannay [1] and later addressed by Cook, Fearn, and Milonni [1] who have suggested that light propagation at a fluid vortex is analogous to the Aharonov-Bohm (AB) effect, where charged matter waves (electrons) encircle a localized magnetic flux [3]. Generally, in quantum effects of the AB type [3]-[4] matter waves undergo an em interaction as if they were propagating in a flow of em origin that acts as a moving medium [4] and modifies the wave velocity. This analogy has led to the formulation of the so-called *magnetic model of light propagation* [1], [2].

According to Fresnel [13], light waves propagating in a transparent, incompressible moving medium with uniform refractive index n, are dragged by the medium and develop an interference structure that depends on the velocity $\mathbf{u}$ of the fluid ($u << c$). At the time of Fresnel the preferred inertial frame was that at rest with the so-called aether, which here may be taken to coincide with the laboratory frame. The speed achieved in the aether frame is

$$v = \frac{c}{n} + (1 - \frac{1}{n^2})u \qquad (2)$$

as later corroborated by Fizeau [13]. Because of the formal analogy between the wave equation for light in slowly moving media and the Schrödinger equation for charged matter waves in the presence of the external vector potential $\mathbf{A}$ (i.e., the magnetic Aharonov-Bohm effect), both equations contain a term that is generically referred to as the interaction momentum $\mathbf{Q}$. Thus, the Schrödinger equation for quantum effects of the AB type (with $\hbar = 1$) [4] and the wave equation for light in moving media can be written [1], [2] as

$$(-i\boldsymbol{\nabla} - \mathbf{Q})^2 \Psi = p^2 \Psi. \qquad (3)$$

Eq.(3) describes matter waves if the momentum p is that of a material particle, while, if p is taken to be the momentum $\hbar k$ of light (in units of $\hbar = 1$), Eq.(3) describes light waves.

a) All the effects of the AB type discussed in the literature [3]-[4] can be described by Eq.(3), provided that the interaction momentum $\mathbf{Q}$ is related [4], [12] to $\mathbf{P}_e$, the momentum of the em fields. The AB term $\mathbf{Q} = (e/c)\mathbf{A}$ of the magnetic AB effect is obtained by taking $\mathbf{Q} = \mathbf{P}_e = \frac{1}{4\pi c} \int (\mathbf{E} \times \mathbf{B}) d^3 x'$ where $\mathbf{E}$ is the electric field of the charge and $\mathbf{B}$ the magnetic field of the solenoid. A general proof that this result holds in the *natural* Coulomb gauge, has been given by several authors [14]. For these quantum effects, the solution to Eq. (3) is given by the matter wave function

$$\Psi = e^{i\phi}\Psi_0 = e^{i \int \mathbf{Q}\cdot d\mathbf{x}} \Psi_0 = e^{i \int \mathbf{Q}\cdot d\mathbf{x}} e^{i(\mathbf{p}\cdot\mathbf{x} - Et)} \mathcal{A} \tag{4}$$

where Ψ_0 solves the Schrödinger equation with $\mathbf{Q} = 0$.

b) Calculations of the quantity $\mathbf{Q} = \mathbf{P}_e$ (1) for light in slowly moving media show [12] that the interaction term yields the Fresnel-Fizeau momentum [2]

$$\mathbf{Q} = -\frac{\omega}{c^2}(n^2 - 1)\mathbf{u}, \tag{5}$$

and that a solution of the type described in (4) may assume the forms

$$\Psi = e^{i\phi}\Psi_0 = e^{i \int \mathbf{Q}\cdot d\mathbf{x}} e^{i \int (\mathbf{k}\cdot d\mathbf{x} - \omega\, dt)} \mathcal{A}; \quad \Psi = e^{i \int (\mathbf{K}(\mathbf{x})\cdot d\mathbf{x} - \omega\, dt)} \mathcal{A} \tag{6}$$

where $\mathbf{k}$ and $\mathbf{K}(\mathbf{x})$ are wave vectors, $\omega = k\, c/n$ the angular frequency, and n the index of refraction, while Ψ_0 solves Eq.(3) with $\mathbf{Q} = \mathbf{u} = 0$.

The fact that the interaction momentum $\mathbf{Q}$ is related to $\mathbf{P}_e$ [4], [12] for both matter waves of effects of the AB type [4] and light waves in moving media [12], definitely reinforces the existing analogy between the two wave equations. Two theoretical possibilities arise [2]:

- By incorporating the phase ϕ in the term $\int \mathbf{K}(\mathbf{x})\cdot d\mathbf{x}$, the last expression on the rhs of Eq.(6) keeps the usual invariant form of the solution as required by special relativity and one finds [12] for the speed of light the result $\mathbf{v} = (c/n)\widehat{\mathbf{c}} + (1 - 1/n^2)\mathbf{u} = (c/n)\widehat{\mathbf{c}} - \mathbf{Q}(c^2/n^2\omega)$ in agreement with Eq.(2) and Special Relativity.

- Maintaining instead the analogy with the AB effect, the solution can be chosen to be represented by the first term of Eq.(6), $\Psi = e^{i\phi}\Psi_0$. In this case, the phase velocity changes but the speed of light (the particle, or photon) may not change [2]. This result is in total agreement with the analogous result for the AB effect where $\mathbf{Q} = (e/c)\mathbf{A}$ and the particle speed is left unchanged by the interaction with the vector potential $\mathbf{A}$.

The established relation (5) will be used in the next sections to tentatively express in a quantitative way the hypothesis of Consoli and Costanzo [8] referring to v, the speed of light in a moving rarefied media. With a quantitative expression for v it is then possible to formulate a dedicated experiment that tests Consoli and Costanzo's hypothesis.

2.1 Propagation of em waves in rarefied moving media

Duffy [15] has noted that the concept of an aether-like preferred frame has always incited controversy, even in modern scientific investigations aimed at exploring the less understood aspects of relativity theory. Within this scenario, Consoli and Costanzo [8], Cahill and Kitto [9], and Guerra and de Abreu [7], after a re-analysis of the optical experiments of the Michelson–Morley type, claim that the available data point towards a consistency of non-null results when light in the arms of the interferometer propagates in a rarefied gas, for instance air at normal pressure and temperature. The possibility of maintaining the existence of a preferred frame, and parallel interests in the Michelson-Morley, Trouton-Noble and related effects, arises because the coordinate transformation used, the Tangherlini transformations [16] foresee the same length contraction and time dilation as the Lorentz transformations. However, they contain an arbitrariness in the determination of the time synchronization parameter, with the consequence that there are quantities which eventually cannot be measured, such as the one-way speed of light, its measured value depending on the synchronization procedure adopted [16]. Different synchronization procedures are possible [6]-[8], fully compatible with Einstein's relativity in practice, but with very different assertions in fundamental and philosophical terms.

The original important assumption made by Consoli et al. to corroborate their claims of a non-null result and open a window for the possible existence of a preferred frame, is that light in a moving rarefied gas of refractive index n very close to 1 propagates with speed c/n, isotropically, in the preferred frame, as if the medium were not moving. Obviously, this hypothesis is in contrast with special relativity that foresees the speed (2), but it is not ruled out by the known optical tests. Thus, this assumption needs justification and experimental corroboration.

In the following, we explore possible modifications of the form of the present Fresnel-Fizeau momentum when the moving medium is composed of rarefied gas. It is not unconceivable that the effectiveness of the light delay mechanism in a compact moving medium differs, and perhaps even substantially so, from that of a non-compact moving medium, such as a rarefied gas, even if they have the same index n. As an *ad hoc* hypothesis or a tentative model of a light delay mechanism, it has been supposed [17] that its effectiveness e_f arises from the relative spatial extension V_i of the interaction em momentum $\mathbf{Q}(\mathbf{u})$ with respect to the extension V of the total em momentum. Introducing then the ratio $e_f = V_i/V$, the effective em interaction momentum, to be used in determining the speed of light in a moving media, will be assumed to be given by the effective Fresnel-Fizeau term $e_f \mathbf{Q} = (V_i/V) \mathbf{Q}$, while the resulting velocity of light in

13

moving rarefied media is

$$\mathbf{v} = \frac{c}{n}\hat{\mathbf{c}} - \frac{c^2}{n^2\omega}e_f \mathbf{Q} = \frac{c}{n}\hat{\mathbf{c}} + e_f\left(1 - \frac{1}{n^2}\right)\mathbf{u}. \tag{7}$$

The hypothesis of Consoli et al. of the speed c/n in the preferred frame for moving rarefied gases, will be justified by our model if $e_f = V_i/V$ turns out to be very small and, in this case, negligible. Calculations leading to a rough estimate of V_i/V for air at room temperature yield [17] $e_f = N_a(a^3/R^3)\,22.9 = 6.1 \times 10^{-3}$, which indeed can be neglected. Thus, our model foresees that the speed of light in moving media is actually not c/n but, quantitatively, the changes found do not alter significantly the basic hypothesis and resulting analysis by Consoli et al. [8], [9] and Guerra et al. [7].

3   Optical test in the first order in v/c

The main consequence is that, with the present hypothesis of negligible drag-like effect for moving rarefied gases, aether drift experiments of the order v/c become meaningful again. Let us consider for example the following experiment which is a variant of the Mascart and Jamin experiment of 1874 [18].

A ray of light travels from point A to point B of a segment A–====–B representing an optical interferometer. The original ray is split in two at A, which results in separate propagations through both arms (1 and 2) of the interferometer. The rays then recombine at B where the interference pattern is observed. The arms 1 and 2 are made of a transparent rarefied gases or materials with indices of refraction n_1 and n_2 and wherein the speeds are c/n_1 and c/n_2 in the preferred frame, respectively, in agreement with Consoli's et al. hypothesis [8] of the velocity expression (7) with $e_f = 0$. The laboratory frame with the interferometer and the rarefied gas is moving with speed u with respect to the preferred frame. We could be using the expressions for the speed in the moving laboratory frame resulting from the Tangherlini transformation, which can be found in [16], [7]. The calculation can also be done using the standard velocity addition from the Lorentz transformation, i.e., using the definition of *Einstein speed* as detailed in [7]. Both approaches yield the same result. The speed of light in arm 1 in the frame of the interferometer, moving with speed u with respect to the preferred frame, is respectively

$$w_1 = \frac{c/n_1 - u}{1 - u^2/c^2} \quad \text{or} \quad w_1 = \frac{c/n_1 - u}{1 - u/(c\,n_1)}, \tag{8}$$

and analogously for w_2. If L is the length of the arms, the time delay, or optical path difference, for the two rays yields, to first order in u/c,

$$\Delta t(0^o) = L\left(\frac{1}{w_1} - \frac{1}{w_2}\right) \simeq \frac{L}{c}(n_1 - n_2)\left[1 + \frac{u}{c}(n_1 + n_2)\right]. \tag{9}$$

14

In order to observe a fringe shift, the interferometer needs to be rotated, typically by 90 or 180 degrees. The time delay for 180 degrees is the same as that of Eq.(9) with u replaced by $-u$. The observable fringe shift upon rotation of the interferometer does not vanish in the first order in u/c and is related to the time delay variation

$$\delta t = \Delta t(0^o) - \Delta t(180^o) \simeq 2\frac{u}{c}(n_1^2 - n_2^2)\frac{L}{c}. \tag{10}$$

Choosing two media with different refractive index such that $n_1^2 - n_2^2$ is not too small ($> 10^{-3}$), the resulting fringe shift should be easily observable if the preferred frame exists and its speed u is not too small. Knowing the sensitivity of the apparatus, one could set the lower limit of the observable preferred speed u. Interferometers, used in advanced Michelson-Morley's type of experiments, can detect a speed u as small as $1km/s$ (a few m/s for He-Ne maser tests). Thus, this optical experiment, in passing from second order (u^2/c^2) to first order tests, should be able to improve the range of detectability of u by a factor $(c/u)(n_1^2 - n_2^2) \simeq 3 \times 10^5 \times 10^{-3} = 3 \times 10^2$, i.e., detect with the same interferometer speeds 3×10^2 smaller.

New, more refined versions of the Michelson-Morley type of experiment (including the tests using He-Ne masers.) are not suitable to test the hypothesis of Consoli et al. [8] because of the relatively low sensitivity of these experimental approaches for rarefied gases. However, as shown above, an optical test of first order in v/c becomes meaningful in this case and can provide important advantages over the second order experiments of the Michelson-Morley type.

4   Effects of the Aharonov-Bohm type and the photon mass

We have shown in the previous sections that all the effects of the AB type can be described in a unified way by the wave equation (3) where, for each one of the effects, the quantity $\mathbf{Q}$ represents the em interaction momentum (1). Both the interaction energy and momentum appear in the expression of the phase of the quantum wave function. Through the phenomenon of interference, phase variations can be measured and the observable quantity can be related to variations of the interaction em momentum or energy. In the following sections we show how the photon mass can be determined by measuring its effect on the observable phase variation via the related changes of em momentum or energy.

The possibility that the photon possesses a finite mass and its physical implications have been discussed theoretically and investigated experimentally by several researchers [19], [20]. Originally, the finite photon mass m_γ (measured in *centimeters*$^{-1}$) has been related to the range of validity of Coulomb law [19]. If $m_\gamma \neq 0$ this law is modified by the Yukawa potential $U(r) = e^{-m_\gamma r}/r$, with $m_\gamma^{-1} = \hbar/m_{ph}c = \lambda_C/2\pi$ where m_{ph} is expressed in *grams* and λ_C is the Compton wavelength of the photon.

There are direct and indirect tests for the photon mass, most of them based on classical approaches. Recalling some of the classical tests, we mention the results of Williams, Faller and Hill [19] yielding the range of the photon rest mass $m_\gamma^{-1} > 3 \times 10^9 cm$, and of Luo, Tu, Hu, and Luan [20] yielding the range $m_\gamma^{-1} > 1.66 \times 10^{13} cm$ and corresponding photon mass $m_{ph} < 2.1 \times 10^{-51} g$.

Several conjectures related to the Aharonov-Bohm (AB) effect have been developed assuming electromagnetic interaction of fields of infinite range, i.e., zero photon mass. The possibility that any associated effects become manifest within the context of finite-range electrodynamics has been discussed by Boulware and Deser (BD) [21]. In their approach, BD consider the coupling of the photon mass m_γ, as predicted by the Proca equation $\partial_\nu F^{\mu\nu} + m_\gamma^2 A^\mu = J^\mu$, and calculate the resulting magnetic field $\mathbf{B} = \mathbf{B}_0 + \hat{\mathbf{k}} \, m_\gamma^2 \, \Pi(\rho)$, that might be used in a test of the AB effect. Because of the extra mass-dependent term, BD obtained a nontrivial limit on the range of the transverse photon from a table-top experiment yielding $m_\gamma^{-1} > 1.4 \times 10^7 cm$.

After the AB effect, other quantum effects of this type have been developed, such as those associated with neutral particles that have an intrinsic magnetic [22] or electric dipole moment [4], and those with particles possessing opposite electromagnetic properties, such as opposite dipole moments or charges [4], [23]-[25]. The impact of some of these new effects on the photon mass has been studied by Spavieri and Rodriguez (SR) [26].

Based on theoretical arguments of gauge invariance, SR point out that, in analogy with the AC effect for a coherent superposition of beams of magnetic dipoles of opposite magnetic moments $\pm\mu$ [24] and the effect for electric dipoles of opposite moments $\pm d$ [25], the Spavieri effect [23] of the AB type for a coherent superposition of beams of charged particles with opposite charge state $\pm q$ is theoretically feasible. Using this effect, SR evaluate its relevance in eventually determining a bound for the photon mass m_{ph}. SR consider a coherent superposition of beams of charged particles with opposite charge state $\pm q$ passing near a huge superconducting cyclotron. The $\pm$ charges feel the effect of the vector potential $\mathbf{A}$ created by the intense magnetic field of the cyclotron and the phases of the associated wave function are shifted, leading to an observable phase shift [26]. For a cyclotron of standard size, SR show that the limit

$$m_\gamma^{-1} = 10^6 \, m_{\gamma BD}^{-1} \simeq 2 \times 10^{13} cm$$

is achievable. With their table-top experiment, BD obtained the value $m_{\gamma BD}^{-1} \simeq 1.4 \times 10^7 cm$ that is equivalent to $m_{ph BD} = 2.5 \times 10^{-45} g$. With SR approach, the new limit of the photon mass is $m_{ph} \simeq 2 \times 10^{-51} g$ which is of the same order of magnitude of that found by Luo *et al.* [20]. Of course, by increasing the size of the cyclotron a better limit could be obtained. With the standard technology available, we expect that the limit $m_{ph} \simeq 2 \times 10^{-52} g$ is not out of reach.

4.1 The scalar Aharonov-Bohm effect and the photon mass

Having exploited the magnetic AB effect in the previous section, we consider now the scalar AB effect. In this effect charged particles interact with an external scalar potential V. The standard phase φ_s acquired during the time of interaction is $\varphi_s = \frac{1}{\hbar} \int eV(t)\, dt$.

In the actual test of the scalar AB effect, a conducting cylinder of radius R is set at the potential V during a time τ while electrons travel inside it. Since no forces act on the charges it is a field-free quantum effect. If the photon mass does not vanish the potential is modified according to the Proca equation. Gauss' law is modified and the potential Φ obeys the equation $\nabla^2 \Phi - m_\gamma^2 \Phi = 0$, with the boundary condition that the potential on the cylinder be V. In cylindrical coordinates the solutions are the modified Bessel functions of zero order, $I_0(m_\gamma \rho)$ and $K_0(m_\gamma \rho)$ which are regular at the origin and infinite, respectively. It follows that the acceptable solution is

$$\Phi(\rho) \simeq V \left[1 + \frac{m_\gamma^2}{2} \left(\rho^2 - R^2 \right) \right] \tag{11}$$

where the first two terms of the expansion of $I_0(m_\gamma \rho)$ have been considered [27].

For two interfering beams of charges passing through separate cylinders, the relative phase shift is

$$\delta\varphi_s = \frac{1}{\hbar} \int e \left[V_1(t) - V_2(t) \right] dt \tag{12}$$

where $V_1(t)$ and $V_2(t)$ are the potentials applied to cylinder 1 and 2, respectively. Consequently, according to (11), the contribution of the photon mass to the relative phase shift is

$$\delta\varphi = \delta\varphi_s + \Delta\varphi = \delta\varphi_s + \frac{m_\gamma^2}{4} \left(\rho^2 - R^2 \right) \delta\varphi_s. \tag{13}$$

Obviously, this additional phase shift term vanishes if m_γ vanishes and the standard result is recovered. The last term of (13) is useful for determining the photon mass in a table-top experiment. We consider the simple case of one beam travelling inside cylinder 1 and the other travelling outside it ($V_2(t) = 0$) for a short time interval τ. It follows that $\Delta\varphi = \delta\varphi - \delta\varphi_s$ reads

$$\Delta\varphi = -\frac{em_\gamma^2}{4}\left(\rho^2 - R^2\right)V\frac{\tau}{\hbar} \tag{14}$$

where $V = V_1(t) - V_2(t)$. This is our main result for determining the photon mass limit. Interferometric experiments may be performed with a precision of up to 10^{-4}, therefore, following the approaches of BD and SR we set $\Delta\varphi = \varepsilon, \varepsilon = 10^{-4}$. Also, we suppose that the beam 1 travels nearly at the centre of the cylinder ($\rho \ll R$) so that

$$m_\gamma^{-1} = \frac{R}{2}\sqrt{\frac{\pi V\tau}{\varepsilon(h/2e)}} \tag{15}$$

The following values may be used to estimate m_γ^{-1}: $V = 10^7 V$, $h/2e = 2.067 \times 10^{-15} Tm^2$, $\tau = 5 \times 10^{-2}s$ and $R = 27cm$. The corresponding range of the photon mass is

$$m_\gamma^{-1} = 3.4 \times 10^{13} cm \tag{16}$$

which yields the improved photon mass limit $m_{ph} = 9.4 \times 10^{-52}g$, but we are left to justify the values used above for τ and R, which are both quite high.

It is interesting to compare the strength of the AB phase of the scalar AB effect with that of the magnetic AB effect. The scalar AB phase may be expressed as $eV\,\tau/\hbar$, while the magnetic AB phase is $eAL/(c\hbar)$, and the link between the particle's classical path is $L = \tau v$ with its speed v assumed to be uniform. According to special relativity, magnetism is a second order effect of electricity, therefore in normal conditions the strength of the coupling eA/c is smaller than the coupling eV. As a consequence of this, the phase variation due to the finite photon mass should be smaller in the magnetic than in the scalar AB effect. In other words, the scalar AB effect should be yielding a better limit for the photon mass than the magnetic AB effect. However, the above consideration is valid if in the actual experiments we have comparable path lengths, i.e., if $\tau \simeq L/v$. In the table-top experiment by SR [26] L is of the order of several meters. Choosing as charged particles heavy ions, for example $^{133}Cs^+$, their speed could be $27m/s$ [28]. With this speed and $L = 1.35\,m$ for the cylinder length, we get $\tau = 5 \times 10^{-2}s$ for the time of flight inside the cylinder. Since $\tau \simeq L/v$, the improved result (16) obtained by exploiting the scalar AB effect is justified.

However, the high values chosen for R and L imply that the charged particle beams will have to keep their state of coherence through an extended region of space $L = 1.35$ m during the interferometric measurement process, while in standard interferometry the path separations are of the order of at most a few cm. Thus, technological advances are needed in this respect, as also mentioned in the article by SR [26] and the references cited therein.

Nevertheless, the feasibility of testing the photon mass with the scalar AB effect has been confirmed by the recent work of Neyenhuis, Christensen, and Durfee [27], lending support to the quantum approach. Actually, one can conceive the possibility of extending to the case of the scalar AB effect the techniques of Refs. [24] and [25] for a coherent superposition of beams of charged particles with opposite charge state $\pm q$, as suggested by SR in Ref. [26]. This might lead to even better limits for the photon mass. These and other technical aspects of our table-top experimental approach will be elaborated elsewhere.

5 Conclusions

We have recalled that the interaction momenta $\mathbf{Q}$ of the effects of the AB type and of light in moving media have the same physical origin, i.e., are given by the variation of the momentum of the interaction em fields $\mathbf{P}_e$. Expecting that the effectiveness of the light delay mechanism in a rarefied gas differs from that of a compact transparent fluid or solid, we consider a tentative model of light propagation that validates the analysis made by Consoli et al. [8] and Guerra et al. [7]. As a test of the speed of light in moving rarefied media and of the preferred frame velocity, we propose an improved first order optical experiment that is a variant of the historical Mascart-Jamine experiment.

Finally, we have considered the table-top approach of Boulware and Deser for determination of the photon mass and verified its applicability to other effects of the AB type, concluding that the new effect using beams of charged particles with opposite charge state $\pm q$ for the magnetic AB effect, and the scalar AB effect are good candidates for setting limits on the photon mass. Using a quantum approach to evaluate the limit of m_{ph} with these effects, we consider realistic table-top experiments capable of yielding the limit $m_{ph} = 9,4 \times 10^{-52} g$, an important result that could either match or improve the limits achieved with recent classical and quantum approaches. In conclusion, advances in this area indicate that quantum approaches to the photon mass limit are feasible and may compete with and perhaps even surpass the traditional classical methods.

6 Acknowledgments

This work was supported in part by the CDCHT (Project C-1413-06-05-A), ULA, Mérida, Venezuela.

References

[1] J. H. Hannay, unpubl., Cambridge Univ. Hamilton prize essay (1976); R. J. Cook, H. Fearn, and P. W. Milonni, Am. J. Phys. **63** (1995) 705.

[2] G. Spavieri and G. T. Gillies, Chin. J. Phys., **45** (2007) 12; G. Spavieri, G. T. Gillies et al., in *Ether, Spacetime & Cosmology*, Vol. 1, 305-356, PD Publications, Liverpool, UK (2008).

[3] Y. Aharonov and D. Bohm, Phys. Rev. **115** (1959) 485; Y. Aharonov and A. Casher, Phys. Rev. Lett. **53** (1984) 319; G. Spavieri, Phys. Rev. Lett. **81** (1998) 1533, Phys. Rev. A **59** (1999) 3194; V. M. Tkachuk, Phys. Rev. A **62** (2000) 052112-1.

[4] G. Spavieri, Phys. Rev. Lett. **82**, 3932 (1999); Phys. Lett. A, **310**, 13 (2003); Eur. J. Phys. D, **37** (2006) 327.

[5] G. Spavieri and G. T. Gillies, Nuovo Cimento, **118** B (2003) 205; G. Spavieri, L. Nieves, M. Rodriguez, and G. T. Gillies. *Has the last word been said on Classical Electrodynamics?-New Horizons*, Rinton Press, USA (2004) 255.

[6] J. S. Bell, Speakable and Unspeakable in Quantum Mechanics, Cambridge Univ. Press, Cambridge, 1988; C. Leubner, K. Aufinger, P. Krumm, Eur. J. Phys. **13** (1992) 170. F. Selleri, Found. Phys. **26** (1996) 641; Found. Phys. Lett. **18** (2005) 325.

[7] R. de Abreu, V. Guerra, Relativity–Einstein's Lost Frame, 1st ed., Extra]muros[, Lisboa, 2005. V. Guerra and R. de Abreu, Found. Phys. **36** (200691826; V. Guerra and R. de Abreu, Phys. Lett. A **333** (2004) 355.

[8] M. Consoli, E. Costanzo, Phys. Lett. A **333** (2004) 355; astro-ph/0311576; M. Consoli, A. Pagano and L. Pappalardo, Phys. Lett. A **318** (2003) 292; M. Consoli, Phys. Rev. D **65** (2002) 105017; Phys. Lett. B **541** (2002) 307; M. Consoli and E. Costanzo, Phys. Lett. A **361** (2007) 513.

[9] R.T. Cahill, K. Kitto, physics/0205070; Apeiron **10** (2003) 104; R.T. Cahill, Apeiron **11** (2004) 53.

[10] D.C. Miller, Rev. Mod. Phys. **5**, 203 (1933) .

[11] L. Indorato and G. Masotto, Annals of Science, **46**, 117-163 (1989).

[12] G. Spavieri, Eur. Phys. J. D, **39**, 157 (2006).

[13] A. J. Fresnel, Ann. Chim. (Phys.) **9**, 57 (1818). H. Fizeau, C. R. Acad. Sci. (Paris) **33**, 349 (1851).

[14] See: T. H. Boyer, Phys. Rev. D **8** (1973) 1667; X. Zhu and W. C. Henneberger, J. Phys. A **23** (1990) 3983; G. Spavieri, in Refs. [4].

[15] M. Duffy, private comm., Int. Conf. *Physical Interpretation of Relativity Theory* 2006. See also M. Duffy and J. Levy (p. 13 and 69, respectively) in *Ether, Space-time & Cosmology*, Vol. 1, PD Publications, Liverpool, UK (2008).

[16] F. R. Tangherlini, Supp. Nuovo Cimento **20** (1961) 1; T. Sjodin, Nuovo Cimento, B **51** (1979) 299; T. Sjodin and M. F. Podlaha, Lett. Nuovo Cimento, **31** (1982) 433; R. Mansouri and R. V. Sexl, Gen. Rel. Grav., **8**, (1977) 497, 515, 809.

[17] G. Spavieri, G. T. Gillies, V. Guerra, and R. De Abreu, Eur. Phys. J. D **47**, 457–463 (2008).

[18] E. Mascart and J. Jamine, Ann. Éc. norm. **3** (1874) 336.

[19] E. R. Williams, J. E. Faller and H. A. Hill, Phys. Rev. Lett., **26**, 721 (1971); L. Davis, A.S. Goldhaber and M. M. Nieto, Phys. Rev. Lett., **35**, 1402 (1975); P. A. Franken and G. W. Ampulski, Phys. Rev. Lett, **26**, 115 (1971); J. J. Ryan, F. Accetta, and R. H. Austin, Phys. Rev. D, **32**, 802 (1985); R. Lakes, Phys. Rev. Lett. **80**, 1826 (1998).

[20] J. Luo, L.-C. Tu, Z. K. Hu, and E.-J. Luan, Phys. Rev. Lett., **90**, 081801-1 (2003); L.-C. Tu, J. Luo, and G. T. Gillies, Rep. Prog. Phys. **68**, 77 (2005).

[21] D. G. Boulware and S. Deser, Phys. Rev. Lett., **63**, 2319 (1989).

[22] G. Spavieri, Phys. Lett. A, **310**, 13 (2003).

[23] G. Spavieri, Eur. J. Phys. D, **37**, 327 (2006).

[24] K. Sangster, E.A. Hinds, S.M. Barnett, E. Riis, Phys. Rev. Lett. **71**, 3641 (1993); K. Sangster, E.A. Hinds, S. M. Barnett, E. Riis, A.G. Sinclair, Phys. Rev. A **51**, 1776 (1995); see also R.C. Casella, Phys. Rev. Lett. **65**, 2217 (1990).

[25] J.P. Dowling, C.P. Williams, J.D. Franson, Phys. Rev. Lett. **83**, 2486 (1999).

[26] G. Spavieri and M. Rodriguez, Phys. Rev. A **75**, 052113 (2007).

[27] B. Neyenhuis, D. Christensen, D. S. Durfee, Phys. Rev. Lett. **99**, 200401 (2007).

[28] Z. T. Lu, K. L. Corwin, M. J. Renn, M. H. Anderson, E. A. Cornell, and C. E. Wieman, Phys. Rev. Lett. **77**, 3331 (1996).

Relativistic Physics in Complex Minkowski Space, Nonlocality, Ether Model and Quantum Physics

E.A. Rauscher* and R.L. Amoroso[#]

*Tecnic Research Laboratories, 3500 S. Tomahawk Rd, Bld #188, Apache Junction, AZ 85219 USA
Email: bvr1001@sbcglobal.net
[#]The Noetic Advanced Studies Institute, 608 Jean St., Oakland, CA 94610-1422 USA
Email: cerebroscopic@mindspring.com

Abstract.

Many naturally occurring phenomena require theoretical treatment utilizing complex analysis by methods such as the Cauchy-Riemann relations using hyper-geometrical spaces which treat inherently nonlinear, non-dispersive, collective nonlocal resonant states of a quantum system, so as to be consistent with the nonlinearity inherent in General Relativity. Typical quantum approaches form linear approximations limiting the ability to formulate a quantum consistent Relativity Theory. The fundamental nature of remote connectedness is exemplified by Young's double slit experiment, Bell's Theorem, nonlocality, Mach's Principle and operation of a Foucault pendulum, which may imply the existence of an aether. We demonstrate that a geometric aether is not precluded by the structure of Relativity, although Einstein excluded a fixed reference aether frame. In fact, certain observable phenomena, such as Mach's Principle, Bell's Theorem and Young's double-slit experiment imply the existence of a fixed geometric spacetime aether. A basic tenet of this aether is the quantum principle of nonlocality understood in terms of the soliton-solitary wave solutions of the Schrödinger equation solved in complex relativistic Minkowski space. Formulation of the complex modified relativistic multidimensional aether allows us to understand the fundamental nature and mechanism of nonlocality allowing experimental designs to further evaluate the properties of nonlocal coherent collective phenomena. The structure of quantum theory using the Schrödinger equation, covariant Dirac equation and sine-Gordon equation are solved in a complex hyper-eight dimensional relativistic geometric space. The symmetry of this space possesses relativistic Lorentz invariance for nonlinear hyper-dimensional geometry, nonlocality, and nonlinear coherent states which are expanded in terms of quantum soliton solutions.

Keywords: Modified Relativity, Complex Minkowski spaces, Nonlocal aether and Quantum Theory

23

1. Introduction: Remote Connectedness and Coherent Collective Phenomena

The interpretation of the extremely successful quantum theory, beyond the Copenhagen Theory, carries within it the vital need for the interpretation of what it means to make a measurement, primarily in the micro-domain. The rapid and major development of the structure, content and interpretation of quantum theory in the 1920s and 1930s, as exemplified by the Heisenberg Uncertainty Principle and Schrödinger Cat Paradox and EPR Paradox [1], led to conceptual paradoxes beyond the practical application of quantum theory. The Schrödinger Cat Paradox arises over the issue of the collapse of the wave function. For two equally probable states arising from a microscopic process, only observation can determine which state exists. Heisenberg and Bohr demonstrated that the act of observation necessarily leaves the system in a new state through what Wheeler terms "participation" [2].

The Copenhagen interpretation of quantum mechanics (that is that quantum theory can only predict the probability of the outcome of a specific experiment) was an attempt to dismiss the observer's participation, but by this dismissal, we can no longer build models of reality. The test of the universality of quantum theory's experimental validity demands that nonlocality is a fundamental property of the quantum domain. The issue of nonlocality as a fundamental property of space-time has been thoroughly proven by experimental verification. If quantum theory is universally valid, nonlocality is necessarily true [3] Bohm termed the nonlocal correlations.

Einstein's dissatisfaction with the lack of determinism of quantum theory, and its probabilistic nature, led him to write the Einstein, Podolsky, and Rosen, EPR, paper. He had hoped to find a flaw in the quantum theory that would allow a way around the Heisenberg Uncertainty Principle and the probabilistic nature of the quantum theory [1] He was not the only physicist to be discontented with the, "spooky action of a distance." Bell reformulated the EPR Paradox into a rigorous that could be experimentally tested. In more recent years, the formation of the EPR Paradox terms of Bell's theorem [4], and its extensive tests which demonstrate that quantum theory holds in all known quantum experiments which necessarily demands the properties of nonlocality on the space-time manifold.

What are some of the possible implications from the quantum description, if we choose to pursue the development of models of reality and perhaps relax the pure objectivity constraint in physical theory? This issue is well exemplified by the Bell's theorem formulation of the Einstein, Podolsky, Rosen Paradox [1]. An indication that non-locality is a principle in Nature is contained in Bell's theorem, which asserts that no deterministic local "hidden variable" theories can give all the predictions of quantum theory [5]. However, most physicists believe that Nature is

24

non-deterministic and that there are no hidden variables. The prevailing view is that Bell's theorem merely confirms these ideas, rather than that it is an indication of a fundamental statement of nonlocality. However, in recent years this view has changed.

Stapp demonstrates that determinism and hidden variables occupy no essential role in the proof of Bell's theorem, which Stapp has reformulated [6]. Stapp asserts that no theory which predicts the outcome of individual observations which conform to the predictions of quantum theory can be local. A less restrictive interpretation of Bell's theorem is that either locality or realism fail [7]. Realism is a philosophical view in which external reality is assumed to exist and have definite properties fundamentally independent of an observer [7,8] Stapp presents reasonable and comprehensive models of reality in which nonlocality, as implied by Bell's theorem, is inconsistent with "objective reality," in which observable attributes can become definite, independent of the observer, the so-called "collapse of the wave function".

In Young's double slit experiment, photons from a source can go through one of two slits or openings of a slit interference arrangement. Through which slit did the photon go that blackens a photographic plate at the other end of the apparatus. The answer is not yet defined because of the Heisenberg Uncertainty Principle. One can observe interference fringes when both slits are open, but at the cost of not knowing through which slit the photon went. Or, one can know through which slit the photon went when one slit is closed, but at the cost of not having any interference fringes. Again, the choice appears to be that of the observer [9]. This experiment also brings the role of the observer into consideration and may also involve nonlocality and anticipation [10]. Modifying relativity theory by geometric complexification allows the accommodation of quantum theory within its context. Certainly, one of the most desirable consequences of scientific discovery is the ability to discover and refine our concepts of reality.

2. Complex Eight Space and the Formation of Nonlocality

We have introduced a complex multi-dimensional geometry of the four real dimensions of space, X_{Re} of x_{Re}, y_{Re}, z_{Re}, and t_{Re} and four imaginary dimensions X_{Im} of ix_{Im}, iy_{Im}, iz_{Im} and it_{Im}, such that we can describe nonlocal macroscopic connections of events that do not violate causality [11]. There are several motivations for introducing such a model; one of which relates to a possible macroscopic formulation of a Bell's theorem-like nonlocal correlation function that may have macroscopic implications, leading to a new interpretation of the Bell's theorem experimental results and to a more fundamental interpretation of the quantum measurement issue. The complex Minkowski Space M^4 is constructed so as to maintain causality and analytic continuation in the complex manifold [11-13]. The four real dimensional space can be considered a slice though the hyperdimensional complex eight space 13]. Such a model is consistent with an ubiquitous aether and with the de Sitter algebra.

Events that appear remote in four space, M^4, are contiguous in the complex eight space. We have demonstrated a fundamental relationship between the complex eight space and the topology of the Penrose twistor algebra [8,14,15]. In this model, spacetime events can become contiguous in the complex eight space, demonstrating that the remoteness of the observer and observed can become contiguous in the complex eight space in which causality conditions are preserved and the acquisition of apparent remote information is allowed.

We have solved the Schrödinger equation in the complex eight dimensional space and, with the inclusion of a relatively small, but significant, non-linear term, $g^2(\tau)$, we find soliton and solitary wave solutions. The non-linear term, which depends on the imaginary time component, overcomes dispersion giving the non-dispersive soliton waves. The coherence over remote space and time of these wave solutions relates to macroscopic-related phenomena as it does for Bell's theorem, Young's double-slit experiment and other nonlocal anticipatory phenomena. The non-linear form of the Schrödinger equation may be related formally to the non-linear gravitational phenomena and relativity theory [15] and also has implications for the quantum measurement problem [16]. Resolution of the observer-participant problem may be at hand as demonstrated by a new interpretation of the Schrödinger equation. In this formation, remote spacetime events are contiguous so that the observer has direct acquisition to remote observable information, in such a manner as to preserve causality.

3. Space-Like Remote Connectedness, Bell's Theorem and Experimental Tests

A most significant theorem about the nature of physical systems is Bell's formulation [4] of the Einstein, Podolsky and Rosen (EPR) "completeness" formulation of quantum mechanics [1]. The EPR paper was written in response to Bohr's proposal that the non-commuting operators (Heisenberg uncertainty principle) comprise a *complete* theory. (Copenhagen quantum mechanics view). Einstein, Podolsky and Rosen define a complete theory as one in which every element of the theory corresponds to an element of "reality". Bohm introduced additional quantum non-observable variables or "hidden variables," as we presented in the last section, in order to make the EPR quantum interpretation consistent with *causality* and *locality* [17]. In 1964, Bell "formalized" the EPR statement and showed mathematically that locality is incompatible with the statistical predictions of quantum mechanics. The locality or separability assumption states that the result of a measurement on one system is *unaffected* by operations on a *distant* system with which it may have previously interacted or had become entangled.

Bell discusses a specific experiment, Stern-Gerlach measurements of two spin one-half particles in the singlet spin state moving freely in opposite directions. The spins are called s_1 and s_2; we make our component spin measurements remote from each other at position P_1 and P_2, such that the Stern-Gerlach magnet at P_1 does not affect one at P_2 and vice versa. Since we can predict, in advance, the result of measuring any chosen component of s_2 at P_2 by previously measuring the same component of s_1 and P_1, this implies that the result of the second measurement must actually be predetermined by the result of the first (remote from P_2) measurement. In Bell's proof, he introduces a more complete specification of the parameters of a system by introducing parameters which in essence are hidden variables. Bell's proof is most eloquent and clear. He calculates the conditions on the correlation function for measurements at P_1 and P_2, as an inequality.

Bell's precise statement made it possible for Clauser and Horne to test the predicted statistical distribution of quantum processes and demonstrate a laboratory instance of quantum connectedness and nonlocality [18,19]. Indeed, in Clauser's calcium two photon cascade system, two photodetectors remote from each other are each preceded by independent, randomly-oriented polarizers. The statistical predictions of quantum mechanics is borne out in the measurements made at the two photomultiplier tubes (PMT). In Bell's words, "there must mechanism whereby the setting of one measuring device can influence the reading of another instrument, however remote." Moreover, the signal involved must propagate instantaneously so that a theory could not be Lorentz invariant. Lorentz invariance, in the usual sense, implies $v \leq c$. Feinberg discusses the relationship between Lorentz invariance and superluminal signals which he found not to be incompatible). It is not completely clear that superluminal signals must be invoked to drive Bell's theorem [20] but this author has demonstrated that indeed Bell's theorem demands $v \leq c$ [21].

The conclusion from Bell's theorem, then, is that any hidden variable theory that reproduces all statistical predictions of quantum mechanics must be nonlocal (implying remote connectedness) and the possibility of an aether-like structure. Of course, thus far, all these formulations involve microproperties only, but some recent formulations seem to imply possible macroscopic consequences of Bell's theorem as well. It is believed that the key lies in the formulation of the correlation function which represents the interconnectedness of previously correlated events. Stapp has demonstrated that hidden variable theory is not necessary to the formalism of Bell's theory [22]. Stapp has recently expanded the pragmatic view of

Bell's theorem and discusses the role of the macroscopic detection apparatus as well as the possible role of superluminal signals. He explores both cases for superluminal propagation or subluminal connection issuing from the points in common to the backward light cones coming from the two regions.

We can write a general correlation function, for example, for an angle θ between polarization vectors in two polarizers as $C(\theta) = \frac{1}{2} - \frac{1}{2}\cos 2\theta = \cos^2 \theta$ for J. Clauser's experiment, or for odd integers we can write $nC(\theta) - C(n\theta) - (n - 1) \leq 0$, which is Bell's inequality – specifically for n = 3; $3C(\theta) - C(3\theta) - 2 \leq 0$. We can write in general $C(\theta) = \frac{1}{2} + g\cos 2\theta$ where g is determined by the particular experiment under consideration. The magnitude correlation function constant, g, relates to the type of nonlocal correlation experiment. For $g = \frac{1}{2}$, we have the Bell's theorem photon-photon correlation.

An exciting and extremely important test of Bell's inequality was designed and implemented by Clauser et al. in the early 1970s at the University of California, Berkeley, which author (EAR) had the privilege to observe [7,18], as well as the work of Aspect, et. al. at the University of Orsay, France [23]. These extremely well designed and implemented experiments unique remote causal connections and nonlocality on the spacetime manifold. Photon correlations have been observed over meter distances in the Aspect experiment. More recently, Gisin et al. has tested Bell's inequality over kilometer distances [24,25]. Rauscher and Targ apply the complex eight space and its description of nonlocality, such as exemplified in the Bell's theorem tests, to the nonlocal aspects of consciousness [26,27]. Precognition and retrocognition comprise an anticipatory system. Clauser expressed his impression of these nonlocality experiments to the above authors. He stated that quantum experiments have been carried out with photons, electrons, atoms, and even 60-carbon-atom Bucky balls. He said that, "it may be impossible to keep anything in a box anymore." Bell emphasizes, "no theory of reality compatible with quantum theory can require spatially separate events to be independent." That is to say, the measurement of the polarization of one photon determines the polarization of the other photon at their respective measurement sites. Stapp also stated to one of us (EAR) that these quantum connections could be the, "most profound discovery in all of science" [26].

Bohm has conducted research on the concept of the undivided nonlocal whole, and Bohm and Hiley [3], having extensive discussions with one of us (EAR). Also Wheeler's fundamental explanations on the concept of nonlocal interactions and the foundations of the quantum theory in publications and discussion with author (EAR) are fundamental to anticipatory systems [2]. Wheeler's design of his delayed choice experiment demonstrates that, according to quantum theory, the choice to measure one or another pair of complimentary variables at a given time can apparently affect the physical state of events for considerable periods of time *before* such a decision is made. Such complimentary variables are typically momentum and distance, or in Wheeler's experiment refer to the dual wave and particle nature of light, as observed in a two slit interference apparatus.

Wigner attempted to formulate a nonlinear quantum theory and stated support of the complex Minkowski eight space which has macroscopic nonlocal consequences [28]. The fundamental issue he addressed is when are where does the measurement observation occur for a stochastic causal system. Earlier, von Neumann had suggested a sequence of observations, or von Neumann chain. Wigner also addresses the issue of multiple observers of a quantum generated event [28].

4. Complex Eight Space and Nonlocal Anticipatory Systems

Within the context of a fundamental observation and theoretical formalism of nonlocality and anticipation, such a theory must be consistent with the main body of the principles of physics. The major universal principles are used to determine the structure and nature of physical laws and act as constraints on physical phenomena. These are *Poincaré invariance* and its corollary, *Lorentz invariance* (which expresses the space-time independence of scientific laws in different frames of reference), *analyticity* (which is a general statement of causality conditions in the complex space), and *unitarity* (which can be related to the conservation of physical quantities such as energy and momentum).

These principles apply to microscopic as well as macroscopic phenomena. The quantum description of elementary particles has led to the formulation of the analyticity principle in the complex momentum plane [29]. Complex geometries occupy a vital role in many areas of physics and engineering. Analyticity relates to the manner in which events are correlated with each other in the space-time metric (that is, causality). When we apply this critical principle to the complex eight-dimensional space we can reconcile nonlocality and anticipatory systems with physics, without violating causality. It has been mathematically demonstrated that the equations of Newton, Maxwell, Einstein, and Schrödinger are consistent with the eight-dimensional complex space described here [12-14,20,30-34]. In addition, nondispersive solitary wave solutions are obtained for the complex eight space Schrödinger equation [21].

The least number of dimensions that has the property of nonlocality and that is consistent with Poincaré invariance or Lorentz invariance is eight dimensions. In this space, each physical *spatial* distance has an imaginary temporal counterpart, such that there is a zero *spatial* separation in the higher dimensional space. Likewise for every real physically temporal separation, there is a counterpart imaginary *spatial* separation that subtracts to zero on the metric, allowing access to future information and bringing it into the present, which acts as an anticipatory system. Consistency with the relativity theory is assured and a relativistic aether model can be, in part, formulated.

We have also demonstrated the properties of nonlocality with the formalism of Maxwell's equations in complex eight space [29-31]. In the next section, we present a brief description of the complex Minkowski eight space and its properties and implications. Then we present in section 4 the solution to the Schrödinger equation in complex eight space and nonlinear recursive solutions which are consistent with and explanatory of Bell's nonlocality and the general principles of nonlocality and anticipatory phenomena in the quantum domain.

Both special and general relativistic forms of the complex eight space have been formulated and examined in applications [11,13,15]. We present a brief description of the formalism which we utilize to solve the Schrödinger equation. We express the solution of the Schrödinger equation in complex eight space. In the usual four dimensional Minkowski space, where Einstein considered time as the fourth dimension of three space, this is formulated as a four dimensional light cone diagram displayed in two dimensions, in which the ordinate is the time coordinate and the abscissa is the space coordinate, representing the three dimensions of space as $X = x,y,z$. the sides of the forward and backward light cone form signal connections at the velocity of light, c, and the apex of the cone represents "now" space-time. Inside the forward, future time, and backward, past time, light cone event connections are represented by signaling for $v < c$ called time-like signaling. The space-like signaling outside of the light cone represents greater than light speed, or space-like signaling, $v > c$.

Bell's nonlocality test implies space-time signaling and hence, even though experimentally well-verified, some physicists find nonlocality unsatisfactory. However, as we know, the truth is in what Nature shows us, not in our particular biased beliefs. The complex eight space formalism not only yields a mathematical description of nonlocality, but the complexified Schrödinger formalism gives a detailed picture of the quantum nonlocality that is consistent with the statistical nature of the quantum theory, but is also consistent with the formalism of relativity. Apparent superluminal signaling can occur for the connection of correlated past time events that remain correlated for present measurement and are related by luminal velocity of light signaling in the complex eight space. Also, this formalism allows anticipatory measurements such as in the Aspect, Gisin experiments and Wheeler's delayed choice experimental proposal.

The conditions for causality in the usual four space, distance ds^2 is invariant and given as the relativistic metrical space $ds^2 = g_{ab}dx^a dx^b$ where the indices a and b run 1 to 4. We use the metrical signature (+,+,+,-) for the three spatial and one temporal component in the metric g_{ab}. This metric is expressed as a sixteen element four by four matrix which represents a measure of the form and shape of space. This is the metric defined on the light cone, connecting time-like events. A second four imaginary dimensional space light cone can be constructed, which intersects with the usual four dimensional Minkowski space, can be constructed. These two light cones coincide in there "now" space-time realities. The complexified eight space metric is denoted as M^4 because it represents the complexification of four space-time dimensions. The complex space is expressed in terms of the complex eight space variable Z^μ, where $Z^\mu = X_{Re}^\mu + iX_{Im}^\mu$, and $Z^{*\nu}$ is the complex conjugate of Z^μ so that $Z^\nu = X_{Re}^\nu - iX_{Im}^\nu$. We now form the complex eight space differential line element $dS^2 = \eta_{\mu\nu}dZ^\mu dZ^{*\nu}$ where the indices run 1 to 8 and $\eta_{\mu\nu}$ is the complex metric of eight space. The generalized complex metric in the previous equation is analogous to the usual Einsteinian four-space metric. In our formalism, we proceed by extending the usual four dimensional Minkowski space into a four complex dimensional space-time. This new manifold (or spacetime structure) is analytically expressed in the complexified eight space.

As stated before we represent X_{Re} by x_{Re}, y_{Re}, z_{Re} and t_{Re} i.e. the dimensions of our usual four space. Likewise, X_{Im} represent the four additional imaginary dimensions of x_{Im}, y_{Im}, z_{Im}, and t_{Im}. Hence, we represent the dimensions of our complex space as Z^μ or x_{Re}, y_{Re}, z_{Re}, t_{Re}, x_{Im}, y_{Im}, z_{Im}, and t_{Im}. These are all real quantities. It is the i before the x_{Im}, etc. that complexifies the space. We write the expression showing the separation of the real and imaginary parts of the differential form of the metric in general relativistic notation: $dZ^\mu dZ^{*\mu} = \left(dX_{Re}^\mu\right)^2 + \left(dX_{Im}^\mu\right)^2$. We can write for real and imaginary space and time components in the special relativistic formalism.

$$ds^2 = \left(dx^2_{Re} + dx^2_{Im}\right) + \left(dy^2_{Re} + dy^2_{Im}\right) + \left(dz^2_{Re} + dz^2_{Im}\right) - c^2\left(dt^2_{Re} + dt^2_{Im}\right) \quad (1)$$

We now use lower case x and t for the three dimensions of space and on of time. We represent the three real spatial components dx_{Re}, dy_{Re}, dz_{Re} as dx_{Re} and the three imaginary spatial components dx_{Im}, dy_{Im}, dz_{Im} as dx_{Im} and similarly for the real time component $dt_{Re} = dt$, the ordinary time and imaginary time component dt_{Im} remains

dt_{Im}. We then introduce complex space -time coordinates as a space-like part x_{Im} and time-like part t_{Im} as imaginary parts of x and t. Now we have the invariant line elements as,

$$s^2 = |x'|^2 - c^2|t'|^2 = |x'|^2 - |t'|^2 \qquad (2)$$

again where we take the units $c^2 = c = 1$ which is made for convenience

$$x' = x_{\text{Re}} + ix_{\text{Im}} \qquad (3a)$$

and

$$t' = t_{\text{Re}} + it_{\text{Im}} \qquad (3b)$$

as our complex dimensional components so that [11,27]

$$x'^2 = |x'|^2 = x_{\text{Re}}^2 + x_{\text{Im}}^2 \qquad (4a)$$

and

$$t^2 = |t|^2 = t_{\text{Re}}^2 + t_{\text{Im}}^2 . \qquad (4b)$$

Recalling that the square of a complex number is given as,

$$|x'|^2 = x'x'^* = (x_{Re} + ix_{Im})(x_{Re} - ix_{Im}) \qquad (5)$$

where the modulus of a complex number squared is $|x'|^2 = x_{\text{Re}}^2 + x_{\text{Im}}^2$ so that x_{Re} and x_{Im} are real numbers. This is a very important point, as we can only measure events described in terms of the mathematics of real numbers. Therefore, we have the eight-space line element where spatial and temporal distances are taken from the origin.

$$s^2 = x_{\text{Re}}^2 - c^2 t_{\text{Re}}^2 + x_{\text{Im}}^2 - c^2 t_{\text{Im}}^2 \qquad (6a)$$

$$s^2 = x_{\text{Re}}^2 - t_{\text{Re}}^2 + x_{\text{Im}}^2 - t_{\text{Im}}^2 \qquad (6b)$$

Causality is defined by remaining on the right cone, in real space-time as,

$$s^2 = x_{\text{Re}}^2 - c^2 t_{\text{Re}}^2 = x_{\text{Re}}^2 - t_{\text{Re}}^2 \qquad (7)$$

using the units of $c = 1$. Then the generalized causality in complex space – time is defined by

$$s^2 = x_{\text{Re}}^2 - t_{\text{Re}}^2 + x_{\text{Im}}^2 - t_{\text{Im}}^2 \qquad (8)$$

where the coordinates in complex eight space can be represented by $x_{\text{Re}}, t_{\text{Re}}, x_{\text{Im}}, t_{\text{Im}}$ on two generalized light cones eight dimensional space [11,12,31].

We calculate the interval separation between two events or occurrences, Z_1 and Z_2 with real separation $\Delta x_{Re} = x_{Re,2} - x_{Re,1}$ and imaginary separation $\Delta x_{Im} = x_{Im,2} - x_{Im,1}$. Then the distance along the line element is $\Delta s^2 = \Delta\left(x_{Re}^2 + x_{Im}^2 - t_{Re}^2 - t_{Im}^2\right)$ and it must be true that the line interval is a real separation. The spatial and temporal distances that are generalized and are not taken only from the origin, but from any two points in space and time. Then,

$$\Delta s^2 = \left(x_{Re,2} - x_{Re,1}\right)^2 + \left(x_{Im,2} - x_{Im,1}\right)^2 - \left(t_{Re,2} - t_{Re,1}\right)^2 - \left(t_{Im,2} - t_{Im,2}\right)^2 \qquad (9a)$$

Or we can write equation 9a as:

$$\Delta s^2 = \left(x_{Re,2} - x_{Re,1}\right)^2 + \left(x_{Im,2} - x_{Im,1}\right)^2 - \left(t_{Re,2} - t_{Re,1}\right)^2 - \left(t_{Im,2} - t_{Im,1}\right)^2 \qquad (9b)$$

In equation (9b), the upper left diagonal term $\left(x_{Re,2} - x_{Re,1}\right)^2$ is be offset or "cancelled" by the lower right diagonal term $-\left(t_{Im,2} - t_{Im,1}\right)^2$, and the lower left diagonal term $-\left(t_{Re,1} - t_{Re,1}\right)^2$ is off set by the upper right diagonal term $\left(x_{Im,2} - x_{Im,1}\right)^2$. Because of the relative signs of the real and imaginary space and time components, and in order to achieve the causality connectedness condition between the two events, or $\Delta s^2 = 0$, we must "mix" space and time. That is, we use the imaginary time component to effect a zero space separation. We identify $\left(x_{Re,1}, t_{Re,1}\right)$ with a subject receiver remotely perceiving information from an even target $\left(x_{Re,2}, t_{Re,1}\right)$.

The nonlocality of Bell's theorem and its experimental test involves a real physical separation $\Delta x_{Re} = x_{Re,2} - x_{Re,1} \neq 0$ and can either involve a current time observation such that $\Delta t_{Re} = t_{Re,2} - t_{Re,1} = 0$ or a anticipatory time interval $\Delta t_{Re} = t_{Re,2} - t_{Re,1} > 0$. The case where there is no anticipatory time element $\Delta t_{Re} = 0$. The simplest causal connection then is one in which $\Delta x_{Im} = 0$, and we have,

$$\Delta s^2 = 0 = (x_{Re,2} - x_{Re,1})^2 - (t_{Im,2} - t_{Im,1})^2 .\qquad(10)$$

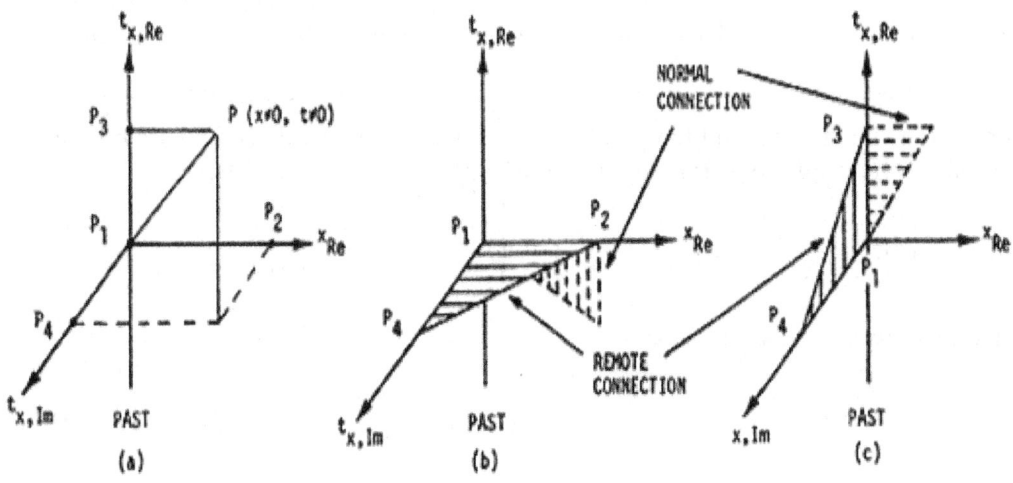

Figure 1. We represent the location of four points in the complex manifold. In figure 1a, point P_1 is the origin, and P is a generalized point which is spatially and temporally separated from P_1. In figure 1b, the Points P_1 and P_2 are separated in space but synchronous in time. This could be a representation of real-time nonlocal spatial separation.. In figure 1c, points P_1 and P_3 are separated temporally and spatially contiguous. This represents an anticipatory temporal connection.

These conditions are illustrated in figure 1. In figure 1a we represent a generalized point $P(x_{Re}, t_{Re}, t_{Im})$, displaced from the origin which is denoted as P_1. This point can be projected on each dimension x_{Re}, t_{Re} and t_{Im} as points P_2, P_3, and P_4 respectively. In Figure 1b, we denote the case where a real-time *spatial* separation exists between points, P_1 and P_2 on the x_{Re} axis, so that $\Delta x_{Re} \neq 0$, and there is no anticipation, so that $t_{Re} = 0$, and access to imaginary time t_{Im}, nonlocality can occur between the P_1 to P_4 interval, so that $\Delta t_{Im} \neq 0$. Then, our metric gives us $\Delta s^2 = 0$, where nonlocality is the contiguity between P_1 and P_2 by its access to the path to P_4. By using this complex path, the physical spatial separation between P_1 and P_2 becomes equal to zero, allowing direct nonlocal connectedness of distant spatial locations, observed as a fundamental nonlocality of remote connectedness on the spacetime manifold.

Figure 1c represents the case where anticipation occurs between P_1 and an apparent future anticipatory accessed event, P_3 on the t_{Re} axis. In this case, no physical spatial separation between observer and event is represented in the figure. Often such separation on the x_{Re} exists. In the case where $x_{Re} = 0$, then access to anticipatory information, along t_{Re} can be achieved by access to the imaginary temporal component, t_{Im}. Hence, remote, nonlocal events in four space or the usual Minkowski space, appear contiguous in the complex eight space and nonlocal temporal events in the four space appear as anticipatory in the complex eight space metric. Both nonlocality and anticipatory systems occur in experimental tests of Bell's Theorem and perhaps in all quantum measurement processes.

5. Solitary Wave Coherent Non-dispersive Solutions in Complex Geometries

The properties and some of the implications of complex Minkowski spaces hold fundamental significance. We have presented the formalism for complex geometries in the previous section and also for superluminal x direction boosts in these geometries and the possible implications for remote connectedness through the nonlocal aether field [11]. Also the symmetry relations of the vector and scalar electromagnetic potential and other properties of Maxwell's equations, the x-directional superluminal boost, have been formulated [18]. A generalized form of Maxwell's equations formulated in complex Minkowski space are given in [30,35]. The relationship of this approach to the Schrödinger equation in this work is of interest.

In this section we determine solutions to the Schrödinger equation formulated in a complex Minkowski space and demonstrate the relationship of the solutions to inter-connectedness and the nonlocality principle. The solutions are solitary or soliton waves which exhibit little or no dispersion over long distances. We present several implications of this formalism, for the test of Bell's Theorem, anticipatory processes and an explanation for some coherent, nonlinear, non-dispersive phenomena, such as nonlinear plasma phenomena [37,38].

We examine the relationship between our multi-dimensional remote connectedness geometry and possible coherent, non-dispersive solutions to the Schrödinger equation. These non-dissipative or non-dispersive solutions are termed soliton solutions, or solitary wave solutions, and are well known in macroscopic hydrodynamic phenomena. There has been some recent interest in the use of the soliton or instanton model to describe the gluon quark structure for "infinitely" bound quarks, in part, to explain the lack of experimentally observed free quarks.

The solution to linear wave equations are dispersive in space and time, that is, their amplitude diminishes and width at half maximum becomes larger as a function of time. The term soliton is commonly used to define a wave which retains its amplitude and "half width" over space and can interact and remain intact with other solitons. The term instanton, or evanescent wave, is used to describe a structure which experiences both spatial and temporal displacement. The term instanton seems to imply a short-lived structure but actually instantons can retain their spatial and temporal configuration indefinitely and interact with other instantons in a particle-like manner as do solitons. These unique solutions can explain the existence of long spatial and temporal phenomena such as Bell's remote connectedness phenomenon, Young's double slit experiment, plasma coherent collective states and other coherent phenomena.

We solve the Schrödinger equation in complex spacetime. As seen in previous papers [8,11], complex geometries have properties which are consistent with the above mentioned phenomena. We proceed from the time-dependent Schrödinger equation in a vacuum with no potential term, $V\psi$. We consider this term later [21]. In real spacetime, we have

$$\frac{\hbar \nabla^2 \psi}{2m} = \frac{1}{i} \frac{\partial \psi}{\partial t} .$$

(11)

Monochromatic plane wave solutions for one dimension of space, or x-direction, such as

$$\psi = e^{\frac{i(kx - \omega t)}{\hbar}} \quad \text{or} \quad \psi^* = e^{\frac{-i(kx - \omega t)}{\hbar}}$$

(12)

which comprise the usual solutions. We can also write (12) as

$$\psi = e^{i\alpha} \quad for \quad \alpha = \frac{kx - \omega t}{\hbar}$$

(13)

and we can write (13) as

$$\psi = e^{i\alpha} = \cos \alpha + i \sin \alpha$$

(14a)

and also

$$\psi = e^{i\alpha} = \sinh i\alpha + \cos i\alpha .$$

(14b)

Equation (11) is the usual linear form of the Schrödinger equation in which the superposition principle holds and the quantum measurement issue arises.

We proceed to formulate the Schrödinger equation in complex spacetime. The form of complex derivative utilized here is given in [8,11]. Only one-dimensional forms of the derivative are considered in the del operator. We consider x-directional spatial dependence only for the real component of x as x_{Re}

$$\hbar \frac{\nabla^2}{2m} \psi \rightarrow \frac{\hbar}{2m} \frac{\partial^2}{\partial x_{Re}^2} \psi \,. \tag{15}$$

Using the imaginary components of space and time x_{Im} and t_{Im}, we have

$$\hbar \frac{\nabla_{Im}^2}{2m} \psi \rightarrow \frac{\hbar}{2m} \frac{\partial^2}{\partial x_{Im}^2} \psi \,. \tag{16}$$

Note that the sign change occurs for the spatial second derivative for $ix \rightarrow x_{Im}$. The imaginary time derivative yields

$$\frac{\partial}{\partial it_{Im}} \rightarrow \frac{1}{i} \frac{\partial}{\partial t_{Im}} \tag{17}$$

which is an imaginary term derivative.

The imaginary form of the Schrödinger equation becomes

$$\frac{\hbar}{2m} \nabla_{Im}^2 \psi = \frac{\partial \psi}{\partial t_{Im}} \,. \tag{18}$$

Because the Schrödinger equation is second order in space and first order in time and no imaginary term occurs in Eq. (18), the harmonic solutions in Eqs. (13), (14a), and (15) are not solutions to the imaginary components of the Schrödinger equation. Since the Dirac equation is first order in space and time, and the Klein-Gordon equation and classical wave equation are second order in space and time, quite a different picture emerges.

Starting from a real solution, which is a plane exponential growth function

$$\psi = e^{\alpha} \quad for \quad \alpha \equiv \frac{kx + \omega t}{\hbar} \tag{19}$$

we then have from Eq. (18),

$$\frac{\partial \psi}{\partial x_{Im}} = \frac{k}{\hbar} \psi \quad or \quad \frac{\partial^2 \psi}{\partial x_{Im}^2} = \frac{k^2}{\hbar^2} \tag{20}$$

and

$$\frac{\partial \psi}{\partial t_{Im}} = \frac{\omega}{\hbar} \,. \tag{21}$$

and Eq. (19) satisfies Eq. (18). Note that $\alpha \equiv (kx - \omega t)/\hbar$ does not satisfy Eq. (18) because of the minus sign which then occurs in Eq. (21). All quantities $k^2, \hbar^2, \omega^2$ are real as is x_{Im} and t_{Im}.

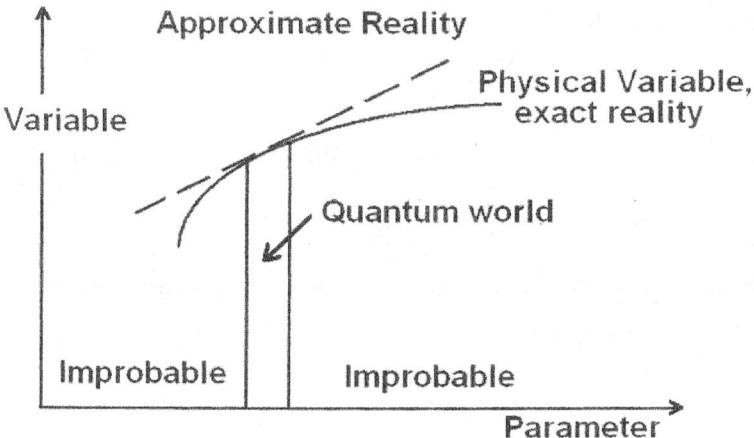

Figure 2. We approximate the quantum domain as a linear variable dependent on a parameter. The full "space" of exact reality is nonlinear.

The form of the solution in Eq. (19) for α positive definite, for all quantities greater than zero, yields an undamped growth function, that is we find that solutions in an imaginary spacetime geometry yield growth equations. Equation (19) is of a linear form. We also have another solution in Eq. (22a), but Eq. (22b) is not a solution where Eq. (19) is the same as (22b):

$$\psi = e^{-\alpha} \quad for \quad \alpha \equiv \left(\frac{kx - \omega t}{\hbar} \right) \tag{22a}$$

and

$$\psi = e^{+\alpha} \quad for \quad \alpha \equiv \left(\frac{kx + \omega t}{\hbar} \right) \tag{22b}$$

where in kx, x is x_{Im} and standing wave solutions cannot occur. Before we examined the solution of the Schrödinger equation in complex spacetime for $x' = x_{Re} + ix_{Im}$ and $t' = t_{Re} + it_{Im}$. Let us briefly discuss the introduction of a nonlinear term with a small coupling constant.

A. Nonlinear Schrödinger Equation for Complex Temporal Perturbation

We introduce a 'potential' like term which is coupled by a small coupling constant, g^2, and is associated with an attractive force. If the coupling term is small, then solutions can be determined in terms of a perturbation expansion. A $g^2 > 0$ implies an attractive force when it is regarded as a second quantized Fermi field which comprises in part our aether model which is in progress [33]. This field satisfies the Dirac equation and introduces an additional term in the Lagrangian. In reference [11] we detail this formalism, in which causality conditions in terms of analytic continuation in the energy plane gives motivation for identifying the nonlinear coupling term with the imaginary temporal coordinate, as $t^* = it_{Im}$.

By analogy to this form of the Dirac equation, we can write

$$\frac{\hbar \nabla_{Im}^2 \psi}{2m} + g^2 (\psi^+ \psi)\psi = 0 \tag{23}$$

for the time-dependent equation where ψ^+ is the Hermitian conjugate of ψ or $\psi^+ \equiv \tilde{\psi}^*$ that is the complex conjugate and transpose. For the real time-dependent equation, we have

$$\frac{\hbar \nabla_{Im}^2 \psi}{2m} + g^2 (\psi^+ \psi)\psi = \frac{1}{i}\frac{\partial \psi}{\partial t_{Im}}. \tag{24}$$

For the Schrödinger and Dirac equation, we can find solutions which we can identify in a field theory, in which each point is identifiable with a kinetic, potential and amplitude function. Linearity can be approximated for $g^2 \sim 0$, for g^2 expressed in terms of it_{Im}. In the following subsection we examine the complexification of the Schrödinger equation.

B.   The Schrödinger Equation in Complex Space and Time

Returning to our definition of complex space and time,

$$x' = x_{Re} + ix_{Im}, \quad t' = t_{Re} + it_{Im} \tag{25}$$

where x_{Re} and t_{Re} are the real parts of space and time and x_{Im} and t_{Im} are the imaginary parts of space and time and are themselves real quantities. In the most general case we have functional dependencies $x_{Im}(x,t)$ and $t_{Im}(x,t)$ where x and t are

39

x_{Re} and t_{Re}. With the quantum superposition principle, we can combine real and imaginary parts. For the x-directional form of Eq. (11), we have

$$\frac{\hbar}{2m}\frac{\partial^2 \psi_1}{\partial x_{Re}^2} = \frac{1}{i}\frac{\partial \psi_1}{\partial t_{Re}}.$$

(26)

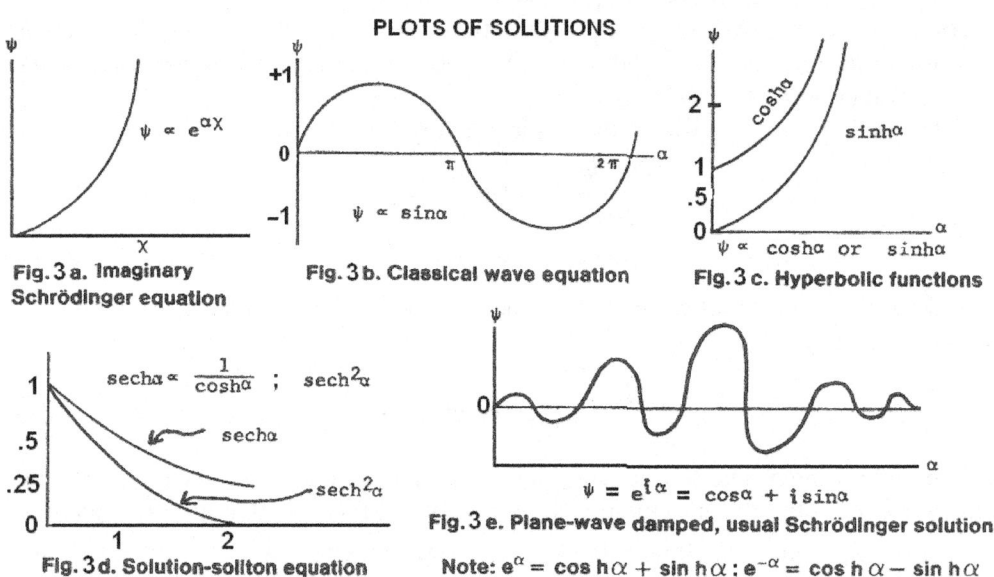

PLOTS OF SOLUTIONS

Fig. 3 a. Imaginary Schrödinger equation

Fig. 3 b. Classical wave equation

Fig. 3 c. Hyperbolic functions

Fig. 3 d. Solution-soliton equation

Fig. 3 e. Plane-wave damped, usual Schrödinger solution

Note: $e^{\alpha} = \cos h\alpha + \sin h\alpha$; $e^{-\alpha} = \cos h\alpha - \sin h\alpha$

Figure 3. Plots of various solutions.

For the imaginary part, we have from Eq. (18)

$$\frac{\hbar}{2m}\frac{\partial^2 \psi_1}{\partial x_{Im}^2} = \frac{\partial \psi_2}{\partial t_{Im}}.$$

(27)

By linear superposition which, in some cases may be an approximation, we can combine the above equation, as

$$\frac{\hbar}{2m}\left(\frac{\partial^2}{\partial x_{Re}^2} + \frac{\partial^2}{\partial x_{Im}^2}\right)\psi = \left(\frac{1}{i}\frac{\partial}{\partial t_{Re}} + \frac{\partial}{\partial t_{Im}}\right)\psi.$$

(28)

40

Note that we make an assumption that the mass in Eq. (26) is the same as in Eq. (27). We discuss this assumption and tachyonic implications in [11]. We now form solutions $\psi(x_{\mathrm{Re}}, x_{\mathrm{Im}}, t_{\mathrm{Re}}, t_{\mathrm{Im}})$ in terms of linear combinations of $\psi_1(x_{\mathrm{Re}}, t_{\mathrm{Re}})$ and $\psi_2(x_{\mathrm{Im}}, t_{\mathrm{Im}})$.

Equation (27) is defined on a four-dimensional space $(x_{\mathrm{Re}}, x_{\mathrm{Im}}, t_{\mathrm{Re}}, t_{\mathrm{Im}})$. In the first approximation, we will choose $\partial^2 \psi / \partial x_{\mathrm{Im}}^2 = 0$ so that we have

$$\frac{\hbar^2}{2m} \frac{\partial^2}{\partial x_{\mathrm{Re}}^2} \psi = \left(\frac{1}{i} \frac{\partial}{\partial t_{\mathrm{Re}}} + \frac{\partial}{\partial t_{\mathrm{Im}}} \right) \psi . \tag{29}$$

Motivation for this approximation can be found in our discussion of remote connectedness properties, represented in Figs. 2 and 3.

Let us rewrite Eq. (26) as

$$\frac{\hbar^2}{2m} \frac{\partial^2}{\partial x_{\mathrm{Re}}^2} \psi - \frac{\partial}{\partial t_{\mathrm{Im}}} \psi = \frac{1}{i} \frac{\partial}{\partial t_{\mathrm{Re}}} \psi \tag{30}$$

where ψ is a function of $(x_{\mathrm{Re}}, t_{\mathrm{Re}}, t_{\mathrm{Im}})$. From examination of the forms of Eq. (24) and (29), we can identify the g^2 term with the imaginary time derivative $\partial / \partial t_{\mathrm{Im}}$. This result is similar to the more comprehensive field theoretic argument for the Dirac equation. The associated metric space for $(x_{\mathrm{Re}}, t_{\mathrm{Re}}, t_{\mathrm{Im}})$ defines a remote connectedness geometry. We then have

$$\frac{\hbar}{2m} \frac{\partial^2 \psi}{\partial x_{\mathrm{Re}}^2} + G^2 \psi = \frac{1}{i} \frac{\partial \psi}{\partial t_{\mathrm{Im}}} \tag{31}$$

where $G^2 = g^2(\psi^+ \psi)$ is identified with the $\partial / \partial t_{\mathrm{Im}}$ term. We proceed from the assumption that $(x_{\mathrm{Re}}, t_{\mathrm{Re}}, t_{\mathrm{Im}})$ are independent variables of each other.

We can define three cases for the right side of Eq. (31), that is, the real time-dependent case, (a) zero, time dependent cases, (b) $\frac{1}{i} E_n \psi$, and (c) $\frac{1}{i} \frac{\partial \phi}{\partial t}$. In determining the coupling constant G^2, we define solutions $\phi(x_{\mathrm{Re}}, t_{\mathrm{Re}}, t_{\mathrm{Im}})$ for the third case. We have, in general,

$$\frac{\partial^2 \phi}{\partial x^2} + G^2(\tau) \frac{\partial \phi}{\partial \tau} = \frac{1}{i} \frac{\partial \phi}{\partial t} . \tag{32}$$

41

We define the quantity $\xi = kx_{Re} - \omega t_{Re} - \alpha t_{Im}$. For case (a) above we have solutions

$$\phi = \phi_0 + A \sec h^2 a\xi \qquad (33)$$

where

$$G^2(t_{Im}) = \frac{a\hbar^2 k^2}{2\alpha m} \tanh a\xi \qquad (34)$$

where k is the wave number or k_{Im}. The constant, a can be expressed in terms of $\hbar$ and m where $m' = im = m^* = m_{Im}$ which is the tachyonic mass, which we formulate in complex eight space. For case (c), we find a similar solution for ϕ for

$$G^2(t_{Im}) = \frac{\dfrac{a\hbar^2 k^2}{2m}}{(\alpha + \frac{1}{2}\omega)} \tanh a\xi . \qquad (35)$$

Solutions and the form of G^2 (t_{Im}) is more complicated for case (b). Note the analogy to the solutions for the Korteweg-deVries equation [21] for

$$u(x,t) = A \sec h^2 k \quad for \quad K = x - ct/L \qquad (36)$$

where L is a characteristic length dimension of a soliton wave which is expressed in terms of the amplitude A and the hydrodynamic media depth h or $L = \sqrt{h^3/3A}$ [39].

The form of $G^2(t_{Im})$ is nonlinear and is compatible with the soliton solutions. The non-dispersive nature of the solutions may be associated with a complex space "signal" which defines the connection of remote parts of the multi-dimensional geometric space [11]. Several types of solutions are displayed in Fig. (3). See Fig. 4 for the implications of the Quantum Theory and Bell's theorem.

C. Discussion and Application of Coherent State Solutions

The soliton solution is a unique solution in that it is non-dispersive. All other solutions to the Schrödinger equation are dispersive to various degrees. Each state solution has a particular amplitude at a specific point in space and instant in time. One can calculate the probability of this existence of a specific amplitude as a function of x and t. A unique feature of the soliton is that it retains its amplitude in space and time and therefore we have a reasonable certainty in our measure of it for each space and time.

42

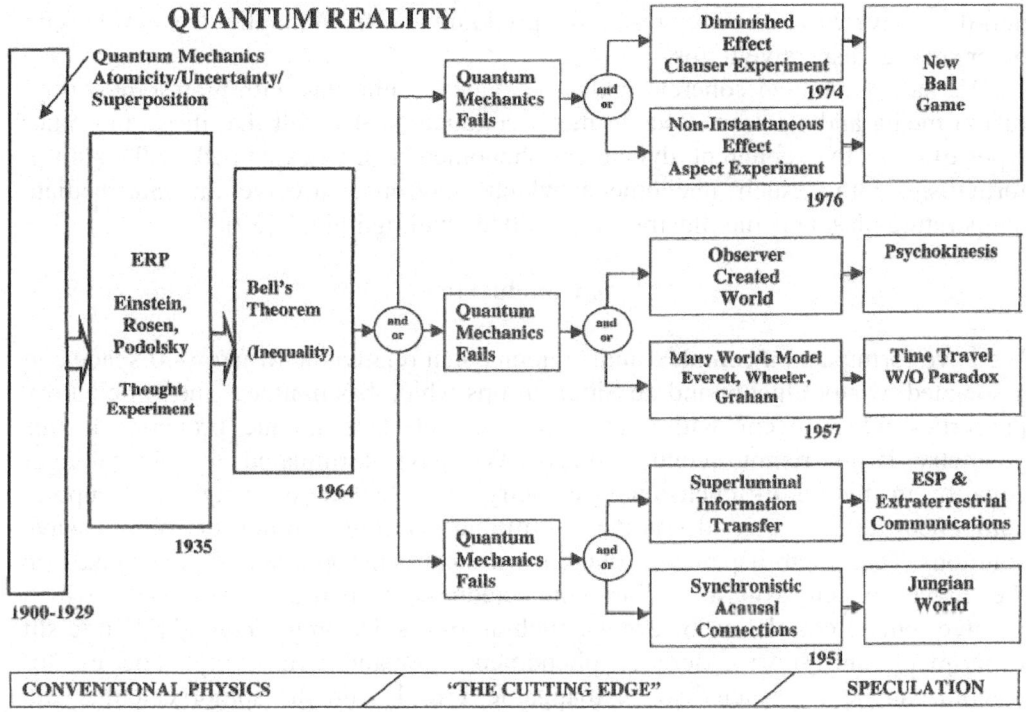

Figure 4. Historical development of the quantum theory. See [40].

In practice of course, there are no completely non-dispersive waves but we define the soliton solutions in terms of a coherent, non-dispersive state that retains its identity and amplitude value over many iteration times. Hence the soliton acts like a particle, in that soliton solution collisions do not disrupt the wave form or amplitude in elastic processes [38].

In hydrodynamics, the interpretation of the soliton or solitary wave is not completely clear [38]. One possible interpretation of this particular type of solution to the wave equation in this particular complex geometry, including the small coupling nonlinear term, is that the geometry selects the particular wave function. Note that this possible interpretation may have deep implications for the quantum measurement issue or the "collapse of the wave function". In the usual nuclear energy levels, a particular state may be composed of a sum of various states of angular momentum and spin which sum to the total I and l values. The amplitude of these states vary, with one predominant term [37]. In the current case, the soliton

43

non-dispersive wave could represent the predominant, fixed amplitude solution with other smaller dispersive terms.

We have examined coherent collective states in plasmas with high temperature fusion media and electron gases in metal conductors. It is felt that these and other types of collective, coherent, dynamical phenomena can be explained by the soliton formalism. Other such phenomena which may also involve an intermediate temperature plasma is the illustrative so-called "ball lightning" [39].

6. Conclusion

We have formulated a complex multi-dimensional relativistic Minkowski space and associated twistor algebra and de Sitter groups which has nonlocal and anticipatory properties which occur within an aether model. One unique property of this geometry is its remote connectedness. We have formulated the Schrödinger equation in this multi-dimensional geometry. We identify the imaginary temporal component term as a small nonlinear term and determine soliton or solitary wave solutions. These non-dispersive, coherent waves are appropriate to define signals, in the space, which exhibit remote connectedness properties. Phenomena which involve remote correlation of events, such as Bell's Theorem, Young's double slit experiment, and super-coherence phenomena, demand nonlocality. The twistor algebra can be constructed to be mappable 1 to 1 with the spinor calculus and allows us to develop a unique formalism of Bell's inequality.

We also speculate that the nonlinear quantum model with coherent non-dispersive solutions to the Schrödinger equations, which is an expression of the remote nonlocality property of the space, may lend insight into the quantum measurement problem [33, 34] which is consistent with the relativistic formalism. A mechanism may be formulated which defines a connection between the observer and the observed. The properties of certain systems appear to demand a nonlinear, nonlocal anticipatory description.

Acknowledgements

Extensive discussions and collaboration with R. Targ is most appreciated. The interest and discussions with Eugene Wigner, David Bohm and John A. Wheeler were very fruitful. A Michrowski brought useful references to EAR's attention.

References

[1] Einstein, A., Podolsky, B., & Rosen, N. (1935) Can a quantum mechanical description of physical reality be considered complete? Physical Review, 47, pp. 777-780.

[2] Misner, C.W., Thorne, K.S. & Wheeler, J.A. (1973) Gravitation, San Francisco: W.H. Freedman; and private communications with EAR 1976-1979.

[3] Bohm, D. & Hiley, B. (1993) The Undivided Universe, London: Routledge; and private communications with EAR 1976-1979.

[4] Bell, J.S. (1964) On the Einstein, Podolsky, Rosen paradox, Physics, 1, pp. 195-200.

[5] Bell, J.S. (1966) On the problem of hidden variables in quantum theory, Review of Modern Physics, 38, 447.

[6] Stapp, H.P. (1971) Phys. Rev. 3D, 1303; and private comm. with EAR 1964-1993.

[7] Clauser, J.F. & Shimony, A. (1978) Rep. Prog. Phys. 41, 1881; and private communications with EAR 1971-1992.

[8] Rauscher, E.A. (1979) Some physical models potentially applicable to nonlocal measurement, in The Iceland Papers: Frontiers of Physics Conference, pp. 50-93, Amherst: Essentia Research Associates and 2nd publication by Planetary Association for Clean Energy, Altoma, Ontario, Canada.

[9] Amoroso, R.L., Vigier, J-P, Kafatos, M. & Hunter, G. (2002) Comparison of near and far-field double-slit interferometry for dispersion of the photon wave packet, in R.L. Amoroso, G. Hunter, M. Kafatos & J-P Vigier (eds.) Gravitation and Cosmology: From the Hubble Radius to the Planck Scale, Dordrecht: Kluwer Academic.

[10] Antippa, A.F. & Dubois, D.M. (2004) Anticipation, orbital stability and energy conservation in the discrete harmonic oscillator, in D.M. Dubois (ed.) AIP Conference Proceedings 718, CASYS03, Liege, Belgium, pp. 3-44, Melville: American Inst. Physics.

[11] Ramon, C. & Rauscher, E.A. (1980) Superluminal transformations in complex Minkowski spaces, LBL Report 9752 ; Foundations of Physics (1980) 10, 661.

[12] Hansen, R.O. & Newman, E.T. (1975) A complex Minkowski approach to twistors, General Relativity and Gravitation, 6, 361-385.

[13] Newman, E.T. (1976) H-space and its properties, General Relativity and Gravitation, 7, 107-111.

[14] Newman, E.T., Hansen, R.O., Penrose, R. & Ton, K.P. (1978) The metric and curvature properties of H-space, Proceedings of the Royal Society of London, A363, 445-468.

[15] Haramein, N. & Rauscher, E.A. (2008) Complex Minkowski space formalism of the Penrose twistor and the spinor calculus, in D. Dubois (ed.) Proceedings of CASYS07, Liege, Belgium, in press; Rauscher, E.A. (1971) A unifying theory of fundamental processes, LBNL/UCB Prress, UCRL 20808, June, and references therein.

[16] Rauscher, E.A. (1972) A set of generalized Heisenberg relations and a possible new form of quantization, Letters il Nuovo Cimento, 4, 757.

[17] Bohm, D. (1952) Physical Review, 85, 166; and private communication with EAR.

[18] Clauser, J.F. & Horne, W.A. (1974) Physical Review, 10D, 526.

[19] Freedman, S. & Clauser, J.F. (1972) Experimental test of local hidden variable theories, Physical Review Letters, 28, 934-941.

[20] Feinberg, G. (1967) Possibility of faster-than-light particles, Physical Rev., 159, 1089.

[21] Rauscher, E.A. (2007) A detailed formalism of Bell's theorem in complex eight space, in progress.

[22] Stapp, H.P. (1975) Theory of reality, LBL Report 3837.

[23] Aspect, A., Grangier, P. & Roger, G. (1992) Experimental tests of Bell's inequalities using time-varying analyzers, Phys. Rev. L. 49, 1804-1907; and private comm. with EAR

[24] Gisin, N., Tittel, W. Brendel, J. & Zbinden, H. (1998a) Violation of Bell's inequalities by photons more than 10 km apart, Phys. Rev. Let. 81, 3563-3566.

[25] Gisin, N., Tittel, W. Brendel, J. & Zbinden, H. (1998b) Quantum correlation over more than 10 km, Optics and Photonics News, 9, 41.

[26] Rauscher, E.A. & Targ, R. (2001) The speed of thought: Investigations of a complex space-time metric to describe psychic phenomena, J. Sci. Exploration, 15, 331.

[27] Rauscher, E.A. & Targ, R. (2006) Investigation of a complex space-time metric to describe precognition of the future, in D.P. Sheehan (ed.) Frontiers of Time: Retrocausation – Experiment and Theory, Melville: AIP Conference Proceedings.

[28] Wigner, E. (1967) Symmetries and Reflections: Scientific Essays, Bloomington: Indiana Univ. Press; and private comm. with EAR.

[29] Chew, G. (1964) The Analytic S-Matrix, Benjamin: Frontiers of Physics; and private comm. with EAR.

[30] Rauscher, E.A. (2002) Non-Abelian gauge groups for real and complex amended Maxwell's equations, in R.L. Amoroso, G. Hunter, M. Kafatos & J-P Vigier (eds.) Gravitation and Cosmology: From the Hubble Radius to the Planck Scale, pp. 183-188, Dordrecht: Kluwer Academic; Also (1992) Bulletin Am. Phys. Society, 47, 167.

[31] Rauscher, E.A. & Amoroso, R.L. (2006) The physical implications of multidimensional geometries and measurement, in D. Dubois (ed.) Int. J. Computing Anticipatory Systems, vol. 19, Liege: University of Liege, CHAOS, Institute of Mathematics.

[32] Rauscher, E.A. (1983) Electromagnetic Phenomena in Complex Geometries and Nonlinear Phenomena and Non-Hertzian Waves, Milbrae: Tesla Books; 2nd Edition (2008) Oakland: The Noetic Press.

[33] Amoroso, R.L. & Rauscher, E.A. (2009) The Holographic Anthropic Multiverse: Formalizing the Geometry of Ultimate Reality, Singapore: World Scientific.

[34] Amoroso, R.L. (2009) Probability $\equiv 1$: An Empirical Protocol for Surmounting Uncertainty, submitted.

[35] Sewell. G.L. (2002) Quantum Mechanics and its Emergent Macrophysics, Princeton: Princeton Univ. Press.

[36] Landau, L.J. (1987) Experimental tests of general quantum theories, Lett. In Math. Physics, 14, 33-40.

[37] Rauscher, E.A. (1968) Electron interactions and quantum plasmas, J. Plas. Phys., 2, 517.

[38] Lonngren, K.& Scott, A. (1978) Solitons in Action, New York: Academic Press.

[39] Osborne, A.R. & Burke, T. (1980) Science, 208, 451.

[40] The Fundamental Physics Group (1973-1979) was organized and chaired by E,A, Rauscher at the Lawrence Berkeley National Laboratory with forty physicists, when she was on the Theoretical Physics Department staff (1971-1979) and LBNL staff (1963-1979).

A Classical Dirac Equation

R. A. Close
Email: robert.close@classicalmatter.org

Abstract

The time evolution of electrons and other fermions is described by the first-order Dirac equation. Although typically interpreted probabilistically, the Dirac equation is fundamentally a deterministic equation for the evolution of physical observables such as angular momentum density. The Dirac equation can be considered as a second-order wave equation if the wave function is a representation of the first derivatives. The conventional Dirac formalism has two serious flaws. First, the conventional derivation of the parity operator is incorrect. Conventional theory holds that the wave function of a Dirac particle is its own mirror image but certain mirrored interactions do not occur. Such mirror particles have never been observed. Experimental evidence, such as beta decay, supports the alternative hypothesis that the mirror image of matter is antimatter. This problem is solved by identifying a flaw in the conventional derivation of the parity operator, then deriving a new parity operator based on the algebraic properties of vectors. Second, the conjugate momenta (p_i) in the free-particle Hamiltonian (H) do not have the proper relation ($p_i = \delta H / \delta \dot{q}_i$) to the time derivatives of coordinates ($\dot{q}_i$). This problem is solved by replacing the mass term with convection and rotation terms. We then show that the resultant bispinor equation of evolution is equivalent to a classical second-order wave equation for angular momentum density in an elastic solid. The co-existence of forward- and backward-propagating waves along a single axis is the basis of half-integer spin. Wave interference produces both the Lorenz force and the Pauli Exclusion Principle. Mass is associated with radially inward acceleration of the wave such as occurs in a soliton. Angular correlations between spin states are equal to the quantum correlations. Bell's Theorem is not applicable to classical bispinors. Matter and anti-matter are related by spatial inversion, consistent with experimental observations. The classical wave formulation therefore provides a conceptually clear interpretation of fermion dynamics.

1. Introduction

In classical electrodynamics the electron is generally regarded as a point-like particle. This view became untenable in 1927 when de Broglie's hypothesis[1] that matter behaves like waves was confirmed in electron diffraction experiments by Davisson and Germer,[2] and independently by Thomson and Reid.[3] However, the quantum mechanical equations developed to describe these 'matter waves' are first order equations rather than classical second-order wave equations. Although these waves are commonly interpreted as probability waves, the quantum mechanical Dirac equation is also fundamentally a deterministic equation for the evolution of angular momentum density and other physical observables. As such, it should correspond to classical wave theory. Others have reformulated the Dirac theory in terms of deterministic relations between local physical observables.[4,5] However, these investigators did not construct a corresponding classical wave theory describing evolution of a field variable entirely in terms of its own derivatives.

1.1. Factoring Wave Equations

The Klein-Gordon (or relativistic Schrödinger) operator can be factored into a product of two Dirac operators acting on the wave polarization (or amplitude) $\mathbf{a}$:

$$\left\{\partial_t^2 - c^2\nabla^2 + M^2\right\}\mathbf{a} = \left\{\gamma^0\partial_t + c\gamma^i\partial_i + \mathrm{i}\,M\right\}\left\{\gamma^0\partial_t + c\gamma^i\partial_i - \mathrm{i}\,M\right\}\mathbf{a} \tag{1}$$

where the gamma factors are related to the Minkowski metric $g^{\mu\nu}$ by:

$$\gamma^\mu\gamma^\nu + \gamma^\nu\gamma^\mu = 2g^{\mu\nu} \tag{2}$$

The factors γ^μ have traditionally been regarded as matrices. However, they can also be interpreted geometrically using multivariate vectors.[5-7] The wave polarization $\mathbf{a}$ is a classical 3-vector in Galilean space-time. The Minkowski metric of relativity is introduced through the operators and applies to the space of measurements made with waves.

If we define a wave function:

$$\Psi \equiv \left\{\gamma_0\partial_t + c\gamma_i\partial_i - \mathrm{i}\,M\right\}\mathbf{a} \tag{3}$$

then the resultant first-order Dirac equation is equivalent to the original Klein-Gordon equation:

$$\left\{\gamma_0\partial_t + c\gamma_i\partial_i + \mathrm{i}\,M\right\}\Psi = 0 \tag{4}$$

In the above case the two Dirac operators have different sign for the mass term. Rowlands[8,9] and Rowlands and Cullerne[10] used a combination of multivariate 4-vectors and quaternions to write the Dirac equation in a nilpotent form in which the two successive Dirac operations are identical. This formulation yields an elegant classification of particle states within the Standard Model.

Standard solutions of the Klein-Gordon equation (with additional potentials) yield different energy eigenvalues than the Dirac equation.[11] This result is quite peculiar given the fact that each component of the Dirac wave function actually satisfies the Klein-Gordon equation! Factoring the Klein-Gordon equation cannot change its eigenvalues. The problem is that in the usual analysis of Klein-Gordon, the angular functions are chosen to be eigenfunctions of the squared orbital angular momentum L^2, whereas in the analysis of the Dirac equation the angular functions are eigenvalues of the squared total angular momentum $(L+S)^2$. The difference is not in the equations, but in the choice of angular eigenfunctions. The usual analysis of the Klein-Gordon equation assumes no spin contribution to angular momentum. This choice corresponds to a scalar boson. Rowlands uses angular eigenvalues obtained from Dirac theory, corresponding to a spin one-half fermion.

1.2. Parity Conservation

Historically, parity conservation was a fundamental assumption of physics (we use here a narrow definition of parity conservation meaning symmetry of physical laws, rather than the requirement of equal incidence of left-handed and right-handed particles). Any physical bias toward right- or left-handed processes would be completely arbitrary and therefore unjustifiable. However, Lee and Yang[12] proposed that weak interactions may violate parity, and experiments by Wu[13] demonstrated that beta decay exhibits left-right asymmetry. This asymmetry has been interpreted as implying parity violation, although Lee and Yang mention that their theory could be consistent with parity conservation if protons are not identical (within a rotation) to their mirror images.

If Wu's experiment could be constructed using antimatter, such an experiment would behave exactly like a mirror image of the original. This property is typically called "*PC*" conservation, although the experimental observation is independent of the theoretical operators. Even the decay of neutral kaons, which supposedly exhibits *PC* violation, is experimentally consistent with mirror symmetry between matter and antimatter (the *PC* violation is attributed to a temporal change in parity, not to an asymmetry of the *PC* transformation itself). All experimental evidence is consistent with the hypothesis that spatial inversion exchanges matter and antimatter, whereas the conventional parity operator results in unexplained left-right asymmetries. Therefore a reformulation of the

Dirac parity operator is necessary. The logical arguments for the new parity operator are as follows:

The Dirac equation for a free particle is:

$$\partial_t \psi + c\gamma^5 \sigma_i \partial_i \psi = -iM\gamma^0 \psi \tag{5}$$

The corresponding Hamiltonian operator is:

$$H = -ic\gamma^5 \boldsymbol{\sigma} \cdot \nabla + M\gamma^0 \tag{6}$$

where $\boldsymbol{\sigma}$ is the matrix operator for spin and $M = mc^2/\hbar$ has units of frequency. There exists an oriented set of basis vectors $(\gamma^5, \tilde{i}\gamma^5\gamma^0, \gamma^0)$ with commutation relations equivalent to those of Pauli matrices ($\tilde{i}$ is the unit imaginary pseudoscalar). These matrices may be represented in terms of the 2×2 identity matrix I_2:

$$\gamma^5 = \begin{pmatrix} 0 & I_2 \\ I_2 & 0 \end{pmatrix}; \qquad \gamma^4 = \begin{pmatrix} 0 & -\tilde{i}I_2 \\ \tilde{i}I_2 & 0 \end{pmatrix}; \qquad \gamma^0 = \begin{pmatrix} I_2 & 0 \\ 0 & -I_2 \end{pmatrix} \tag{7}$$

These represent directions relative to velocity ($c\gamma^5\boldsymbol{\sigma}$) and should all be inverted by the parity operator. The conventional parity operator $P_0\psi(\mathbf{r}) \equiv \gamma^0\psi(-\mathbf{r})$ inverts only two of these (i.e. $\gamma^a \to \gamma^0\gamma^a\gamma^0$).

The unit imaginary associated with spin is a true scalar:

$$\tilde{i} \equiv \sigma^1\sigma^2\sigma^3 \tag{8}$$

The roles of the different imaginaries can be clarified by factoring the Dirac wave function:[11]

$$\psi(\mathbf{r}) \equiv a^{1/2} \exp(\tilde{i}\sigma_i\varphi_i)\exp(\gamma^5\sigma_i\alpha_i)\exp(\tilde{i}\gamma^5\zeta)\psi_0 \tag{9}$$

It is clear that $\tilde{i}\sigma_i = \varepsilon_{ijk}\sigma_j\sigma_k/2$ is associated with rotation in the plane orthogonal to the x_i axis. Similarly, $\tilde{i}\gamma^5$ is associated with rotation of relative velocity axes in the plane orthogonal to the wave velocity. The matrices $\tilde{i}\gamma^4$ and $\tilde{i}\gamma^0$ may also be used to rotate velocity independently of spin. These matrices do not appear in the factorization because the velocity orientation is contained in the velocity parameters α_i.

Next we define a new wave function in which all imaginary pseudoscalar factors are inverted: $\psi^{\#}(\tilde{i}) \equiv \psi(-\tilde{i})$. This pseudoscalar conjugation operation differs from complex conjugation, which inverts both scalar and pseudoscalar imaginaries. Pseudoscalar conjugation inverts observables associated with γ^4 since:

$$\psi^{\#\dagger}\gamma^4\psi^{\#} = \left[\psi^{\dagger}\gamma^{4\#}\psi\right]^{\#} = \left[\psi^{\dagger}\left[-\gamma^4\right]\psi\right]^{\#} = \psi^{\dagger}\left[-\gamma^4\right]\psi \tag{10}$$

The parity operator, which inverts all of the relative velocity vectors, is then (within an arbitrary phase factor):

$$P\psi(\mathbf{r}) \equiv \psi_P(\mathbf{r}) = \gamma^4\psi^{\#}(-\mathbf{r}) \tag{11}$$

This operator inverts observables computed from γ^5, γ^4, and γ^0 independently of the change in sign of $\mathbf{r}$. We will see below that this parity operator changes the sign of the electron energy eigenvalue, consistent with an exchange of matter and anti-matter.

1.3. Conjugate momenta

Another problem with Dirac theory is that the conjugate momenta in the free particle Hamiltonian (6) do not have the proper form. For example, the angular momentum operator is conventionally taken as:

$$\mathbf{J} = -i\,\mathbf{r} \times \nabla + \frac{1}{2}\boldsymbol{\sigma} \tag{12}$$

The spin term is justified by the fact that conservation of angular momentum requires commutation with the Hamiltonian. However, using (6), the Hamiltonian expression for the angular momentum operator (conjugate to coordinate φ) is:

$$\mathbf{J} \equiv \frac{\delta H}{\delta(\partial_t\varphi)} = 0 \tag{13}$$

(The conjugate momentum is properly defined as a derivative of the Lagrangian, but in classical mechanics the same relationship holds for the Hamiltonian since the kinetic energy contribution is the same for both). Clearly the spin angular momentum operator should appear explicitly in the Hamiltonian as the coefficient of an angular velocity. Hestenes[7] has interpreted mass as the scalar product of spin $\mathbf{S}$ and its associated angular velocity $\dot{\varphi} \equiv \partial_t\varphi$:

$$mc^2 \equiv \frac{1}{2}\dot{\boldsymbol{\varphi}}\cdot\mathbf{S} \tag{14}$$

but cannot explain what is rotating. We will assume that the classical Dirac equation $\partial_t\psi = -iH\psi$ contains convection terms with velocity $\mathbf{u}$ and vorticity $\mathbf{w} = \dot{\boldsymbol{\varphi}}$ in order to yield the correct conjugate momenta:

$$H = -ic\gamma^5\boldsymbol{\sigma}\cdot\nabla - i\mathbf{u}\cdot\nabla + \mathbf{w}\cdot\frac{\boldsymbol{\sigma}}{2} \tag{15}$$

It is possible that this Hamiltonian is incomplete, or may not yield masses and potentials consistent with experiment. Nonetheless, this formulation is sufficient to explain the general dynamical behavior of electrons. It also offers a possible means for deriving physical quantities which at present are only determined empirically. We will interpret the motion represented by $\mathbf{u}$ and $\mathbf{w}$ as referring to an elastic solid, which we take as a model of the vacuum.

1.4. Classical Interpretation

A classical interpretation of Dirac spinors was presented by Close in describing torsion waves.[14] In this paper we use a classical approach to factoring a wave equation for angular momentum density. The one-dimensional wave equation is generalized to three dimensions under the assumption that the wave polarization and velocity rotate together. This procedure yields a first-order bispinor wave equation. Wave interference is shown to produce both the Lorenz force and the Pauli Exclusion Principle. Angular correlations between different states are also derived.

2. Spinorial Representation of Waves

Consider a scalar quantity (a) which satisfies a wave equation with wave speed (c) in one spatial dimension (z):

$$\partial_t^2 a = c^2 \partial_z^2 a \tag{16}$$

This equation can be factored:

$$[\partial_t + c\partial_z][\partial_t - c\partial_z]a = 0 \tag{17}$$

The general solution is a sum of forward (a_F) and backward (a_B) propagating waves:

$$a(z,t) = a_F(z - ct) + a_B(z + ct) \tag{18}$$

This form of the solution to the one-dimensional wave equation can be found in any elementary textbook on waves. We can write the equations for forward and backward waves in matrix form:

$$\left[\partial_t + \begin{pmatrix} 1 & 0 \\ 0 & -1 \end{pmatrix} c\partial_z \right]\begin{pmatrix} a_F(z-ct) \\ a_B(z+ct) \end{pmatrix} = 0 \tag{19}$$

The spatial derivatives are related to the temporal derivatives:

$$c\partial_z\begin{pmatrix} a_F(z-ct) \\ a_B(z+ct) \end{pmatrix} = \partial_t\begin{pmatrix} -a_F(z-ct) \\ a_B(z+ct) \end{pmatrix} = -\begin{pmatrix} 1 & 0 \\ 0 & -1 \end{pmatrix}\partial_t\begin{pmatrix} a_F(z-ct) \\ a_B(z+ct) \end{pmatrix} \tag{20}$$

Let $\dot{a} \equiv \partial_t a$ and $a' \equiv \partial_z a$. We define a wave function in terms of the time derivatives:

$$\Psi \equiv \begin{pmatrix} \dot{a}_F(z-ct) \\ \dot{a}_B(z+ct) \end{pmatrix} \tag{21}$$

The wave equation for the forward and backward waves is now:

$$\left[\partial_t + \begin{pmatrix} 1 & 0 \\ 0 & -1 \end{pmatrix} c\partial_z \right]\Psi = \partial_t\begin{pmatrix} \dot{a}_F(z-ct) \\ \dot{a}_B(z+ct) \end{pmatrix} - c\partial_z\begin{pmatrix} a'_F(z-ct) \\ a'_B(z+ct) \end{pmatrix} = 0 \tag{22}$$

We have now reduced the second-order wave equation to a first-order matrix equation.

2.1. Spinors and Bispinors

If we regard the z-axis as one of three orthogonal axes, then the two independent components $\dot{a}_F$ and $\dot{a}_B$ differ by a 180 degree rotation. This is the definitive property of independent states in spin one-half systems. Unfortunately, this property is de-emphasized in the physics literature in favor of the more exotic property that complex spinors change sign upon 360 degree rotation. This latter property does not apply to physical observables which are computed from bilinear products of spinors. However, the separation of independent states by 180 degrees does apply to wave velocity, implying that solutions of the wave equation generally form spin one-half systems. Note that unlike positive and negative scalars or vector components (which can also be expressed as bilinear products of spinors), waves with positive and negative velocity are not related by a multiplicative factor of minus one. The forward and backward waves are independent states. The mathematical basis of this property is that wave velocity is a property of the functional arguments and is not simply an amplitude.

The relationship between waves and spinors can be made explicit by further decomposition into positive-definite components $(\dot{a}_{F+}, \dot{a}_{B-}, \dot{a}_{B+}, \dot{a}_{F-})$ or $(a'_{F+}, a'_{B-}, a'_{B+}, a'_{F-})$ representing positive $(+)$ or negative $(-)$ contributions to the wave derivatives: [14]

$$\dot{a}(z,t) = \dot{a}_{F+}(z-ct) - \dot{a}_{B-}(z+ct) + \dot{a}_{B+}(z+ct) - \dot{a}_{F-}(z-ct) \tag{23}$$

and

$$ca'(z,t) = c[a'_{F+}(z-ct) - a'_{B-}(z+ct) + a'_{B+}(z+ct) - a'_{F-}(z-ct)]$$
$$= -\dot{a}_{F+}(z-ct) - \dot{a}_{B-}(z+ct) + \dot{a}_{B+}(z+ct) + \dot{a}_{F-}(z-ct) \tag{24}$$

From here on the functional arguments will not be written explicitly. Note that the positive-definite components may have discontinuous derivatives where the original signed quantities pass continuously through zero. For example, to make the time derivatives continuous requires matching conditions for $\dot{a}$:

$$\partial_t \dot{a}_{F+}\big|_{\dot{a}_{F+}=\dot{a}_{F-}=0} = -\partial_t \dot{a}_{F-}\big|_{\dot{a}_{F+}=\dot{a}_{F-}=0}$$
$$\partial_z \dot{a}_{F+}\big|_{\dot{a}_{F+}=\dot{a}_{F-}=0} = -\partial_z \dot{a}_{F-}\big|_{\dot{a}_{F+}=\dot{a}_{F-}=0} \tag{25}$$

Similar relations hold for the backward wave components. Such discontinuities do not affect the validity of the first order equations. However, higher derivatives may be undefined at some points.

Since each component has a unique sign, we can express $\dot{a}$ and a' in spinorial form with the one-dimensional wave function ψ_v (the subscript 'v' refers to the velocity axis):

$$\dot{a} = \begin{pmatrix} \dot{a}_{F+}^{1/2} \\ \dot{a}_{B-}^{1/2} \\ \dot{a}_{B+}^{1/2} \\ \dot{a}_{F-}^{1/2} \end{pmatrix}^T \begin{pmatrix} 1 & 0 & 0 & 0 \\ 0 & -1 & 0 & 0 \\ 0 & 0 & 1 & 0 \\ 0 & 0 & 0 & -1 \end{pmatrix} \begin{pmatrix} \dot{a}_{F+}^{1/2} \\ \dot{a}_{B-}^{1/2} \\ \dot{a}_{B+}^{1/2} \\ \dot{a}_{F-}^{1/2} \end{pmatrix} \equiv \psi_v^T \sigma \psi_v \tag{26}$$

$$ca' = \begin{pmatrix} \dot{a}_{F+}^{1/2} \\ \dot{a}_{B-}^{1/2} \\ \dot{a}_{B+}^{1/2} \\ \dot{a}_{F-}^{1/2} \end{pmatrix}^T \begin{pmatrix} -1 & 0 & 0 & 0 \\ 0 & -1 & 0 & 0 \\ 0 & 0 & 1 & 0 \\ 0 & 0 & 0 & 1 \end{pmatrix} \begin{pmatrix} \dot{a}_{F+}^{1/2} \\ \dot{a}_{B-}^{1/2} \\ \dot{a}_{B+}^{1/2} \\ \dot{a}_{F-}^{1/2} \end{pmatrix} \equiv -\psi_v^T \beta \psi_v \tag{27}$$

where the superscript T indicates transposition of the column matrix and the matrix $\beta\sigma$ tabulates the forward and backward velocities (v):

$$v\psi_v = c\begin{pmatrix} 1 & 0 & 0 & 0 \\ 0 & -1 & 0 & 0 \\ 0 & 0 & -1 & 0 \\ 0 & 0 & 0 & 1 \end{pmatrix}\begin{pmatrix} \dot{a}_{F+}^{1/2} \\ \dot{a}_{B-}^{1/2} \\ \dot{a}_{B+}^{1/2} \\ \dot{a}_{F-}^{1/2} \end{pmatrix} \equiv c\beta\sigma\psi_v \tag{28}$$

This wave function is a one-dimensional bispinor. In one dimension the components of the bispinor may be taken to be real and positive-definite. Extension to three dimensions requires complex components.

Changing the order of terms in the wave function is called a change of 'representation'. A few important points are:

1. The components of the column matrix wave function are real and positive-definite.
2. Only one forward component and one backward component can be non-zero at any given time and place (for one-dimensional waves).
3. The spatio-temporal variation of each component must be consistent with its location in the column matrix.

Since some of the components must be zero, let δ_F and δ_B be either zero or one. Then the wave function is:

$$\psi_v = \begin{bmatrix} \dot{a}_F^{1/2}\delta_F & \dot{a}_B^{1/2}\delta_B & \dot{a}_B^{1/2}[1-\delta_B] & \dot{a}_F^{1/2}[1-\delta_F] \end{bmatrix}^T \tag{29}$$

Using Lorentz boosts, the wave function can be written as:

$$\psi_v = \dot{a}_0^{1/2}\exp(\beta\sigma\,\alpha/2)\begin{bmatrix} \delta_F & \delta_B & [1-\delta_B] & [1-\delta_F] \end{bmatrix}^T/\sqrt{2} \tag{30}$$

This form has two independent continuous parameters and two binary parameters.

The equation of evolution of the wave components is:

$$\partial_t\psi_v + c\beta\sigma\,\partial_z\psi_v = 0 \tag{31}$$

This is the one-dimensional Dirac equation. This equation can be interpreted as a convective derivative with two opposite velocities represented by the matrix $v = c\beta\sigma$.

The relation between one dimensional bispinor equations and scalar wave equations is summarized in Table 1.

Table 1. Corresponding Bispinor and Scalar Wave Equations in One Dimension

Bispinor Equation	Scalar Equation
$\partial_t\left[\psi_v^T \sigma \psi_v\right] + c\partial_z\left[\psi_v^T \beta \psi_v\right] = 0$	$\partial_t^2 a - c^2 \partial_z^2 a = 0$
$\partial_t\left[\psi_v^T \psi_v\right] + c\partial_z\left[\psi_v^T \beta\sigma \psi_v\right] = 0$	$\partial_t\lvert\partial_t a_F\rvert + \partial_t\lvert\partial_t a_B\rvert + c^2\partial_z\lvert\partial_z a_F\rvert - c^2\partial_z\lvert\partial_z a_B\rvert = 0$
$\partial_t\left[\psi_v^T \beta \psi_v\right] + c\partial_z\left[\psi_v^T \sigma \psi_v\right] = 0$	$\partial_t\left[-c\partial_z a\right] + c\partial_z\left[\partial_t a\right] = 0$
$\partial_t\left[\psi_v^T \beta\sigma \psi_v\right] + c\partial_z\left[\psi_v^T \psi_v\right] = 0$	$c\partial_t\lvert\partial_z a_F\rvert - c\partial_t\lvert\partial_z a_B\rvert + c\partial_z\lvert\partial_t a_F\rvert + c\partial_z\lvert\partial_t a_B\rvert = 0$

2.2. Three Dimensional Waves

Extension to three dimensions requires rotation of the wave velocity to arbitrary direction, and, for vector waves, rotation of polarization. For the Dirac wave functions, extension to three dimensions is achieved by requiring that velocity and polarization rotate together as follows.

Let the polarization matrix σ be one component (σ_3) of a vector of matrices ($\sigma_1, \sigma_2, \sigma_3$):

$$\sigma_1 = \begin{pmatrix} 0 & 1 & 0 & 0 \\ 1 & 0 & 0 & 0 \\ 0 & 0 & 0 & 1 \\ 0 & 0 & 1 & 0 \end{pmatrix}, \quad \sigma_2 = \begin{pmatrix} 0 & -i & 0 & 0 \\ i & 0 & 0 & 0 \\ 0 & 0 & 0 & -i \\ 0 & 0 & i & 0 \end{pmatrix}, \quad \sigma_3 = \begin{pmatrix} 1 & 0 & 0 & 0 \\ 0 & -1 & 0 & 0 \\ 0 & 0 & 1 & 0 \\ 0 & 0 & 0 & -1 \end{pmatrix} \tag{32}$$

Another vector of matrices can also be defined ($\beta_1, \beta_2, \beta_3$) with β_3 replacing β :

$$\beta_1 = \begin{pmatrix} 0 & 0 & 1 & 0 \\ 0 & 0 & 0 & 1 \\ 1 & 0 & 0 & 0 \\ 0 & 1 & 0 & 0 \end{pmatrix}, \quad \beta_2 = \begin{pmatrix} 0 & 0 & -i & 0 \\ 0 & 0 & 0 & -i \\ i & 0 & 0 & 0 \\ 0 & i & 0 & 0 \end{pmatrix}, \quad \beta_3 = \begin{pmatrix} 1 & 0 & 0 & 0 \\ 0 & 1 & 0 & 0 \\ 0 & 0 & -1 & 0 \\ 0 & 0 & 0 & -1 \end{pmatrix} \tag{33}$$

These matrices are identical to the relative velocity matrices $(\gamma^5, \tilde{i}\,\gamma^5\gamma^0, \gamma^0)$ defined above, but with β_3 representing the parallel direction. This choice corresponds to the chiral representation except for a different sign convention. Rotations of these β-matrices change the *representation*, because they determine which matrix is associated with the wave velocity direction.

Starting from the one-dimensional wave factorization above, we allow the Lorentz boost to have arbitrary direction:

$$\psi = \dot{a}_0^{1/2} \exp(\beta_3 \boldsymbol{\sigma} \cdot \boldsymbol{\alpha}/2)[\delta_F \quad \delta_B \quad [1-\delta_B] \quad [1-\delta_F]]^T/\sqrt{2} \tag{34}$$

The Lorentz boost determines the wave velocity direction and the ratio of forward and backward waves along that direction.

For a given representation, there is still freedom in the orientation of two of the β-axes, which may be defined by a rotation angle ζ:

$$\psi = \dot{a}_0^{1/2} \exp(\beta_3 \boldsymbol{\sigma} \cdot \boldsymbol{\alpha}/2)\exp(i\beta_3 \zeta/2)[\delta_F \quad \delta_B \quad [1-\delta_B] \quad [1-\delta_F]]^T/\sqrt{2} \tag{35}$$

Next, we allow co-rotation of polarization and velocity (using $\sigma_i \sigma_j - \sigma_j \sigma_i = 2i\varepsilon_{ijk}\sigma_k$):

$$\psi = \dot{a}_0^{1/2} \exp(-i\boldsymbol{\sigma} \cdot \boldsymbol{\varphi}/2)\exp(\beta_3 \boldsymbol{\sigma} \cdot \boldsymbol{\alpha}/2)\exp(i\beta_3 \zeta/2)[\delta_F \quad \delta_B \quad [1-\delta_B] \quad [1-\delta_F]]^T/\sqrt{2} \tag{36}$$

Since Lorentz boosts can exchange forward and backward waves, and rotations can exchange positive and negative components, the delta-functions are no longer necessary. A general wave can be obtained by transforming an initial wave consisting of forward and backward positive components:

$$\psi = \dot{a}_0^{1/2} \exp(-i\boldsymbol{\sigma} \cdot \boldsymbol{\varphi}/2)\exp(\beta_3 \boldsymbol{\sigma} \cdot \boldsymbol{\alpha}/2)\exp(i\beta_3 \zeta/2)[1 \quad 0 \quad 1 \quad 0]^T/\sqrt{2} \tag{37}$$

There are eight free parameters in this factorization, enough to uniquely determine all complex components of the bispinor. For the original Dirac representation, we rotate the basis vectors ($\beta_1, \beta_2, \beta_3$) to let the velocity matrices be represented by $\beta_1 \boldsymbol{\sigma}$:

$$\psi_0 \equiv \exp(-i\beta_2 \pi/4)[1 \quad 0 \quad 1 \quad 0]^T / \sqrt{2} \tag{38}$$

so the factorization is:

$$\psi = \dot{a}_0^{1/2} \exp(-i\boldsymbol{\sigma} \cdot \boldsymbol{\varphi}/2)\exp(\beta_1 \boldsymbol{\sigma} \cdot \boldsymbol{\alpha}/2)\exp(i\beta_1 \zeta/2)\psi_0 \tag{39}$$

This factorization is equivalent to that of Hestenes.[6] The time derivative is:

$$\partial_t \psi = \exp(-i\boldsymbol{\sigma} \cdot \boldsymbol{\varphi}/2)\partial_t \left[\dot{a}_0^{1/2} \exp(\beta_1 \boldsymbol{\sigma} \cdot \boldsymbol{\alpha}/2)\psi_1\right] - \frac{i\boldsymbol{\sigma}}{2} \cdot \dot{\boldsymbol{\varphi}} \psi \tag{40}$$

Now we assume that the first term can be replaced by spatial derivatives corresponding to wave motion and convection by motion of the medium with velocity $\mathbf{u}$ and vorticity $\mathbf{w} = \dot{\boldsymbol{\varphi}}$. This yields:

$$\partial_t \psi = -c\beta_1 \boldsymbol{\sigma} \cdot \nabla \psi - \mathbf{u} \cdot \nabla \psi - \mathbf{w} \cdot \frac{i\boldsymbol{\sigma}}{2} \psi \tag{41}$$

The free-particle Dirac equation is obtained by replacing the convection and rotation terms with $-iM\beta_3\psi$:

$$\partial_t \psi = -c\beta_1 \boldsymbol{\sigma} \cdot \nabla \psi - iM\beta_3 \psi \tag{42}$$

with $M = m_e c^2 / \hbar$.

Although the mass term replaces convection and rotation in the bispinor equation, we do not assume equality between these terms. Rather, we propose that the mass term also includes some spatial derivatives from the term $c\beta_1\boldsymbol{\sigma} \cdot \nabla \psi$. This interpretation would explain why electrons may be treated as plane waves in quantum mechanics in spite of the radially decreasing $1/r$ dependence of the influence of each electron through the electrostatic potential. However, derivation of the exact relationship between the classical bispinor equation (41) and Dirac's equation (42) is beyond the scope of this work.

2.3. Classical Interpretation

We obtain a three dimensional wave equation by multiplying $\psi^\dagger \sigma_j$ by the classical equation (41) and adding the complex conjugate:

$$\frac{\partial}{\partial t}\left[\psi^\dagger\sigma_j\psi\right] = -c\partial_j\left[\psi^\dagger\beta_1\psi\right] + \mathrm{i}\,c\varepsilon_{ijk}\left\{\partial_i\psi^\dagger\beta_1\sigma_k\psi + \psi^\dagger\beta_1\sigma_k\partial_i\psi\right\}$$
$$-\mathbf{u}\cdot\nabla\left[\psi^\dagger\sigma_j\psi\right] + \varepsilon_{kij}w_k\left[\psi^\dagger\sigma_i\psi\right] \tag{43}$$

The classical interpretations of these terms (which define derivatives of the variable $\mathbf{Q}$) are:

$$\partial_t^2 Q_j = \frac{1}{2}\partial_t\left[\psi^\dagger\sigma_j\psi\right]$$

$$c^2\partial_j[\nabla\cdot\mathbf{Q}] = -\frac{1}{2}c\partial_j\left[\psi^\dagger\beta_1\psi\right] \tag{44}$$

$$c^2\left\{\nabla\times\nabla\times\mathbf{Q}\right\}_j = -\frac{\mathrm{i}}{2}c\varepsilon_{ijk}\left\{\partial_i\psi^\dagger\beta_1\sigma_k\psi - \psi^\dagger\beta_1\sigma_k\partial_i\psi\right\}$$

Compared with the one-dimensional case, each component of the time derivative has the same form, but the spatial derivatives now consist of a gradient plus a curl as expected for a vector wave. Using the relation $\nabla^2\mathbf{Q} = \nabla[\nabla\cdot\mathbf{Q}] - \nabla\times\nabla\times\mathbf{Q}$ the corresponding wave equation is:

$$\partial_t^2\mathbf{Q} - c^2\nabla^2\mathbf{Q} + \mathbf{u}\cdot\nabla\dot{\mathbf{Q}} - \mathbf{w}\times\dot{\mathbf{Q}} = 0 \tag{45}$$

This is a classical wave equation for a vector wave with polarization $\mathbf{Q}$ with convection and rotation.

If we assume that the wave represents rotational oscillations in an elastic solid with $\rho\mathbf{u} = \left(\nabla\times\dot{\mathbf{Q}}\right)/2$, then we can interpret $\mathbf{Q}$ as an angular potential and derive the following dynamical variables:

$$\mathbf{S} = \dot{\mathbf{Q}}$$

$$\mathbf{a}_\mathrm{T} = \frac{1}{2\rho}\nabla\times\mathbf{Q}$$

$$\mathbf{q} = \rho\mathbf{u} = \rho\dot{\mathbf{a}}_\mathrm{T} = \frac{1}{2}\nabla\times\mathbf{S} = \frac{1}{2}\nabla\times\dot{\mathbf{Q}} \tag{46}$$

$$\varphi = -\frac{1}{4\rho}\nabla^2\mathbf{Q}$$

$$\mathbf{w} = \dot{\varphi} = -\frac{1}{4\rho}\nabla^2\mathbf{S} = -\frac{1}{4\rho}\nabla^2\dot{\mathbf{Q}}$$

61

where **S** is the angular momentum density of rotations of the medium, $\mathbf{a}_T$ is the (transverse) displacement, ρ is the inertial density of the medium, **u** is the velocity of the medium, **q** is the linear momentum density of motion of the medium, ($\boldsymbol{\varphi}$) is the rotation angle (proportional to torque density), and **w** is the vorticity.

If the vorticity is computed from the angular potential as above, the wave equation (45) is nonlinear, as is the corresponding bispinor equation (41). Such nonlinear equations typically have quantized solutions, and we propose that these soliton solutions correspond to elementary particles. Several investigators have attempted to explain quantization using nonlinear Dirac bispinor equations.[15-19] This appears to be the first time that the form of the proposed nonlinearity has been derived from a physical model. However, solution of this soliton equation is beyond the scope of this paper.

3. Electron Waves

3.1. Free Electron Equation

The free-particle wave equation for a stationary state with energy eigenvalue E is:

$$\left[-iE + c\beta_1 \boldsymbol{\sigma} \cdot \nabla\right]\psi = -iM\beta_3\psi \tag{47}$$

The operator $\boldsymbol{\sigma} \cdot \nabla \psi$ can be factored:

$$\boldsymbol{\sigma} \cdot \nabla \psi = \sigma_r \left[\partial_r + i\frac{\boldsymbol{\sigma}}{r} \cdot [\mathbf{r} \times \nabla]\right]\psi = \sigma_r \left[\partial_r - \frac{\boldsymbol{\sigma} \cdot \mathbf{L}}{r}\right]\psi \tag{48}$$

Letting $\kappa \equiv l+1$, the two-component angular solutions of the eigenvalue equations $\boldsymbol{\sigma} \cdot \mathbf{L}\Phi_{l,m}^{(+)} = l = -1+\kappa$ and $\boldsymbol{\sigma} \cdot \mathbf{L}\Phi_{l,m}^{(-)} = -[l+2] = -1-\kappa$ are well known.[20] These two angular solutions are related by $\sigma_r \Phi_{l,m}^{(+)} = \Phi_{l,m}^{(-)}$ and yield opposite eigenvalues of the parity (spatial inversion) operation.

Denote two wave functions as:

$$\psi_{l,m}^{(+)} = \begin{bmatrix} \tilde{i} \, G\Phi_{l,m}^{(+)} \\ F\Phi_{l,m}^{(-)} \end{bmatrix} \quad \text{or} \quad \psi_{l,m}^{(-)} = \begin{bmatrix} \tilde{i} \, F\Phi_{l,m}^{(-)} \\ G\Phi_{l,m}^{(+)} \end{bmatrix} \tag{49}$$

Each of these is an eigenfunction of the conventional parity operator, but they are exchanged by the correct parity operator:

$$P\psi_{l,m}^{(+)}(\mathbf{r}) = \tilde{i}\,\gamma^0\gamma^5\psi_{l,m}^{(+)\#}(-\mathbf{r}) = (-)^l\psi_{l,m}^{(-)}(\mathbf{r})$$
$$P\psi_{l,m}^{(-)}(\mathbf{r}) = \tilde{i}\,\gamma^0\gamma^5\psi_{l,m}^{(-)\#}(-\mathbf{r}) = (-)^{l+1}\psi_{l,m}^{(+)}(\mathbf{r})$$

(50)

Using $\psi_{l,m}^{(+)}$ yields the coupled radial equations:

$$[E - M]G + c\left[\partial_r + \frac{1}{r} + \frac{\kappa}{r}\right]F = 0$$
$$[E + M]F - c\left[\partial_r + \frac{1}{r} - \frac{\kappa}{r}\right]G = 0$$

(51)

$\psi_{l,m}^{(-)}$ yields similar coupled equations with the sign of E reversed. Therefore the new parity operator is associated with exchange of matter and anti-matter. This result means that electrical charges are associated with left- and right-handedness. Therefore charge neutrality may be interpreted as a balance between left- and right-handed physical processes.

3.2. Velocity Rotation and Mass

It is instructive to compute the effect of mass on the wave velocity:

$$\frac{d}{dt}\left[\psi^{(+)\dagger}\beta_1\sigma_i\psi^{(+)}\right] = \psi^{(+)\dagger}\beta_1\sigma_i\left[-iM\beta_3\psi^{(+)}\right] + \left[-iM\beta_3\psi^{(+)}\right]^{\dagger}\beta_1\sigma_i\psi^{(+)}$$

$$= -2\frac{M}{r^2}FG\left\{\Phi_{l,m}^{(+)\dagger}\sigma_i\Phi_{l,m}^{(-)} + \Phi_{l,m}^{(-)\dagger}\sigma_i\Phi_{l,m}^{(+)}\right\}$$

$$= 2M\frac{FG}{r^2}\left\{\Phi_{l,m}^{(+)\dagger}[\sigma_i\sigma_r + \sigma_r\sigma_i]\Phi_{l,m}^{(+)}\right\}$$

$$= 4M\frac{FG}{r^2}\delta_{ir}\left\{\Phi_{l,m}^{(+)\dagger}\Phi_{l,m}^{(+)}\right\}$$

(52)

The mass term represents a radial acceleration of the wave, which is inward provided that the appropriate sign is chosen for M. This is consistent with an interpretation of electrons as soliton waves.

3.3. Hamiltonian Formulation

Hamilton's equations of motion have the form:[21]

$$\partial_t \psi = \frac{\partial H}{\partial p_\psi} \tag{53}$$

where ψ is a field variable and p_ψ is the conjugate momentum to the field defined by:

$$p_\psi = \frac{\partial H}{\partial [\partial_t \psi]} \tag{54}$$

We can fit the bispinor equation to this form by taking the momentum conjugate to the wave function to be:

$$p_\psi = \frac{i}{2} \psi^\dagger \tag{55}$$

and the Hamiltonian is:

$$H = \frac{i}{2} \left\{ \psi^\dagger \partial_t \psi - [\partial_t \psi]^\dagger \psi \right\} = \frac{1}{2} \psi^\dagger \left\{ -i c \beta_1 \boldsymbol{\sigma} \cdot \nabla - i \mathbf{u} \cdot \nabla + \mathbf{w} \cdot \frac{\boldsymbol{\sigma}}{2} \right\} \psi + c.c. \tag{56}$$

From here on we will remove the factor of ½ and simply discard the imaginary part. The Hamiltonian H will have units of energy if the wave polarization has units of angular momentum.

We can also define a Hamiltonian operator with $\partial_t \psi = -i H \psi$:

$$H = -i c \beta_1 \boldsymbol{\sigma} \cdot \nabla - i \mathbf{u} \cdot \nabla + \mathbf{w} \cdot \frac{\boldsymbol{\sigma}}{2} \tag{57}$$

Using $\mathbf{w} = \nabla \times \mathbf{u}/2$, the conjugate momentum for $\mathbf{r}$ is:

$$\mathbf{p_r} = \frac{\delta H}{\delta [\mathbf{u}]} = \psi^\dagger \left\{ -i \nabla - \frac{\boldsymbol{\sigma}}{4} \times \nabla \right\} \psi = \mathbf{p} + \mathbf{q} \tag{58}$$

The quantity $\mathbf{p} = \psi^\dagger \{ -i \nabla \} \psi$ is familiar from quantum mechanics. The quantity $\mathbf{q} = (\nabla \times \mathbf{S})/2$ is associated with rotations of the medium and is assumed to have an average value of zero.

64

Defining the rotational component of velocity by $\mathbf{u}_R = \mathbf{r} \times \dot{\boldsymbol{\varphi}}$, the conjugate momentum for rotation is:

$$\mathbf{p}_\varphi = \frac{\delta H}{\delta \dot{\boldsymbol{\varphi}}} = \psi^\dagger \left\{ -i\mathbf{r} \times \nabla + \frac{1}{2}\boldsymbol{\sigma} \right\} \psi = \mathbf{L} + \mathbf{S} \tag{59}$$

This angular momentum operator includes both orbital and spin components just as in Dirac theory. The spin component is the angular momentum associated with motion of the medium, while the orbital component is the angular momentum associated with motion of the wave.

The conjugate momentum for time is simply H itself:

$$\mathbf{p}_t = H \tag{60}$$

3.4. Wave Interference and Potentials

Next we investigate the origin of electromagnetic potentials. Observables (scalars and vectors) should be additive when two waves are superposed. This implies that when two waves ψ_A and ψ_B are superposed, the total wave ψ_T has the property that:

$$\psi_T^\dagger G \psi_T = \psi_A^\dagger G \psi_A + \psi_B^\dagger G \psi_B \tag{61}$$

for any Hermitian operator G. If we simply added the two wave functions, we would have instead:

$$\left[\psi_A^\dagger + \psi_B^\dagger \right] G \left[\psi_A + \psi_B \right] = \psi_A^\dagger G \psi_A + \psi_B^\dagger G \psi_B + \psi_A^\dagger G \psi_B + \psi_B^\dagger G \psi_A \tag{62}$$

The additional terms are clearly not zero in general. Suppose we wish to treat ψ_A as being independent of ψ_B. We can introduce phase shifts to ψ_B in order to cancel the interference terms. For $G=1$ a scalar phase is sufficient: $\psi_B' \equiv \exp(-i\chi/2)\psi_B$. The new wave function may now be regarded as independent of ψ_A.

$$\psi_A^\dagger \psi_B' + \psi_B'^\dagger \psi_A = 0 \tag{63}$$

This assumption of independent waves is just the Pauli Exclusion Principle. However, some observables computed from the independent wave $\psi_B' \equiv \exp(-i\chi)\psi_B$ may differ from those of the free particle wave:

$$\psi_B^\dagger G \psi_B = \psi_B'^\dagger [\exp(-i\chi) G \exp(i\chi/2)]\psi_B' = \psi_B'^\dagger G' \psi_B' \tag{64}$$

Hence the effect of wave interference is to change the operator for wave packet ψ'_B from G to G':

$$G' = \exp(-i\chi/2)G\exp(i\chi/2) \tag{65}$$

The equation of motion of wave packet ψ'_B is:

$$\{\partial_t + [\exp(-i\chi/2)\partial_t\exp(i\chi/2)]\}\psi'_B = \exp(-i\chi/2)iH\exp(i\chi/2)\psi'_B \tag{66}$$

Substituting the general form of the Hamiltonian:

$$\begin{aligned}
&\left[\partial_t + \frac{i}{2}\partial_t\chi\right]\psi'_B + \left[c\beta_1\boldsymbol{\sigma}\cdot\nabla + \frac{i}{2}c\beta_1\boldsymbol{\sigma}\cdot\nabla\chi\right]\psi'_B \\
&+ \left[(\mathbf{u}_A + \mathbf{u}_B)\cdot\nabla + \frac{i}{2}(\mathbf{u}_A + \mathbf{u}_B)\cdot\nabla\chi\right]\psi'_B + \frac{i}{2}[\boldsymbol{\sigma}\cdot(\dot{\boldsymbol{\varphi}}_A + \dot{\boldsymbol{\varphi}}_B)]\psi'_B = 0
\end{aligned} \tag{67}$$

Substituting the mass term for the free electron:

$$\begin{aligned}
&\left[\partial_t + \frac{i}{2}\partial_t\chi\right]\psi'_B + \left[c\beta_1\boldsymbol{\sigma}\cdot\nabla + \frac{i}{2}c\beta_1\boldsymbol{\sigma}\cdot\nabla\chi\right]\psi'_B \\
&+ \left[\mathbf{u}_A\cdot\nabla + \frac{i}{2}(\mathbf{u}_A + \mathbf{u}_B)\cdot\nabla\chi\right]\psi'_B + \frac{i}{2}[\boldsymbol{\sigma}\cdot\dot{\boldsymbol{\varphi}}_A]\psi'_B + iM\beta_3\psi'_B = 0
\end{aligned} \tag{68}$$

Since we are interested in the effects of the phase shift, we will neglect the extra terms which are independent of χ (without explicit justification). We then define the electromagnetic potentials as:

$$\begin{aligned}
e\mathbf{A} &\equiv -\frac{c}{2}\nabla\chi \\
e\Phi &\equiv \frac{1}{2}\partial_t\chi - \mathbf{u}\cdot\frac{e}{c}\mathbf{A}
\end{aligned} \tag{69}$$

The curl of $\mathbf{A}$ (magnetic field) may be nonzero because χ is a phase angle which may be multi-valued. For example, the multi-valued function $\chi = \arctan(x_2/x_1.)$ has gradient components:

$$\partial_1 \chi = -\frac{x_2}{\left[x_1^2 + x_2^2\right]^{(1/2)}}$$

$$\partial_2 \chi = \frac{x_1}{\left[x_1^2 + x_2^2\right]^{(1/2)}}$$

(70)

The curl of this gradient is clearly non-zero. See Kleinert[22] for a discussion of multi-valued potentials in electromagnetism.

With these definitions, the electron equation in the presence of another wave becomes:

$$[\partial_t + i e\Phi]\psi'_B + [c\beta_1\boldsymbol{\sigma}\cdot\nabla - i e\beta_1\boldsymbol{\sigma}\cdot\mathbf{A}]\psi'_B + i M\beta_3\psi'_B = 0$$

(71)

Electromagnetic potentials result from wave interference under the assumption that different wave packets are independent. These potentials also cancel interference terms when adding spins (at least in directions for which either wave function is an eigenfunction). Other potentials may arise from the addition rules for other observables.

3.5. Lorenz Force

In terms of electromagnetic potentials, the modified Hamiltonian is:

$$H = e\Phi - i c\beta_1\boldsymbol{\sigma}\cdot\nabla - c\beta_1\boldsymbol{\sigma}\cdot\frac{e}{c}\mathbf{A} + i\mathbf{u}\cdot\nabla + \dot{\boldsymbol{\varphi}}\cdot\frac{\boldsymbol{\sigma}}{2}$$

$$= e\Phi - i c\beta_1\boldsymbol{\sigma}\cdot\nabla - c\beta_1\boldsymbol{\sigma}\cdot\frac{e}{c}\mathbf{A} + M\beta_3$$

(72)

Using $\dot{\boldsymbol{\varphi}} = \nabla \times \mathbf{u}/2$ and noting the $\mathbf{u}$-dependence of the electrostatic potential above, the conjugate momentum for $\mathbf{r}$ is now:

$$\mathbf{p_r} = \frac{\delta H}{\delta[\mathbf{u}]} = \psi^\dagger\left\{-i\nabla - \frac{e}{c}\mathbf{A} - \frac{\boldsymbol{\sigma}}{4}\times\nabla\right\}\psi = \mathbf{p} + \mathbf{q}$$

(73)

The time derivative of any observable Q is:

$$\partial_t[\psi^\dagger Q\psi] = [\partial_t\psi^\dagger]Q\psi + \psi^\dagger Q\partial_t\psi + \psi^\dagger[\partial_t Q]\psi = \psi^\dagger i[Q,H]\psi + \psi^\dagger\partial_t Q\psi$$

(74)

An example of this is the force density. Substituting the linear wave momentum for Q yields the Lorenz force law:

$$\partial_t \mathbf{p} = \psi^\dagger \left\{ \nabla \left[c\beta_1 \boldsymbol{\sigma} \cdot \frac{e}{c} \mathbf{A} \right] - \nabla e\Phi - c\beta_1 \boldsymbol{\sigma} \cdot \nabla \left[\frac{e}{c} \mathbf{A} \right] - \frac{e}{c} \frac{\partial}{\partial t} \mathbf{A} \right\} \psi$$

$$= \psi^\dagger \left\{ c\beta_1 \boldsymbol{\sigma} \times \left[\nabla \times \frac{e}{c} \mathbf{A} \right] - \nabla e\Phi - \frac{e}{c} \frac{\partial}{\partial t} \mathbf{A} \right\} \psi = \psi^\dagger \left\{ c\beta_1 \boldsymbol{\sigma} \times \frac{e}{c} \mathbf{B} + e\mathbf{E} \right\} \psi$$

$$(75)$$

where $\mathbf{E}$ and $\mathbf{B}$ are the usual electric and magnetic fields, respectively. Hence the Lorenz force has a straightforward interpretation in terms of classical wave interference.

3.6. Measurement Correlations

It is widely believed that the correlations between polarization measurements of entangled particles cannot be predicted classically. This belief is based on correlation predictions using an equation of the form described by John Bell:[23]

$$P(\mathbf{a}, \mathbf{b}) = \int A(\mathbf{a}, \lambda_1, ..., \lambda_n) B(\mathbf{b}, \lambda_1, ..., \lambda_n) \rho(\lambda_1, ..., \lambda_n) d\lambda_1 .. d\lambda_n \qquad (76)$$

where λ_i represent variables which describe the state of the system, $\rho(\lambda_1, ..., \lambda_n)$ is the probability distribution of these variables, $\mathbf{a}$ and $\mathbf{b}$ are the measured polarization directions for the two entangled particles, A and B are the theoretical outcomes of the measurement (± 1), and $P(\mathbf{a}, \mathbf{b})$ is the correlation.

Bell proved that quantum correlations cannot be represented in this form.[23] In particular, he proved that for three different measurements $A(\mathbf{a}, \lambda_1, ..., \lambda_n)$, $B(\mathbf{b}, \lambda_1, ..., \lambda_n)$, and $C(\mathbf{c}, \lambda_1, ..., \lambda_n)$:

$$1 + P(\mathbf{b}, \mathbf{c}) \geq |P(\mathbf{a}, \mathbf{b}) - P(\mathbf{a}, \mathbf{c})| \qquad (77)$$

This condition is violated by quantum mechanical (and physically observed) correlations, which can be measured using two or more particles whose spins are constrained. For example, if a pair of spin ½ particles is produced with opposite spin, the observed correlation between their spin measurements by detectors oriented with relative angle φ is:

$$C_{\text{pair}}(\varphi) = -\cos\varphi \qquad (78)$$

This correlation violates Bell's condition. For example, if the detectors $\mathbf{a}$, $\mathbf{b}$, and $\mathbf{c}$ are oriented at angles 0, $\pi/4$, and $3\pi/4$, respectively, then:

$$C(\mathbf{a},\mathbf{b}) = C(\pi/4) = -1\big/\sqrt{2}$$
$$C(\mathbf{b},\mathbf{c}) = C(\pi/2) = 0$$
$$C(\mathbf{a},\mathbf{c}) = C(3\pi/4) = 1\big/\sqrt{2} \qquad\qquad (79)$$
$$1 + C(\mathbf{b},\mathbf{c}) < \big|C(\mathbf{a},\mathbf{b}) - C(\mathbf{a},\mathbf{c})\big|$$

The key assumption of Bell's Theorem is that the correlation is computed by multiplying the theoretical measurement results, A and B, in the integral. This assumes that for a given set of parameters, the measurement result essentially propagates to the detector along with the particle. In fact, however, it is the bispinor wave function ψ which propagates from place to place since the first order Dirac equation is a kind of convection equation. Therefore Bell's Theorem does not generally apply to classical waves. Since ψ has spin one-half, correlation between states related by rotation $R(\varphi)$ about an axis perpendicular to the spin is:

$$C = \frac{\big|\psi^{\dagger} R(\varphi)\psi\big|^2}{\big|\psi^{\dagger}\psi\big|^2} = \cos^2\frac{\varphi}{2} \qquad\qquad (80)$$

The correlation for angle $(\pi\hat{\varphi} - \varphi)$ is $\cos^2[(\pi - \varphi)/2] = \sin^2[\varphi/2]$.

Assuming that spin measurements are coincident or anti-coincident in proportion to the correlations between the spinor wave functions, the correlation C_s between spin measurements separated by angle φ is:

$$C_s(\varphi) = C_{\psi}(\varphi) - C_{\psi}(\pi - \varphi) = \cos^2\frac{\varphi}{2} - \sin^2\frac{\varphi}{2} = \cos\varphi \qquad\qquad (81)$$

In the case of pair production in EPR-type experiments, the spins of the two electrons (or electron and positron) are opposite (changing φ to $\pi - \varphi$ above), thereby changing the sign of the correlation. Hence the classical correlations are in agreement with the quantum correlations.

3.7   Quantum Mechanics

In the preceding section we computed the correlation between two states related by rotation. The two states may be denoted by $\psi(\mathbf{r},t)$ and $R(\varphi)\psi(\mathbf{r},t)$. The correlation at a given position and time is given by (80). A more global correlation between two wave functions $\psi_1(\mathbf{r},t)$ and $\psi_2(\mathbf{r},t)$ at a given time is obtained by integrating over space:

$$C = \frac{\left| \int \psi_1^\dagger \psi_2 \, d^3 r \right|^2}{\left| \int \psi_1^\dagger \psi_1 \, d^3 r \right| \left| \int \psi_2^\dagger \psi_2 \, d^3 r \right|} = \frac{\cdot \left| \int \psi_2^\dagger \psi_1 \, d^3 r \right|^2}{\left| \int \psi_1^\dagger \psi_1 \, d^3 r \right| \left| \int \psi_2^\dagger \psi_2 \, d^3 r \right|} \tag{82}$$

The correlation between spin one-half states is non-negative, and the correlation of a wave function with itself is unity. These properties provide the basis for a probabilistic interpretation of the wave functions. If a wave function is decomposed into multiple wave functions (states), then the correlation between the wave function and each 'state' may be interpreted as the probability of occurrence of that state for the given wave function. Temporal evolution of the wave function is expressed as:

$$\psi(\mathbf{r}, t_2) = \exp\left(-i \int_{t_1}^{t_2} H(\mathbf{r}, t) \, dt \right) \psi(\mathbf{r}, t_1) \tag{83}$$

Therefore the probability that an initial state $\psi_1(\mathbf{r}, t_1)$ will evolve into a final state $\psi_2(\mathbf{r}, t_2)$ is:

$$C(\psi_2(t_2) \,|\, \psi_1(t_1)) = \frac{\left| \int \psi_2^\dagger \exp\left(-i \int_{t_1}^{t_2} H(\mathbf{r}, t) \, dt \right) \psi_1 \, d^3 r \right|^2}{\left| \int \psi_1^\dagger \psi_1 \, d^3 r \right| \left| \int \psi_2^\dagger \psi_2 \, d^3 r \right|} \tag{84}$$

In quantum mechanics, the states are normalized to one:

$$\psi' = \frac{\psi}{\left| \int \psi^\dagger \psi \, d^3 r \right|^{1/2}} \tag{85}$$

Dropping the primes, the correlation integrals are then written in the form:

$$C(\psi_2(t_2) \,|\, \psi_1(t_1)) = \left| \langle \psi_2 \| \psi_1 \rangle \right|^2 = \left| \langle \psi_1 | \exp\left(-i \int_{t_1}^{t_2} H(\mathbf{r}, t) \, dt \right) \| \psi_1 \rangle \right|^2 \tag{86}$$

This means that correlations between physical states are equal to the square of a complex amplitude. This statistical property of matter is due to the simple fact that independent wave states are 180 degrees apart.

4. Wave Properties of Matter

We have shown that classical wave theory provides a qualitatively correct description of electron dynamics. This result lends support to recent efforts to revive the classical aether (or ether) as a medium of propagation of matter waves. Duffy[24] has surveyed modern aether theory.

The model of vacuum as an ideal elastic solid was quite successful in explaining classical properties of light in the 19th century.[25] Quantum effects are only apparent in interactions with matter, which evidently consists of classical soliton waves. Our classical equation for the angular potential is essentially empirical, although all of the terms have intuitive interpretations in terms of rotations in an elastic solid. At present there appears to be no rigorous description of rotational waves in an ideal elastic medium. Kleinert[26] attempted to include rotations in the elastic energy but was compelled to introduce new elastic constants dependent on an arbitrary scale length. Close[14] showed that torsion waves (with rotation axis parallel to wave velocity) can be described by a Dirac equation, but did not derive a general equation (and also incorrectly related mass and velocity rotation).

Many physical properties of matter can be derived from a wave model of matter. The Uncertainty Principle applies to all classical waves. Lorentz invariance is also a property of waves, and Special Relativity is therefore a consequence of any wave theory of matter.[27] Parity conservation in physical processes, as demonstrated above, is consistent with matter waves propagating in a Galilean space-time, and provides a rationale for charge neutrality.

5. Conclusions

In this paper we interpret the Dirac equation as a second-order wave equation whose first order spatial and temporal derivatives are represented by a bispinor wave function. The parity transformation is corrected and the free electron Hamiltonian is generalized with a classical interpretation modeling waves in an elastic solid. Mass is assumed to be derived from spatial derivatives, convection, and rotation operators. Half-integer spin is attributable to the co-existence of waves traveling in opposite directions along the gradient axis. The wave function in a given representation can be factored into constant matrix, an amplitude, a three-dimensional Lorentz velocity boost, and a rotation operator. Wave interference yields both the Pauli Exclusion Principle and the Lorenz force. Mass is related to angular velocity and associated with radially inward acceleration of the wave, implying a soliton. The theory is consistent with parity conservation for physical laws, which is the simplest explanation for the mirror symmetry ordinarily associated with exchange of matter and anti-matter. Correlations between rotated states are the same as for quantum theory. Bell's Theorem does not apply to classical electron waves because it is the bispinor wave functions, and not the measurement values, which convect from place to place. Hence classical wave theory can correctly describe electron dynamics and offers the possibility of a simple mechanical model of the vacuum.

Acknowledgments

The author is grateful to Damon Merari for his interest and encouragement. Thanks also to Peter Rowlands and Lorenzo Sadun for helpful discussions during the course of this research. The author also thanks a referee for suggesting some clarifications.

References

1. de Broglie L.V.: Recherches sur la Theorie des Quanta. Ph.D. thesis, University of Sorbonne, Paris (1924)
2. Davisson, C., Germer, L.H.: Diffraction of Electrons by a Crystal of Nickel. Phys. Rev. 30, 705–740 (1927)
3. Thomson, G.P., Reid, A.: Diffraction of cathode rays by a thin film. Nature 119, 890-895 (1927)
4. Takabayashi, Y.: Relativistic hydrodynamics of the Dirac matter. Suppl. Prog. Theor. Phys. 4(1), 1-80 (1957)
5. Hestenes, D.: Local observables in the Dirac theory. J. Math. Phys. 14(7), 893-905 (1973)
6. Hestenes, D.: Real Spinor Fields. J. Math. Phys. 8(4), 798-808 (1967)
7. Hestenes, D.: The Zitterbewegung Interpretation of Quantum Mechanics. Found. Phys. 20(10), 1213-1232 (1990)
8. Rowlands, P.: The physical consequences of a new version of the Dirac equation. In: Hunter, G. et al. (eds.) Causality and Localaity in Modern Physics and Astronomy: Open Questions and Possible Solutions. Fundamental Theories of Physics, vol. 97, pp. 397-402. Kluwer Academic Publishers, Dordrecht (1998)
9. Rowlands, P.: Removing redundancy in relativistic quantum mechanics. arXiv:physics/0507188 (2005)
10. Rowlands, P., Cullerne, J.P.: The connection between the Han-Nambu quark theory, the Dirac equation and fundamental symmetries. Nucl. Phys. A 684, 713-715 (2001)

11. Schiff, L.I.: Quantum Mechanics, Third Edition pp. 467-471. McGraw-Hill, New York (1968)

12. Lee, T.D. and Yang, C.N.: Question of Parity Conservation in Weak Interactions. Phys. Rev. 104, 254-258 (1956)
13. Wu, C.S., et al.: Experimental test of parity conservation in beta decay. Phys. Rev. 105, 14 (1957)

14. Close, R.A.: Torsion Waves in Three Dimensions: Quantum Mechanics with a Twist. Found. Phys. Lett. 15, 71-83 (2002)

15. Rañada, A.F.: Classical Nonlinear Dirac Field Models of Extended Particles. In: Barut, A.O. (ed.) Quantum Theory, Groups, Fields, and Particles, pp. 271-288. Reidel, Amsterdam (1983)

16. Fushchych, W., Zhdanov, R.: Symmetries and Exact Solutions of Nonlinear Dirac Equations. Mathematical Ukraina, Kyiv (1997, Russian version 1992)

17. Gu, Y.Q.: Some Properties of the Spinor Soliton. Adv. Appl. Clifford Algebras 8(1), 17-29 (1998)

18. Bohun, C.S., Cooperstock, F.I.: Dirac-Maxwell solitons. Phys. Rev. A 60(6), 4291-4300 (1999)

19. Maccari, A.: Nonlinear field equations and solitons as particles. Electron. J. Theor. Phys. 3(10), 39–88 (2006)

20. Bjorken, J.D., Drell, S.D.: Relativistic Quantum Mechanics, p 53. McGraw-Hill, New York (1964)

21. Morse, P.M., Feshbach, H.: Methods of Theoretical Physics, vol. I, pp. 304-306. McGraw-Hill, New York (1953).

22. Kleinert, H.: Mutivalued Fields in Condensed Matter, Electromagnetism, and Gravitation, p. 119. World Scientific, Singapore (2007)

23. Bell, J. S.: On the Einstein-Podolsky-Rosen paradox. Physics 1, 195-200 (1965)

24. Duffy, M.C.: The Ether Concept in Modern Physics. In: Dvoeglazov, V. (ed.) Einstein and Poincaré: the Physical Vacuum, pp. 11-34. Apeiron, Montreal (2006)

25. Whittaker, E.: A History of the Theories of Aether and Electricity, vol. 1, pp. 128-169. Thomas Nelson and Sons Ltd., Edinburgh (1951)

26. Kleinert, H.: Gauge Fields in Condensed Matter, vol. II, p. 1259. World Scientific, Singapore (1989)

27. Close, R.A.: The Other Meaning of Special Relativity. Http://www.classicalmatter.org/ClassicalTheory/OtherRelativity.doc (2001)

Mass and energy in the light of aether theory

Joseph Levy

4 Square Anatole France, 91250, St Germain-lès-Corbeil, France
E-mail: levy.joseph@orange.fr

Abstract

The laws of physics dealing with mass and energy are reviewed in the light of the assumption of a preferred aether frame, which relies today on weighty theoretical and experimental arguments [1A,1B,1F,29] and [25-27]. The existence of such a privileged aether frame makes sure that clocks slow down and meter sticks contract, as a function of their speed with respect to this frame. These real physical processes are supported by their ability to rationally account for the *apparent* isotropy of the speed of light. Yet, the information they provide being dependent on their absolute speed, and therefore not invariant, moving standards and clocks give a distorted view of reality. Therefore the physical data are subjected to alterations and need to be corrected. As a result of these corrections, they assume a different mathematical form, which is the real value. In the text which follows we propose to highlight the corrected value of the basic laws dealing with mass and energy. This concerns the mass-energy equivalence law and the variation of mass with speed. The real proper mass of moving bodies is shown to vary as a function of their absolute speed, and the kinetic energy is shown not to be observer dependent. The compatibility of the relativity principle with mass-energy conservation is discussed, and the mass, is shown not to be an intrinsic property of matter, it depends on the presence of the aether. In the appendices, we show by which mechanisms the standard measurement procedures alter the physical data.

I. Introduction

In previous publications [1A, 1B] we saw that the measured values of the co-ordinates in the transformations of space and time, result from the distortions caused by length contraction, clock retardation and arbitrary clock synchronization which affect these measurements.

Consequently, the use of these transformations can not allow estimating the physical data on the basis of criteria independent of absolute velocity, which means that the expressions of the physical data derived from them are not reliable as such and must be corrected.

In the developments which follow, we shall study the consequences of this fact as regards the laws dealing with mass and energy and compare them to conventional relativity.

Conventional relativity assumes that, whatever the 'inertial frame' S considered, the energy content of a body measured by an observer at rest in this frame, is $E_0 = m_0 C^2$, where m_0, the rest mass, is the same in all 'inertial frames'. This implies that when a body moves from one 'inertial frame' S_1 to another S_2 receding from S_1 at speed v, its rest energy remains unchanged. The fact that the kinetic energy of the body has increased has, nevertheless, not changed its rest mass and its rest energy.

This viewpoint is simply untenable. Of course we are aware that, if we use a standard that also moves from S_1 to S_2 to measure the mass of the body by comparison, the rest mass will be (erroneously) found identical in S_1 and S_2. This is because the rest mass of the standard will have changed in the same ratio as the body's rest mass. But the real rest masses in S_1 and in S_2 are different.

This untenable point of view, results from the fact that according to relativity mass-energy is observer dependent. It can be overcome if one assumes the existence of a preferred aether frame.

Thus, even though they are difficult to estimate, the rest mass and the rest energy of a body increase when the speed of the body increases with respect to the fundamental frame. These issues will be studied in detail in the text which follows.

If the conventional space-time transformatons result from measurement distortions, they cannot be used as such to demonstrate the fundamental laws of physics, because, of course, they give a distorted view of reality. Therefore, the laws of physics determined from them, also need a correction. This is the case for the mass variation $m = m_0 \gamma$ which is generally derived from the conventional space-time transformations.

We shall nevertheless demonstrate, by means of arguments independent of relativity, that the law $m = m_0 \gamma$ applies, but, contrary to relativity, it applies as such only when a body is carried from the fundamental frame to any other frame. This is because m_0 is the rest mass in the fundamental frame, it is not the rest mass in other reference frames.

Insofar as a body is carried from one co-ordinate system, not at rest in the aether frame, to another, the law will take a different mathematical form contrary to what conventional relativity asserts. But this result requires that the measurement distortions are corrected. *It cannot be obtained with the usual measurement procedures which, in contrast, give rise to the conventional laws.* It therefore enables to correct some illogical consequences of relativity as the example studied below will show.

Some other issues will be addressed showing that the existence of a preferred frame gives rise to a number of significant differences between relativity and aether theory: among them the fact that the kinetic energy is not observer dependent, and that the mass is not an intrinsic property of matter, it depends on the presence of the aether.

The role played by lenth contraction, clock retardation and arbitrary clock synchronization in the distortion of the measured physical data will be highlighted in the appendices.

In appendix 1 we will show that, assuming the existence of a preferred aether frame, length contraction is a necessary condition so that the *measured* speed of light along a rigid path assumes the value C in any direction of space and independently of the absolute velocity of the reference frame where it is measured. This *measured* speed is actually different from the real one-way speed of light which depends on the angle as the demonstration will show.

The issue relative to the synchronization of clocks will be addressed in the appendix 2. This question is often ignored by the physics community. Yet, the usual synchronization procedures play an essential part in the alteration of the measured parameters, in addition to the alterations entailed by length contraction and clock retardation.

II. <u>A classical derivation of mass-energy equivalence based on aether theory</u>

Let us consider a body at rest in a co-ordinate system S_0 that emits N identical photons simultaneously in two opposite directions ($+x$ and $-x$,). See Figure1.
We assume that S_0 is firmly linked to the fundamental aether frame.
(For this demonstration, we will follow arguments given by Rohrlich, [2] but with different assumptions.)

Consider now another system S moving along the x-axis at constant speed v, (with $(v/C)^2 \ll 1$.) In S_0, the total momentum is conserved. This must also be true for any observer moving with respect to S_0. With respect to the system S, we have:

$$P_0 = P_1 + N\frac{hv}{C}\left(1+\frac{v}{C}\right) - N\frac{hv}{C}\left(1-\frac{v}{C}\right),$$

where P_0 is the initial momentum, and P_1 the final momentum of the body. The other terms are the momenta of the photons altered by the Doppler shift.
(Note that the relation $p=E/C$ which relates the energy and the quantity of motion was known before the formulation of relativity theory and does not depend on it. The formula can be derived on the basis of classical electrodynamics arguments [28]. Using the relation $E = hv$, it is easy to verify that the quantity of motion mv transmitted to a perfectly absorbing surface by any quantum of light is given by:

$$mv = h\frac{v}{C}$$

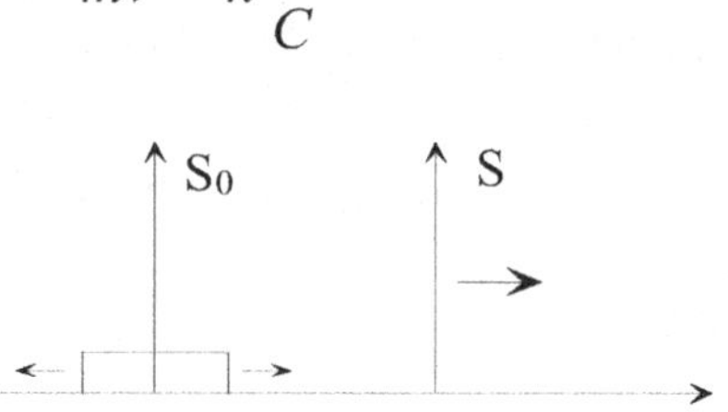

Figure 1. The body at rest in frame S_0 emits N identical photons in two opposite directions.

where v is the frequency of the light quantum.)

Since in the system S_0 the speed of ligt is isotropic, the role of the aether is identical in both directions, and therefore we can ignore it. This would be different if the body was standing in a frame different from the aether frame, particularly at high speed, contrary to the assumptions of Rohrlich based on special relativity.

Viewed from the system S, the momentum $\Delta(mv)$ lost by the body will be:

$$P_0 - P_1 = 2N\frac{hv}{C^2}v .$$

Since, obviously, the source is at rest in S_0 both before and after emission, it is clear that, with respect to the system S, it must have the speed v both before and after emission, thus:

$$\Delta(mv) = v\Delta m = \frac{2Nhv}{C^2}v . \tag{1}$$

Now, according to the energy conservation law:

$$E_0 = E_1 + Nhv\left(1+\frac{v}{c}\right) + Nhv\left(1-\frac{v}{c}\right) = E_1 + \Delta E \tag{2}$$
$$= E_1 + 2Nhv$$

$\Delta E = 2Nhv$ is the variation of energy resulting from the emission of the photons. From (1) and (2) we obtain

$$\Delta E = \Delta mC^2 . \tag{3}$$

Note that this mass-energy equivalence formula has been obtained without the help of the Lorentz-Poincaré transformations.

III. <u>Variation of mass with speed from the fundamental frame</u>

Let us consider a body *initially at rest in the fundamental frame*, which is subjected to a force F. The elementary expression of the kinetic energy acquired by the body in the displacement $d\ell$ is:

$$dE_C = Fd\ell = \frac{d(mv)}{dt}d\ell , \tag{4}$$

where $Fd\ell$ is the work carried out by the force F during the displacement. (We suppose that F and $d\ell$ are aligned.)

Now, as seen previously, the equivalence of mass and energy takes the form

$$E = mC^2 = E_C + m_0C^2$$

where m_0 is the rest mass assumed by the body in the fundamental frame.

Therefore

$$dE = dE_C = C^2dm . \tag{5}$$

From Equation (4) and (5) we have

78

$$C^2 dm = (v\frac{dm}{dt} + m\frac{dv}{dt})vdt,$$

which gives

$$\frac{dm}{dv}C^2 = \frac{dm}{dv}v^2 + mv,$$

and

$$\frac{dm}{m} = \frac{v}{C^2 - v^2}dv.$$

Denoting $C^2 - v^2$ as u so that $v\,d\,v = -du/2$, we then find

$$Log\ m = -\frac{1}{2}Log\left(C^2 - v^2\right) + Log\ k$$

$$= Log\ k\left(C^2 - v^2\right)^{-1/2}$$

and

$$m = \frac{k}{C\sqrt{1 - v^2/C^2}}.$$

For $v = 0 \Rightarrow m = k/C = m_0$, thus:

$$m = \frac{m_0}{\sqrt{1 - v^2/C^2}}. \qquad (6)$$

(Another equivalent demonstration has been given by Selleri [3], based on a work by Lewis)

As we shall see, in contrast to what conventional relativity asserts, expression (6) is completely exact only if m_0 represents the rest mass of the body in the fundamental frame. But this result can be revealed only after the alterations which affect the measurements have been corrected.

IV. <u>Different conceptions about mass increase with speed</u>

In relativity, since no fundamental frame exists, whatever the 'inertial frame' considered, the mass of a body attached to this frame, as seen by an observer at rest in it, is always the same. This mass is defined as the proper mass or the rest mass of the body.

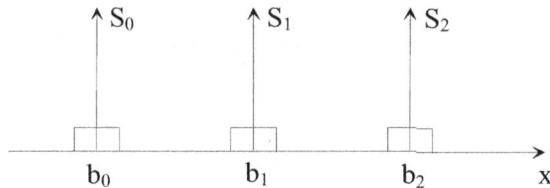

Figure 2. According to relativity, the rest masses of the three bodies are identical. This point of view is not shared by the fundamental aether theory.

If the body moves with respect to an 'inertial system' S with velocity v, its mass with respect to S is assumed to be:

$$m = \frac{m_0}{\sqrt{1 - v^2/C^2}},$$

whatever the system S may be. Therefore, the body is supposed to possess at the same time an infinite number of masses different from its rest mass, depending on the speed relative to it of the observer who measures the mass.

The point of view of the fundamental aether theory is quite different. Let us consider a body having mass m_0 in the fundamental system S_0. Since this body needs to acquire kinetic energy E_C in order to go from S_0 to any other system S moving at constant speed, the rest mass of the body in S will be $m_0 + E_C/C^2$. This means that a hierarchy of rest masses exists, each a function of the absolute speed of the body.

(Note that it is necessary to distinguish the real mass from the measured mass, which can be incorrectly determined. If we measure the mass m_0 of a body in the fundamental system S_0 by comparison with a standard μ_0 and if m_0 and μ_0 are transported into another system S, they are changed in the same ratio. As a result, the mass m_0 appears not to have changed, which is incorrect.)

In other words, the real mass m of the body in S, cannot be measured by an observer at rest in this co-ordinate system. In all cases, the measurement gives the value m_0, which is the mass of the body in the aether frame.

-Let us now examine the consequences of these results in the following example. Consider three co-ordinate systems S_0, S_1 and S_2, receding from one-another at constant speed along the common x-*axis,* and let us assume that a body is at rest in each of the three systems. The masses were initially identical in S_0 and equal to m_0, before being transported into their respective reference system. We propose to determine the effect of motion on these masses. (See Figure 2.)

iV.1. Mass increase with speed speed according to the conventional theory of relativity

Measured by an observer at rest with respect to one of the bodies, the mass remains equal to m_0 in all cases. Therefore, for observer S_1, we have

$$m_2^1 = \frac{m_0}{\sqrt{1 - v_{12}^2 / C^2}} \qquad (7)$$

where m_2^1 refers to the relativistic mass of body b_2 measured by observer S_1, and v_{12} refers to the relative speed of the reference systems S_1 and S_2.

If we suppose that $v_{12} \ll C$, expression (7) can be written to first order, as follows:

$$m_2^1 \approx m_0(1 + \frac{1}{2} v_{12}^2 / C^2).$$

So that, viewed by observer S_1, the energy of body b_2 is:

$$m_2^1 C^2 \approx m_0 C^2 + \frac{1}{2} m_0 v_{12}^2.$$

(This corresponds to the sum of the rest energy and the kinetic energy needed by b_2 to move from S_1 to S_2.) Therefore for observer S_1 the kinetic energy is reduced to:

$$(m_2^1 - m_0)C^2 \approx \frac{1}{2} m_0 v_{12}^2. \qquad (8)$$

Given that the usual measurements ignore the absolute velocity and cannot highlight the increase of the rest mass when a body moves from the aether frame to another frame, the kinetic energy assumes (erroneously) the form of equation (8) with respect to any frame not subjected to perceptible external forces (when $v/C \ll 1$), a fact which *seems* in agreement with the relativity principle, (in contrast, as we shall see, to aether theory.)

For an observer at rest in the reference system S_0, the energy of b_2 is different. Indicating per m_2^0 the mass of body b_2 measured by the observer at rest in S_0, we have, (for

$v_{02} \ll C$): $\qquad\qquad m_2^0 C^2 \approx m_0 C^2 + \frac{1}{2} m_0 v_{02}^2,$

and the energy of body b_1 is assumed to be

$$m_1^0 C^2 \approx m_0 C^2 + \frac{1}{2} m_0 v_{01}^2,$$

thus, for observer S_0 the kinetic energy needed by the body b_2 to move from S_1 to S_2 is:

$$(m_2^0 - m_1^0)C^2 \approx \frac{1}{2} m_0 (v_{02}^2 - v_{01}^2).$$

This result is different from the measurement made by observer S_1, $m_0 v_{12}^2 / 2$ although, obviously, it should be the same. *This is a serious internal contradiction that affects special relativity.*

IV.2. Mass increase with speed according to aether theory

As we shall see, the results below are the *theoretical* results that are obtained when the measurements do not vary with speed. (In contrast, the relativity principle *seems* to apply only when the measurements are subjected to distortions and therefore it gives a distorted view of reality.)

We now go back to the figure with the three bodies, and suppose that S_0 is a co-ordinate system at rest in the fundamental inertial frame, and S_1 and S_2 two systems receding from S_0 at constant speed, along the common x-axis. According to the fundamental aether theory, m_2^0 and m_2^1 have no meaning. A body at rest in one of these systems has only one real mass. The mass of the body b_2 is:

$$m_2 = \frac{m_0}{\sqrt{1 - v_{02}^2/C^2}}, \tag{9}$$

and the mass of b_1:

$$m_1 = \frac{m_0}{\sqrt{1 - v_{01}^2/C^2}} \tag{10}$$

where m_0 is the mass assumed by by the bodies when they are at rest in the aether frame. From (9) and (10) we obtain

$$m_2 = m_1 \frac{\sqrt{1 - v_{01}^2/C^2}}{\sqrt{1 - v_{02}^2/C^2}}, \tag{11}$$

and

$$m_2 C^2 = m_1 C^2 \frac{\sqrt{1 - v_{01}^2/C^2}}{\sqrt{1 - v_{02}^2/C^2}}.$$

(In the same way, from $\ell_2 = \ell_0 \sqrt{1 - v_{02}^2/C^2}$ and $\ell_1 = \ell_0 \sqrt{1 - v_{01}^2/C^2}$ we have:

$$\ell_2 = \ell_1 \frac{\sqrt{1 - v_{02}^2/C^2}}{\sqrt{1 - v_{01}^2/C^2}}.)$$

We see that expression (11), which connects any pair of co-ordinate systems moving at constant speed assumes a mathematical form different from (9) and (10).

If we now suppose that $v_{02} << C$, m_2 reduces to:

$$m_2 \approx m_1 + \frac{m_1}{2C^2}(v_{02}^2 - v_{01}^2) \tag{12}$$

$$\approx m_1[1 + \frac{1}{2}v_{12}^2/C^2 + v_{01}v_{12}/C^2].$$

In this case, m_1 is hardly different from m_0.

Expression (12) is different and obviously greater than the relativistic expression of the mass m_2^1 viewed from observer S_1 which is:

$$m_2^1 = \frac{m_0}{\sqrt{1 - v_{12}^2 / C^2}} \approx m_0 (1 + \frac{1}{2} v_{12}^2 / C^2).$$

We can see that, according to aether theory, the increase of kinetic energy of the body b_2 in the transfer from S_1 to S_2 is:

$$(m_2 - m_1)C^2 \approx \frac{m_1}{2} (v_{02}^2 - v_{01}^2)$$

$$\approx \frac{m_1}{2} (v_{12}^2 + 2 v_{01} v_{12}). \tag{13}$$

This expression (13) is different from the relativistic expression which is:

$$(m_2^1 - m_0)C^2 = \frac{1}{2} m_0 v_{12}^2,$$

indeed expression (13) contains a term depending on v_{01} which vanishes when S_1 is at rest with respect to S_0.

These results are incompatible with the exact application of the relativity principle.

We also note that, when $v_{12} \to 0$ or in other words when $v_{02} \to v_{01}$, the terms depending on v_{01} and v_{02} in expression (12) cancel. Thus, m_1 represents the real rest mass assumed by the aforementioned bodies when they are at rest in reference frame S_1. (Actually, there is no distinction between the real mass and the real rest mass.) This is a different result from special relativity.

Nevertheless, we must distinguish the absolute rest mass m_0, from the other rest masses measured in reference frames that are in motion with respect to the aether frame.

We see that the relativity principle, does not apply to real values of the physical variables. But we have shown in ref [1B] that, with the usual measurements which are performed with contracted meter sticks and clocks slowed down by motion synchronized with light signals, the experimental space-time transformations assume a mathematical form identical to the Lorentz transformations, (although their meaning is quite different) and, therefore, with these transformations, the *apparent* laws of physics, (including $m = m_0 \gamma$ and $\ell = \ell_0 / \gamma$ and the expression for the kinetic energy), take an identical mathematical form whatever the platform from which the measurement is made, provided that the platform is not subjected to *perceptible* external forces.

With these measurement distortions therefore, the relativity principle *seems* to apply.

This argument, which enables to surmount the objections raised to the Lorentz approach, merely confirms, the coexistence of the Lorentz assumptions and the experimental (apparent) law of mass increase, despite what differentiates them.

83

(Of course in order to obtain the exact values of the physical data, the experimental results must be corrected in order to supress the measurement distortions.)

Note, however, that when $v_{12} \gg v_{01}$, and $v_{01} \ll C$ expression (11) reduces to:

$$m_2 \approx \frac{m_1}{\sqrt{1 - v_{02}^2 / C^2}} \approx \frac{m_1}{\sqrt{1 - v_{12}^2 / C^2}}$$

and since, $m_1 \approx m_0$ we obtain:

$$m_2 \approx \frac{m_0}{\sqrt{1 - v_{12}^2 / C^2}} .$$

This applies, for example, to particles moving at a significant fraction of the speed of light with respect to the Earth frame, *while the Earth moves at relatively low speed with respect to the aether frame* (a value whch is estimated at $\cong 400$ km/sec.). In such cases, the Earth can be regarded as almost at rest with respect to the Cosmic Substratum. So, the relativistic approach and the fundamental approach lead to practically equivalent results.

IV.3. Critique of the concept of reciprocity

This question makes a crucial distinction between relativity and fundamental aether theories. According to relativity, when a body is transported from one 'inertial system' S_0 to another S_1, viewed from S_0, its mass is supposed to be

$$m_1 = \frac{m_0}{\sqrt{1 - v_{01}^2 / C^2}} .$$

But conversely, if the body comes back to S_0, viewed from S_1 its mass will also appear equal to m_1.

For the treatment in the fundamental aether theory, let us assume that S_0 is the fundamental frame. If the body is at rest in frame S_1, we also have

$$m_1 = \frac{m_0}{\sqrt{1 - v_{01}^2 / C^2}} ,$$

where $m_1 > m_0$. Indeed we have been compelled to supply energy to the body in order to move it from S_0 to S_1; but if the body returns to S_0, the energy is restored. All observers (including the observer in frame S_1) will conclude that the real mass in frame S_0 is equal to m_0.

This conclusion is in total contradiction with relativity, but it is the only one in agreement with mass-energy conservation.

Of course this result applies only to real masses whose measurement is not subject to alterations.

Important remark

In the fundamental aether theory, we must distinguish the total available energy of a body (which is equal to the sum of the rest energy m_0C^2 and the kinetic energy with respect to the fundamental frame), from the available energy of the body with respect to any other frame, which is smaller than the previous energy, and takes another mathematical form.

In the example discussed earlier, the total available energy of body b_2 is:

$$m_2\,C^2 = m_0C^2\left(1+\frac{1}{2}v_{02}^2\big/C^2\right) + \text{small terms of higher order.}$$

(This notion has no equivalent in conventional relativity for which the energy of a body is entirely relative and depends on its speed with respect to another body.)

V. Mass energy conservation, inertia and the relativity principle

V.1. Introduction

In contrast to Newtonian physics, special relativity highlighted the fact that the existence of inertial mass does not depend exclusively on the amount of matter, but also on the dynamic properties of bodies, a fact which challenges the Newtonian concept of mass. The more recent developments of physics have demonstrated that the critical examination of the old concept of mass could have other important implications as regards the unification of the physical interactions. This is illustrated by the fact that the unification of the electromagnetic interaction and the weak interaction required a null mass for the W and Z bosons.

Two theories have been proposed to explain the origin of inertial mass which elementary particles *seem* to possess. Although different, these theories assume that the existence of mass is not an intrinsic property of matter, but rather the consequence of the interaction of matter with a physical medium.

According to the Higgs field hyphtesis [9,10], inertial mass results from the interaction of elementary corpuscles with a special kind of particles, referred to as the Higgs bosons, which agglomerate around them.

In contrast, the speculative theory of Puthoff, Haisch and Rueda (P.H.R) [11], based on stochastic electrodynamics, assumes that "it is the interaction of electric charges and the electromagnetic field that creates the appearance of mass".

Among the other issues addressed, we will show in this section that, whatever the assumptions made about the nature of its interaction with a substratum, the inertial mass cannot exist without this mediation.

Although our approach is quite different, we note that the above theories assume some kind of aether, even if they give the concept another appellation. Indeed, in their article Beyond E=mc² [11], P.H.R conclude: "Even if our approach based on stochastic electrodynamics turns out to be flawed, the idea that the vacuum is involved in the creation of inertia is bound to stay".

(Note also that, in his book "The God particle" [12], Lederman refers to the Higgs field as new aether.)

Our approach does not take sides for one approach or for the other; it does not need to make any hypothesis about the nature of the substratum, it is based exclusively on logical arguments and does not postulate some new assumptions such as the Higgs boson. It should be interesting to investigate to what degree the theories of our predecessors are compatible with ours.

V.2. Critical review of usual definitions

Mass-energy conservation and the application of the relativity principle in the physical world are regarded today, by almost all the scientific community, as among the most fundamental principles of physics. It is generally assumed that their compatibility does not require specific conditions, and therefore was not called into question at least by the leading members of the discipline.

Briefly outlined in references [1A, 1D], this subject is nevertheless of the utmost importance and deserves to be reviewed in more detail. In parallel with the role played by a substratum, its impact on the existence of inertial mass will be studied in the following paragraphs.

V.2.1. Mass-energy conservation

With the advent of Lavoisier, physicists in the 18° century realized that matter cannot be destroyed even if this seems to be the case. Lavoisier expressed the idea in the following terms: "Nothing can be lost, nothing can be created, everything can only be transformed". Later, the idea that energy is also conserved became progressively an acknowledged fact. The energy-conservation law was expressed explicitly and accurately by Helmoltz. The final step was taken at the beginning of the twentieth century when mass and energy were regarded as two aspects of the same reality and the law $E = mc^2$ was formulated. The law expresses the fact that mass can be converted into energy and reciprocally. Note that, as we have seen in section II, the equivalence of mass and energy can be demonstrated without resorting to the Lorentz transformations [2]. This is also the case for the law of variation of mass with speed $m = m_0\gamma$ that will be needed for our demonstrations, (as shown in section III). (These two results were also briefly set out in the chapter 8 of ref [1D]). As we have shown, their field of application in aether theory differs from special relativity.

V.2.2. The relativity principle

Aristotle regarded rest and motion as two states of different nature. The Earth was assumed to be in a state of absolute rest, while the bodies moving with respect to it were considered in a state of absolute motion. According to Aristotle, uniform motion needed a motor to be maintained, although for the philosopher, the origin of the motor was not clear.

The idea of relativity departed completely from this viewpoint, considering that rest and uniform motion are only relative, depending on the position of the observer.

It is difficult to give the exact date of the origin of relativity which interested numerous scientists such as Jean Buridan, rector of the University of Paris (1300 - 1358), Giordano Bruno (1568 - 1600), Descartes (1596 -1650), Leibniz (1646 – 1716) and Newton (1642 -1727) among others. But it is Galileo (1564-1642) who deserves credit for having given the idea a clear formulation. We shall discuss the conclusion of Galileo in the following chapters.

Even if it can be challenged, the approach of Galileo represented a progress on that of Aristotle, because, instead of dogmatic claims, he proposed an explanation based on observation.

With the advent of Poincaré and Einstein, the idea has somewhat evolved. In fact there are at least three formulations of the relativity principle whose meaning is a little different. In the following sections we will review these different approaches and examine whether or not the principle strictly applies in the physical world, or if it is reduced to an approximation whose field of application remains limited.

A. Galileo's original idea [4]

Galileo realized that the uniform motion of a vehicle has no detectable influence on the physical processes occurring in it. For example, a pendulum hung on the ceiling of a ship, sailing uniformly on a calm sea, remains vertical (perpendicular to the surface of the sea), a stone released from the top of the ship's mast falls at the foot of the mast, flies and butterflies move in the same way as they do in their normal conditions, in the Earth frame. According to Galileo, if motion was absolute, the stone would fall at a distance from the foot of the mast, the pendulum would adopt a slanting position depending on the speed of the ship etc... Actually, noting that this is not the case, Galileo concluded "...uniform motion is like nothing...," a sentence which can be translated by: there is no absolute uniform motion, rest and uniform motion are only relative.

Actually, the explanation of Galileo is not the only explanation possible. Indeed, even if a preferred aether frame exists relative to which absolute motion can be defined, it remains that the stone possesses a momentum which constrains it to continue its horizontal motion at the same speed as the ship, while its vertical motion is determined by the law of gravitation. Therefore the stone is constrained to fall at the foot of the ship's mast. For the same reason, a pendulum hung on the ceiling of the ship keeps a vertical orientation.

Therefore the observations of Galileo are not enough to corroborate the relativity principle and a principal objection to the existence of absolute motion can be challenged.

B. Poincaré's Relativity Principle and the Lorentz aether[5]

If we assume that rest and uniform motion are only relative, it seems a priori obvious that the laws of physics must be the same in all 'inertial platforms' (not subjected to forces external to the platforms.) However, under a Galilean transformation $x' = x - vt$ and $t' = t$,.the laws of electromagnetism seemed to escape this identity, the Maxwell equations taking a different form in the different platforms, even if they were not subjected to perceptible external forces. In order to bring the Maxwell equations back into line, Poincaré had recourse to a new set of transformations that he called "Lorentz transformations" which did constitute a group. In his approach, Poincaré did not assume that the hidden influence of the aether drift could prevent the physical laws to be strictly invariant, as the sentences below show. He expressed his principle in the following terms:

> "It appears that the impossibility of detecting experimentally the absolute motion of the Earth is a general law of nature. We are naturally inclined to admit this law that we shall call the postulate of relativity, and to admit it without restriction" [5]. Whether or not this postulate, which up to now agrees with experiment, may later be corroborated or disproved by experiments of greater precision, it is interesting in any case to ascertain its consequences."

We note that, in this sentence, Poincaré did not explicitly deny the idea of absolute motion, he simply questioned the possibility of observing it. Although he placed credit in the postulate of relativity, he did not quite exclude the fact that it could be disproved by experiment.

Yet, in other sentences, the rejection of absolute motion and the adhesion to the postulate of relativity were asserted with much conviction although, paradoxically, Poincaré never rejected the concept of preferred aether frame he shared with Lorentz. His agreement with Lorentz is stated in the following declaration:

> The results I have obtained agree with those of Mr. Lorentz in all important points. I was led to modify and complete them in a few points of detail [5].

His belief in the aether is expressed in the sentences below:

> "Does an aether really exist? The reason why we believe in an aether is simple: if light comes from a distant star and takes many years to reach us, it is (during its travel) no longer on the star, but not yet near the Earth. Nevertheless, it must be somewhere, and supported by a material medium" (La science et l'hypothèse chapter 10, p 180 of the French edition, "Les theories de la physique moderne" [6],

and

> "Let us remark that an isolated electron moving through the aether generates an electric current, that is to say an electromagnetic field. This field corresponds to a certain quantity of energy localized in the aether rather than in the electron" [5].

But, as we saw in ref [1] and as we shall confirm in the following paragraphs, *when the systematic measurement distortions are corrected*, the strict validity of the relativity principle proves to be incompatible with the existence of the preferred aether frame which was assumed by Lorentz.

Poincaré's discomfort about the absolute motion, that resulted however from the Lorentz postulates that he admitted, appears obvious in the following sentences which also challenge absolute time and absolute simultaneity. ("La science et l'hypothèse" 1902, chapter VI page 111.)

> There is no absolute space and we only conceive relative motion, nevertheless the mechanical facts are generally expressed as if there were an absolute space to which one could refer.

and,

> There is no absolute time. To say that two times are equal is an assertion which by itself does not have any significance and can only acquire one by convention. Not only we do not have any direct intuition of the equality of two durations, but we do not even have that of the simultaneity of two events.

Moreover Poincaré did not put forward the idea that the relativity principle could be contingent. His attachment to the principle of relativity was too strong for that, as we can see in the following sentence extracted from "La science et l'hypothèse" 1902, Chapter VII page 129:

> The movement of a system must obey the same laws being related to fixed or moving axes entrained in a rectilinear uniform motion. This is the principle of relative motion which imposes to us for two reasons: first, the most vulgar experiment confirms this, second, the opposite hypothesis would be singularly reluctant to our spirit.

This sentence strongly suggests that the principle of relative motion is not for Poincaré a result of distorted measurements. It is rather perceived as something undoubtedly fundamental.

When he says: "the opposite hypothesis would be singularly reluctant to our spirit", he certainly does not take for granted the fact that, behind the measured values of the variables (apparent), exist hidden variables (real) which obey this opposite hypothesis and do not comply with the principle of relative motion.

Even in 1909, long time after the hypothesis of Lorentz contraction (1895), in a conference given in Lille University, he declared:

> There is no absolute space: all the displacements we can observe are relative displacements.

We do not agree with the opinions of Poincaré on this topic. As we saw in ref [1], absolute displacements exist even if it is difficult to highlight them.

We add that, if the relativity principle applied strictly in the physical world, the speed of light would be isotropic because there would be no privileged direction. However, although he assumed the principle without restriction, Poincaré gave credit to the Lorentz assumptions which did not assume light speed isotropy.
Actually, light speed anisotropy implies the negation of the principle of relativity as a fundamental principle of physics.

As for the simultaneity, it is certain for us that it must be absolute. In addition to the arguments given in ref [1A], let us suppose that two events A and B are simultaneous for one observer and not for another. This means that, for the second, a lasting event can happen between A and B, a fact impossible for the first observer.

But an *event cannot exist and also not exist* depending on whoever observes it.

C. Einstein's Relativity Principle [7]

According to Einstein, the laws of physics must be the same in all 'inertial frames'. Contrary to Poincaré, Einstein never acknowledged the existence of a preferred aether frame. In his early period, he categorically denied the concept of aether, but since 1916, he changed his mind in order to formulate the theory of General relativity.
But Einstein's aether is quite different from the concept of aether previously imagined by Lorentz. This aether is not associated with a preferred inertial frame. In his essay, "Ether and the theory of relativity"[8], Einstein expressed his idea of the aether in the following terms:

> "According to the theory of general relativity, space is endowed with physical qualities. In this sense therefore, there exists an ether… But this ether must not be thought of as endowed with the qualities of ponderable media, as consisting of parts which may be tracked through time. The idea of motion may not be applied to it."

If the vacuum was empty, the frames associated to moving bodies could be perfectly inertial, and Einstein's relativity principle would strictly apply. But, as we shall see, this is an abstraction which does not correspond to reality and which is at variance with the existence of massive bodies.

V.2.3. Further examination of the concept of 'inertial frame'

An inertial coordinate system can be defined as a coordinate system in which a body at rest relative to it is not subjected to external forces (*no more hidden than apparent*) that can hinder its state of rest or uniform motion.
An inertial body is a body not subjected to such external forces. If such a body is the place for experiments, it can be described as inertial platform.

The set of coordinate systems at rest with respect to the previous one constitutes an inertial frame of reference.

All inertial frames move with respect to one another with rectilinear uniform motion. The first question worth asking is the following: is the existence of bodies strictly inertial compatible with the Lorentz aether? (This kind of aether implies the existence of a drift acting on all bodies except in the preferred frame and, as we shall see in the paragraph V.3.1, weighty arguments in favour of the Lorentz aether exist. These arguments are added to those developed in Ref [1])

We must be aware that the concept of 'inertial frame', which is sanctioned by use, ignores the aether drift. However, except for the preferred frame, real frames associated to bodies not subjected to perceptible external forces, are never perfectly inertial, because they are subjected to the hidden influence of the aether drift whose magnitude depends on the absolute speed of the frame.

One can therefore conclude that, if the existence of the aether drift is proved, real frames moving with respect to one another are never exactly equivalent for the description of the physical laws. This demonstrates that the application of the relativity principle is not compatible in all generality with the Lorentz aether. Actually, it only *seems* compatible, because of the alterations that affect the measurements [1B].

Yet, in practice, the term 'inertial frame' can be used to describe reference frames whose absolute speed is low relative to the speed of light, ($v/C \ll 1$) because, as will be explained below, the effects of aether that result from the pressure exerted on moving bodies, are negligible at low speed..

But, as we shall see, this rule suffers from exceptions since the presence of aether can account for the existence of inertial mass, a fact which makes a neat difference between the concept of inertial frame assumed by conventional relativity, and frames surrounded by aether, even at low speed.

Of course this point of view asks a very fundamental question. If the aether interacts with matter, it should give rise to a resistance to the movement of material bodies. As Einstein said, the planets for example move through the aether without encountering the resistance that such a medium should cause. This is the reason why in his essay, "Ether and the theory of relativity" [8], he envisaged the case of an aether which does not oppose to motion.

However it is inconceivable that mass increase or length contraction can exist in the absence of a substratum interacting with matter.

The response to the argument raised by Einstein is that the pressure exerted by the aether is negligible at current speeds, and becomes effective only when absolute speeds exceed 10^5 km/sec. Below this value the ratio L/Lo falls in the range $0.95 < L/Lo < 1$.

Above this value, its effect should be effective, but not necessarily observable, because the field of application where the process occurs is considerably limited, and because the physical processes occurring at very high speeds which seem to mask gravity, may also make it difficult to highlight the effects of the aether drift.

However, in this field, our knowledge is limited by much uncertainty and some notions such as dark energy or an accelerating universe, not only have not yet found a definitive explanation, but are contested by several authors[24, 25].

In any cases, whether real or apparent, these phenomena are capable to mask the effect of the aether drift at high speed.

(We will check in the following sections whether the kind of aether described by Einstein in Ref [8] is (or not) in agreement with other well established laws of physics.)

V.2.4. Critique of the conventional concept of kinetic energy

The conventional concept of kinetic energy is closely related to the relativity principle and to the assumed absence of aether drift. Indeed, insofar as there is no preferred frame, the kinetic energy has no absolute character, it is observer dependent. The following example will put forward the paradoxes raised by this concept of kinetic energy.

When a spaceship travels from one 'inertial frame' S_1 to another S_2, a part of its fuel provides the chemical energy which is converted into kinetic energy K. According to relativity, for an observer attached to frame S_1, when the spaceship reaches frame S_2, its kinetic energy has increased, but viewed from the observer attached to frame S_2, it has decreased. However, a chemical energy has been used during the travel, and this is true for all observers. This energy is not dependent on whoever observes it. Chemical energy cannot give rise to a decrease of kinetic energy, because in such a case, the mass-energy conservation law would be transgressed. We must add that for observer S_2, the part of chemical energy K mentioned above could not be converted into heat and exhaust energy because heat and exhaust energy are related to the environment and not to the spaceship, and the two observers cannot draw opposite conclusions.

Such a paradox results from the fact that the kinetic energy of a body in relativity is regarded as observer dependent. Indeed, the relativity principle implies that there is no preferred frame where a body at rest has zero kinetic energy and from which the kinetic energy of moving bodies should be measured. If one assumes that the relativity principle applies in all generality in the physical world, a body is viewed as having zero kinetic energy for any observer at rest in the same frame as the body, and therefore, there is no storage of a definite amount of kinetic energy, identical for all observers, when a body moves from one 'inertial frame' to another.

There is no paradox any more if we assume that the kinetic energy is defined with respect to a privileged aether frame in which its value for any body at rest is zero. In this case, the total kinetic energy of a moving body has a well defined value and is not observer dependent. In the above example, the increase of kinetic energy in the transfer from frame S_1 to frame S_2 will be absolute, and recognized as the same by all observers. Conversely, in the transfer from S_2 to S_1, the decrease will be also recognized as the same by all. This implies that rest and motion are not only relative and that absolute speeds do exist.

The same considerations can be applied to the mass of a body which is transferred from one frame A to another frame B. According to relativity, viewed from frame A, the mass of the body increases, but viewed from frame B, it decreases. This conclusion is at variance with logic. If we need to supply energy to the body to transfer it from frame A to frame B, its mass cannot decrease.

V.3. Mass-energy conservation and the relativity principle

V.3.1. Is the relativity principle compatible with the mass-energy conservation law?

As we have seen, reference frames associated to bodies not subjected to perceptible external forces cannot be strictly inertial if an aether drift is present. But, until now we have not given complete arguments in favour of the aether drift. To this end, we will make use of the criterion expressed in the paragraph V.2.4 which is required by logic and we will reason by contradiction and put forward the consequences of the absence of aether drift. (These arguments complete those which were developed in Ref [1].)

Suppose that two perfect inertial frames S_1 and S_2 really exist. A spaceship at rest in one of them would not be subjected to any external force (no more hidden than apparent.) Now, suppose that the spaceship at rest in S_1 leaves frame S_1, and after acceleration becomes firmly attached to frame S_2. If we adopt the point of view of relativity, for an observer at rest in frame S_1, the initial kinetic energy of the spaceship is zero. The transfer from S_1 to S_2 requires an amount of fuel F capable of supplying the chemical energy Q. We assume that the mass of the fuel is negligible relative to the mass of the spaceship. When the spaceship reaches frame S_2, the chemical energy has been converted, on the one hand into the kinetic energy K and, on the other hand, into the heat and exhaust energy h which is released in the environment.

In frame S_2, the fuel tank is filled up again with the same amount of fuel F as in frame S_1. If the frames S_1 and S_2 are assumed to be perfectly inertial (equivalent), they differ from one another only by their relative speed. This means that if there were no aether drift, no difference could be observed in the physical properties of the transfers from S_1 to S_2 and from S_2 to S_1. Therefore, in order to come back to frame S_1 the spaceship should use the same amount of fuel as it does going from S_1 to S_2, a fact at the origin of the paradox. The observer of frame S_2 should note that the chemical energy used in the transit from S_2 to S_1 is Q = K + h, the same as in the reverse direction. Assuming that $v/C \ll 1$, the observers of both frames should agree on that, and for the same reason, they should agree on the fact that the amount of fuel which was to be converted into kinetic energy had to be the same in the two reverse directions (see the discussion below in section V.3.2).

Therefore, on a round trip, the spaceship should have used an amount of fuel equal to 2F corresponding to the energy $2Q = 2K + 2h$. We realize that, since the heat and exhaust energy has been released in the environment, it is conserved, but the kinetic energy is not, because with the assumed hypotheses a part of the fuel has been used to this end, while the final kinetic energy has not increased. We are therefore led to a contradiction. Finally the total mass-energy involved in the process is not conserved.

This assertion asks a very fundamental question because mass-energy conservation is a very basic principle which can in no way be ignored. (We note that we have neglected the variation of mass with speed which for $v \ll C$ is imperceptible. In any case, in this problem, it has no impact on the conclusions drawn, since, with the questionable hypotheses assumed, the amount of fuel used to be converted into kinetic energy is not null, while the final kinetic energy has not increased.)

Impact on the origin of mass
In order to reconcile the relativity principle with mass-energy conservation, the only way would be to assume that one can transfer the spaceship from S_1 to S_2 (or from S_2 to S_1) without changing the kinetic energy. Such a paradoxical result can be easily explained when we know that in the absence of aether drift, there is no hierarchy between frames

Theoretically, *in the absence of aether drift*, the expression of the kinetic energy gained by the spaceship when it moves from S_1 to S_2 would be:

$$(m - m_0)c^2 = m_0 c^2 [(1 - v^2/c^2)^{-1/2} - 1], \tag{14}$$

where m_0 is the rest mass of the spaceship, m is the mass in frame S_2 viewed from frame S_1, and v the relative speed of frames S_1 and S_2. Insofar as the kinetic energy remains the same on a round trip, expression (14) must be null. Since v is not null, we obtain $m_0 = 0$.

We therefore conclude that, if the relativity principle could exactly apply in the physical world, the mass-energy of bodies would be null. Since this is not the case, we must infer that the mass-energy is not an intrinsic property of bodies, it results from their interaction with the aether. The mass is minimum in the aether frame where there is no drift; it increases with absolute velocity as the aether drift increases.

This implies that the aether exerts its influence as well inside the bodies as outside of them.
On the contrary, a fundamental theory which assumes the Lorentz aether, implies a different influence of the aether on frames S_1 and S_2. Let us assume that the aether drift is greater in S_2 than in S_1. During the transfer of the spaceship from S_1 to S_2, the chemical energy Q is converted into kinetic energy K and heat and exhaust energy h. The heat and exhaust energy is released in the environment and therefore is conserved. Now, if we suppose that v_1 is the speed of frame S_1 with respect to the aether frame

S_0 and v_2 the speed of frame S_2 with respect to S_0, the extra kinetic energy acquired by the spaceship when it moves from S_1 to S_2 will be:

$$m_0 C^2 [(1 - v_2^2 / C^2)^{-1/2} - (1 - v_1^2 / C^2)^{-1/2}]. \tag{15}$$

Upon its return, this extra kinetic energy will be restored to the environment (aether), in agreement with the mass-energy conservation law. We are not constrained to assume that expression (15) is null, and therefore $m_0 \neq 0$.

(The situation is similar to that of a body which acquires potential energy E when it moves from one level A to another B. Upon its return to A, the body must give up the same energy E.)

In all probability, the amount of fuel, converted into heat and exhaust energy h' will be slightly lower in the transit from S_2 to S_1, than in the transit from S_1 to S_2. Therefore the total energy, h' – K needed for the transit $S_2 \rightarrow S_1$ will require a smaller amount of fuel.

Of course the process described above can be easily highlighted only if the frames under consideration are not subjected to additional physical forces, and are not surrounded by an atmosphere. In such cases, we should take account of these additional physical influences, which can mask the process.

Note that this result (mass resulting from the interaction with the aether) would not be obtained if we assumed the concept of aether defined by Einstein [8] since, according to Einstein, "the idea of motion may not be applied to this aether", and therefore it does not create an obstacle to the motion of bodies. This concept of aether denies the existence of an aether drift, (and therefore, in this problem, the same conclusions must be drawn as if the aether did not exist.)

Similar considerations concern the mass of a body which is transferred from one frame to another [1A]. According to the relativity principle, we should use the same chemical energy E to increase the mass of the body, in the transfer from one 'inertial system' A, to another B, as we do in the reverse direction, which for a round trip yields a total energy of 2E. Yet, the mass of the body would not have changed and the mass-energy conservation law would not be obeyed.

V.3.2. Discussion

In response to our arguments we received the following response which tried to challenge them on the basis of relativistic arguments:
" The correct statement of the law of conservation of energy is that an observer, in a fixed, inertial reference frame, will observe that the total energy (understood to be that of an isolated system) is constant in time: i.e, d(total energy)/dt = 0. The error in your (K+h) + (K+h)= 2K + 2h equation is that you are adding the energy observed in A's frame (the first K+h) to energy observed in B's frame (the second K+h); this is not allowed; in any case, it has nothing to do with conservation of energy, which refers to the total energy in either A's frame OR in B's frame, but not some from one and some

from the other. Energy is the 4th component of a 4-vector and is observed to be different in different frames; when A sees an increase/decrease in the object's kinetic energy, B sees a decrease/increase.

The correct equations in A's frame are:

Acceleration: $Q = K + h1$,

chemical energy released is converted into an increase in the object's kinetic energy plus heat energy

and then:

Deceleration: $Q + K = h2$ (i.e. $Q = -K + h2$)

chemical energy released plus the object's kinetic energy is converted in heat energy.

The equations in B's frame are the same but reversed; first deceleration and then acceleration. Note that the heats are not the same; while burning the same amount of fuel will generate the same amount of heat in the rest frame of the object, during deceleration, A is no longer in the object's rest frame and will see a different amount of heat energy".

Analysis and response to these arguments

The above arguments can be refuted for the following reasons

1. As we saw in paragraph 5.2.4 the kinetic energy gained in the transfer from one reference frame to another cannot be observer dependent.
2. The criticisms made by our contradictor lead to absurd consequences for the following reasons:
According to the relativity principle, viewed from observer A, the burning of fuel in the transit from A to B is $Q=K+h1$. Now, since, *if the principle were true*, nothing would distinguish frame A from frame B, we infer that, viewed from observer B, the burning of fuel in the transit from B to A would be the same, i.e. $Q=K+h1$, the energy released in the environment being therefore $h1= Q-K$.
Yet, in the transit back from B to A, our contradictor asserts that for observer A, the energy released in the environment is $h2=Q+K$.
As a consequence of this reasoning, the energy released in the transit from B to A is viewed different by the two observers, the difference being $h2-h1=2K$. This cannot be true; indeed, even if the contradictor assumes that, due to the variation of mass with speed, the energy released can be evaluated differently by the two cosmonauts, (a difference which in most usual cases where $v/C<<1$ would be hardly perceptible), this consideration does not explain the difference (=to 2K) observed between h1 and h2.

V.3.3. Consequences

The fact that bodies possess inertial mass, demonstrates that the aether exerts a hidden influence on them. This influence, being dependent on their absolute speed, leads us to conclude that, *provided that the measurement of the space and time co-ordinates is reliable*, the laws of physics, including electromagnetism, must somewhat vary as a function of this velocity.

Yet the measurements are usually made with contracted standards and with clocks whose ticking is slowed down by motion and which are synchronized with light signals. As we demonstrated in ref [1B], only when these measurement distortions act, and only if we replace the real speeds by apparent speeds, the space-time transformations assume a mathematical form identical to the conventionnal transformations and the relativity principle *seems* to apply.

Being established, these facts make it possible to find a solution to the paradoxes generated by the reciprocity of observations, between frames which are supposed to be inertial (even at very high speed), that affect special relativity [1F].

V.3.4. Concluding remarks

I would like to insist on the fact that it is not the validity of the relativity principle as an abstract concept which is called into question, it is the fact that it does not strictly apply to the true laws of physics (not subjected to measurement distortions), given that the reference frames in which these laws apply, suffer from the *hidden influence* of the aether drift.

It is clear that if this influence was offset, or if it was compensated, the relativity principle would strictly apply. But this condition is never fulfilled in the experiments. Only a theoretical correction is possible.

Of course, these remarks concern the true laws of physics not altered by arbitrary measurements. As we saw in ref [1B], due to the distortions, the relativity principle *seems* to apply even at very high speed, despite the fact that the interaction with the aether drift is not offset. But this does not concern the true laws.

VI. <u>Principle of inertia</u>

In previous writings, we declared that any objection to the applicability of the relativity principle, also challenges the principle of inertia. It is important here to give further information in order to specify what we mean. In its original formulation, the principle was expressed in concrete terms, "a marble *sliding* on a perfectly smooth horizontal surface (without any friction) in vacuum, remains perpetually in its state of rest or uniform motion".

Of course, if, in agreement with the Galilean relativity principle, rest and uniform motion are only relative, we can view the marble as at rest in its reference system and, as a result, it must remain in this state of rest.

But, in the fundamental aether theory proposed here, absolute rest exists and is distinct from motion. The difference results from the existence of the aether. Under the action of the aether, and even if we assume that the friction of the air and of the ground are balanced, the marble will experience a gradual slowing down, which at low absolute speed is hardly perceptible, but which increases with absolute velocity. The Galilean principle of inertia is therefore challenged.

(It is clear that if the pressure exerted by the aether were offset, the principle of inertia would strictly apply, a necessary condition for the mass-energy conservation law is obeyed.)

Note that, in the usual experimental situations, the pressure exerted by the aether is extremely low, as shown by the fact that the ratio L/L_0 hardly varies when absolute speeds are below 10^5 km/sec. Therefore the obstacle to inertia is negligible.

When the speed of the platform, where it has to be verified, reaches an important fraction of the speed of light, the principle of inertia is no longer strictly valid. However due to the measurement distortions, the relativity principle *seems* to apply, so that, a body at rest in such a platform remains at the same place, although the platform is not inertial. Therefore for an observer at rest in the platform, the principle of inertia still *appears* to be verified exactly.

Note that the physical processes occurring at very high speed are likely to be made complicated because of the mechanisms which seem to oppose gravity, that have not found a rational and definitive explanation (see the discussion in paragraph V.2.3.)

VII. _Conservation of momentum_

Insofar as the particles interacting in a collision are slowed down by the aether drift, their total quantity of motion cannot be the same before and after the collision. A part of the impulse is transferred to the aether. This effect, imperceptible at low speed, should not be ignored at high speed, ($v > 10^5$ km/sec.)

Of course, if the impulse transferred to the aether was taken into account, the total quantity of motion would be conserved.

Note also that, as we saw in refs [1A and 1B], in the usual experiments, because of the alterations suffered by the measuring instruments during the movement, the laws of physics *appear* invariant. Therefore, the law of conservation of the relativistic momentum *seems* to apply as such in all frames not subjected to perceptible physical forces. *As a result, the relativistic law of variation of mass with speed, which is derived on the basis of the conservation of the relativistic momentum, takes the form conferred by conventional relativity.*

In the following appendices, we will show how the alterations suffered by the measuring instruments because of their absolute movement, as well as the arbitrary synchronization procedures, alter the experimental results.

Appendix I

How Lorentz-FitzGerald's contraction (L.C) explains the *apparent* light speed invariance.

In this appendix we will show that L.C is necessary to account for the invariance of the two way transit of light time along a rigid path, irrespective of its orientation in space, in conformity with the experiment.

But this is not all. Combined with clock retardation, L.C will make it possible to explain why the *measured* value of the speed of light is found isotropic and equal to C, whereas its real value is not, and this in any co-ordinate system not subjected to perceptible physical forces.

The demonstration is based on Builder and Prokhovnik's studies [13] whose importance is indisputable but, as we shall see, some of the conclusions of Prokhovnik were questionable and could not enable to demonstrate that this *apparent* velocity is found to be C.

Let us consider two co-ordinate systems, S_0 and S. S_0 is at rest in the cosmic substratum (aether frame) and S is attached to a body which moves with rectilinear uniform motion along the x_0-axis of the S_0 system and suppose that a rod AB making an angle θ with the x_0, x-axis, is at rest with respect to the system S (see figure 3.)

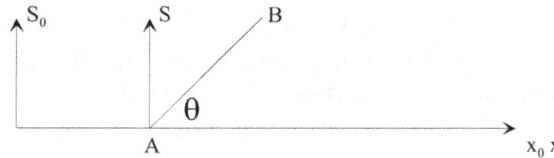

Figure 3. The rod AB is at rest with respect to the system S.

At the two ends of the rod, let us place two mirrors facing one another by their reflecting surface, which is perpendicular to the axis of the rod $\ell = AB$. At the initial instant, the two systems S_0 and S are coincident. At this very instant a light signal is sent from the common origin and travels along the rod towards point B. After reflection the signal returns to point A.

We do not suppose a priori that $\ell = \ell_0$ (where ℓ_0 is the length of the rod when it is at rest in the aether system S_0). We remark that the path of the light signal along the rod is related to the speed C_1 by the relation:

$$C_1 = \frac{AB}{t}$$

where t is the time needed by the signal to cover the distance AB. (see figure 4).

In addition, when the signal reaches point B, the system S has moved away from S_0 a distance $AA' = vt$, so that:

$$v = \frac{AA'}{t}.$$

Now, from the point of view of an observer at rest in S_0, the signal goes from point A to point B' (see figure 4).

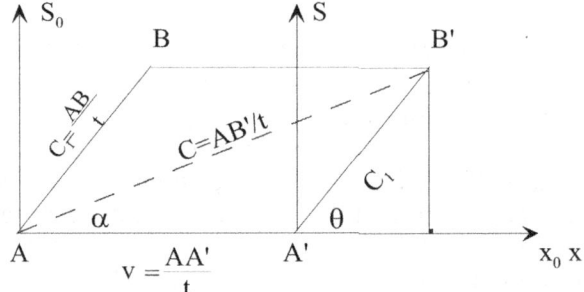

Figure 4. The speed of light is equal to C from A to B', and to C_1 from A' to B'.

C being the speed of light in S_0, we have:

$$\frac{AB'}{t} = C$$

and hence, the projection along the x-axis of the speed of light C_1 relative to the system S, will be equal to $C\cos\alpha - v$. So that:

$$C \cos \alpha - v = C_1 \cos \theta.$$

The three speeds, C, C_1 and v being proportional to the three lengths AB', AB and AA' with the same coefficient of proportionality, we have

$$C^2 = (C_1 \cos \theta + v)^2 + C_1^2 \sin^2 \theta.$$

Therefore:

$$C_1^2 + 2vC_1 \cos \theta - (C^2 - v^2) = 0. \qquad (16)$$

(We must emphasize that equation (16) implies that the three speeds C, C_1 and v have been measured with the help of the same clock, which obviously is a clock whose ticking is not slowed down by motion.)

Resolving the second degree equation, we obtain:

$$C_1 = -v\cos\theta \pm \sqrt{C^2 - v^2 \sin^2 \theta}.$$

The condition $C_1 = C$ when $v = 0$ compels us to only retain the + sign so:

$$C_1 = -v\cos\theta + \sqrt{C^2 - v^2 \sin^2 \theta}. \qquad (17)$$

Now, the return of light can be illustrated by the figure 5 below:

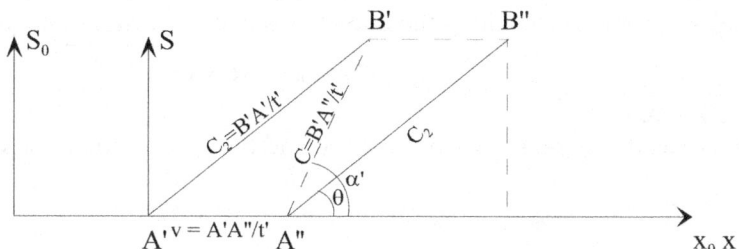

Figure 5.The speed of light is equal to C from B' to A" and to C_2 from B" to A".

From the point of view of an observer attached to the system S, the light comes back to its initial position with the speed C_2.
Therefore we can write:

$$C_2 = \frac{B'A'}{t'}.$$

For the observer attached to S_0 the light comes from B'to A'' with the speed C, so that:

$$C = \frac{B'A''}{t'}.$$

During the light transfer, the system S has moved from A' to A'' with the speed v therefore:

$$v = \frac{A'A''}{t'}.$$

The projection of the speed of light relative to S along the x-axis will be:

$$C\cos\alpha' + v = C_2\cos\theta.$$

We easily verify that:

$$(C_2\cos\theta - v)^2 + (C_2\sin\theta)^2 = C^2,$$

therefore,

$$C_2 = v\cos\theta + \sqrt{C^2 - v^2\sin^2\theta}. \qquad (18)$$

The two-way transit time of light along the rod AB, measured with clocks not slowed down by motion, is:

$$2T = \frac{\ell}{C_1} + \frac{\ell}{C_2}. \qquad (19)$$

According to the experiment, T must be essentially independent of the angle θ. Therefore, $2T$ must be equal to:

$$\frac{2\ell_0}{C\sqrt{1 - v^2/C^2}}$$

101

which is the two way transit time of light along the y direction (transversal arm) of the Michelson interferometer. We will see that, in order for this condition to be satisfied, the projection of the rod along the x-axis must shrink in such a way that:

$$\ell \cos\theta = \ell_0 \cos\varphi \sqrt{1 - v^2/C^2} \qquad (20)$$

(see figure 6)

where φ was the angle separating the rod and the x_0-axis when the rod was at rest in S_0.

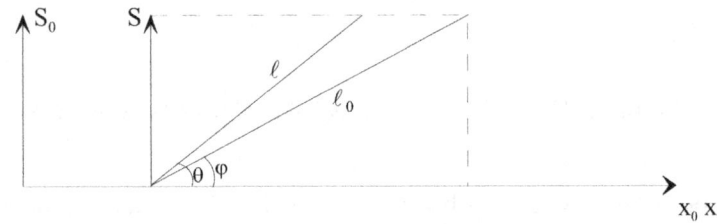

Figure 6. Along the x_0, x-axis, the projection of the rod ℓ_0 contracts, along the y-axis it is not modified.

from:

$$\ell_0 \cos\,\varphi = \frac{\ell \cos\theta}{\sqrt{1 - v^2/C^2}}$$

and

$$\ell_0 \sin\,\varphi = \ell \sin\,\theta,$$

we easily verify that:

$$\left(\frac{\ell \cos\theta}{\sqrt{1 - v^2/C^2}}\right)^2 + \left(\ell \sin\theta\right)^2 = \ell_0^2.$$

Finally:

$$\ell = \frac{\ell_0 \left(1 - v^2/C^2\right)^{1/2}}{\left(1 - v^2 \sin^2\theta/C^2\right)^{1/2}}. \qquad (21)$$

Replacing ℓ with this expression (21) in (19) we obtain, as expected:

$$2T = \frac{2\ell_0}{C\sqrt{1 - v^2/C^2}}. \qquad (22)$$

We conclude that length contraction along the x_0, x-axis is a necessary condition so that the two-way transit time of light along a rod, (given by formula (22)), is independent of the orientation of the rod.

We shall now show that the same conditions combined with clock retardation, will enable us to explain why the *apparent* (measured) two way speed of light is found equal to C in any direction of space.

Clock retardation is an experimental fact. Let us denote by 2ε the *apparent* (measured) two way transit time of light along the rod in frame S. We will have (from (22)):

$$\varepsilon = T\sqrt{1 - \frac{v^2}{C^2}} \tag{23}$$

$$= \frac{\ell_0}{C}.$$

Now, the length of the rod, measured with a contracted meter stick, is always found equal to ℓ_0, so that the two-way speed of light is (erroneously) found to be C in any direction of space and independently of the speed v. (As we have seen in Ref[1A], this is also the case for the *apparent* one-way speed of light measured with clocks synchronized by means of the Einstein-Poincaré procedure or by slow clock transport) (see appendix 2.)

This result is highly meaningful. It demonstrates that, assuming the existence of a preferred aether frame, length contraction is a necessary condition to account for the invariance of the *apparent* (measured) two way speed of light along a rigid path. This value is found equal to C independently of its orientation in space and of its absolute speed in agreement with the experiment.

Note

In our demonstration, although we are indebted to Prokhovnik, we differ with some of his conclusions [13]; indeed, since $C = AB'/t$ and $C = B'A''/t'$, it is obvious that t and t' are the real transit times of light along the rod (measured with clocks not slowed down by motion.)

Now, since $C_1 = \dfrac{AB}{t}$ and $C_2 = \dfrac{B'A'}{t'}$ there is no doubt that C_1 and C_2 are also measured with the help of clocks not slowed down by motion. This is also the case for

$$2T = \frac{\ell}{C_1} + \frac{\ell}{C_2}.$$

Nevertheless, in his book "The logic of special relativity" [13] chapter "The logic of absolute motion", Prokhovnik identifies the time

$$2T = \frac{2\ell_0}{C\sqrt{1 - v^2/C^2}}$$

with the two-way transit time of light along the rod, measured with clocks attached to the moving co-ordinate system. This cannot be true for the reason indicated above.

(Note that in our notation the moving co-ordinate system is designated as S, while in Prokhovnik's notation, S denotes the aether system and A the moving system. We will continue the demonstration with our own notation.)

In addition, if Prokhovnik's approach were true, the *apparent* two-way speed of light in S would not be C. Indeed, since the standard used for the measurement is also contracted, observer S would find ℓ_0 for the length of the rod.

Therefore, the *apparent* (measured) two way speed of light in the system S would have been:

$$\frac{2\ell_0}{2\ell_0/(C\sqrt{1-v^2/C^2})} = C\sqrt{1-v^2/C^2}$$

which is not in agreement with the experimental facts.

The two-way transit time of light along the moving rod, measured with clocks not slowed by motion is actually:

$$2\ell_0/(C\sqrt{1-v^2/C^2}),$$

and the *apparent* two-way transit time, measured with clocks attached to the system S, is $2\ell_0/C$. This corresponds to the experimental facts, since, with these values, the *apparent* two-way speed of light in S is found equal to

$$2\ell_0\left/\frac{2\ell_0}{C}\right. = C.$$

Note also that according to aether theory, the real two-way speed of light (measured with non-contracted standards and with clocks not slowed down by the movement) can be easily determined from (21) and (22). Along the x_0, x-axis we obtain:

$$\frac{2\ell_0\sqrt{1-v^2/C^2}}{2\ell_0/(C\sqrt{1-v^2/C^2})} = C(1-v^2/C^2).$$

As expected, this expression tends to 0 when $v \Rightarrow C$)

Appendix 2

Clock Synchronization and Light Velocity

We will now show how the *usual* clock synchronization procedures are affected by *systematic* errors generating a distorted vision of reality. We shall examine successively the Einstein-Poincaré procedure with light signals and the slow clock transport method.

1. Clock synchronization with light signals

In order to measure the speed of light with this method, we can use one or two clocks. When we use one clock, the signal is sent from the clock toward a mirror, and, after reflection, comes back to its initial position. In this case, what we determine actually is the *apparent* average round trip velocity of the light signal (measured with contracted meter sticks and clocks slowed down by motion.)

As we saw in the appendix 1, formula (23), even if we subscribe to the Lorentz postulates, which assume the anisotropy of the one-way speed of light in the Earth frame, the theory demonstrates that this average round trip velocity along a rod is (erroneously) found equal to C irrespective of the orientation of the rod. It also appears independent of the relative speed of the frame in which it is measured with respect to the aether frame. (These results follow from the systematic measurement distortions already mentioned.)

As a result, the method does not enable to distinguish aether theory from special relativity.

Therefore, *a priori*, the use of two clocks seems justified in order to accurately measure the one-way speed of light. With this goal in sight, we need first to synchronize two distant clocks A and B.

In the Einstein-Poincaré procedure, this requires two steps. First, we send a light signal from clock A to clock B at instant t_0; after reflection the signal comes back to A at instant t_1. Then we send another signal at instant t'_0. The clocks will be considered synchronous if, when the signal reaches clock B, the display of clock B is:

$$t'_0 + \frac{t_1 - t_0}{2} = t'_0 + \varepsilon$$

where ε is equal to half the *'apparent'* two-way transit time of the signal measured with the clocks slowed down by motion attached to the Earth frame. But this clock reading is usually (improperly) identified with the one way transit time of light.

As we saw in formula (23), $\varepsilon = \ell_0 / C$ in any direction of space; and since the distance AB is always found equal to ℓ_0, the speed of light is found equal to C in the same way as when we use one clock. Thus, even though the speed of light is given by formulas (17) and (18) of the appendix 1, the Einstein-Poincaré procedure yields C.

It is therefore justified to test another method, *i.e.*, the slow clock transport procedure.

2.   The slow clock transport method

Many physicists believe that an exact measurement of the speed of light can be obtained by the slow clock transport method. The procedure consists of setting two clocks A and B to the value zero at a point O' in the Earth frame, and then transporting clock B to a distance from A at low speed ($v \ll C$.) The problem has been envisaged in various ways by different authors [14-22].

A priori, it would appear that, since the transport is very slow and $v \to 0$, the motion would have no perceptible influence on the time displayed by clock B, and that the two clocks would remain almost synchronized all the time. But is this really the case?

2.1 Point of view of the conventional theory of relativity

If we regard the assumptions of special relativity as indisputable, then absolute speeds have no meaning: only relative speeds exist. On this basis, clock B will display:

$$t' = t\sqrt{1 - v^2/C^2} \approx t(1 - \frac{1}{2}\frac{v^2}{C^2})$$

where t is the reading displayed by clock A. (Note that for convenience we have supposed that the two clocks display $t_0 = 0$ at the initial instant.)

Once clock B has stopped (at point P), its lag behind clock A will remain constant. The synchronism discrepancy between clocks A and B is then, to first order

$$\Delta t = \frac{1}{2}\frac{v^2}{C^2}T \,,$$

where T denotes the time displayed by clock A when clock B reaches point P.

The speed of light will thus appear to be:

$$\frac{O'P}{T - \Delta t} \approx \frac{O'P}{T} + \frac{O'P\Delta t}{T^2}, \tag{24}$$

where $O'P$ refers to the path covered by the light signal.

Since $v \to 0$, expression (24) reduces to

$$\frac{O'P}{T}.$$

Note that the value of the speed of light is assumed to be known. The measurement therefore consists in verifying whether the results obtained by this method are in agreement with the premises.

The experimental value of the speed of light obtained is C.

Since the measurements of $O'P, T$ and Δt are supposed to be exact, special relativity concludes that the method is reliable. Actually if the method gives a value of the speed of light in agreement with the assumed hypotheses, it does in no way enable these hypotheses to be justified.

2.2 The fundamental aether theory point of view

It is interesting to see whether the above results can be obtained with basic hypotheses different from those of special relativity. Today, we have strong arguments in favour of the Lorentz assumptions. Several of them have been reviewed in ref [1].

According to Lorentz, the speed of light is constant exclusively in the aether frame. If we denote this speed by C, the speed of light in moving frames is given by the formulas (17) and (18) of the appendix 1.

If the slow clock transport method is reliable, it should give a value for the one-way speed of light in agreement with the assumed hypotheses.

We will see in reality that contrary to what many authors think, the method does in no way allow synchronizing the clocks exactly. Yet it presents a great interest since, even if we assume the Lorentz postulates, the speed of light is found equal to C in contradiction with these postulates but in agreement with the Einstein-Poincaré procedure. It therefore enables to give a rational explanation of the experimental results: due to measurement distortions, both methods yield the same value C.

We will now verify this point. Two cases will be considered.

2.2.1 The light ray travels in the direction of motion of the Earth frame

Consider two co-ordinate systems S_0 and S_1. S_0 is at rest in the Cosmic Substratum, and S_1 is firmly linked to the Earth frame. Initially the two systems are coincident. At this instant, a vehicle equipped with a clock starts from the common origin and moves slowly and uniformly along the x-axis of S_1 toward a point P of this co-ordinate system. We suppose that during the time of the experiment, the x_0-axis and the x-axis are aligned along the direction of the Earth orbital motion (See Figure 7.) v_{01} is the real speed of the Earth with respect to the fundamental frame S_0, v_{02} is the real speed of the vehicle with respect to S_0, and v_{12} the real speed of the vehicle with respect to S_1.

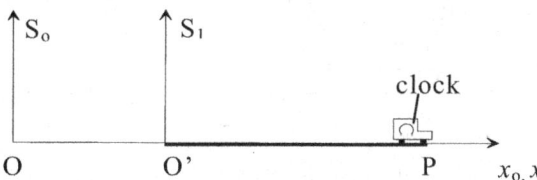

Figure 7. Synchronization of two clocks placed at O′ and P by the
slow clock transport method.

(Note that, for a short time, the orbital motion of the Earth can be considered rectilinear and uniform. If this were not the case, the bodies standing on the Earth platform would be submitted to perceptible accelerations.) The duration of the transport should be short enough so that the orbital and rotational motions of the Earth do not significantly affect the measurement. When the vehicle reaches point P, it stops. The real time needed to reach point P is given by

$$t_r = \frac{\ell}{v_{02} - v_{01}} = \frac{\ell_0 \sqrt{1 - v_{01}^2 / C^2}}{v_{02} - v_{01}}$$

where ℓ is the length of O′P (which is contracted because of the motion of the Earth with respect to the Cosmic Substratum), ℓ_0 is the length that O′P would assume if it were at rest in the aether frame, t_r is the vehicle's real transit time from O′ to P. It is the time that a clock attached to the aether frame, opposite the vehicle at the instant when it reaches point P, would display. But the clock in the vehicle (B) is slow relative to the clock of S_0, and will display the reading:

$$\frac{\ell_0\sqrt{1-v_{01}^2/C^2}\,\sqrt{1-v_{02}^2/C^2}}{v_{02}-v_{01}}.$$

(Let us recall that, in the fundamental aether theory, real speeds obey the Galilean law of composition of velocities. The relativistic law aplies only to apparent speeds [1].)

Now the clock placed at the origin O′ of the Earth system (A) slows down relative to a clock attached to S_0 opposite it. When the vehicle reaches point P, it will display the reading:

$$\frac{\ell_0\sqrt{1-v_{01}^2/C^2}\,\sqrt{1-v_{01}^2/C^2}}{v_{02}-v_{01}}.$$

(This implies that, for an instantaneous event occurring at point P, all the clocks attached to the Cosmic Substratum display the same time.) Thus, between clock B and clock A, we find a synchronism discrepancy equal to:

$$\frac{\ell_0\sqrt{1-v_{01}^2/C^2}}{v_{02}-v_{01}}(\sqrt{1-v_{01}^2/C^2}-\sqrt{1-v_{02}^2/C^2})$$

$$\approx\frac{\ell_0\sqrt{1-v_{01}^2/C^2}}{v_{02}-v_{01}}(1-\frac{1}{2}v_{01}^2/C^2-1+\frac{1}{2}v_{02}^2/C^2) \qquad (25)$$

$$\approx\frac{\ell_0}{2C^2}\sqrt{1-v_{01}^2/C^2}\,(v_{02}+v_{01}).$$

We can see that, once the vehicle has stopped, the discrepancy will remain constant.

As shown by Prokhovnik [13] the synchronism discrepancy between clocks synchronized by the Einstein-Poincaré method is equal to $v_{01}\ell_0/C^2$ (see also Ref [1D] chapter 3.) The difference with expression (25) is actually quite negligible if one considers that the transport is very slow and therefore that $v_{02}\approx v_{01}$. Regarding the gamma factor, for the usual measures ($v/C\!\ll\!1$), it is hardly different from 1. This is the case for the absolute velocity of the Earth frame (which is estimated at about 400 Km/sec.)

We can therefore conclude that for the usual measurements, the two methods yield similar results.

Speed of light along O'P

If we assume the Lorentz postulates, the real time of light transit along the distance ℓ is theoretically:

$$\ell_0\frac{\sqrt{1-v_{01}^2/C^2}}{C-v_{01}}.$$

We suppose here, *a priori*, that the speed of light with respect to frame S_1 is $C - v_{01}$. This is intentional, since we want to check whether the results are in agreement with the premises.

Now, as a result of clock retardation, (and without making allowance for lack of synchronism) the display of a clock in frame S_1 placed at point P when the signal reaches this point should be:

$$\frac{\ell_0 \sqrt{1 - v_{01}^2/C^2}\, \sqrt{1 - v_{01}^2/C^2}}{C - v_{01}} = \frac{\ell_0}{C - v_{01}} \left(1 - v_{01}^2/C^2\right).$$

If, in addition, we take into account the synchronism discrepancy given by formula (25), the apparent (measured) light transit time will be:

$$\frac{\ell_0}{C - v_{01}} \left(1 - v_{01}^2/C^2\right) - \frac{\ell_0}{2C^2} \sqrt{1 - v_{01}^2/C^2}\, \left(v_{02} + v_{01}\right). \tag{26}$$

Ignoring the terms of high order, expression (26) reduces to

$$\frac{\ell_0}{C} \left(1 + \frac{v_{01} - v_{02}}{2C}\right) = \frac{\ell_0}{C} \left(1 - \frac{v_{12}}{2C}\right).$$

Now, since the measured length of $O'P$ is always found equal to ℓ_0, the apparent speed of light will be

$$\frac{\ell_0}{\dfrac{\ell_0}{C}\left(1 - \dfrac{v_{12}}{2C}\right)} = \frac{C}{1 - \dfrac{v_{12}}{2C}} \approx C\left(1 + \frac{v_{12}}{2C}\right) = C + \frac{v_{12}}{2}.$$

Since v_{12} is taken as small as possible, the apparent speed of light is found equal to C. Therefore, even if the real speed of light is $C - v_{01}$, the slow clock transport method will (erroneously) yield C in the same way as the Einstein-Poincaré method.

Therefore the two methods can be considered equivalent.

2.2.2 General case

We now measure the speed of light along a rigid path O'B making an angle θ with the *x*-axis of the co-ordinate system S_1, firmly tied to the Earth frame (See Figure 8.)

We need, to this end, to synchronize two clocks placed at O' and B.

Initially the two systems S_0 and S_1 are coincident. At the initial instant, a vehicle leaves the common origin, and moves slowly and uniformly along the rigid path toward point B. We suppose that during the time of the experiment, the x_0-axis and the *x*-axis are aligned along the direction of the Earth orbital motion. For a short period of time this motion can be considered rectilinear and uniform.

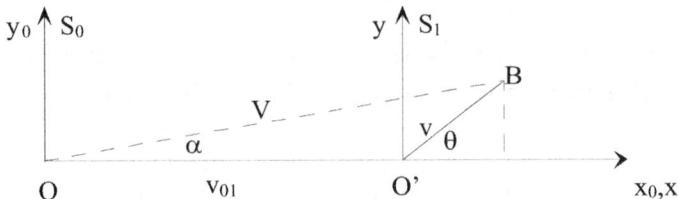

Figure 8. Synchronization of two clocks placed at O' and B by the
slow clock transport method.

(Note also that the rigid path is in the x, y plane, but obviously, provided θ remains the same, the following reasoning would be identical in any plane passing by the x_0, x-axis.)

As we saw in formula (21) of the appendix 1, due to length contraction along the x_0, x-axis, the length of the rigid path is given by

$$\ell = \frac{\ell_0 \sqrt{1 - v_{01}^2/C^2}}{\sqrt{1 - v_{01}^2 \sin^2 \theta/C^2}},$$

where v_{01} is the speed of the Earth with respect to the fundamental co-ordinate system S_0. We will denote by v the real speed of the vehicle with respect to S_1, and V its real speed with respect to S_0. (See Figure 8.)

The real time needed by the vehicle to reach point B is ℓ/v, but the apparent time in frame S_1, after allowance is made for clock retardation, is:

$$\frac{\ell}{v} \sqrt{1 - v_{01}^2/C^2} .$$

The apparent time measured with a clock placed inside the vehicle is

$$\frac{\ell}{v} \sqrt{1 - V^2/C^2} .$$

As a result, the synchronism discrepancy between the apparent time displayed by a clock attached to frame S_1 placed at point O' and the clock of the vehicle is:

$$\Delta = \frac{\ell}{v} \left(\sqrt{1 - v_{01}^2/C^2} - \sqrt{1 - V^2/C^2} \right).$$

We easily find that

$$V^2 = v^2 \sin^2 \theta + \left(v_{01} + v \cos \theta \right)^2,$$

and

$$V = v_{01} \sqrt{\frac{v^2}{v_{01}^2} + 1 + \frac{2v}{v_{01}} \cos \theta} . \tag{27}$$

If the inequality $v \ll v_{01}$ is taken into account, expression (27) reduces to

$$V \approx v_{01}(1 + \frac{v}{v_{01}}\cos\theta) = v_{01} + v\cos\theta.$$

Therefore, to first order, Δ becomes

$$\frac{\ell}{C^2}\left(v_{01}\cos\theta + \frac{1}{2}v\cos^2\theta\right).$$

Speed of light along O'B

Let us now suppose that we place, in O' and B, two clocks that have been (apparently) synchronized by the slow clock transport method. Actually, there is a synchronism error equal to Δ. The real speed of light along the rigid path from O' to B is (as seen in appendix 1 formula (17))

$$C_1 = -v_{01}\cos\theta + \sqrt{C^2 - v_{01}^2\sin^2\theta}.$$

As a result of clock retardation, but without the synchronism discrepancy effect, the *apparent* time needed by the light ray to reach point B should be

$$T_L = \frac{\ell}{C_1}\sqrt{1 - v_{01}^2/C^2}.$$

However, we must allow for the synchronism discrepancy, so that the *'apparent'* (measured) transit time of light will be:

$$\frac{\ell}{C_1}\sqrt{1 - v_{01}^2/C^2} - \Delta. \tag{28}$$

Ignoring terms of high order, expression (28) reduces to

$$\ell_0\frac{\left(1 - \frac{v_{01}}{C}\cos\theta - \frac{1}{2}\frac{v}{C}\cos^2\theta\right)}{C\left(1 - \frac{v_{01}}{C}\cos\theta\right)}.$$

Since the rigid path O'B is measured with a contracted meter stick, it appears equal to ℓ_0.

The apparent speed of light is then:

$$C_{app} = \frac{\ell_0}{T_L - \Delta} = \frac{C\left(1 - \frac{v_{01}}{C}\cos\theta\right)}{1 - \frac{v_{01}}{C}\cos\theta - \frac{1}{2}\frac{v}{C}\cos^2\theta}.$$

Since $v \to 0$, $C_{app} \to C$, which is different from the real value C_1 as shown earlier.

Therefore, contrary to what is often claimed, the slow clock transport method does not allow exact measurement of the speed of light [23]. It is approximately equivalent to

111

the Einstein-Poincaré method, and like this method, gives the erroneous value C for all measurements.

It is interesting to note that, even if the speed of light is not constant, it is found constant when standard methods of synchronization are used. Consequently, these methods must be seen as inadequate.

References

[1] J. Levy, A. Basic concepts for a fundamental aether theory, in *"Ether space-time & cosmology"* Volume 1, (Modern ether consepts relativity and geometry), Michael C. Duffy and Joseph Levy Editors, (PD Publications, Liverpool, UK), March 2008, ArXiv: *Physics*/0604207.
B. Aether theory and the principle of relativity, in *"Ether space-time & cosmology"* Volume 1, (Modern ether consepts relativity and geometry), Michael C. Duffy and Joseph Levy Editors, (PD Publications, Liverpool, UK), March 2008, ArXiv: *physics*/0607067.
C. "Extended space-time transformations for a fundamental aether theory", *Proceedings of the International Conference* "Physical Interpretations of Relativity Theory VIII", Imperial College London 6-9 September 2002, pp 257.
D. *From Galileo to Lorentz and beyond* (Apeiron, Montreal, 2003) web site http://redshift.vif.com
E. "Experimental and real coordinates in space-time transformations", *Found. Phys*, vol 34, N° 12, December 2004, pp 1905.
F. Aether theory clock retardation vs. special relativity time dilation, in *Ether space–time & cosmology, Volume 2* (New insights into a key physical medium), Apeiron, Montreal, Canada, 2009, pp 37-51.

[2] F. Rohrlich, *Am. J. Phys*, 58 (4), pp 348 (1990)

[3] G.N. Lewis, *Phil. Mag*, 16, pp 705 (1908), F. Selleri, On the meaning of special relativity if a fundamental frame exists, in *Progress in new Cosmology,* pp 269-284, Ed H. Arp et al, Plenum, New York London, 1993.

[4] G. Galilei, *Dialogo sopre I due massimi sistemi del mondo* (Opere Italiani, vol VII)

[5] H. Poincaré, Sur la dynamique de l'électron, and Lecture given in Lille, France, 1909, in *La mécanique nouvelle,*(Jacques Gabay, Sceaux, France, 1989)

[6] H. Poincaré, *La science et l'hypothèse* (Paris, Champs, Flammarion, 1968)

[7] A. Einstein, *Annalen. der. Physik*, 17, pp 891 (1905)
The principle of relativity, (Dover, New York, 1952)

[8] A. Einstein, Ether and the theory of relativity, Address delivered in the university of Leyden, May 5th, 1920, in *Sidelights on relativity*, (Dover, New York, 1983)

[9] P.W. Anderson, Plasmons, Gauge invariance and mass, *Phys. Rev*, 130, pp 439 (1963)

[10] P.W. Higgs, Broken symmetries, massless particles and gauge fields, *Phys. Lett*, 12, pp 132 (1964), Broken symmetries and the masses of gauge bosons, *Phys. Rev. Lett*, 13, pp 508 (1964)

[11] B. Haisch, A. Rueda and H.E. Puthoff, *Phys. Rev A*, vol 49, pp 678 (1994), B. Haisch, A. Rueda and H.E. Puthoff, Beyond E=mc², The sciences, November December 1994, pp 26, A. Rueda and B. Haisch, *Phys. Lett. A* vol 240, n° 3, pp 115, (1998), B. Haisch, A. Rueda and Y. Dobyns, *Annalen. der. Physik* 10, pp 393, (2001), A. Rueda and B. Haisch *Annalen. der. Physik*, 14, pp 479 (2005)

[12] L. Lederman and D. Teresi, *The God particle* ,(Houghton and Mifflin, New York, 1993)

[13] S. J. Prokhovnik, 1 - *The logic of special relativity*, (Cambridge University press, 1967). 2 - *Light in Einstein's Universe*, (Reidel, Dordrecht 1985), -references to the arti cles of G. Builder.

[14] A.S. Eddington, *The mathematical theory of relativity*, 2nd ed. (Cambridge Univer sity Press, Cambridge 1924)

[15] H. Reichenbach, *The philosophy of space and time* (Dover, New York, 1958)

[16] A. Grünbaum, *Philosophical problems of space and time* (A. Knopf, New York, 1963)

[17] P. W. Bridgman, *A Sophisticate's primer of relativity* (Wesleyan University Press, Middletown, 1962)

[18] B. Ellis and P. Bowman, Conventionality in distant simultaneity, *Phil. Sci* 34 (1967) pp 116-136.

[19] A.Grünbaum, Simultaneity by slow clock transport in the special theory of rela tivity, *Phil. Sci*, 36 (1969) pp 5-43.

[20] Yu. B. Molchanov "On a permissible definition of simultaneity by slow clock transport," in: *Russian Einstein Studies*, Nauka, Moskow, 1972.

[21] J. A. Winnie, Special relativity without one-way velocity assumptions *Phil. sci*, 37, pp 81-89 and pp 223-238 (1970).

[22] R.G. Zaripov, Convention in defining simultaneity by slow clock transport," *Galilean. Electrodynamics* 10, May June 1999, pp 57.

[23] R. Anderson *et al.*, *Physics. reports* 295 (1998) p 93-180. See in particular pp 100, where the authors criticize attempts to measure the one-way speed of light by means of the slow clock transport procedure. References to Krisher *et al.*, Nelson *et al.*, Will Haughan *et al.* and Vessot.

[24] T. Suntola, *Ether space–time & cosmology, Volume 2* (New insights into a key physical medium), Apeiron, Montreal, Canada, 2009, pp 67-134.

[25] R. T. Cahill, *Ether space–time & cosmology, Volume 2* (New insights into a key physical medium), Apeiron, Montreal, Canada, 2009, pp 135-200.

[26] G.F. Smoot, Cosmic microwave background radiation anisotropies, their discovery and utilization. *Nobel Lecture* December 8th 2006.
Aether drift and the isotropy of the universe: a measurement of anisotropies in the primordial blackbody radiation, *Final report* 1 November 1978 - 31 Oct 1980, University of California, Berkeley.

[27] G.F. Smoot, M.V. Gorenstein and R.A. Muller, Detection of anisotropy in the cosmic blackbody radiation *Phys. Rev. Lett*, vol **39**, p 898-901, (1977).
M.V. Gorenstein and G.F. Smoot, *Astrophys.J*, **244**, 361, (1981).

[28] F.S Crawford Jr, *Berkeley physics course Volume 3*, (Mc-Graw-Hill 1965-1968)

[29] F. Selleri, Relativistic physics from paradoxes to good sense-1 in *Ether space– time & cosmology, Volume 2* (New insights into a key physical medium), Apeiron, Montreal, Canada, 2009, pp 201-265

Relativistic Physics from Paradoxes
to Good Sense - 2

Franco Selleri

Dipartimento di Fisica, Università di Bari
INFN, Sezione di Bari

Abstract

This second part of "Relativistic physics from paradoxes to good sense" complements the topics covered in the second volume of "Ether space-time & cosmology" P 201, which review the results obtained in recent years by the author in relativistic physics. Historically the two theories of relativity were born from the clash of positivism and realism. The former current of thought used relativism as a weapon against ideas of realistic inclination, like Lorentz's. Paradoxes were the consequence in the new relativistic paradigm of emarginating realism. The recent understanding of the role of the conventional definition of simultaneity in relativistic physics has opened the doors to new lines of thought. Epistemologists have stressed that the coefficient of the space variable x in the Lorentz transformation of time (we call it e_1) has a nonphysical ("conventional") nature. Therefore, it should be possible to modify e_1 without touching the empirical predictions of the theory. Given that Einstein's principle of relativity leads necessarily to the Lorentz transformations, such a modification implies however a reformulation of the relativistic idea itself. With respect to this ideal picture, the concrete development of the research has produced some exciting surprises. Nature does not seem to be so indifferent about the value of e_1, given that several phenomena, in particular those taking place on a rotating platform (Sagnac effect, and all that) converge in a strong indication of the value $e_1 = 0$. This implies absolute simultaneity and a new type of space and time transformations which we call "inertial". Today we count on six proofs of absolute simultaneity, which are essentially independent of one another (three are contained in this second part of the paper). The cosmological consequences of the new structure of space and time go against the big bang model. After our results relativism, although weakened, is not dead and keeps proposing itself under milder forms.

1. The general transformations

In many a problem it turns out to be useful to work with a theoretical scheme even more general than that of the equivalent transformations. It must then be the scheme of the "general transformations". Given the inertial frames S_0 and S one can set up Cartesian coordinates [1] and make the following standard assumptions:

(i) Space is homogeneous and isotropic and time homogeneous, at least from the point of view of observers at rest in S_0;

(ii) Relative to the isotropic system S_0 the velocity of light is "c" in all directions, so that clocks can be synchronized in S_0 and the one way velocities relative to S_0 can be measured;

(iii) The origin of S, observed from S_0, moves with velocity $v < c$ parallel to the $+x_0$ axis, that is according to the equation $x_0 = v t_0$;

(iv) The axes of S and S_0 coincide for $t = t_0 = 0$.

The general transformations from S_0 to S are usually written as [2]

$$\left\{ \begin{aligned} x &= f_1(x_0 - vt_0) \\ y &= g_2 y_0 \quad ; \quad z = g_2 z_0 \\ t &= e_1 x_0 + e_4 t_0 \end{aligned} \right. \tag{1}$$

where f_1, g_2, e_4 and e_1 are v dependent parameters. The transformations inverse of (1) can easily be found

$$\left\{ \begin{aligned} x_0 &= \frac{(e_4 / f_1)x + v t}{e_4 + e_1 v} \\ y_0 &= \frac{1}{g_2} y \quad ; \quad z_0 = \frac{1}{g_2} z \\ t_0 &= \frac{t - (e_1 / f_1)x}{e_4 + e_1 v} \end{aligned} \right. \tag{2}$$

An electromagnetic spherical wave front, born at time $t_0 = 0$ in the origin of the isotropic reference frame S_0 satisfies the condition

$$x_0^2 + y_0^2 + z_0^2 = c^2 t_0^2 \tag{3}$$

116

The velocity of light relative to the moving system S can be found by introducing the transformations (2) into Eq. (3), which then becomes a second degree equation in t. The physically acceptable solution is

$$t = \frac{ce_1 + \beta e_4}{c(1-\beta^2)f_1}x + \frac{e_4 + c\beta e_1}{cf_1}\left[\frac{x^2}{(1-\beta^2)^2} + \frac{f_1^2}{g_2^2}\frac{y^2 + z^2}{1-\beta^2}\right]^{1/2} \tag{4}$$

where $\beta = V/c$. Polar coordinates with the Cartesian x axis as polar axis, give

$$x = r\cos\theta \quad ; \quad y = r\sin\theta\sin\phi \quad ; \quad z = r\sin\theta\cos\phi \tag{5}$$

and Eq. (4) becomes

$$\frac{t}{r} = \frac{1}{cf_1(1-\beta^2)}\left\{(e_1 c + e_4\beta)\cos\theta + (e_4 + e_1\beta c)\left[\cos^2\theta + q^2\sin^2\theta\right]^{1/2}\right\} \tag{6}$$

if

$$q = \frac{f_1\sqrt{1-\beta^2}}{g_2} \tag{7}$$

Clearly θ is the angle, in S, locally on the spherical wave front, between the light propagation direction and the absolute velocity $\vec{V}$ **of** S. Introducing the one way velocity of light $c_1(\theta) = r/t$, we have

$$\frac{1}{c_1(\theta)} = \frac{e_4 + e_1\beta c}{cf_1(1-\beta^2)}\left[\cos^2\theta + q^2\sin^2\theta\right]^{1/2} + \frac{e_1 c + e_4\beta}{cf_1(1-\beta^2)}\cos\theta \tag{8}$$

One can use the more compact form

$$\frac{1}{c_1(\theta)} = \frac{p}{c}\left[\cos^2\theta + q^2\sin^2\theta\right]^{1/2} + \frac{\Gamma}{c}\cos\theta \tag{9}$$

where

$$p = \frac{e_4 + e_1\beta c}{f_1(1-\beta^2)} \quad ; \quad \Gamma = \frac{e_1 c + e_4\beta}{f_1(1-\beta^2)} \tag{10}$$

117

Particular cases of (9) are

$$\frac{1}{c_1(0)} = \frac{p + \Gamma}{c} \qquad ; \qquad \frac{1}{c_1(\pi)} = \frac{p - \Gamma}{c} \qquad (11)$$

whence

$$\frac{c_1(\pi)}{c_1(0)} = \frac{(e_4 + e_1 c)(1 + \beta)}{(e_4 - e_1 c)(1 - \beta)} \qquad (12)$$

In order to obtain the two way velocity of light $c_2(\theta)$ we notice that the inverse two way velocity is the average of the inverse forward and backward (one way) velocities. From (9) it follows then

$$\frac{1}{c_2(\theta)} = \frac{p}{c} \left[\cos^2\theta + q^2 \sin^2\theta \right]^{1/2} \qquad (13)$$

where

$$p = \frac{e_4 + e_1 \beta c}{f_1 \left(1 - \beta^2\right)} \qquad ; \qquad q = \frac{f_1 \sqrt{1 - \beta^2}}{g_2} \qquad (14)$$

The Galilei transformations can be seen as a particular case of (1) with

$$f_1^G = g_2^G = e_4^G = 1 \qquad ; \qquad e_1^G = 0 \qquad (15)$$

Therefore

$$p^G = \frac{1}{1 - \beta^2} \qquad ; \qquad q^G = \sqrt{1 - \beta^2} \qquad ; \qquad \Gamma^G = \frac{\beta}{\sqrt{1 - \beta^2}} \qquad (16)$$

and the Galilean velocities of light are:

$$\frac{1}{c_1^G(\theta)} = \frac{1}{c} \frac{1}{1 - \beta^2} \left\{ \left[1 - \beta^2 \sin^2\theta\right]^{1/2} + \beta\cos\theta \right\} \qquad (17)$$

and

118

$$\frac{1}{c_2^G(\theta)} = \frac{1}{c}\frac{1}{1-\beta^2}\left[1 - \beta^2 \sin^2\theta\right]^{1/2} \tag{18}$$

These results will be useful in the following sections.

2. Revival of the MM experiment

We will now apply to practical problems some consequences of the general transformations. The expression for the inverse $c_2(\theta)$, the two way velocity of light relative to S, at an angle θ with respect to the "absolute" velocity, is given by (18). For small β^2 this can be approximated as

$$\frac{1}{c_2^G(\theta)} \approx \frac{1}{c}\left[1 - \frac{1}{2}\beta^2 \text{sen}^2\,\theta\right] \tag{19}$$

where a practically unobservable constant term $1-\beta^2$ has been approximated to 1. But now it is useful to review the situation concerning the Michelson-Morley (MM) experiment

In 1925 Miller [3] and in 1932 Kennedy and Thorndike (KT) [4] detected a possible anisotropy in the propagation of light in their interference studies of the absolute motion of earth. When the anisotropy was analyzed by means of the Galilean composition of velocities the following values for the Earth absolute velocity were obtained: about 9 km/sec (Miller), (24 ± 19) km/sec (KT from the daily effect), (15 ± 4) km/sec (KT from the annual effect). Miller insisted on the physical nature of his observations and on their compatibility with the results of the Michelson-Morley experiment [5], while KT discarded their results for the following (almost unbelievable) reason: "In view of relative velocities amounting to thousands of kilometers per second known to exist among the nebulae, this can scarcely be regarded as other than a clear null result." [6]

The first modern experiment on the foundations of relativity, performed by Jaseja et al (JJMT) [7], used two masers mounted with axes perpendicular on a rotating table. Rotations of the table through 90° produced repeatable variations of the frequency difference of about 275 kc/sec, while 3000 kc/sec was the result predicted by considering only the earth orbital motion in the Galilean theory. Given that the latter motion has a velocity of 30 km/sec and that the frequency difference turns out to be proportional to the squared earth velocity, one can say that JJMT observed again a velocity of 9 km/sec. However JJMT stated that the observed effect was "presumably" due to magnetostriction in the metallic parts of their apparatus due to the earth's magnetic field and promised to repeat the experiment by eliminating those parts. This promise has never been kept, as far as I know.

119

More recent experiments [8] - [10] were designed in such a way as to avoid what by then were considered the "spurious effects" detected by Miller, KT, and JJMT, so that today a doubt actually remains concerning the exact invariance of the two way velocity of light. The doubt has been reinforced by very recent papers published by M.
Allais [11], J. De Meo [12], R. Cahill [13], H. Munera [14] and M. Consoli [15].

As far as the two way velocity of light is concerned all the equivalent theories give the same prediction, the TSR included. Remembering (18) and (16) we see that

$$ f_1^E = 1/R \quad ; \quad g_2^E = 1 \quad ; \quad e_4^E + e_1^E v = R \quad \rightarrow \quad p^E = q^E = 1 \tag{20} $$

Of course, if the equivalent transformations hold fully, the MM experiment must give a null result, which is not what we wish to explore. Thus we consider the possibility of a **small breakdown** of the "equivalent" q and set

$$ q \approx 1 - \frac{\varepsilon}{4} \beta^2 \tag{21} $$

with $\varepsilon \ll 1$ as our "small breakdown" has to be such for all values of β^2 ($0 < \beta^2 < 1$). A factor ¼ has been included for convenience. We will see that

$$ \varepsilon \approx 10^{-3} \tag{22} $$

would be optimal for our thesis. Inserting (22) in (18) we get

$$ \frac{1}{c_2(\theta)} \approx \frac{1}{c}\left[1 - \frac{\varepsilon}{2}\beta^2 \operatorname{sen}^2 \theta\right] \tag{23} $$

Now we compare the consequences of (19) with those of (23). Measurements of $c_2(\theta)$ have produced data compatible with a constant $c_2(\theta)$ with a percent error of 10^{-9}. Eventual variations should not be larger than such an error. To mime the convictions of Miller and of some contemporary authors we assume an "experimental" two-way velocity of light given by

$$ \frac{1}{c_2(\theta)} \approx \frac{1}{c}\left[1 - \frac{\Delta}{2}\operatorname{sen}^2 \theta\right] \quad \text{with} \quad \Delta \approx 10^{-9} \tag{24} $$

A factor 1/2 has been introduced for convenience. The comparison of this "experimental result" with the Galilean velocity of light (19) gives

120

$$\beta^2 \approx \Delta \quad \rightarrow \quad \beta^2 \approx 10^{-9} \quad \rightarrow \quad \beta \approx 3 \cdot 10^{-5} \quad \rightarrow \quad v = 9 \; km/s$$
$$(25)$$

Next we compare the "experimental result" (24) with the broken equivalent transformations formula (23):

$$\varepsilon\beta^2 \approx \Delta \quad \rightarrow \quad \beta^2 \approx 10^{-6} \quad \rightarrow \quad \beta \approx 10^{-3} \quad \rightarrow \quad v = 300 \; km/s \quad (26)$$

This is a very reasonable value for the planet of a star in an arm of a spiral galaxy like ours! *Passing from the classical Newtonian approach to a slightly broken relativistic approach (with the same experimental data) increases radically the absolute velocity of the laboratory.*

3. Lorentz contraction deduced

In this section we explore some interesting connections between three fundamental properties of relativistic physics: Lorentz contraction, clock retardation, and the two way velocity of light [16].

3.1 Lorentz contraction of moving objects. P_1 and P_2 are any two points of a body completely at rest in the system S (no translation, no rotation). Seen from S_0 the body is in a state of uniform translation parallel to the x_0 axis. The Lorentz contraction implies

$$\begin{cases} x_{02}(0) - x_{01}(0) = \sqrt{1 - V^2/c^2}\,(x_2 - x_1) \\[2mm] y_{02}(0) - y_{01}(0) = y_2 - y_1 \end{cases} \quad (27)$$

where $x_{01}(0)$, $y_{01}(0)$ $\left[x_{02}(0),\ y_{02}(0)\right]$ are the coordinates of P_1 $\left[P_2\right]$ seen from S_0 at time zero; x_1, y_1 $\left[x_2, y_2\right]$ are the coordinates of P_1 $\left[P_2\right]$ seen from S. The time $t_0 = 0$ in the left hand side of (27) was chosen for simplicity, but any other value of t_0 would be acceptable. In other words, the distance between P_1 and P_2 measured in S_0 is the same for all t_0. The comparison of (27) with similar formulae easily deducible from the system (1) gives

121

$$f_1 = \frac{1}{\sqrt{1 - V^2 / c^2}} \qquad ; \qquad g_2 = 1 \qquad (28)$$

Obviously, a set of transformations of the space and time variables of the type (1) contains the Lorentz contraction if and only if the conditions (28) are satisfied.

3.2 Moving clock retardation. Let Q be a clock at rest in S marking the time t. The equation of motion of Q seen from S_0 is $x_0(t_0) = vt_0 + x_0(0)$. Substituting into the 4th eq. (1) one has

$$t = \left(e_1 V + e_4\right)t_0 + e_1 x_0(0) \qquad (29)$$

Considering any two times t_1 and t_2 marked by the moving clock Q and the corresponding S_0 times t_{01} and t_{02}, one easily gets from (29)

$$t_2 - t_1 = \left(e_1 V + e_4\right)(t_{02} - t_{01}) \qquad (30)$$

As we know, the retardation of moving clocks in the case considered is given by

$$t_2 - t_1 = \sqrt{1 - V^2 / c^2}\,(t_{02} - t_{01}) \qquad (31)$$

The comparison with (30) finally gives

$$e_1 V + e_4 = \sqrt{1 - V^2 / c^2} \qquad (32)$$

3.3 Invariance of the two way velocity of light. A flash of light propagating forth and back on any segment AB at rest in S does so with a two-way velocity

$$c_2(\theta) = c \qquad (33)$$

independent of S and of the angle θ formed by the light propagation direction with the x axis. Now let us see the consequences. First of all, as we saw, $c_2(\theta)$ as given by (13) is independent of θ if and only if $q = 1$. Once $q = 1$ is satisfied, one has

$$\frac{1}{c_2(\theta)} = \frac{p}{c} \qquad (34)$$

122

At this point it becomes clear that condition (33) is satisfied if and only if

$$p = 1 \quad ; \quad q = 1 \qquad (35)$$

Notice that the conditions (35) can also be written

$$g_2 = f_1 \sqrt{1 - V^2 / c^2} \quad ; \quad e_1 V + e_4 = f_1 \left(1 - V^2 / c^2\right) \qquad (36)$$

A set of transformations of the space and time variables of the general type (1) contains an invariant two-way velocity of light if and only if the conditions (36) are satisfied.

3.4 Lorentz contraction as a consequence of other phenomena. At this point it becomes rather easy to prove the following theorem: "If the Larmor clock retardation and the invariance of the two way velocity of light hold in nature, then the Lorentz contraction holds necessarily as well." The proof is easy if we set

$$R = \sqrt{1 - V^2 / c^2} \qquad (37)$$

and summarize the previous results as follows:

(i)   **Lorentz contraction**   is a consequence of the GT (1) if and only if

$$f_1 = 1 / R \quad ; \quad g_2 = 1 \qquad (38)$$

(ii)   **Clock retardation** is a consequence of the GT (1) if and only if

$$e_1 V + e_4 = R \qquad (39)$$

(iii) $c_2(\theta) = c$ is a consequence of the GT (1) if and only if

$$g_2 = f_1 R \quad ; \quad e_1 V + e_4 = f_1 R^2 \qquad (40)$$

Our argument can be formalized as follows

$$(ii) + (iii)_1 + (iii)_2 \quad \Rightarrow \quad (i) \qquad (41)$$

so that the proof is obtained in two steps:

$$(ii) \ + \ (iii)_2 \quad \Rightarrow \quad R \ = \ f_1 R^2 \ \Rightarrow$$

$$; \tag{42}$$

$$\Rightarrow \quad f_1 \ = \ 1 \ / \ R$$

$$(1 = f_1 R) \quad + \quad (iii)_1 \quad \Rightarrow \quad g_2 = 1. \tag{43}$$

Thus the Lorentz contraction is obtained as a consequence of (*ii*) and (*iii*), that is, of two firmly established facts. This shows that the lack of empirical evidence for the Lorentz contraction is more seeming than real. In fact, all the evidence collected for the clock retardation and for the invariance of the two way velocity of light can be considered evidence, albeit indirect, for the Lorentz contraction as well.

4. General proof of absolute simultaneity

Given the inertial frames S_0 and S one can set up Cartesian coordinates and make only the usual four assumptions seen at the start of Section 1. The general transformations are given by eq. (1), and in (12) it was also shown that

$$\frac{c_1(\pi)}{c_1(0)} \ = \ \frac{(e_4 + e_1 c)(1 + \beta)}{(e_4 - e_1 c)(1 - \beta)} \tag{44}$$

The continuity condition of the physical quantities in passing from slowly accelerated systems to inertial systems leads to a simple result, obtained also in the second part of the present section, which is:

$$\frac{c_1(\pi)}{c_1(0)} \ = \ \frac{1 + \beta}{1 - \beta} \tag{45}$$

By comparing (44) and (45) it necessarily follows the vanishing of the synchronization parameter. It must be stressed that in the TSR, in which $c_1 = c$ is isotropic, the result (45) gives rise to a discontinuity at zero acceleration.

We can conclude that the famous synchronization problem is solved by nature itself: it is not true that the synchronization procedure can be chosen freely as the usually adopted choice leads to an unacceptable discontinuity in the physical theory.
Absolute simultaneity is a logical necessity not only in relativistic physics, but also in the broader domain of the "general transformations". In fact we can so prove that $e_1 = 0$ is a characteristic property of all theories treating inertial systems in a way

continuous with the accelerated systems. In view of our results absolute simultaneity seems to be the only serious way of dealing with space and time physics. Seen in this way, absolute simultaneity looks very much like a fundamental property of nature.

Next we review earlier arguments showing that the condition $e_1 = 0$ has to be adopted also in he framework of the general transformations. Consider an inertial reference system S_0 and assume it is isotropic so that the one-way velocity of light relative to S_0 has the usual value c in all directions.
By recalling the inverse Galilean velocity of light

$$\frac{1}{c_1^G(\theta)} = \frac{1}{c\left(1 - \beta^2\right)}\left\{\left[1 - \beta^2 \sin^2\theta\right]^{1/2} + \beta\cos\theta\right\} \qquad (46)$$

we see that Eq. (46) contains the term $\beta\cos\theta$, characteristic of all theories treating inertial systems in a way continuous with the accelerated ones. The absence of this term in the TSR in which the velocity of light is isotropic relative to all inertial systems, leads to an unacceptable discontinuity in the physical theory.
Absolute simultaneity is a logical necessity not only in relativistic physics, but also in the broader domain of the "general transformations" given by (1). In fact we are going to prove that

$$e_1 = 0 \qquad (47)$$

is a characteristic property of all theories treating in a continuous way the step from inertial to very slowly accelerated systems.
Thus the most general transformations of the space and time variables allowed by the continuity condition contain the absolute simultaneity. In fact, according to the fourth Eq. (1), if (47) is satisfied, two point-like events with coordinates x_{10} and x_{20} ($x_{01} \neq x_{02}$), taking place at the same time t_0, are judged to happen at the same time also in S. Once more the absolute simultaneity is seen to be unavoidable. With $e_1 = 0$ the velocity of light, as given by equation (8), becomes

$$\frac{1}{c_1(\theta)} = \frac{e_4}{cf_1\left(1 - \beta^2\right)}\left\{\left[\cos^2\theta + q^2 \sin^2\theta\right]^{1/2} + \beta\cos\theta\right\} \qquad (48)$$

showing that the $\beta\cos\theta$ term in the one way velocity of light is <u>a fixed ingredient in all theories of inertial systems satisfying the continuity condition with the accelerated ones</u>. Such a term is present also in Galilean physics, as we saw.
In a laboratory there is a circular platform (radius r and center constantly at rest in S_0) which rotates uniformly around its axis with angular velocity ω and

peripheral velocity $v = \omega r$. On its rim there is only a single clock Σ (marking the time t). We assume it to be set as follows: When a clock of the laboratory momentarily very near Σ shows time $t_0 = 0$ then also Σ is set at time $t = 0$. When the platform is not rotating, Σ constantly shows the same time as the nearby laboratory clocks. When it rotates, however, motion modifies the pace of Σ and the relationship between the times t and t_0 is taken to have the general form

$$t_0 = t\, F(v) \tag{49}$$

where F is a function of velocity. Eq. (49) is a consequence of the isotropy of S_0. Its validity can be shown in three elementary steps:

1.   In the inertial system S_0 all directions are physically equivalent. If a clock is moving on a straight line with a certain speed v relative to S_0, the change in the rate of advancement of its hands cannot depend on the orientation of the line.

2.   Similar is the case of the clock Σ at rest on a uniformly rotating platform, with centre at rest in S_0. If S_0 is isotropic the rate of advancement of the Σ hands cannot depend on the orientation of the clock instantaneous velocity, but only on speed v.

3.   This conclusion, clearly correct by symmetry reasons, was confirmed experimentally by the 1977 CERN measurements of the anomalous magnetic moment of the muon [17]. The decay of muons was followed very closely in different parts of the storage ring and the results showed a decay rate constant in the different points of the ring.

Thus we have every reason to believe (49) to be correct. We are of course far from ignorant about the function F. There are strong experimental indications that $F(v, \ldots) = 1/R$, with R given by (37). This is however irrelevant for our present needs as the results obtained below hold for all possible factors F.

On the rim of the platform besides clock Σ there is a light source placed in a fixed position very near Σ Two light flashes leave Σ at the time t_1 marked by Σ itself and are forced to move on a circumference, by "sliding" on the internal surface of a cylindrical mirror placed at rest on the platform, all around it and very near its border. Mirror apart, the light flashes propagate in the vacuum. The mirror behaves like a source ("virtual") and a source motion never changes the velocity of the emitted light signals. Thus, relative to the laboratory, the light flashes propagate with the usual velocity c.

The description of light propagation given by the laboratory observers is the following: two light flashes leave at time t_{01}. The first one propagates on a circumference, in the sense discordant from the platform rotation, and comes back to Σ at time t_{02} after circling around the platform. The second flash propagates on the same circumference, in the sense concordant with the platform rotation, and comes back to Σ time t_{03} after circling around the platform. These laboratory times, all relative to events taking place in a fixed point of the platform very near Σ, are related to the corresponding platform times as follows:

$$t_{0i} = t_i \, F(v) \qquad (i = 1,2,3) \qquad (50)$$

The circumference length is assumed to be L_0 and L, measured in the laboratory and on the platform, respectively. If $\tilde{c}(0)$ and $\tilde{c}(\pi)$ are the light velocities, relative to the disk, for the flash propagating in the direction of the disk rotation and in the opposite direction, respectively, one can show with a few elementary steps using the very definition of velocity:

$$\begin{cases} \dfrac{1}{\tilde{c}(\check{s})} - \dfrac{t_2 - t_1}{L} - \dfrac{L_0}{FL} \dfrac{t_{02} - t_{01}}{L_0} - \dfrac{L_0}{FL} \dfrac{1}{c + v} \\[2ex] \dfrac{1}{\tilde{c}(0)} = \dfrac{t_3 - t_1}{L} = \dfrac{L_0}{FL} \dfrac{t_{03} - t_{01}}{L_0} = \dfrac{L_0}{FL} \dfrac{1}{c - v} \end{cases} \qquad (51)$$

In both equations the last step is justified by considering the speed with which a circular length L_0 closes. From (51) it follows equation (45) as anticipated. Notice that the functions F, L, L_0 have disappeared in the ratio.

Clearly, Eq. (45) gives us not only the ratio of the two global light velocities for full trips around the platform, but *the ratio of the instantaneous velocities as well*. Refusing this last point means accepting space anisotropy. In fact the isotropy of the inertial system S_0 implies, by symmetry, that the instantaneous velocities of light are the same in all points of the rim of the rotating circular disk whose center is at rest in S_0. There is no reason why the light instantaneous velocities relative to the disk in the different points of the rim should not be equal to one another.
Thus the crucial quantity is

$$\rho \equiv \frac{\tilde{c}(\pi)}{\tilde{c}(0)} \qquad (52)$$

127

which, owing to (45), is larger than unity. Therefore the light velocities parallel and antiparallel to the disk peripheral velocity are different. For the TSR this is not acceptable, because a set of platforms with the same peripheral velocity locally approximates an inertial system. The situation is shown in Fig. 1.

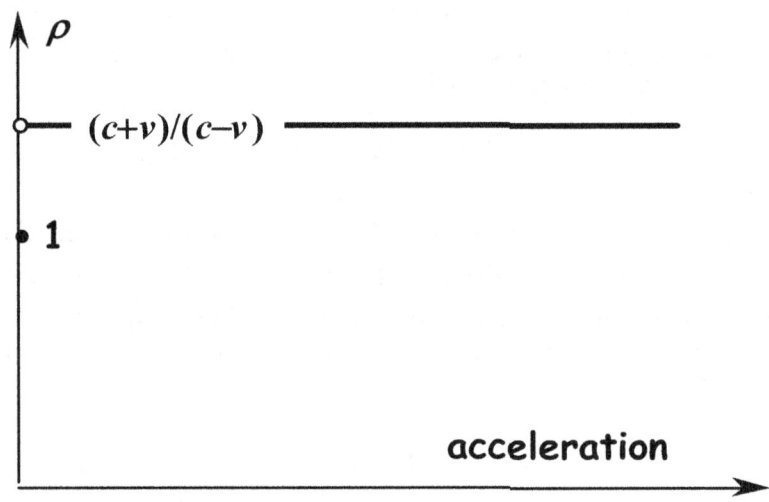

Figure 1. The ratio $\rho = \tilde{c}(\pi)/\tilde{c}(0)$ plotted as a function of acceleration a for rotating platforms of constant peripheral velocity and decreasing radius (increasing acceleration). The prediction of the TSR is $\rho = 1$ for $a = 0$ (black dot on the ρ axis) and is not continuous with the ρ value of the rotating platforms.

Thus the TSR predicts for ρ a discontinuity at zero acceleration. While all the experiments are performed in the real physical world [$a \neq 0$, $\rho = (1+\beta)/(1-\beta)$], the theory has gone out of the world ($a = 0$, $\rho = 1$)!

Notice that the velocity of light given by (48) with $e_1 = 0$ is required for all inertial systems but one, the isotropic system S_0. In fact, for every small region AB of every such system it is possible to imagine a large rotating platform with center at rest in S_0 and rim locally co-moving with AB and the result (48) can be applied. Thus the velocity of light depends on direction in all inertial systems with the exception of the privileged one, S_0.

5   Lifetimes in particle physics

We are going to show that several famous experiments on the velocity dependent modification of the lifetime of unstable objects share the property of being explained equally well by all the equivalent transformations (ET). In other words the theoretical predictions do not depend on the synchronization parameter e_1. We start from the relativistic modification of the lifetime of unstable objects. Three examples in particle physics [18] are

π (pion) disintegrates mostly: $\pi^+ \to \mu^+ + \nu$ with $\tau = 2.197 \times 10^{-8}$ s;

N (neutron) mostly: $N \to P + e^- + \nu$ with $\tau = 886$ s ;

. Λ (Lambda) mostly $\Lambda \to P + \pi^-$ or $N + \pi^\circ$,   with $\tau = 2.63 \times 10^{-10}$ s.

Unstable particles moving at high velocity are known to have a correspondingly longer lifetime. It has been checked many times and very accurately that the following formula holds without exception

$$\tau = R \tau_0 \tag{53}$$

where τ_0 is the lifetime measured in the laboratory crossed by the beam of the unstable particles with velocity u ; τ is the lifetime of the same particles measured by an observer at rest with them; finally $R = \sqrt{1 - u^2 / c^2}$.

The experimentalists up to now trusted completely the TSR. Therefore they measured the time delay Δt of the pions (say) in a beam with respect to a light pulse. If ℓ / c and ℓ / u are the times spent by the light pulse and the particle, respectively, to cover the distance ℓ , the time delay is by definition

$$\Delta t = \frac{\ell}{u} - \frac{\ell}{c} \tag{54}$$

The previous equation contains the unknown velocity u. Everything else is either known or easily measurable. It can be solved to obtain u. Once this is done the square root factor is easily calculated. The results are

$$u = \frac{\ell c}{c\Delta t + \ell} \qquad ; \qquad R = \left[1 - \frac{\ell^2}{\left(c\Delta t + \ell\right)^2} \right]^{1/2} \tag{55}$$

129

In this way, the value of R in (53) is numerically predicted. It turns out to be in full agreement with the observed velocity dependent lifetime increase. Obviously this represents an application of the theory of special relativity, given that the one-way velocity of light is assumed to be c, consistently with the Lorentz transformations. Does the excellent agreement with experiments imply that alternative transformations (such as the inertial transformations) have difficulties in trying to explain the modified lifetimes of unstable moving particles? The answer is negative, the great successes scored by the TSR in this field do not imply a choice in favor of the relativistic theory essentially because a time delay, in principle measurable with a single clock placed near the point of arrival, cannot depend on the way clocks are synchronized. In fact, if the calculation is repeated with the ET formula for inverse velocities, one gets a result identical to (54). In fact in the theory of the equivalent transformations (TET) it is easy to show that the inverse velocities which the TSR indicates by 1/c and 1/u, in the TET are instead

$$\frac{1}{c} + \frac{1}{c}\Gamma\cos\theta \qquad \text{and} \qquad \frac{1}{u} + \frac{1}{c}\Gamma\cos\theta \qquad (56)$$

respectively. Notice that the Γ - dependent term is the same in the two cases.

$$\Delta t = \left(\frac{\ell}{u} + \frac{\ell}{c}\Gamma\cos\theta\right) - \left(\frac{\ell}{c} + \frac{\ell}{c}\Gamma\cos\theta\right) = \frac{\ell}{u} - \frac{\ell}{c} \qquad (57)$$

In conclusion, all the theories "equivalent" to the TSR lead to the same prediction, namely to the validity of (53) with the same R. The triumphs of the TSR are triumphs of all TET, in particular, of course, of the TIT.

6.   The differential retardation effect

The differential retardation effect consists of two clocks which separate and reunite later again: of the two clocks will mark a smaller proper time that one which has used the infra-meeting time interval to visit places, to bend its trajectory, to run here and there. It all depends on a more elementary property of nature, the velocity dependent retardation of moving clocks, a well established physical phenomenon. A large amount of experimental evidence, in agreement with Einstein's 1905 statements, shows that the clock proper time variation $d\tau$ equals the inertial system time variation dt_0 times the usual velocity dependent square root factor. The formula is

$$d\tau = dt_0\sqrt{1 - u^2(t_0)/c^2}$$

130

where $u(t_0)$ is the clock instantaneous velocity relative to the inertial frame S_0. In the TSR S_0 is arbitrary, while in the theory of the ET S_0 has to be the privileged isotropic frame, otherwise the right expression for $d\tau$ is slightly more complicated.

After all it is only natural that the rectilinear uniform motion of a clock between two events (P_a, t_{oa}) and (P_b, t_{ob}) corresponds to the maximum proper time of all the trajectories connecting the same two events. In fact, given that the times of departure from the point P_a and of arrival at point P_b are the same for all trajectories, the clock following the rectilinear trajectory will be slower than the clocks moving over longer curvilinear trajectories. Thus the square root factor will be larger, on the average, with obvious consequences for the proper time interval. Clearly the clock paradox is a non classical phenomenon: in classical pre-relativistic physics the above formula would be $d\tau = dt_0$ and proper times would last $t_{ob} - t_{oa}$ seconds irrespective of the trajectory followed to go from the event (P_a, t_{oa}) to the event (P_b, t_{ob}).

The differential retardation effect ("clock paradox") has been simplified very much by contemporary research. Now we recognize that velocity and nothing else is the cause of this phenomenon. Inertial observers disagree, however, on the numerical value of the velocity of either clock at any instant of time. In relativity all potential observers (forming an infinite set) are completely equivalent, so that, if the TSR is right, one can say that the clock velocity assumes at any time all conceivable values. But a quantity having at the same time infinitely many values is totally undefined. In this way our presumed cause of the differential retardation would seem to vanish into nothingness. This is not physically reasonable, however, obviously the cause of a real physical effect has to be concrete as well, in spite of the evasive description coming from the theory. We recognize also from this that the relativistic theory needs to be modified.

Clocks are real, and so are the numbers appearing on their screens. When two clocks for a very short time enter in direct physical contact and numbers are written on their screens, there is no space for disagreement between different observers. If the proper time is used, the differential retardation effect (read on the clocks at separation and at reunification) appears to be the same to observers in all states of motion, and in this sense it is not only absolute but especially well defined. The emergence of a well defined effect consequence of velocity implies the existence of a privileged inertial frame, in the sense that motion relative to this frame assumes a particular significance for a well defined velocity which can be associated to well defined effects.

Therefore causality implies that velocity itself should be well defined, that is, relative to a physically active reference background (ether) that constitutes at the same time the privileged inertial reference frame.

131

To escape from conclusions favorable to the ether the original formulation of the clock paradox was completed with a later one based on the theory of general relativity (TGR). The work of many people, including Builder [20], Prokhovnik [21], and Unnikrishnan [22], has shown that the 1918 proposal of Albert Einstein is untenable. To understand the clock paradox we need to consider only the velocities defined by the ether. And this ether has to be itself concrete and well defined. Therefore also the "weak relativity" conclusion of our research should not be taken as final: we cannot do better for the moment, but the day of overcoming all forms of relativity is approaching.

7. The accelerations do not act.

The empirical evidence shows that accelerations are not the cause of any phenomenon. For example, the CERN muon storage ring experiment proves that an acceleration as big as 10^{18} g does not modify the muon lifetime in the slightest. In the same experiment was instead very visible the effect of velocity, with an increase of the muon lifetime $\tau \rightarrow \tau_0$ by a factor of about 28 according to the formula

$$\tau_0 = \tau / \sqrt{1 - v^2 / c^2}$$

and exactly equal to the lifetime growth of a rectilinear beam of muons having the same speed. It is evident that accelerations are no protagonists of the game of physics. One could jokingly say that accelerations do not exist, that they are only variations of velocities.

Very different is the case of velocities, which are easily recognized as cause of important phenomena:
1)   Determination of the shape in space of the electric and magnetic fields generated by any structure in motion containing charges and currents.
2)   Corresponding determination of the structure of atoms in motion, as their electrons feel the modified em fields (point 1) . This is the Lorentz contraction.
3)   Determination by the velocities, besides the structure, also of the evolution of atomic systems. Periodic motions undergo the famous "clock retardation".
4)   As a consequence of 3), the rhythm with which a macroscopic clock oscillates turns out to be determined, another example of clock retardation.
Isotropy of space implies that the direction of velocity is irrelevant for the size of the produced effects. Therefore, if a velocity changes direction only, all its consequences remain the same. Then it is better to say that the effects listed above are consequence of the speed of the moving body. In the CERN muon storage ring experiment the effect of motion on the lifetime is exactly equal to the effect generated by motion on a linear beam of muons having the same speed.

Having excluded the existence of direct consequences of accelerations, we can extend the negative conclusions to indirect consequences. In particular the resolution of the clock paradox proposed by the TGR, is invalid. There is no gravitational potential of the fictitious forces. As is well known, the general relativistic effect hypothesized by the TGR is not on the accelerating system, but on the other systems participating in the ideal experiment with a static role. If this idea were correct there should be also an effect of the acceleration of the storage ring on other low energy muons eventually present in the laboratory. But nothing of the type has ever been seen and the very proposal of the idea makes it sound very unlikely.

This conclusion had been anticipated by Builder in the fifties [20]. More recenty C.S. Unnikrishnan of the Tata Institute (Mumbai) has carried out a new detailed analysis of the clock paradox [22]. The Indian author remarks that Einstein's paper on *Die Naturwissenschaften* [19] is entitled "Dialogue about objections to the theory of relativity" and has the form of a dialogue between a critic and an "adherent" of the theory of relativity. The discussion starts with a complaint of the critic that none of the relativists discussions has adequately responded to the many published critical remarks. This suggests that Einstein considered that none of the earlier adequately addressed the problem and that it was necessary to respond.

Unnikrishnan's conclusion is that Einstein's resolution of the clock paradox suffers also from logical and physical flaws and gives incorrect answers in a general setting. But before discussing issues of logic and consistency in Einstein's resolution of the clock paradox it must be noticed that this resolution cannot stand up to the counter examples that do not involve any accelerations. Since Einstein invoked the equivalence principle and a homogeneous gravitational field equivalent to the acceleration as the physical cause of asymmetrical clock retardation, acceleration is absolutely essential for his analysis to work. But the clock paradox can be stated without acceleration by invoking a third inertial observer.

If U's calculation invokes a differential clock retardation due to a homogeneous gravitational field, then it is illogical and inconsistent to ignore it in U_1's calculations. It is true that U_1 has no way of determining locally whether there was a homogeneous gravitational field present when U was reversing in motion, since a homogeneous gravitational field cannot be detected while in free fall. But during the comparison of notes if U_1 agrees to U's claim that there was indeed a homogeneous gravitational field, then he should also calculate the total time retardation incorporating this fact. This means that U_1 would overestimate grossly the clock retardation of U and would not find agreement with the observations.

The results obtained by Prokhovnik [21], Builder, Unnikrishnan,[22] lead us to conclude that the gravitational potential of the fictitious forces exerts no action on the clocks, contrary to Einstein's 1918 opinion. The gravitational fields in the accelerated systems are not ordinary static fields like that of the Earth, but arise from the accelerations of bodies. Einstein assumed that these fields had on clocks the same action as ordinary fields, but we can now say that on this particular point he was not right, in spite of the very probable correctness of the general idea of equivalence between fictitious and gravitational forces. One finds an analogy in the magnetic field, which can be considered a dynamical manifestation of the electric field, but has quite different interaction properties.

8. No synchronization jumps

Two identical clocks, C_A and C_B are at rest on the x_0 axis of the isotropic reference frame S_0 at a distance D_0 from one another. Like the other clocks of S_0, C_A and C_B have been synchronized on the basis of the velocity of light (Einstein's method). Therefore a new flash of light propagating in the $+x_0$ direction and touching C_A and C_B at times t_A and t_B, respectively, satisfies the relationship $D_0 / (t_B - t_A) = c$, meaning that the new measurement of the velocity of light is bound to give the result "c".

What happens if one modifies the times shown by the two clocks with two equal re-settings? For example, one subtracts 33'45'' from the time of C_A and 33'45'' from the time of C_B. What happens if after doing this, one uses C_A and C_B for a new determination of the velocity of light? Of course nothing happens, and the result "c" is found again, because the equal corrections of the two clocks cancel in the time difference.

We can also try to implement the resetting of time by physical means, that is by using the time retardation of moving clocks. We can do it, because C_A and C_B are on board of two spaceships, A and B, respectively. At a certain common time marked by the two clocks, A and B ignite their engines and actuate a pre-established program of acceleration such that at every instant of time the velocity is the same for A and B. But then, also the usual velocity dependent square root factor of time retardation is the same. Integrating over time, the retardation of the two clocks at the end of the acceleration period is the same. Incidentally C_A and C_B at the end of the acceleration period are at rest in a different inertial system S. Anyway, if they are used for a new measurement of the velocity of light they do not produce "c", but the value predicted by the inertial transformations (based, as you will remember, on $e_1 = 0$). Does this mean that during the trip there has been a jump in the reciprocal synchronization of C_A and C_B?

During the parallel trip of the two spaceships there seems to be a change in the
synchronization of C_A and C_B [23,24]. Before departure the two clocks were *synchronized à la Einstein*, at the end of acceleration they had adopted the absolute synchronization. Moreover, the calculation performed allows me to claim that the acceleration is not the cause of the seeming change of synchronization

Actually there was no synchronization change. The observer in S_0 is a witness of the perfect identity of the motions of the two clocks. Therefore he comes to the firm conclusion that if motion has consequences for the moving clock, then the consequences must be identical for C_A and C_B. It is enough that this conclusion is arrived at by an observer to conclude that it must hold for all observers, for the conclusions of the first observer cannot be negated. Thus the effects of motion on C_A and C_B are exactly of the trivial type described above when 33'45'' were subtracted from the time of C_A and another 33'45'' from the time of C_B. In these circumstances no change of synchronization can arise for the two clocks. The truth obviously is that the IT provide the natural continuation to moving systems of the physics holding in S_0. *In other words the absolute synchronization method provides the natural continuation to moving systems of the Einstein synchronization valid in S_0.*

No, there is no synchronization change for the two clocks.

9.   Time, space, what is left after positivism?

According to the positivistic thinkers of the beginning of the XXth century (Mach, Avenarius, Ostwald, ...) our general conceptions of space, time, cause, effect, ... are metaphysical, that is beyond the reach and the interests of physics and, more generally, science. Therefore a serious scientific approach is possible only if such ideas are left out of our elaborations. Einstein, influenced by positivism in that first part of his scientific activity which produced the two relativistic theories, tried hard to get rid of the metaphysical ideas by giving a central role to the "observer" and to the inertial frames (considered equivalent, as nothing physical could select one of the set) in which different observers could be at rest. Over this landscape positivism gave birth to relativism.

The 1905 theory of special relativity was simple and beautiful from the point of view of the "modern" positivistically oriented philosophy of science, but was judged to be little less than a disaster by the old Galilei, Descartes, Leibnitz, Newton lines of thought. Time, space, and of course velocity, had been modified in depth, with the objective simultaneity of events declared nonexistent, the velocity of light taken to be invariant after the expulsion from physics of one way velocities. In whatever was left of space and time other distortions arose from the exclusive dependence on relative velocity of the contraction of moving rods and the retardation of moving clocks. Time was considered to be of no interest in itself, and it had been demoted simply to the fourth dimension of Minkowski's space-time continuum. For a classical physicist the new theory introduced a nightmare of paradoxes, but a modern physicist could retaliate that relativity was mathematically simple and elegant and the paradoxes existed only in the attempts of the old theory to give physical meaning to metaphysical conceptions. Anyway, the word paradox had entered in the scientific debate. It is still there more than a century later.

The way out of this complex situation has been found thanks to the contributions of several workers: Reichenbach [25], Jammer [26], Tangherlini [27], Mansouri and Sexl [28] for example. It is based on new transformations of space and time in which time is independent of the space variables. The inertial transformations, based on absolute simultaneity, imply that a privileged (isotropic) inertial reference

135

system exists. One can show, however, that it is possible to resynchronize clocks in all inertial frames in such a way as to select a different, arbitrarily chosen frame as "privileged". Such a resynchronization of clocks (ROC) does not modify any empirical consequence of the theory, which is thus compatible with a new form of relativity principle. Einstein based the theory of special relativity on two principles which together lead necessarily to the Lorentz transformations. In an important sense we can consider Einstein's relativity as a strong principle. When it says that the physical laws "are not affected" by a change of reference system, it requires the laws of nature to have <u>exactly</u> the same form in all inertial reference frames.

From such a point of view the inertial transformations, based on absolute simultaneity and alternative to the Lorentz transformations, have milder implications. For example, Buonaura [29] showed that Maxwell's equations, transformed from the isotropic inertial system S_0 (where they retain the usual form) to another inertial system S, acquire a generalized form, which depends on the velocity of S relative to S_0 ("absolute velocity"). From such a point of view Maxwell's equations are "affected" by a change of reference frame given that the velocity of the frame to which they are referred is modified. A recent book [2] contains six independent proofs that the physics of space and time has to be based on $e_1 = 0$.

In interstellar space there is a very good vacuum, which can reach an atomic density as low as 0.1 atom/cm3. One can make a safe guess: if even the few atoms left were taken away, none of the fundamental properties of space would change.

The property which here interests us most, is that space in some sense is a container of distances. We see in space bright objects and our science has been able to measure their distances. Three examples: the moon is 1.2 light seconds away; the nearest star (Proxima Centauri) is at about 4.3 light years; the great spiral galaxy of Andromeda at about 2.2 million light years.

For a positivist the vacuum is pure emptiness: take away those few atoms and nothing remains. The vacuum is the physical zero. The vacuum is nothing, it does not exist.

But an answer can be given to the positivistic rejection of space. If the vacuum is nothing how is it possible that the distances separating cosmic objects be so different from case to case? These distances are made of vacuum. If the vacuum is the physical zero, one would expect that 0+0=0, or a unicity of distances, even the vanishing of all distances. In the real world, instead, there are small distances, large distances, immense distances, ... Correspondingly one can have small, large, huge quantities of space. We can measure space on a line, in a surface, and in a volume. It is not possible that something so easily quantified be nonexistent.

Our conclusion about the reality of space allows us to move another step forward. In fact space is not only real, but is homogeneous. We do not know how far this property can be extrapolated, but for our concrete physics, limited to the interior of the solar system, homogeneity is certainly at least an excellent approximation. The line of thought based on space homogeneity has been developed in the quoted book [2] and the result has been a remarkable strengthening of the realistic point of view. But this would be too long to review here. Instead we pass to the last argument of the paper, the overcoming of relativistic paradoxes.

10.   Overcoming the relativistic paradoxes

The adoption of the inertial transformations implies essentially a violation of Einstein's relativity principle. Clearly the point needs to be discussed further. It is useful to distinguish two formulations of the relativity principle:

R1.   **Strong relativity**, according to which the laws of physics are exactly the same in all inertial systems. This is Einstein's formulation.

R2.   **Weak relativity**, stating merely the impossibility to measure the absolute velocity of the Earth. This principle does not demand necessarily the validity of the Lorentz transformations and opens a logical space for new theories, such as the one based on the inertial transformations. This formulation is essentially the original one given by Galilei extended, of course, to all physical phenomena.

Experimental evidence based on the clock paradox and on stellar aberration shows that absolute velocities exist in nature. The weak relativity principle accepts this, but maintains that they nevertheless remain not measurable. The conclusions of this book are not in disagreement with some of those obtained by Rizzi, Ruggiero and Serafini, who wrote: "… the 'privileged role' played by S_0 … is a merely artificial element, S_0 being just the IRF [inertial reference frame] in which, by stipulation, Einstein synchronization has been performed: as a matter of fact, any IRF S can play the role of S_0."

In spite of this conclusion, with which it is possible to agree, I must insist that the statements made in the initial part of the present book are correct. The theory with the free e_1 applied to the rotating platform and to the Sagnac effect shows that only $e_1 = 0$ gives a rationally acceptable formulation of the physics on the disk. Similarly, only with $e_1 = 0$ one can obtain a reasonable description of aberration. The paradoxes of the special theory of relativity disappear if $e_1 = 0$. The growing evidence for the existence in nature of superluminal signals can easily be accommodated if $e_1 = 0$, while it is incompatible with standard relativity due to the presence of a famous causal paradox. We will next check that the paradoxes of the special theory of relativity disappear if $e_1 = 0$. Therefore a possible theory of space and time seems to be the one based on absolute simultaneity.

P1.   The velocity of a light signal, considered equal for observers at rest and observers pursuing it with very high velocities. The answer of a theory based on the ITs is as follows. After having established that $e_1 = 0$, the velocity of light relative to a moving reference frame is given by Eq. (18.2). Therefore, the speed of the light signal (absolute velocity c) relative to an inertial frame which is running after it (then, $\theta = 0$) with absolute velocity almost equal to that of light ($v \cong c$) has a denominator $\cong 1 + 1 = 2$ and the limit velocity is $c/2$. This is a 50% reduction with respect to the TSR prediction.

P2.   The retardation of moving clocks, phenomenon for which the theory of relativity does not provide a description in terms of objective causation. It was possible to show that with the inertial transformations objectivity can be restored in terms of action of the ether on all the periodic phenomena which are used to measure time, very much in the realistic line of thought of Hendrik Lorentz.

P3.   The contraction of moving objects, phenomenon for which, once more, the theory does not provide a causal description. The objectivity can be restored with the inertial transformations in terms of action of the ether on every atom, with reduction of the atomic length in the direction of motion. Also this follows the realistic line of thought of Lorentz.

P4.   The (relativistic) idea that the simultaneity of spatially separated events does not exist in nature and must therefore be established with a human choice was accepted by Mansouri and Sexl, who fully believed in the conventionality of clock synchronization. In spite of the broad diffusion of this type of expectation, we have established that a rational description of physical phenomena (Sagnac effect, linearly accelerating systems, objective reality of inertial observers, superluminal propagations) can be obtained only if absolute simultaneity is adopted: $e_1 = 0$.

P5.   The relativity of simultaneity, according to which two events simultaneous for an observer in general are no more such for a different observer was overcome when it was shown that assuming $e_1 = 0$ all inertial observers have the same reality, where reality is defined by the set of events simultaneous with a given event (e.g., the "here-now" event establishing the local present).

P6.   The conflict between the reciprocal transformability of mass and energy and the ideology of relativism. The TSR declares all inertial observers perfectly equivalent so depriving energy of its full reality. The retrieval of the objectivity of energy and of the other physical quantities is possible due to the inequivalence of the different reference frames and to our present conviction that there is one at rest in the ether, which has a more fundamental role. The idea is developed and the objectivity of energy is fully recovered by working with the inertial transformations.

P7.   The hyperdeterministic block universe of relativity, fixing in the least detail the future of every observer, is now out, having been overcome thanks to the conclusion that $e_1 = 0$. In a two dimensional diagram the reality line of a theory based on the inertial transformations is the same for all observers, independently of their state of motion.

P8.   The TSR predicts a discontinuity between the inertial systems and systems endowed with a very small acceleration. The discontinuity is in (see equation (12)) the variable ρ, ratio of the velocities of light along two opposite directions. It turns out to be a very serious problem for all $e_1 \neq 0$. If one takes $e_1 = 0$, however, the discontinuity does not exist anymore and the relative difficulty is completely overcome. The existence of this discontinuity is related to the great difficulties met by the TSR with the Sagnac effect.

P9.   The propagations from the future towards the past, generated in the TSR by the eventual existence of superluminal signals. It has been shown that the essence of the causal paradox lies in the impossible requirement that a superluminal propagation may overtake a set of clocks marking a progressively decreasing physical (then, not conventional) time. The particular choice   , which, as we saw, is selected by several phenomena, remains by far the best one also in the present context, being the only one not leading to the causal paradox. The same choice avoids the complications of the TSR describes all the propagations as forwards in time for all observers.

P10.   The asymmetrical aging of the twins in relative motion in a theory waving the flag of relativism. We discussed the differential retardation effect between separating and reuniting clocks ("clock paradox"). A variational method was used to show, both in the TSR and in more general theories with arbitrary e_1, that among all possible trajectories of a clock connecting two given points at two given times the rectilinear uniform motion requires the longest proper time. A complete resolution of the clock paradox is so obtained by giving an exhaustive unified description of all possible situations.

Relativism does not apply and must be considered obsolete. Velocity (and nothing else) is seen to be responsible for the differential retardation effect. Of course it must be an absolute velocity! Hidden behind the relativism of Einstein's theory there is a physically active background.

In spite of the overcoming of the relativistic paradoxes we must admit that our results may seem somewhat contradictory. On the one hand they point to a theory of space and time in which such conceptions as absolute velocity, privileged frame and absolute simultaneity have a central role, while, on the other hand, relativism comes back in the arbitrariness of the choice of the "privileged" inertial reference frame. In spite of the fact that our results are mixed, we must stress that both sides of the contradiction (namely, $e_1 = 0$ and weak relativity) are absolutely correct if the assumptions made in the first section are correct. I can add that from the point of view of the inertial transformations the validity of weak relativity appears accidental, more than fundamental. It would be enough to discover a very small non invariance of the two way speed of light to make the whole game of ROC impossible. Therefore weak relativity is provisional, as it is the whole idea of relativism that soon will probably be fully overcome [33].

REFERENCES

[1]   A. Einstein, H.A.Lorentz ..., THE PRINCIPLE OF RELATIVITY, Dover, NY (1952).
[2]   F. Selleri, WEAK RELATIVITY, Apeiron, Montreal (2009).
[3]   D. C. Miller, *Rev.Mod.Phys.* **5**, 203 (1933).
[4]   R. J. Kennedy and E. M. Thorndike, *Phys.Rev.* **42**, 400 (1932).
[5]   A.A.Michelson and E.W. Morley, *Am. J. Science*, **34**, 333 (1887).
[6]   Ref. [4], p. 416.
[7]   T.S. Jaseja, A. Javan, J. Murray and C.H. Townes, *Phys. Rev.* **133**, A1221 (1964).
[8]   A. Brillet and J. L. Hall, *Phys. Rev. Lett.* **42**, 549-552 (1979).
[9]   E. Rijs, L.-U. Aaen Andersen, N. Bjerre, O. Poulsen, S.A. Lee and J.L. Hall, *Phys. Rev. Lett* **60**, 81 (1988).
[10]   D. Hils e J.L. Hall, *Phys. Rev. Lett.* **64**, pp. 1697-1700 (1990).
[11]   M. Allais, *C.R. Acad. Sci. Paris, Série IV* **1**, 1205 (2000).
[12]   R. De Meo, *Pulse of the Planet,* **5**, 114 (2002).
[13]   R. Cahill, "The Einstein Postulates 1905-2005. A Critical Review of Evidence." In V. Dvoeglazov, ed., *Einstein and Poincaré: The Physical Vacuum*, Apeiron, Montreal (2006).
[14]   H. Mùnera, D. Hernàndez-Deckers, G. Arenas and E. Alfonso, *Electromagnetic Phenomena* **6**, 100 (2006).

[15]   M. Consoli, A. Pagano and L. Pappalardo, *Phys. Lett.* **A318**, 292 (2003).

[16]   F. Selleri, *Apeiron* **4**, 100 (1997).

[17]   J. Bailey, et al. *Nature*, **268**, 301 (1977).

[18]   Particle Data Group, Review of Particle properties, *Phys. Lett.* **B 239**, 1-516 (1990).

[19]   A. Einstein, *Die Naturwissenschaften* **6**, 697 (1918).

[20]   G. Builder, *Aust. J. Phys.* **10**, 246 (1957).

[21]   S.J. Prokhovnik, *Spec. Science and Techn.* **2**, 225 (1979).

[22]   C.S. Unnikrishnan, *Current Science* **89**, 2009 (2005).

[23]   S.P. Boughn, *Am. J. Phys.* **57**, 791 (1989).

[24]   S. K. Ghosal, S. Nepal and D. Das, *Twin paradox: a classic case of 'like cures like'* , paper presented at the International Conference MATHEMATICS, PHYSICS AND PHILOSOPHY IN THE INTERPRETATIONS OF RELATIVITY THEORY, Budapest, September 7-9, 2007.

[25]   H. R. Reichenbach, THE PHILOSOPHY OF SPACE AND TIME, Dover, New York (1958).

[26]   M. Jammer, CONCEPTS OF SIMULTANEITY, The Johns Hopkins University Press, Baltimore (2006).

[27]   F.R.Tangherlini, *Nuovo Cim. Suppl.* 20, 351 (1961).

[28]   R. Mansouri and R. Sexl, *General Relat. Gravit.* **8**, 497, 515, 809 (1977).

[29]   B. Buonaura, *Found. Phys. Letters* **17**, 627-644 (2004).

[30]   See the chapters 13 (aberration) and 14 (clock paradox) of ref. [2].

[31]   G. Rizzi, M.L. Ruggiero and A. Serafini, *Found. Phys.* **34**, 1835 (2005).

[32]   See p. 1840 of ref. [31].

[33]   Presently the two way velocity of light in vacuum is known with an error of about 20 cm/s and has the value $c = \left(299\ 792\ 458.8 \pm 0.2\right)$ m/s. See: P.T. Woods, K.C. Sholton, W.R.C. Rowley, Appl. Opt. **17**, 1048 (1978); D.A. Jennings, R.E. Drullinger, K.M. Evenson, C.R. Pollock, J.S Wells, J. Res. Natl. Bur. Stand. **92**, 11 (1987). J.S Wells, J. Res. Natl. Bur. Stand. **92**, 11 (1987).

Relativity Based on Physical Processes, not on Space-Time

Albrecht Giese

Postal Address: Taxusweg 15, D 22605 Hamburg, Germany

E-mail: note@ag-physics.de

Abstract

At the beginning 20th century, Einstein developed the relativity theory as one of the prominent topics of physics. Einstein gave this subject a very specific direction, when he explained the relativistic phenomena in a geometric way, by specific assumptions about space and time. This approach was accepted as a fact by the physical community. It was never seriously questioned at a later date.

However, relativity theory does not depend on Einstein's assumptions. There are good reasons to go another way. We can explain relativistic effects as a consequence of physical processes, which means as properties of fields and of elementary particles. We then achieve great benefits.

(1) The formalism to describe relativity, special relativity as well as general relativity, turns out to be much easier.

(2) The paradoxes and logical conflicts, which come along with Einstein's way of relativity, are avoided.

(3) We achieve further understanding of other areas of physics. Unresolved open points of present physics get a solution. - Relativity is no longer a separate topic within the total range of physics, but it follows from general physical structures, and it is thus integrated into physics as a whole. This also explains open problems in present physics like the incompatibility between relativity and quantum physics.

About the ether: The model of relativity presented here has to accept an ether in the sense of an absolute frame of reference. This is the only compelling assumption of an ether following from our model.

The exclusive reference of relativity to space-time according to Einstein has historically blocked the further development of physics to a great extent. The aim of this article is to provide a contribution to a way out of this trap.

1. Overview

This article is structured into the following parts:

Chapter 2 shows in a historical review, how Einstein came to his understanding of relativity and to his way to develop the theory. Connected with it we present some considerations to his way of the philosopher Hans Reichenbach, who supported him.

Chapter 3 explains how special relativity can be derived from the properties of fields and of particles. Here we will tie in the considerations of Hendrik A. Lorentz, who took the first step towards this direction. We will then refer to the insights provided by Louis de Broglie, Paul Dirac, and Erwin Schrödinger to derive dilation. We will see that dilation can be logically connected to the internal oscillations and to the wave properties of elementary particles.

In chapter 4 we further develop the consequences of dilation and internal oscillation, and we build a particle model from them to unify those aspects. This will also explain the origin of mass and derive the relativistic properties of mass. This general particle model will be explained to some necessary detail.

In chapter 5 we will see how also general relativity, that is gravitation, can be understood without reference to Einstein's assumptions about space-time. Gravitation will be shown to be a consequence of a varying speed of light. The result is an easy way to general relativity, which will be developed here up to the Schwarzschild solution.

In chapter 6 some implications of our model of relativity as regards questions of cosmology are presented. If gravitation is a mass-independent phenomenon, photons fulfil the properties of dark matter particles in an almost ideal way. And if we accept that the speed of light is not a constant given by space and time but controlled by classical physical processes and can vary, then the problem of dark energy disappears.

In chapter 7 the inferences of the model as regards the speed of light and the question of an ether is discussed. We present an idea about the physical cause of the constancy of the speed of light. This will also show that the assumptions of some physicists these days, that the speed of light is decreased during the development of the universe, can be justified. This gives also some hint, what we can understand to be the ether.

In general the question of exchange particles is of importance for the model presented here. Exchange particles have been invented by quantum mechanics; however, not many details are given by quantum mechanics. We will, when referring to exchange particles, at the places we use them indicate which characteristics are used. So, it should be visible, which properties are needed by the particle model used here and which conclusions we can draw from the properties used.

2. History of Special Relativity

2.1 The Conclusions from the Michelson-Morley Experiment

The basic experiment, which started all activities about relativity, was the Michelson-Morley experiment. For the observed null-result several explanations were offered at that time by physicists. Of those we shall only describe here two of them, which have relevance until today.

Quite early after the experiment, an explanation was presented by the Dutch physicist Hendrik A. Lorentz [3]. It was understood that electric fields contract, if at motion with respect to the ether. This was found as a consequence of the theory of Maxwell about electromagnetism. As a further consequence, objects also contract at motion. This contraction is also assumed for the apparatus of Michelson and Morley, and so the null result is a logical consequence.

Most physicists, however, in the end followed the position of Einstein, that the speed of light is in fact physically the same for every inertial system. This position went along with the opinion that no ether of any kind existed. Einstein himself discounted the position of Lorentz as an "ad hoc explanation" without a physical meaning.

2.2 Einstein's Solution

Einstein based his physical understanding on the direct impression that the speed of light was apparently constant in every inertial system. In his deduction he referred to an experimental situation, in which a light signal is sent from a source to a mirror and back to the location of the source (Figure 2.1). He took it as a given fact that the time, needed for the light signal to move to the mirror, is equal to the time for the way back. In his understanding, this assumption is even true for every experimental set up moving at a constant speed vector.

2.2.1 Einstein Trapped by his Principle

In his original paper about special relativity [1] Einstein stated (in German)

> Vom Anfangspunkt des Systems k aus werde ein Lichtstrahl zur Zeit τ_0 längs der X-Achse nach x' gesandt und von dort zur Zeit τ_1 nach dem Koordinatenursprung reflektiert, wo er zur Zeit τ_2 anlange; so muß dann sein:
>
> $$\tfrac{1}{2}(\tau_0 + \tau_2) = \tau_1$$

In English (translated by the author):

"From the origin of a system k there shall a light beam be transmitted at the time τ_0 along the X-axis to x' and from there at the time τ_1 reflected back to the origin, where it shall arrive at time τ_2; so must be then:

$$\tfrac{1}{2}(\tau_0+\tau_2) = \tau_1 \text{ "}$$ (2.1)

The following figure is an illustration of Einstein's consideration.

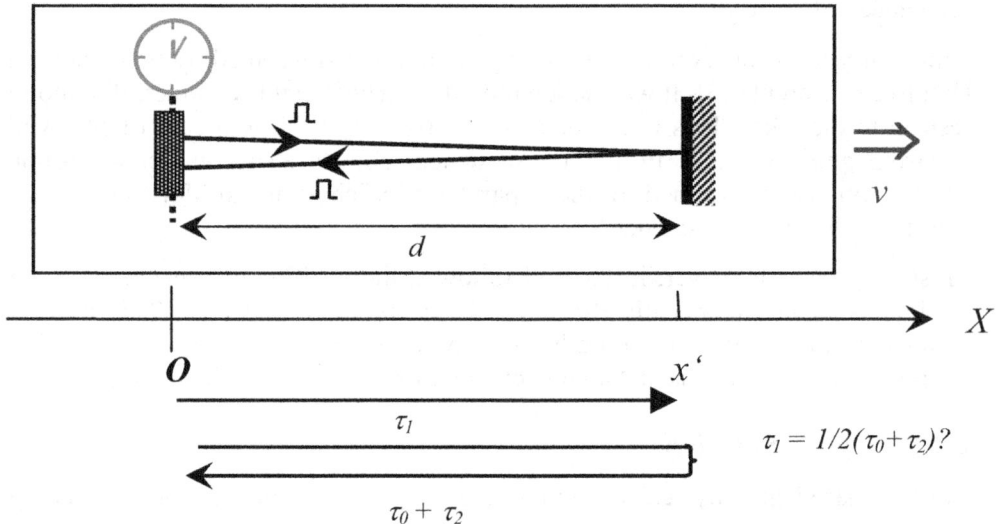

Figure 2.1: Measurement of c in a moving lab

We have to realize that this assumption of Einstein cannot be verified by any experiment. It is not possible to measure the one-way speed of light without certain assumptions. The reason is that any such measurement needs two clocks, one at the starting point and one at the destination point of the signal. These clocks have to be synchronized. Einstein has given in his paper a very general advice how to synchronize clocks: A light signal shall be sent from one clock to the other one and the travel time of the light signal shall be taken into account. - But this needs the knowledge of the speed of light which is to be determined here. So, the use of this method for this measurement results in circular conclusion.

We can assume that Einstein was aware of this fact. He never stated that this assumption can be proven, but he called it a 'principle'. This means a general assumption about our physical world which is accepted without a proof, like a mathematical axiom.

2.2.2 Consequences of the Principle

With the acceptance of his principle, Einstein's concept about space and time is cemented.

From this point no recovery is possible. It is essential to understand that this assumption of Einstein including its consequences for the changes of space and time is on the one hand a mathematically possible solution, but by no means the only possible solution. And also, as we are going to show, it is not at all a practical or easy way.

If we go back to an understanding of relativity, which is based on physical processes, as it was initiated by Lorentz some time before Einstein, we will find that we have on the one hand a much easier understanding of the related phenomena, and on the other hand we get a better understanding about physics in general.

2.2.3 Position of Hans Reichenbach about Einstein

The philosopher Hans Reichenbach supported Einstein in the early years of relativity. In his book "Philosophy of Space and Time" [2] he stated very clearly that Einstein's theoretical treatment of relativistic effects, particularly the constancy of the speed of light in a moving system, is only one possible solution, not the only one.

In his book he says: " ... This definition is essential for the special theory of relativity, but it is not epistemologically necessary. Einstein's definition, too, is only one possible definition. ..."

With respect to Einstein's consideration, shown in figure 2.1, he reasons with respect to eq. (2.1), why the special theory prefers ½ ... :

" it does so on the ground that this definition leads to simpler relations". ...

We will, however, show in the following that Reichenbach's argument of "simpler relations" is not at all applicable.

3. Special Relativity Based on Physical Processes

In this chapter we will show, how the basic phenomena of special relativity, dilation and contraction, can be explained by classical physical processes. Further on we will see that the apparent constancy of the speed of light follows from these two phenomena.

3.1 Contraction

3.1.1 Contraction of Extended Objects

Fields contract at motion. Historically this was found by O. Heaviside in 1888 as a consequence of Maxwell's theory of electromagnetism. We have then to take into account that electric fields determine the size and the shape of macroscopic objects. Those objects are built by atoms and molecules, which are bound to each other by electric multi-pole forces. As fields contract at motion, also all extended objects in our world must contract at motion. So, also the Michelson-Morley apparatus must contract if the laboratory moves in relation to some reference frame.

These conclusions were raised by FitzGerald and Lorentz.

3.1.2 Contraction in General

To restrict field contraction to electric fields is not sufficient to explain special relativity. Also the nucleus of atoms and the behaviour of elementary particles show the phenomena of special relativity. Here other forces are effective, so it is essential that all kinds of fields are subject to this contraction.

The contraction of all different types of fields can conveniently be shown, if we follow the position of modern physics that field forces are realized by the exchange of field-related particles, the so-called 'exchange particles'.

If two particles are in a bind, then this means that they send exchange particles to each other caused by the charges of the binding field. If now both particles get into motion, then there are two effects to the distribution and reception of those particles: On the one hand there is a different travel time. The time for an exchange particle into the direction of motion is extended; the time for the reverse direction is shortened (as seen from the observer at rest). See figure 3.1.

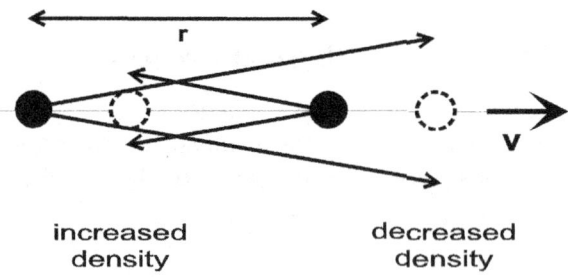

increased   decreased
density   density

Figure 3.1 Path lengths at motion

148

On the other hand there is the deflection of the motion vectors. The beam of exchange particles in the direction of motion is intensified; the beam in the reverse direction is diluted as shown in figure 3.2.

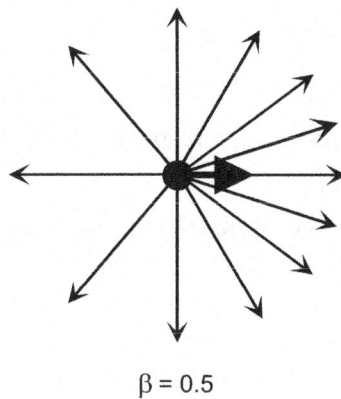

$\beta = 0.5$

Figure 3.2 Vector deflection at motion

We will now determine both effects quantitatively.

At first we calculate the change of path length:

If both charges are moving at a speed v, but the distance of both charges remains r, then the path length to be passed by the exchange particle from the rear particle to the one in front is extended to

$$r' = r \cdot \frac{c}{c - v}. \tag{3.1}$$

Equivalently into the other direction it is reduced to

$$r' = r \cdot \frac{c}{c + v}. \tag{3.2}$$

This means that the density of the exchange particles arriving at the 'destination' charge will be increased or decreased respectively due to the effective change of the path length.

Next we determine the directional deflection:

It is supposed that normally the exchange particles emerging form a charge are equally (i.e. isotropically) distributed into every spatial direction. However, if the emitting

149

charge is moving, there will be an increase of density in the direction of motion and a decrease in the opposite direction.

If an exchange particle is emitted in the direction α in relation to the x-axis, α is described by

$$\tan \alpha = \frac{c_y}{c_x}$$

where c_x and c_y are the x and y component respectively of the exchange particle's speed moving with the speed of light c.

For a charge moving at speed v, the angle α will, due to the geometry of the deflection process, change to α', so that

$$\tan \alpha' = \frac{c'_y}{c'_x} = \frac{c}{\sqrt{c^2 - v^2}} \cdot \frac{c_y}{c_x + v}.$$

From this it follows that:

$$\frac{\tan \alpha'}{\tan \alpha} = \frac{c}{\sqrt{c^2 - v^2}} \cdot \frac{c_x}{c_x + v}.$$

On the other hand:

$$c_x = c \cdot \cos \alpha .$$

This inserted yields:

$$\tan \alpha' = \frac{c}{\sqrt{c^2 - v^2}} \cdot \frac{c \cdot \cos \alpha}{c \cdot \cos \alpha + v} \cdot \tan \alpha .$$

For the case considered here those directions of the exchange particles are essential, that are close either to the direction of motion or to the opposite direction:

1. $\alpha \approx 0$ degrees. Then $c_x \approx c$, $\tan \alpha \approx \alpha$, $\tan \alpha' \approx \alpha'$, $\cos \alpha \approx 1$ and therefore

$$\frac{\tan \alpha'}{\tan \alpha} \approx \frac{\alpha'}{\alpha} \approx \frac{c}{\sqrt{c^2 - v^2}} \cdot \frac{c}{c + v} \tag{3.3}$$

2. $\alpha \approx 180$ degrees, i.e. v in the opposite direction and so with the sign inverted against c_x

$$\frac{\tan \alpha'}{\tan \alpha} \approx \frac{\alpha'}{\alpha} \approx \frac{c}{\sqrt{c^2 - v^2}} \cdot \frac{c}{c - v}. \tag{3.4}$$

In the first case the angle α' will be smaller; this means a higher density of the particle stream, which will be inversely proportional to the square of the angle. In the latter case it means a corresponding dilution.

Now we can get the combined effect of path length and deflection:

The combination of the effects of the change of the path length and of the deflection is the product of both contributions. For the density change G of the exchange particles going from the rear particle to the front particle this means:

$$G = \left(\frac{\alpha}{\alpha'}\right)^2 \cdot \left(\frac{r}{r'}\right)^2 = \left(\frac{c+v}{c}\right)^2 \cdot \frac{c^2}{c^2 - v^2} \cdot \left(\frac{c-v}{c}\right)^2$$

$$G = \frac{(c+v)^2 \cdot (c-v)^2}{c^2 \cdot (c^2 - v^2)} = \frac{(c+v) \cdot (c-v)}{c^2} = \frac{1}{\gamma^2}, \tag{3.5}$$

where

$$\gamma = 1 / \sqrt{1 - \frac{v^2}{c^2}} \ .$$

For the influence of the front particle to the rear particle it means correspondingly:

$$G = \left(\frac{\alpha}{\alpha'}\right)^2 \cdot \left(\frac{r}{r'}\right)^2 = \left(\frac{c-v}{c}\right)^2 \cdot \frac{c^2}{c^2 - v^2} \cdot \left(\frac{c+v}{c}\right)^2$$

with the same result as above.

So the reduction of the field is the same in either direction, in the direction of the general motion and opposite to it. This reduction by γ^2 corresponds to a decrease of the field range by a factor of γ.

This is meant by the statement of the theory of relativity that a field 'contracts' in motion.

With respect to *exchange particles*, the property of them used here is their motion at the speed of light c. This speed refers to a fixed frame of reference; the direction of emission is, however, influenced by the motion of the charge.

3.2 Dilation

The dilation of all time-related processes in physics is the consequence of the internal motion in elementary particles.

3.2.2 The Indications for the Internal Oscillation

When Louis de Broglie detected the wave-particle phenomenon in 1923 [4], it became obvious that there is an oscillation related to every particle.

This detection initiated the quantum mechanical description of a particle by a wave function. Paul Dirac developed this description to the point where he could present a relativistic wave function of the electron in 1928 [5]. When Erwin Schrödinger analyzed this wave function, he concluded in his famous paper of 1930 [6] that there is a permanent motion at the speed of light c in the electron, which he called (in German) 'Zitterbewegung'.

In the scope of the particle model presented here, we assume that not only the electron, but all leptons as well as all quarks have this internal motion at c; this is just a minor extension of our knowledge about the electron.

This assumption not only explains dilation but also some further properties of those particles like the magnetic moment and the spin. Despite of the statements of conventional quantum mechanics these phenomena are explainable in a classical way.

3.2.2 How the Oscillation Works

In anticipation of chapter 4 we assume here that the oscillation has to be a circular motion of two sub-particles. This motion exclusively takes place at the speed of light c in accordance to the particle model.

3.2.2.1 How Dilation Works for a Circular Motion

Every periodic motion, which goes on at the speed of light c, causes necessarily an extension of the period and correspondingly a reduction of the oscillation frequency in the case of a linear motion of the whole configuration. This will be shown here for the simple case of a circular motion.

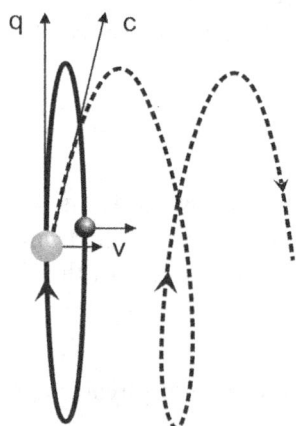

Figure 3.3 Dilation in an elementary particle

Let us begin with the simplest case. The elementary particle is moved into an axial direction at a speed v. This means now that the circular motion changes to a helical motion. The extension of the time period follows from the Pythagorean Theorem:

As the speed on the helix has to be c and the forward motion is v (in relation to an observer at absolute rest), then for the remaining speed on the projected circuit we have

$$q^2 = c^2 - v^2.$$

If the radius of the orbit is R, then the period time T for the configuration at rest is

$$T = 2\pi \cdot R/c. \tag{3.6}$$

the period T' of the projection of the circuit is for the case of motion

$$T' = 2\pi \cdot R/q. \tag{3.7}$$

For the change of the period time we get

$$\frac{T'}{T} = \frac{c}{q} = 1 \bigg/ \sqrt{1 - \frac{v^2}{c^2}} = \gamma$$

which is the known Lorentz factor.

This extension of the period means that a moving clock will indicate a smaller time interval. The indication of the moving clock is conventionally called the "proper time" and denoted by the symbol τ. Using this we get the conventional form of the temporal part of the Lorentz Transformation for an object moving at the speed v.

$$\tau = t \cdot \sqrt{1 - \frac{v^2}{c^2}}. \tag{3.8}$$

3.2.2.2 How Dilation Works in the General Case

There are two generalisations of this process. One more general case is a motion of an elementary particle, which is not into the direction of the axis but in other directions. In this case the result is the same; however the calculation is more complex.

The other generalization is the oscillation of an arbitrary oscillator. In the case of a mechanical oscillator the mass increases at motion as treated in chapter 4. In addition the drive force reduces relativistically during motion. Both influences in combination cause this frequency also to decrease by the Lorentz-Factor.

3.3 Constancy of c

Whereas for Einstein the value of c is a true constant in every inertial system, in the physically based relativity, which we present here, only the *measured* value of c is a constant. This is caused by the contraction of the gauges and by the retardation of clocks. A further consequence of it is the desynchronization of clocks at different positions.

We refer to Figure 3.4. A light pulse is transmitted from a source to a mirror, which is positioned at a distance d, and from there reflected back to the position of the source.

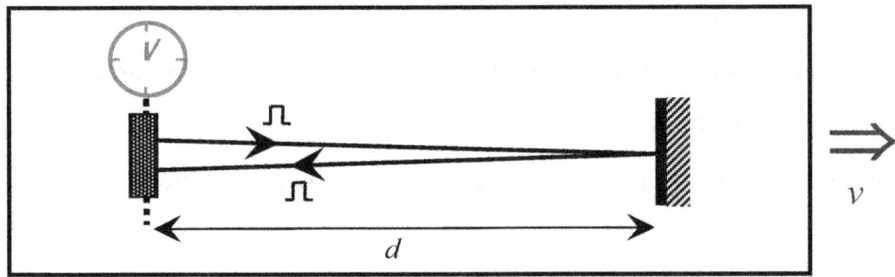

Figure 3.4 Measuring c

For the set up at rest the measured speed of light c is determined from the travel time T of the light pulse and the distance d in the following simple way:

$$c = 2d/T.$$

Or, the other way around, we state that the return time T in this setup at rest is given by

$$T = 2d/c.$$

Now, at motion with speed v, the return time T classically changes to

$$T = d/(c-v) + d/(c+v) = 2cd/(c^2 - v^2) = 2d/c \cdot \gamma^2.$$

But as relativistic effects exist, we have to take into account the length change of the setup and the slow-down of the clock. At first:

$$d \to d' = d/\gamma.$$

From the effect of dilation, there results a different value for the T measured

$$T = 2d'/c \cdot \gamma^2 = 2d/c \cdot \gamma.$$

The slow-down of clock indications is given by:

$$T \to T' = T/\gamma.$$

So, with both effects together, the observer gets the clock indication

$$T = 2d/c \cdot \gamma/\gamma = 2d/c \; ;$$

resolved for c:

$$\boxed{c = 2d/T}$$

which is the *same result as for the situation at rest.*

This fact has misguided Einstein to state that the speed of light is a true constant also in every moving system.

Note: The calculation is done here as a demonstration for one specific direction of motion. It is in fact generalizable for all directions.

4. The Basic Particle Model

The particle model presented in this article is in two ways unavoidable for the understanding of physical phenomena. It is on the one hand the consequence of the facts of special relativity; on the other hand it explains relativity as a whole. Particularly general relativity can easily be understood on the basis of this particle model, whereas this is extremely complex in the way that Einstein has chosen.

4.1 Structure of an Elementary Particle

An elementary particle is composed of two sub-particles, which are called basic particles in the scope of this concept.

The main arguments to understand the structure of elementary particles in this way are the following:

- From the quantum mechanical evaluations of Dirac and Schrödinger it follows that there is in the electron a permanent motion at c. According to de Broglie, the frequency follows from the energy state of the particle as $E = h*v$

- From the dilation there follows that this internal oscillation at c is applicable for all particles, as dilation is a general phenomenon, not restricted to a specific particle type.

We have in addition to take into account that there are further restrictions to the possible structure of an elementary particle:

- An oscillation is only possible, if there are at least two sub-particles in the elementary particle. Otherwise an oscillation would violate the momentum law

- If those sub-particles move permanently at the speed of light c, then they cannot have any mass. It is a general understanding of relativity independent of

155

the specific interpretation of relativity, that c is only possible for a mass-less object.

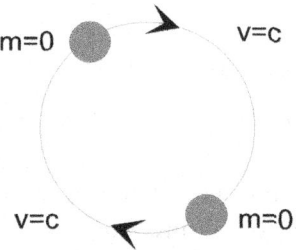

Figure 4.1 Structure of an elementary particle

This particle model is called in the following the '**Basic Particle Model**'.

These assumptions define most of our general particle structure. In addition they raise a new question. If the sub-particles of an elementary particle have no mass, on the other hand the whole particle does have a mass, what is the origin of the particle mass?

4.2 The Mass of an Elementary Particle

The inertia in physics is the direct consequence of the fact that the speed of light c is finite.

As explained in the preceding section, the elementary particle is built by two basic particles. The bind between these basic particles has to be in a way that both particles keep a certain distance. Otherwise the elementary particle could not have any extension. It must have an extension in order to have a spin and a magnetic moment.

4.2.1 Exchange Particles

The model presented here uses the idea, introduced by quantum mechanics, that forces are mediated by specific particles, called exchange particles. These exchange particles are assumed here with the following characteristics:

- An exchange particle moves at the speed of light. This results from the fact that fields propagate at c.

- When an exchange particle meets a charge of the same kind of field, it is absorbed by the charge

- When the receiving charge absorbs the exchange particle, then a fixed momentum is transferred to the particle, which carries the charge. This momentum is directed towards the sending charge in case of an attracting force and off the sending charge in case of a repelling force.

4.2.2 The Bind within an Elementary Particle

The bind between both basic particles can only be a multi-pole bind in the way that the multi-pole field has a potential minimum, which defines the distance of both particles and in this way the extension of the whole particle. A planetary model is not possible as an alternative, as the basic particles do not have a mass.

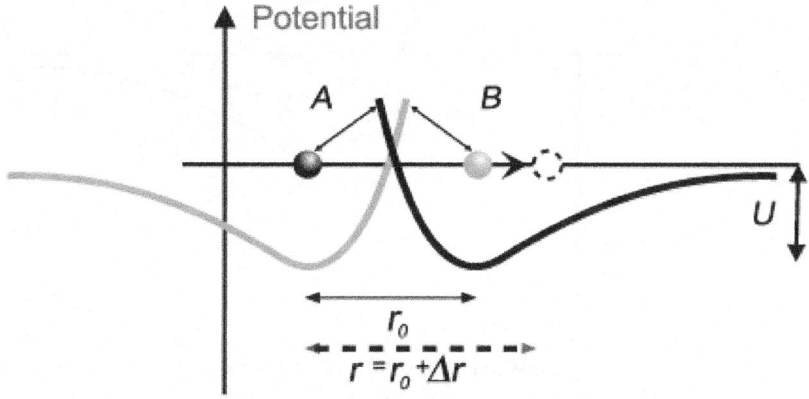

Figure 4.2: The Binding Field

For the multi-pole field a shape of the potential of the field is assumed which causes the following force:

$$F = Kq^2 \cdot \frac{\Delta r}{r^3}. \tag{4.1}$$

This binding force is the strong force by the following consideration.

The multi-pole configuration is achieved by an appropriate arrangement of monopole charges of a different sign. The bind of this arrangement must be strong enough to compensate an additional electric charge in the case of a charged elementary particle (like an electron). This makes it obvious that no other force than the strong force is possible to provide a stable bind.

4.2.3 The Behaviour at Motion

The binding field between the basic particles has this specific shape to cause a bind which keeps a constant distance between both. If a particle is now set to motion, the field follows the changing position with a delay caused by the finiteness of the speed of light. As a consequence the other basic particle stays for a short while at the current position. And, as a further consequence, the field of this other particle will not change at all for a short while, and the displacement of the basic particle under consideration requires a force for a short time.

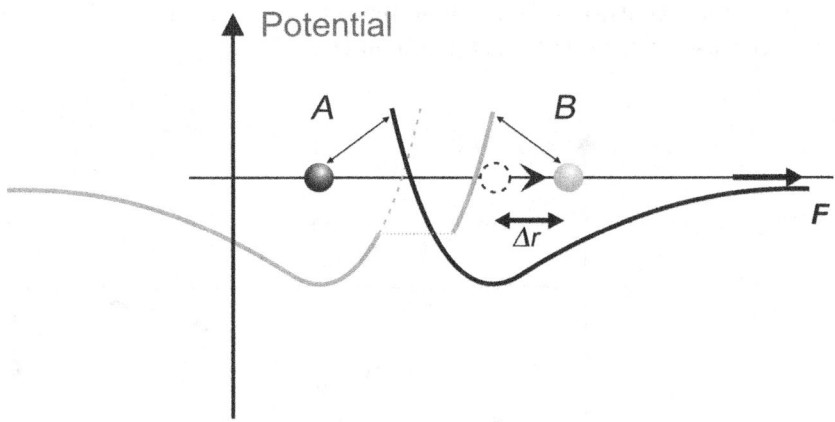

Figure 4.3 Binding field at motion

After a time given by the distance of both basic particles, the field change reaches the other particle, A, which is repositioned; after another time period the field of the repositioned particle A will reach the particle under consideration, B, and no force is necessary any longer.

This fact that a change of the state of motion intermediately needs a force is the physical phenomenon of inertia.

Remark: This displacement of particle B is shown here as a motion *step*. This is done to make the process and the spatial field change better understandable. But in reality it is of course different. The motion of B is a smooth process. During this change the fields change continuously, but with the delay shown here.

4.2.4 The Force in Case of a Constant Acceleration

For the quantitative determination of the force we shall assume that one of the basic particles, B, is accelerated by an external actor. This causes a displacement of the position of this basic particle relative to the other one, A. The displacement results from the time needed for the change of the field of particle B to propagate to particle A:

$$\Delta t_1 = r/c.$$

With a constant acceleration of a, this time means a displacement of:

$$\Delta r = \frac{1}{2} \cdot a \cdot \Delta t_1^2 \qquad (4.2)$$

before particle A can react. From this displacement of Δr there follows a force of

$$F_r = Kq^2 \cdot \frac{1}{r^3} \cdot \Delta r \cdot \qquad (4.3)$$

This means that the force F_r, as a consequence of the first portion of the time delay, Δt_1, has reached the value

$$F_r = Kq^2 \cdot \frac{1}{r^3} \cdot \frac{1}{2} \cdot a \cdot \Delta t_1^2 . \qquad (4.4)$$

After the time Δt_1 particle A will start to move. The change of its field in forward direction will in turn need the time

$$\Delta t_2 = r/c$$

to propagate back to particle B. After this time, the force F_r on the particle B will reach its final value.

So the overall time, until a stationary state is achieved, is

$$\Delta t = \Delta t_1 + \Delta t_2 = 2 \cdot r/c \qquad (4.5)$$

supposed that the actual motion is slow compared to c.

Now Δt (4.5) is inserted to replace Δt_1 in eq. (4.4) which results in

$$F_r = 2 \cdot Kq^2 \cdot \frac{1}{r} \cdot a \cdot \frac{1}{c^2} .$$

Please note for this result that the force F_r is proportional to the acceleration. So, this is already the deduction of Newton's law of motion.

According to the definition of Newton the inertial mass is

$$m_r = \frac{F_r}{a}$$

and therefore:

$$m_r = 2 \cdot Kq^2 \cdot \frac{1}{r} \cdot \frac{1}{c^2} .$$

We now come back to eq. (4.3), and we have to consider that the full force

$$F_r = Kq^2 \cdot \frac{1}{r^3} \cdot \Delta r$$

is only effective if both basic particles are positioned in a line parallel to the direction of the force applied and so to the direction of their motion.

For an arbitrary motion of the elementary particle in the 3-dimensional space, and also caused by the orbital motion inside, the basic particles are positioned to each other at varying angles in relation to the direction of the forced motion, so only a portion of this force is effective.

The magnitude of the portion depends on the 3-dimensional shape of the binding field.

We will at this place not calculate the integral over all directions but use a symbolic factor I as a representation of the integration result.

$$< F > = F_r \cdot I \,.$$

Further down we will present an easy way to determine this factor I.

This integration factor inserted into eq. (4.4) yields now the averaged force $<F>$

$$< F > = I \cdot Kq^2 \cdot \frac{1}{r^3} \cdot \frac{1}{2} \cdot a \cdot \Delta t^2 \,. \tag{4.6}$$

Again (4.5) is inserted to replace Δt. This insertion results in

$$< F > = 2I \cdot a \cdot Kq^2 \cdot \frac{1}{r} \cdot \frac{1}{c^2} \,.$$

And again, we use the definition of Newton for the inertial mass:

$$m = \frac{< F >}{a}$$

and so we get for the effective mass:

$$\boxed{m = 2I \cdot Kq^2 \cdot \frac{1}{r} \cdot \frac{1}{c^2}} \tag{4.7}$$

This now is the inertial mass of an object deduced from the delay, by which field forces between charges are propagated.

Please note that r is the distance between the basic particles in the configuration. So, for an elementary particle built by two constituents, it is the diameter of the particle.

This result has the following remarkable aspects:

1. It yields the fact that the quotient of force and acceleration is constant at non relativistic velocities. Therefore this is a deduction of Newton's law of motion. For Newton, this law had the property of an axiom (or a principle of nature).

2. The result shows, that the mass is inversely proportional to the size of an elementary particle r.

The value of the constant I will be determined next.

4.2.5 Determination of the Constants

We have derived the dependency of the accelerating force from the particle parameters and have shown in this way, why Newton's law of motion is valid. Now we have to determine the internal parameters, i.e. the field parameters of the binding field within the elementary particle. So we will determine the constant I introduced above.

In the Basic Particle Model it is assumed, that the basic particles orbit each other at the orbital speed c and at a certain orbital frequency, which depends on the radius. The field which binds the basic particles to each other propagates into all directions. So outside of this orbit an alternating field exists, the frequency of which is identical to the orbital frequency.

It is now possible to determine this frequency v from the known parameters of the configuration, i.e. the elementary particle.

We use the simple geometric relation for an orbital motion

$$v = \frac{c}{\pi \cdot r} \quad \text{or} \quad r = \frac{c}{\pi \cdot v} \tag{4.8}$$

where r is the distance of the basic particles, i.e. twice the radius of the orbit.

The frequency v is clearly the de Broglie frequency, because it is the frequency of the alternating field.

(Historical remark: Louis de Broglie predicted interference behaviour of all particles at scattering. He assumed a wave surrounding each elementary particle. The existence of this wave follows now as a consequence of the Basic Particle Model.)

In eq. (4.7)

$$m = 2I \cdot Kq^2 \cdot \frac{1}{r} \cdot \frac{1}{c^2} \tag{4.9}$$

we replace r by use of eq. (4.8) and reorder the result; this yields

$$m \cdot c^2 = 2I \cdot \frac{1}{c} \cdot Kq^2 \cdot \pi \cdot v . \tag{4.10}$$

161

In section 4.3.2 we will show that

$$E = mc^2 .$$

We can further use the known equation

$$E = h \cdot v$$

which relates the energy to the frequency of a particle. Both equations combined yield

$$m \cdot c^2 = h \cdot v . \tag{4.11}$$

If this is used to replace the left side of (4.10) we get

$$h = 2I \cdot \pi \cdot \frac{1}{c} \cdot Kq^2 . \tag{4.12}$$

We will in the following use the reduced Planck constant:

$$\hbar = h/2\pi .$$

Eq (4.12) then changes to

$$\hbar = I \cdot \frac{1}{c} \cdot Kq^2 . \tag{4.13}$$

Using now eq. (4.11) through (4.13) and replacing the distance of the basic particles r by the radius of the orbit $R = r/2$, we end up with the formula

$$\boxed{m = \frac{\hbar}{R \cdot c}} \tag{4.14}$$

for the mass of an elementary particle constituted by two basic particles.

This is now a universal relation for the mass of an elementary particle. It is assumed here to be valid for all leptons and all quarks. – Please note that it does not have any free or unknown parameters.

4.2.6 The Relation Mass to Magnetic Moment

We can now use the magnetic moment of electrically charged particles to check the usability of the mass formula deduced above.

First we will recall the classical relation for the magnetic moment of a particle. The magnetic moment μ of a loop current is classically:

$$\mu = i \cdot \pi \cdot R^2 . \tag{4.16}$$

The loop current i within a particle of one elementary charge e_0 at frequency v is simply:

$$i = v \cdot e_0 .\tag{4.17}$$

When using now eq. (4.8) for v as

$$v = \frac{c}{2\pi \cdot R}$$

there follows:

$$\mu = \frac{c \cdot e_0 \cdot R}{2} .\tag{4.18}$$

If now R is inserted from eq.(4.14), the magnetic moment turns out to be

$$\mu = \frac{\hbar \cdot e_0}{2 \cdot m} .\tag{4.19}$$

For the electron this is the 'Bohr Magneton'.

Please note: In textbooks of physics it is stated that the Bohr Magneton can only be derived by using quantum mechanics. The preceding, however, shows that this relation can be derived classically using the Basic Particle Model.

4.3 The Relativistic Mass

4.3.1 The Increase of the Mass at Motion

According eq.(4.14) the mass of a particle is given as

$$m = \frac{\hbar}{R \cdot c}$$

At motion the radius R shrinks by the Lorentz factor

$$\gamma = 1/\sqrt{1 - v^2/c^2} .$$

So, at motion the mass changes in the following way

$$m \rightarrow m' = m \cdot \gamma .\tag{4.21}$$

This is, as we have to admit, a simplified deduction. In the case of motion not only the radius shrinks but the binding field changes, i.e. is reduced. In addition the delay time of the exchange particles on their way from one basic particle to the other one increases, which causes the inertial force to increase. But if we do this calculation in detail, then both latter effects cancel out each other. So the above calculation, which only refers to the contraction effect, describes the process correctly.

4.3.2 The Mass to Energy Relation

From the preceding section, eq. (4.21):

$$m = m_0 / \sqrt{1 - v^2/c^2} \quad \text{or equivalently}$$

$$m = m_0 \sqrt{c^2 / (c^2 - v^2)} \tag{4.23}$$

where m_0 is the rest mass of the particle, it follows that an increase of the velocity of an object will increase its mass. On the other hand an increase of velocity means an increase of its energy. The relation between mass and energy, which is the most famous relation given by Einstein, will now be quantitatively deduced.

Eq. (4.23) is squared and reordered to:

$$m^2 v^2 = m^2 c^2 - m_0^2 c^2 .$$

When using the definition of momentum

$$p = mv$$

there is

$$p^2 = m^2 c^2 - m_0^2 c^2 .$$

Now the change of the momentum p resulting from a change of mass at motion is found by differentiation:

$$2p\,dp = 2mc^2 dm ,$$

which, using again $p = mv$, yields

$$v\,dp = c^2 dm . \tag{4.24}$$

Energy as defined by Newton is:

$$dE = F dx = \frac{dp}{dt} dx = \frac{dx}{dt} dp = v\,dp . \tag{4.25}$$

If this definition of dE is inserted into (4.24), there directly follows:

$$dE = c^2 dm . \tag{4.26}$$

If this is integrated now starting with $E=0$ at $m=0$, we end up with the well known result:

164

$$\boxed{E = mc^2}.\qquad\qquad\qquad (4.27)$$

So, also this famous and important formula is derived from elementary considerations, whereas Einstein has referred to the theory of Maxwell and performed a *gedanken* experiment using the momentum of a reflected light pulse to deduce this formula, originally restricted to light.

Note:

If you follow the idea, that the mass energy equivalence is just a consequence of the structure of an elementary particle, then this has a remarkable further consequence: As a reverse, the mass energy equivalence may not be valid below the level of an elementary particle. That means that for the constituents of an elementary particle and their interactions, energy mass equivalence as well as energy conservation does in general not work!

4.3.3 The Experimental Situation of the Electron

There is an apparent conflict between the model presented here and the experiment.

Present day physics understand the electron as a particle, which is point-like and has no internal structure. This is concluded from scattering experiments. The conclusion is based on an assumption, which has never been questioned up to now. This assumption is that, if an electron had sub-particles, these sub-particles would have a mass. Such a conception of the electron would in fact contradict the measurements.

The Basic Particle Model, however, assumes that the basic particles, which are the sub-particles in our case, do not have a mass. So, no conflict to the experiment exists.

5. General Relativity

General relativity is Einstein's theory to explain gravitation. Einstein explains gravitation on the basis of his geometrical model of space-time.

Our model explains gravitation on the basis of physical processes in contrast to Einstein.

5.1 Gravitation with Einstein

According to Einstein the concept of space-time is also the explanation for gravity. Objects move along geodesics in the four-dimensional space-time. In the vicinity of mass, or according to Einstein equivalently in the vicinity of occurrences of energy, the space-time is curved, and so is also a geodesic.

In order to determine the motion of objects in a gravitational field, it is necessary to determine the shape of the geodesics under question. This requires the use of the multi-dimensional Riemannian geometry, which is a very challenging task.

The calculations according to Einstein are so complex that in the usual cases the more specialized calculation of Schwarzschild, the so-called Schwarzschild Solution, is used.

5.2 Gravitation as a Physical Process

Gravitation based on physics rather than geometry uses the fact, that in a physical understanding, which refers directly to the measurement, the speed of light c is not always constant but varies in the vicinity of matter. This variation of c causes refraction of light-like particles in general. This reflection influences the motion processes within an elementary particle, and causes the particle to accelerate. This process explains gravitation with quantitatively correct results for all phenomena which are treated by general relativity.

Below we will show that general relativity based on the Basic Particle Model is equivalent to the one of Einstein up to the Schwarzschild Solution.

5.2.1 Speed of Light under Gravitation

The speed of light varies in a gravitational field. As a consequence, photons and light-like particles are refracted in this field.

The dependency of c from the position as determined by experiments is:

$$c(r) = c_0 \cdot (1 - 2 \cdot \frac{G \cdot M}{r \cdot c_0^2})^p \qquad (5.1)$$

where c_0 is the speed of light in the gravitation-free space, G is the gravitational constant and M is the mass of the object, which is traditionally said to cause the gravitational potential; r is the distance of the position under investigation to the centre of gravity. The power number p is ½ or 1 depending on the direction of motion with respect to the centre of gravity.

The equation (5.1) is here initially used as an experimental result. Although this dependency is also provided by the theory of Einstein, we do not refer to Einstein in this place. We will explain later how this dependency follows from our model.

This dependency was the first time measured by I. I. Shapiro around the year 1970. The measurement was later repeated by others with increased accuracy.

5.2.2 Gravitational Lensing

Here now follows a collection of the most important equations, which describe the path of a photon in a gravitational field (figure 5.1).

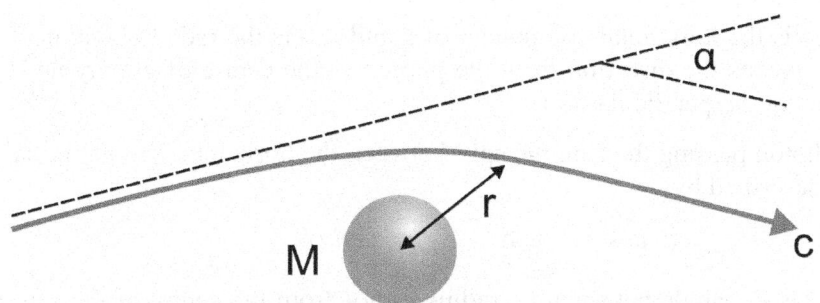

Figure 5.1: Deflection at the sun

At first we have to split the speed of light into a horizontal component and a vertical component as defined above, because the speed reduction depends on the direction of the motion, see eq. (5.1).

The effective speed of light for an arbitrary direction is the vector sum of the components

$$c = \sqrt{c_{hor}^2 + c_{vert}^2} \, .$$

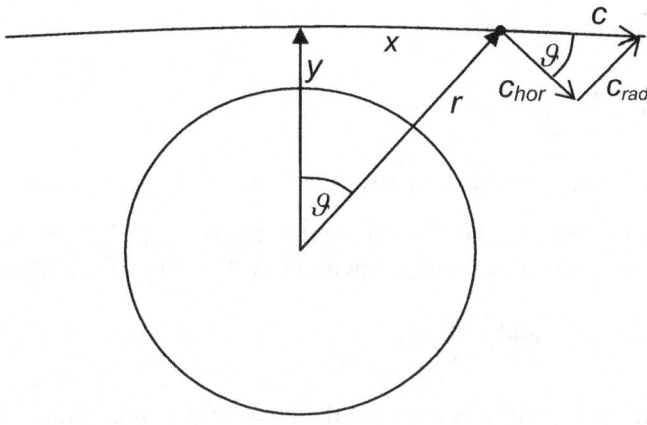

Figure 5.2: Components of c at gravitational lensing

167

Here c_{hor} is the horizontal component of c and c_{rad} is the radial component of c; where "radial" means the direction from the photon to the centre of gravity and "horizontal" the direction perpendicular to it.

For a photon passing the sun, the split between the horizontal and the radial component can be described by

$$c_{hor} = c \cdot cos\,\vartheta \quad \text{and} \quad c_{rad} = c \cdot sin\,\vartheta \tag{5.2}$$

where ϑ is the angle between the radius vector from the centre of the sun to the actual position of the photon on the one hand and the radius vector from the centre of the sun of its closest position (vertex) on the other hand (see fig. 5.2).

To abbreviate the equations, we use the common definition for the so-called Schwarzschild radius r_s

$$r_s = 2 \cdot \frac{G \cdot M}{c_0^{\,2}} \cdot \tag{5.3}$$

(Remark: We describe the gravitational field by $G \cdot M$ following the current conventions. Later we will show that gravity does not depend on M.)

So we can write the components according to (5.1)

$$c_{hor} = c_0 \cdot (1 - \frac{r_s}{r})^{1/2} \cdot cos\,\vartheta \quad \text{and} \quad c_{rad} = c_0 \cdot (1 - \frac{r_s}{r})^1 \cdot sin\,\vartheta. \tag{5.4}$$

Combining (5.2) thru (5.4) we get

$$c = c_0 \cdot \left[1 - \frac{r_s}{2r}(1 + sin^2\,\vartheta) \right] \tag{5.5}$$

taking into account that $r_s \ll r$ is normally fulfilled (i.e. not considering a Black Hole).

We will denote the distance of the photon from the vertex of the light path as x (see fig. 5.2). Further we define the coordinate perpendicular to x as y. Then there is

$$\frac{x}{r} = sin\,\vartheta \quad \text{and} \quad \frac{x}{y} = tan\,\vartheta \quad . \tag{5.6}$$

At the vertex of the light path, y is then the distance of the light path from the centre of the gravitational source, in this case of the sun. Then we can describe r by these two parameters.

$$r^2 = x^2 + y^2. \tag{5.7}$$

Since the speed of light c depends on the location (i.e. the distance to the sun and the angle between the light path and the sun), we get a classical refraction.

Eq. (5.5) can then be presented as

$$c = c_0 \cdot \left[1 - \frac{r_s}{2\sqrt{x^2 + y^2}} \left(1 + \frac{x^2}{x^2 + y^2} \right) \right].$$ (5.8)

Now, as y is perpendicular to x and does so not depend on x, we can perform a straight differentiation with respect to y:

$$\frac{dc}{dy} = c_0 \cdot \left[\frac{r_s}{4} \left(x^2 + y^2 \right)^{-\frac{3}{2}} \cdot 2y + \frac{3}{2} \cdot \frac{1}{2} x^2 r_s \left(x^2 + y^2 \right)^{-\frac{5}{2}} \cdot 2y \right]$$

$$\frac{dc}{dy} = \frac{c_0 r_s y}{2} \cdot \left(\frac{1}{r^3} + \frac{3x^2}{r^5} \right).$$ (5.9)

Next we will determine the deflection angle α. At first we determine the differential deflection angle $d\alpha$. The dispersion dc/dy multiplied by a time differential dt yields the path differential dx according to the refraction. From this we get

$$d\alpha = \frac{dc}{dy} dt \quad \text{or} \quad d\alpha = \frac{1}{c_0} \frac{dc}{dy} dx$$

and then, using eq. (5.9):

$$\frac{d\alpha}{dx} = \frac{r_s y}{2} \left(\frac{1}{r^3} + \frac{3x^2}{r^5} \right).$$ (5.10)

Now setting this equation into relation to the angle ϑ (5.6) we get

$$\frac{d\alpha}{d\vartheta} = \frac{r_s}{2y} \left(\cos \vartheta + \sin^2 \vartheta \cdot \cos \vartheta \right).$$ (5.11)

This equation integrated over $d\vartheta$ from $\vartheta = -\pi/2$ to $\vartheta = +\pi/2$ and using (5.2) yields

$$\alpha = \frac{r_s}{2y} \cdot 4 = 4 \cdot \frac{GM}{c^2 y}.$$ (5.12)

After inserting now the values applicable for the sun

-   $G = 6.674 * 10^{-11} \ m^3 \ kg^{-1} \ s^{-2}$
-   $M = 1.989 * 10^{30} \ kg$ (the mass of the sun)
-   $c = 2.998 * 10^8 \ m \ s^{-1}$
-   $y = 6.95 * 10^8 \ m$ (distance of the light path from the centre of the sun)

we get, after converting to angular units, the correct result of

1.75 arc-sec.

This number corresponds to twice the normal gravitational acceleration and conforms to the observation. This numerical result as well as the analytical result (5.12) conform also to the prediction of General Relativity – however without any use of General Relativity.

Next we will determine the gravitational acceleration for an object at rest.

Geometrically, the acceleration a of a photon - or of every light-like object – across the direction of motion is defined by the change of the cross speed dv:

$$dv = c \cdot d\alpha \tag{5.13}$$

So, for the acceleration a:

$$a = \frac{dv}{dt} = \frac{c \cdot d\alpha}{dt} = c \cdot \frac{d\alpha}{dx} \cdot \frac{dx}{dt} = c^2 \cdot \frac{d\alpha}{dx}. \tag{5.14}$$

Inserting (5.10):

$$\frac{d\alpha}{dx} = \frac{r_s y}{2} \left(\frac{1}{r^3} + \frac{3x^2}{r^5} \right)$$

and using also (5.6) we get

$$a = \frac{c_0^2 r_s}{2r^2} \cos \vartheta (3 \sin^2 \vartheta + 1) = \frac{GM}{r^2} \cos \vartheta (3 \sin^2 \vartheta + 1). \tag{5.15}$$

Inserted now for the vertex, i.e. $\vartheta = 0$ we get:

$$a_{vertx} = \frac{GM}{r^2}, \tag{5.16}$$

i.e. the Newtonian acceleration. – Please note that this value only applies for the vertex. Outside the vertex the acceleration is greater than in the Newtonian case, which is the cause for the known fact that the entire deflection at the sun is twice the Newtonian value.

5.2.3 Gravitational Acceleration for a Particle at Rest

If an elementary particle is placed in a gravitational field, its basic particles are subject to refraction as explained in section 5.2.2. This refraction causes the basic particles to deviate from their circular path. This in turn will cause a movement of the entire elementary particle.

If we take the case, that the elementary particle is oriented such that its orbital axis points towards the source of gravity, then the refraction causes the basic particles to spiral towards the source of gravity. So the entire elementary particle will move into the direction of the source. Figure 5.3 shows the accelerated motion downwards. Due to the refraction, the pitch angle of the basic particles, α in this figure 5.3, will steadily increase. This causes the elementary particle to perform an accelerated motion towards the gravitational source.

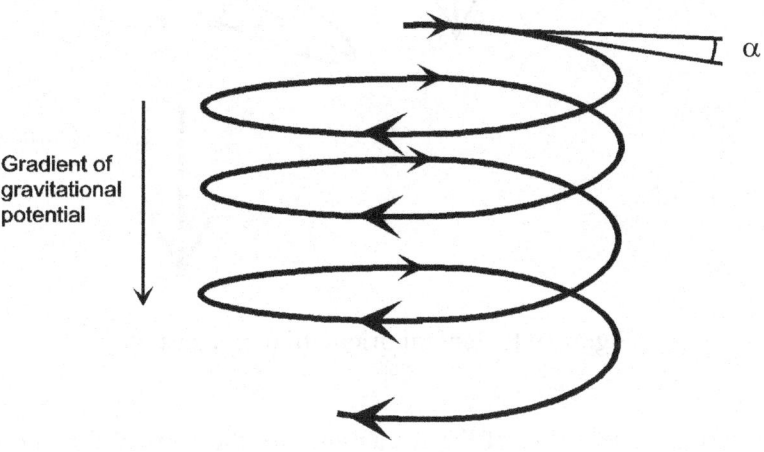

Figure 5.3: Progressive spiralling downwards

Please note that in figure 5.3 only the path of one of the two basic particles is shown to keep the drawing simple.

In this case the acceleration of the (composed) elementary particle is similar to the acceleration given in equation (5.16):

$$a = \frac{G \cdot M}{r^2}$$

which is the Newtonian acceleration.

5.2.4 Gravitational Acceleration at Arbitrary Orientations

In the general case if an elementary particle is placed in a gravitational field the orientation of the axis is at an arbitrary angle θ with respect to the vertical direction (see figure 5.4).

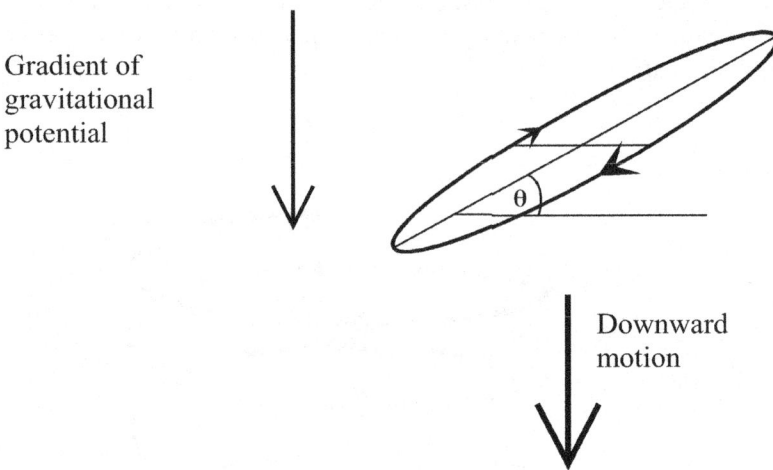

Gradient of
gravitational
potential

Downward
motion

θ

Figure 5.4: General orientation of a particle

In this case only the projection of the refraction into the vertical direction is effective for gravity. This means on the one hand, that the gravitational acceleration is reduced compared to the case above. But on the other hand, the effect of reduction is compensated by the increase of refraction for the vertical component as it is visible in eq (5.15), i.e. the term with a factor of 3.

So, also in the case of an arbitrary orientation of the elementary particle we can get the result eq (5.16)

$$a = \frac{G \cdot M}{r^2}$$

which is the well-known result for the classical case (Newton).

5.2.5 The Equivalence Principle

5.2.5.1 Equivalence Classically

The (weak) equivalence principle treats the fact that every object undergoes the same gravitational acceleration independent of its mass.

In order to explain this fact, Newton has introduced the equivalence principle, saying that there exists an inertial mass and a gravitational mass. Both types of mass are said to be equivalent to a high degree, and as a consequence, the gravitational force onto an object is strictly proportional to the inertial force, and as a further consequence, they undergo strictly the same acceleration in a certain gravitational field.

Einstein has, when he developed general relativity, adopted this principle and has made it to one of the pillars of his theory.

Neither Newton nor Einstein ever made the attempt to explain this phenomenon on a physical basis.

5.2.5.2 Equivalence Based on the Particle Model

If we look to the figure 5.3, the deflection of the path of the basic particles is independent of the radius of the particle and, because of eq. (4.14) independent of the mass of the particle. So, it has a very natural cause that the gravitational acceleration is independent of the mass. No assumptions about any equivalence are needed.

Figure 5.3 shows, why an elementary particle at rest is subject to a gravitational acceleration. It is in fact gravitational lensing on a micro-scale.

This is the physical cause that the gravitational acceleration is independent of the mass of an object.

5.2.6 The Schwarzschild Solution

To work with Einstein's field equations is an extremely challenging task. A short time, after Einstein published general relativity, Karl Schwarzschild presented a solution for the simplified, less general solution of a spherically symmetric field, like the one of the sun, which is a very frequent situation in astronomy. The experiments and observations cited in the literature as proofs for Einstein's general relativity refer usually to the results of the Schwarzschild solution.

The Schwarzschild solution is normally deduced by starting with Einstein's field equations and the use of the Riemannian geometry and then deducing the special solution. We will present here a different deduction. We will start with the physical version of relativity and the Basic Particle Model and demonstrate how easily this solution can be deduced from these physical fundaments.

An elementary particle is, according to the Basic Particle Model, built by two sub-particles orbiting each other. Their temporal behaviour is described by eq. (3.8), from which the following equation results for the proper time of an object in motion:

$$\tau = t \cdot \left(1 - \frac{v^2}{c^2}\right)^{1/2}.$$

This equation is now derived by dt and squared and reordered:

$$c^2 \left(\frac{d\tau}{dt}\right)^2 = c^2 - v^2. \tag{5.18}$$

173

In a gravitational field this time behaviour changes. The understanding of this change directly guides us to the Schwarzschild Solution.

We first split the speed into a radial and a tangential component as the Schwarzschild Solution is normally given with polar coordinates:

$$c^2 \frac{d\tau}{dt} = c^2 - v_r^{\,2} - v_t^{\,2} \qquad (5.19)$$

where v_r and v_t denote the radial und the tangential component of the speed respectively.

Now we have to take into account that c changes in a gravitational field in the following way according to (5.1):

$$c_{rad} \rightarrow c'_{rad} = c_{rad} \cdot \left(1 - \frac{r_s}{r}\right)$$

$$c_{tan} \rightarrow c'_{tan} = c_{tan} \cdot \left(1 - \frac{r_s}{r}\right)^{1/2}.$$

Here again, to abbreviate the equations, we have used the common definition for the so-called Schwarzschild radius r_s

$$r_s = 2 \cdot \frac{G \cdot M}{c_0^{\,2}}. \qquad (5.20)$$

As a consequence of the change of c, also fields contract and so the size of particles in radial direction in relation to the centre of gravity:

$$r \rightarrow r' = r \cdot \left(1 - \frac{r_s}{r}\right)^{1/2}. \qquad (5.21)$$

Now, inside the gravitational field, the following occurs according to (5.1) and (5.21):

1. The circular motion within the elementary particle changes to an ellipsoidal shape. That means, it is compressed in the radial direction of the gravitational field

2. The speed of the basic particles in the orbit changes from c to a value between c_{rad} and c_{tan} depending on the actual direction of motion.

This unfortunately complicates the calculation of the temporal development.

To solve this, we use a trick here in the way that we change to a modified coordinate system. We change

$$y \rightarrow \hat{y} = y \cdot \left(1 - \frac{r_s}{r}\right)^{-1/2}. \tag{5.22}$$

That means for the derivative to the time

$$y' = v_r \rightarrow \hat{y}' = y'\left(1 - \frac{r_s}{r}\right)^{-1/2} = \hat{v}_r = v_r\left(1 - \frac{r_s}{r}\right)^{-1/2}. \tag{5.23}$$

The x-coordinate remains unchanged with

$$x' = v.$$

In the system of $(x, \hat{y})$ we now have the following situation

1. The orbit within the elementary particle is circular again

2. The speed of light c_{gr} is still reduced but now independent of the direction of motion and is

$$c_{gr} = c \cdot \sqrt{1 - \frac{r_s}{r}}$$

where c is here the speed of light outside a gravitational field. c_{gr} now replaces c_{rad} and c_{tan} - The duration of the orbital period of the elementary particle is not changed by this coordinate transformation, so we can use it for our calculation.

Now, with the alternate coordinate system, we can write the Lorentz equation as

$$c^2 \frac{d\tau}{dt} = c_{gr}^2 - \hat{v}_r^2 - v_t^2. \tag{5.24}$$

Please be aware that this equation physically describes the extension of the orbital period of the elementary particle within a gravitational field.

Now inserting

$$c_{gr} = \sqrt{1 - \frac{r_s}{r}} \cdot c,$$

$$\hat{v}_r = \frac{dr}{dt} \cdot \left(1 - \frac{r_s}{r}\right)^{-1/2} = r'\left(1 - \frac{r_s}{r}\right)^{-1/2},$$

$$v_t = \frac{d\varphi}{dt} \cdot r = \varphi' \cdot r$$

into (5.24) we get

$$c^2 \frac{d\tau}{dt} = \left(1 - \frac{r_s}{r}\right) \cdot c^2 - \left(1 - \frac{r_s}{r}\right)^{-1} r'^2 - \varphi'^2 \cdot r^2 \tag{5.25}$$

which is a common form of the Schwarzschild solution.

Note (1):
The reduction of c in the gravitational field is used here as a fact; it will be explained further down.

Note (2):
This deduction of the Schwarzschild solution only uses basic mathematics. Neither the Riemannian geometry nor Einstein's field equations are needed.

5.2.7 The Cause of Gravitation

We have seen that gravity is in fact *not a force but a refraction process*. And the cause of the refraction is the varying speed of light c in the vicinity of matter.

5.2.7.1 Varying Speed of Light

Equation (5.1) is the basis to explain all phenomena of gravitation. Next the question has to be answered, why c is reduced in the vicinity of matter. The answer is that the reduction of c is caused by the effect of the exchange particles, which build the binding field of the basic particles.

According to the Basic Particle Model, the binding field is the field of the strong interaction, which is – also according to the model – the universal force in our world affecting all existing particles.

These exchange particles, emitted by a multi-pole compound and causing attraction or repulsion in a random way, interact as well with every light-like particle. They cause such a particle to be deflected towards the origin of the exchange particle (i.e. the basic particle) or away from it. So the light-like particle performs a random walk as depicted in figure 5.5. As a result, the average speed of the light-like particle is reduced, even though the microscopic speed is still the speed of light c.

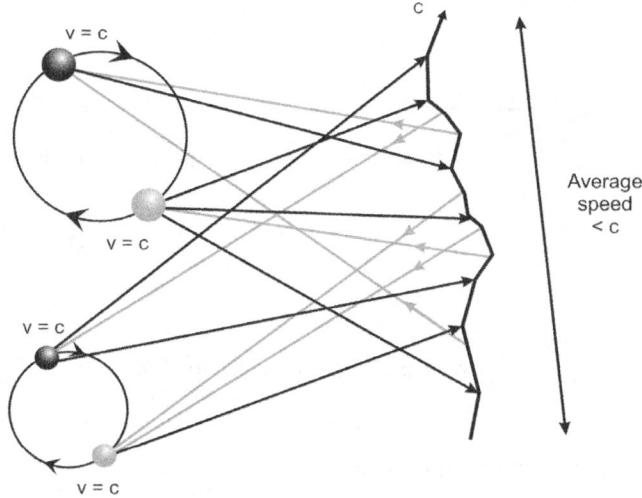

Figure 5.5: Disturbed way of a light-like particle

According to the Basic Particle Model, the sum of charges in every basic particle is the same, irrespective of the elementary particle to which the basic particle belongs, and so is the flow of exchange particles. Consequently the reduction of c and so the gravitational effect is independent of the elementary particle, which means that it is independent of the size and consequently independent of the mass of the elementary particle. Every elementary particle provides the same contribution to the gravitational field.

This fact is in contrast to the conventional physics, but it helps to overcome open problems of present gravitational physics.

5.2.7.2 Speed Reduction in Detail

The speed reduction caused by the permanent deflection of a light-like particle will now be determined in detail. We will treat here the two orthogonal cases in such a way, that every arbitrary motion of such a particle is a vector combination of these two cases.

In case of the tangential motion of a light-like particle at speed c, every interaction with an exchange particle coming from the centre of the gravitational field causes a cross-speed v, which can be into either side depending on the sign of the exchange particle. This reduces the effective speed of the original motion to a projected value c_{eff}. See figure 5.6.

177

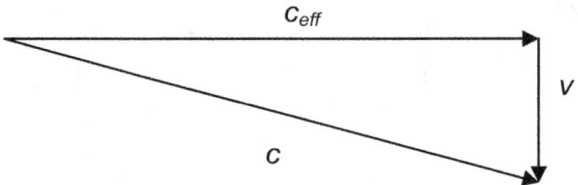

Figure 5.6: Transverse deflection

The projected effective speed is in this case

$$c_{eff,tan} = \sqrt{c^2 - v^2} \quad \text{or}$$

$$c_{eff,tan} = c\sqrt{1 - \frac{v^2}{c^2}} \, . \tag{5.26}$$

The other case is a light-like particle moving into a radial direction in relation to the centre of gravity. The effective speed is in this case given as follows:

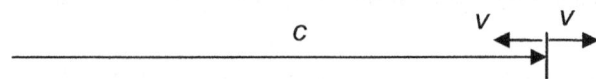

Figure 5.7: Longitudinal deflection

With an attracting impact of an exchange particle we get for the effective speed over an assumed distance s

$$c_{rad} = c - v$$

and for the travel time over this distance s

$$t_1 = s/(c - v);$$

and correspondingly with a repelling impact

$$t_2 = s/(c+v).$$

The time of both cases averaged:

$$t_{avg} = (t_2 + t_2)/2 = s \cdot \frac{c}{c^2 - v^2} \cdot \qquad (5.27)$$

For the effective speed:

$$c_{eff} = \frac{s}{t_{avg}} = \frac{c^2 - v^2}{c} = c \cdot \left(1 - \frac{v^2}{c^2}\right). \qquad (5.28)$$

Now for the quantity of the deflection speed v we calculate the following:

The deflection speed is the sum of all single deflection steps which are caused by the exchange particles. The rate of these steps is proportional to the number of particles N in the configuration which builds the gravitational source. As these events consisting of attracting and repelling pulses are adding up randomly, the resulting deflection is - by the rules of random statistics - the square root of the sum of impacts. So we have

$$v \propto \sqrt{\frac{N}{f(r)}}. \qquad (5.29)$$

Inserted into the eqs. (5.26) and (5.28) we get:

$$c_{eff} = c \cdot \left(1 - g\frac{N}{c^2 \cdot f(r)}\right)^p \qquad (5.30)$$

where $p=1$ for radial motion and $p=1/2$ for tangential motion.

The function $f(r)$ comprises several aspects. A basic particle, which is deflected from its original path by the random process described above, is in the longer term guided back to its original path to keep the particle in the average on its path. Further on it is known in the case of photons that particles, which are originally not correlated to each other, get correlated if they move for some time side by side to each other. This behaviour could also be assumed for the exchange particles. This fact will influence the range dependency of the deflection process. The collection of these influences shall be covered by the function $f(r)$.

As we do not have further information or an actual model about these correlation aspects, we go here the way that we adapt the function to the experimentally known result. That means to assume

$$f(r) = r ; \qquad \text{and so}$$

$$c_{eff} = c \cdot \left(1 - \frac{g \cdot N}{c^2 \cdot r}\right)^p . \qquad (5.31)$$

The parameter g is the proportionality factor for the influence of the flow related to N particles.

The dependency of the extension of multi-pole fields in a gravitational field works in an analogue way to the contraction of fields at motion and is not derived here. The result for the reduced distance is as referred to in section 5.2.6

$$r_{red} = r \cdot \left(1 - \frac{g \cdot N}{c^2 \cdot r}\right)^{(p-1/2)}$$

where again $p=1$ for radial motion and $p=1/2$ for tangential motion.

6. Cosmology

This chapter deals with open problems in astronomy and cosmology, with Dark Matter, Inflation, and Dark Energy.

6.1 Dark Matter

Some decades ago it was detected, that the rotational speed within and around big galaxies is in conflict with the equilibrium speed determined on the basis of standard gravitation. Figure 6.1 shows the discrepancy. As a solution present physics assume a specific type of matter, which is invisible and has almost no interaction with the known matter, but must have a high mass to constitute the missing mass of the calculations. This missing mass was given the name "Dark Matter".

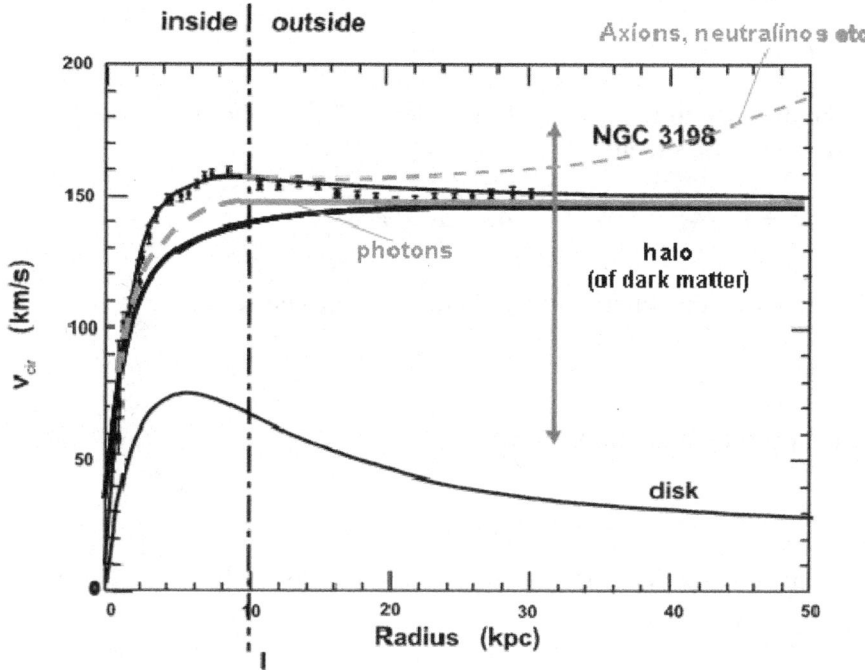

Figure 6.1: Equilibrium conflict at the galaxy NGC 3198
(The radius of the galaxy is 10 kpc)

In figure 6.1 the solid curve labelled "disk" is the rotational speed dependent on the radius as a result of a normal gravitational calculation. The uppermost single values are measurements of the real speed; a curve (also solid) is fitted through these measurements. The dark solid line labelled "halo" describes the required distribution of the assumed "Dark Matter" in order to explain the measured values.

The horizontal grey line, which is very close to the "halo" curve, follows from the assumption described above, that every elementary particle contributes equally to the gravitational field. It is the contribution of light particles, i.e. neutrinos and photons. In the drawing the height of this line was adjusted to fit into this diagram, but it fits within a tolerance of a factor 2-3 to the known data. Its curvature, however, is given by the natural distribution of the light particles and is not parameterised.

Of the light particles mentioned, the photons are mainly generated by the hot, shining stars in the centre of the galaxy. The neutrinos are similarly generated by the nuclear processes within the stars, the sources of which are also mostly in or close to the centre of the galaxy. (However, the flow of neutrinos is too small to contribute considerably to the observed values.) These particles build a continuous flow off the centre with the speed of light c (or almost this speed). This flow causes their spatial density – outside of the core of the galaxy - to be

$$\rho \propto \frac{1}{r^2}$$

where r is the distance to the centre of the galaxy. The number of particles N within a sphere up to a radius r_0 is then

$$N = \int_0^{r_0} \rho \cdot 4\pi r^2 dr \propto \int_0^{r_0} \frac{1}{r^2} \cdot 4\pi r^2 dr \propto r_0 .$$

The acceleration a in the gravitational field towards the centre is for $r = r_0$

$$a \propto \frac{1}{r^2} \cdot N \propto \frac{1}{r^2} \cdot r = \frac{1}{r} .$$

The centrifugal acceleration on the other hand is

$$a = \frac{v^2}{r} .$$

In order to keep both accelerations in a balance, it follows for the orbital speed v that

$$\frac{v^2}{r} \propto \frac{1}{r} \Rightarrow v = cons \tan t .$$

This is the reason for the curvature of the grey line (of photons) in figure 6.1, and so it provides the contribution to the gravitational field, which is normally assigned to the "Dark Matter".

The highest, curved dashed grey line describes the spatial distribution of supersymmetric particles like axions and neutralinos, which are in present main stream physics the candidates for Dark Matter. There is no mechanism known which makes them running in a horizontal line, every plausible distribution causes them to increase at distance from the galaxy.

Another aspect of spatial distribution, which is not shown in the graph, is the three-dimensional spatial distribution of the Dark Matter. According to the current physical theories and according to the particle generation mechanisms, they should follow the flat shape of a galaxy. From the observed effects of Dark Matter, however, one has to conclude that there is an almost spherical distribution around the centre of the galaxy. Also this conforms to the assumption that Dark Matter is constituted of photons. The radiated photons build outside the core of the galaxy an almost isotropic flux in space.

6.2 The Horizon Problem

From the temperature distribution of the Cold Microwave Background (CMB) it is concluded that there must have been a correlation between separate regions of the universe at a very short time after the Big Bang. On the other hand, those regions moved off each other at such a high speed that, in the face of the limited speed of light, a causal connection cannot be understood. To this conflict the name "Horizon Problem" was given.

6.2.1 Inflation as with Einstein

Following the understanding of Einstein, present physics assumes a change of the space as a solution. This assumed process has gotten the name 'Inflation'.

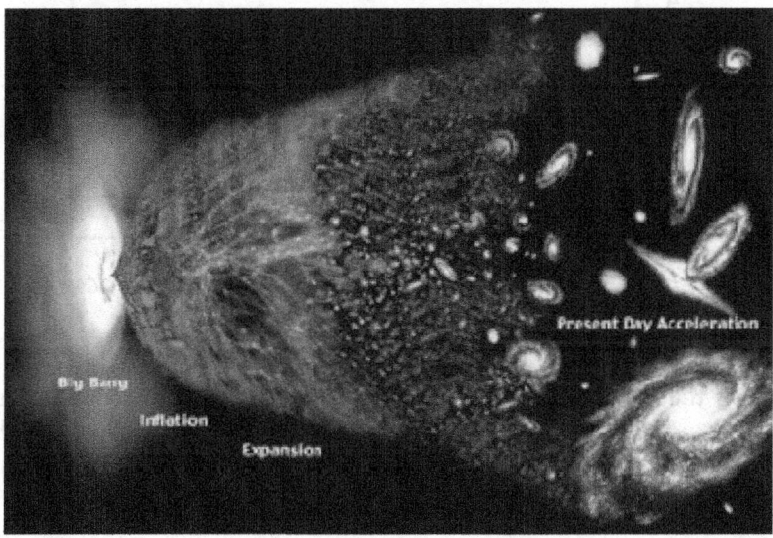

Figure 6.2: Inflation visualized

Figure 6.2 shows how the process is understood in present physics. At a time close to the Big Bang the space was according to this understanding contracted in relation to the present one by a factor of around 10^{50}. For this period of contraction causal exchanges have been possible also in the view of the limited speed of light. Then during a short time the space expanded considerably, and during the following time until now it continued to expand at a low rate.

Present physics does not have a real explanation available for this process of inflation. As an ad-hoc assumption a new field of so-called 'inflatons' is assumed to cause it.

6.2.2 Horizon Problem Explained by Varying Speed of Light

From a logical perspective, the problem of this correlation is a conflict between the spatial extension and the speed of light.

So, as an alternative to an assumed change of the 'space', it may also be assumed that the speed of light changed, namely that it was extremely large during at short period close to the Big Bang. Afterwards the speed of light decreased rapidly at first, and later more slowly to the present value, as depicted in figure 6.3.

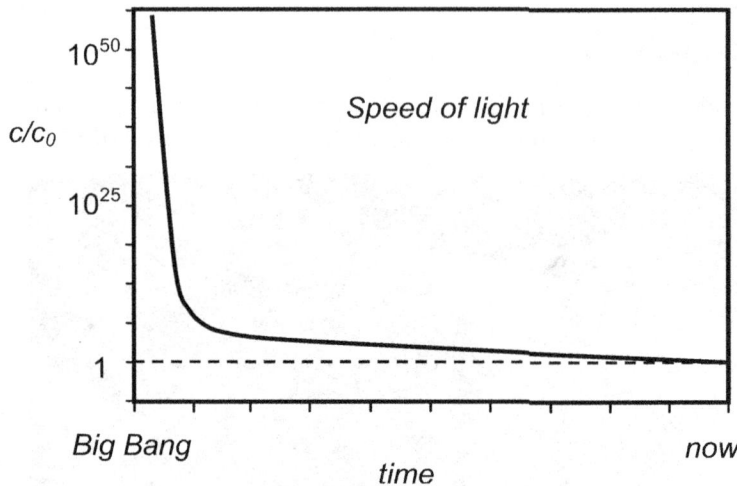

Figure 6.3: Development of the speed of light

The assumption that the speed of light has changed during the development of the universe is anyway attractive, as it would not only solve this causal problem. As a further benefit, it would also solve the presently not understood fine-tuning of basic physical parameters. (Refer e.g. to the work of J. Magueijo [7].)

6.3 Dark Energy

An acceleration of the objects of our universe in recent times has been concluded from

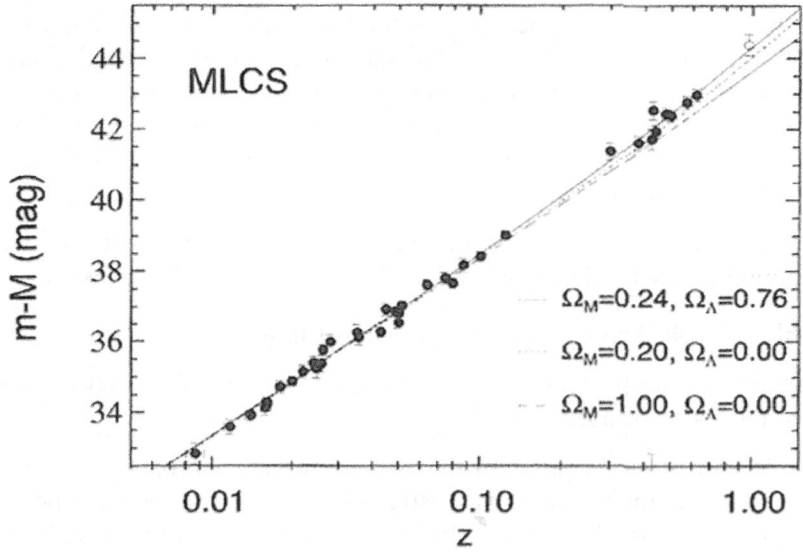

Figure 6.4: Supernova 1a Hubble diagram

the observations of supernovae type 1a. The result of Riess et al. [8] is presented in figure 6.4.

Figure 6.4 shows the apparent magnitude of the observed supernovae (as the ordinate) versus the red shift z (the abscissa), which is identified with the escape speed of the stars.

The red shift z is defined as

$$z = \frac{\Delta \nu}{\nu_{ob}} \tag{6.1}$$

where ν_{ob} is the *observed* frequency, $\Delta \nu$ is the frequency shift. From this follows for the speed V:

$$V = c \cdot \frac{z}{z+1} \tag{6.2}$$

which is used for the evaluation according to figure 6.4, but conventionally with the assumption that c is a constant over all times and the space is unchanged during the time investigated.

According to the Hubble Law, all stars and so also the investigated supernovae should be situated on a straight line, presented in figure 6.4 by the dotted line for the most probable assumption. This means that the escape speed of these objects is proportional to their distance to the observer. However, the measurement points in the upper part can be understood as being too much to the left, which means that the red shift of the older supernovae is too small in comparison to the younger stars, being presented in the lower (left) part. This is commonly interpreted in the way that the younger supernovae are too fast in comparison to the older ones. They are assumed to be accelerated.

The lowest - dashed - line would also be physically acceptable but assumes a higher matter density of the universe indicated by $\Omega_M = 1$, and it does not fit. The highest - solid – line, which fits well the data, is characterized by $\Omega_\Lambda = 0.76$, which means the assumption of an acceleration, here indicated by the denotation of the cosmological constant Λ with reference to the original conception of Einstein.

Presently there are the following models used to explain the phenomenon:

- Dark Energy: This means that the whole universe is filled by some type of energy which causes all objects to accelerate. This energy can, with reference to Einstein's understanding of mass and energy, identified with matter. This leads to the conclusion that – together with the 'mysterious' Dark Matter – only ca. 4% of the matter of the universe are known and described by present physics

- Quintessence: This would represent a new kind of a potential

- The revival of the cosmological constant Λ, which was once introduced by Einstein in order to explain the fact that the universe does not collapse. This idea was abandoned at some later time, after Hubble detected that the universe was in a state of permanent expansion. Einstein called this idea later his worst stupidity.

The assumption that the speed of light c has changed as indicated in figure 6.3 is able to explain the acceleration as an evaluation effect. From equations (6.1) and (6.2) there follows for the speed V in the case of a Doppler shift Δv :

$$V = c \cdot \frac{\Delta v}{\Delta v + v_{ob}}.$$
(6.3)

However, with reference to figure 6.3 it can be assumed that the speed of light c was higher at early times. So, if c is to be replaced by a larger value, then the resulting V will be larger.

In figure 6.5 the – corrected – speed is added to the abscissa scale. That means that in reference to the speed scale the supernovae are to be repositioned to the right to reflect the correction. They can be positioned on the dotted line now without any conflict to the observation. This is indicated as an example by the arrows in the figure.

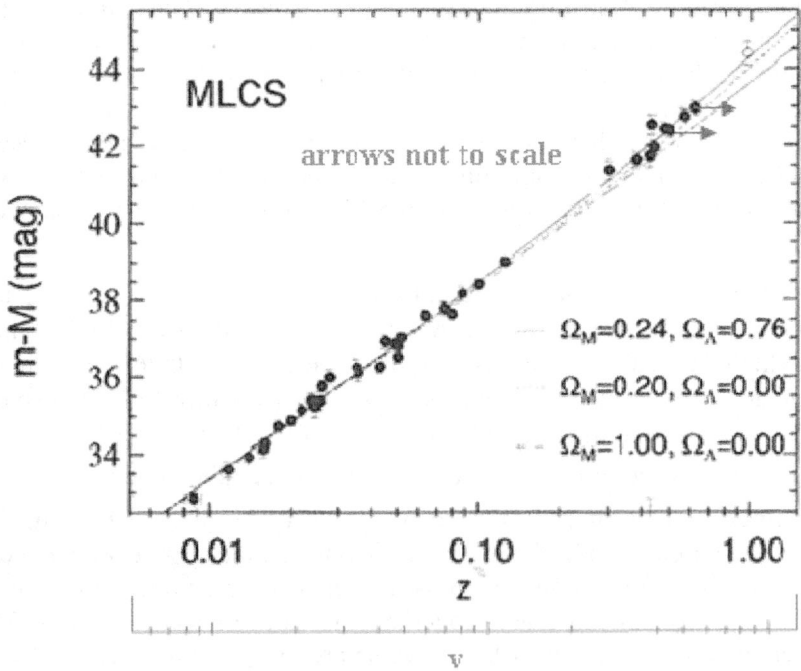

Figure 6.5: Supernova 1a Hubble diagram - corrected

This means physically that the alleged acceleration vanishes.

7. The Speed of Light and the Ether

Both topics, the speed of light and the ether, are related to each other.

As stated above, we do not share Einstein's opinion, that the speed of light is defined on the one hand by properties of the four-dimensional space-time and that on the other hand there is no reason to assume an absolute system of reference, e.g. an ether.

We will present here a model to explain the fact of a normally constant speed of light on the one hand and possible physical reasons for the change of this value during the development of our universe on the other hand.

The following considerations about the speed of light, its causes and mechanisms are in our view plausible and consistent. They are, however, not stringent consequences of our model like the other properties presented in the preceding chapters. Here some work has still to be done.

7.1 The Speed of Light

7.1.2 Constancy of c

Why is c a constant? Has c to be a constant forever or can it change with the time?

With reference to the Basic Particle Model, which is the basis for the considerations in this article, we can state:

There exist two categories of particles

- Basic particles
- Exchange Particles.

Both types of particle are mass-less and both move exclusively at the speed of light c. We will first regard basic particles. The situation about exchange particles will be discussed at the end of this chapter.

All interactions between basic particles are according to this model elastic scattering. That means that interactions can in general only change the direction of the motion of the affected particles, not their speed. We refer here to the classical situation.

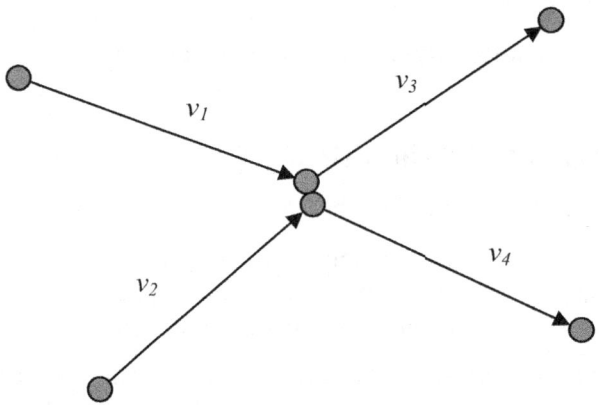

Figure 7.1 Elastic Scattering of Basic Particles

In the classical situation of elastic scattering we have to apply the momentum law, which states for the momentum vector p

$$\vec{p}_{in} = m_1 \cdot \vec{v}_1 + m_2 \cdot \vec{v}_2 = \vec{p}_{out} = m_3 \cdot \vec{v}_3 + m_4 \cdot \vec{v}_4 \ .$$

In the Basic Particle Model the basic particles have no mass. We have to assume that there is a kind of unified reaction similar to the case of unique masses.

So we get for arbitrary speeds

$$\vec{v}_1 + \vec{v}_2 = \vec{v}_3 + \vec{v}_4 \ .$$

If two objects of the same mass have an elastic collision, then the amount of the speed of both is exchanged. This is a known process in classical mechanics and will not be deduced here again. In our case of basic particles these particles have no mass and we assume a similar interaction like in the classical case for a unique mass.

According to this model, such a basic particle nominally has the speed of light c. This means in the case of a collision or a scattering that both maintain their amount of speed c. If they have different speeds (i.e. deviations from the nominal c), then they will exchange the speeds.

However, there is a mechanism to unify different speeds.

Assume three particles, one particle D moving freely, and two particles A and B bound to each other as basic particles in an elementary particle. The particle D may move at a speed c_1, different from the standard speed of light c_0, the particles A and B move both at c_0. If now particle D interacts with A, then D will get the speed c_0 and A will get the speed c_1, *as* described above. On the other hand, A is in a bind with B in their common orbit and both speeds have to adapt to each other. So, finally, A and B will both have the speed

$$c_{new} = 1/2 \cdot (c_0 + c_1) \, .$$

Now the difference between the highest and the lowest speed is

$$\Delta c_{new} = c_0 - 1/2 \cdot (c_0 + c_1) = 1/2 \cdot (c_0 - c_1) \, .$$

Prior to the interaction, the difference was

$$\Delta c_{pre} = (c_0 - c_1) \, .$$

So the interaction has reduced the difference to half.

Interactions between particles occur permanently in the universe. This is able to explain why c is a universal constant – with minor deviations.

7.1.3 Decrease of c

Now we can ask the question whether c can change following a situation where it has been unique. The answer in the scope of this model is that a change occurs in the case, when three particles interact in the same scattering process. This is a rare case but it happens.

This case can be treated in a simplified way by the assumption that two of these particles are connected to each other and the third particle interacts with this compound. In this case major deviations from the original speed c are possible. This is again similar to the classical situation as argued analogously above.

A specific case with important consequences for the development of the universe is the situation, when the compound particle moves off the centre of the universe. This is the statistically more frequent case for an expanding universe as ours. In this case the reflected single particle will have a reduced speed, reduced compared to the original speed c. However, in the next two-particle interactions the speed will be smoothly guided back to the average c. In the sum over all cases there will be a slow decrease of the speed of light in the expanding universe.

7.1.4 The Early Situation after the Big Bang

Early after the Big Bang the density of particles was very high. So the rate of interactions was correspondingly high. This is true on the one hand for the two-particle interactions, which caused the speed of the particles to adapt quickly to each other. On the other hand the probability of three-particle interactions was higher by a disproportionately high factor. As a consequence, in this rapidly expanding situation the speed of the particles decreased dramatically. This easily explains the development which in standard physics is called 'inflation'.

7.2 The Question of the Ether

7.2.1 Ether Related to Basic Particles

In the view of the reactions between basic particles described above, the ether can be seen in a quite minimalist way. Ether can be understood as an absolute frame of reference, which is given by the motion state of the Big Bang. This motion state is, as explained above, conserved by the fact that basic particles interact by elastic processes and in this way did not leave the state space of motion since the Big Bang.

7.2.2 Ether Related to Exchange Particles

The situation of basic particles and exchange particles seems not identical.

In the normal case, i.e. neglecting the generation of additional elementary particles, which is presently not in the scope of this model, every interaction between basic particles keeps the number of them unaltered. And so it is easily understandable, as we have seen above, that the speed of light is a constant here.

Exchange particles are, on the other hand, in the present understanding of fields emitted by a charge. The interaction of a field with another charge goes on in the way that the receiving charge absorbs the exchange particle and receives in this moment an impact towards the emitting charge or away from it, depending on the mutual signs of the charges. Here several questions arise.

- The exchange particles have to move at the speed c. We know that fields of any kind move at c, so also the exchange particles constituting the field. But the question arises, to which motion state the speed c is related. Macroscopically it could not be related to the moving state of the charge, because in that case a moving charge would propagate its information at a different speed than c related to an observer at rest, as could be macroscopically measured. So, the sending charge cannot be the speed reference. That reference can only be provided by a motion state, which is either the original motion state of the Big Bang or the average of the speed of the surrounding basic particles over a sufficiently large area.

- If the speed of light c changes during the development of the universe as we have explained it for the basic particles, then this must also be the fact for the exchange particles. We have to understand in detail how this works.

- The exchange particles, which are emitted from a charge, have the ability to do work on another charge of the same class of field. So they carry energy. On the other hand most of them will never meet another charge but fly into the empty space until infinity. This can only be understood if in this process conservation of energy is invalid. - This point is an unresolved problem in present physics as it is assumed that conservation of energy is a general law in physics. It is, on the other hand, not a problem in the scope of the Basic Particle Model. In the scope of this model energy conservation is a consequence of the structure of elementary articles. So, the violation at the level of exchange particles does not mean a conflict.

Regarding the question of the ether, the mechanism which conserves the speed of exchange particles in relation to an absolute frame is not easily understandable. In our view, more work has to be done to find a mechanism here which can explain this to the necessary detail. Up to that possible result, we have to be open also for the classical understanding, in which the ether is a kind of a medium which guides these particles.

Conclusion

We have shown that relativity can be based on physical processes, i.e. on properties of particles and of fields. We see compelling arguments, that the speed of light is only constant in respect to an absolute frame of reference. Further on we find good arguments that the speed of light changed during the development of the universe – like other physicists assume. That means that the speed of light is only constant in relation to a limited time interval.

On this basis, we gain extraordinary benefits:

- Relativity now fits seamlessly into physics as a whole. The theory has become far more comprehensible and its formalism is much easier to understand. Accounted for in this way, relativity can even be taught at a high school level, yet the results conform to the ones in Einstein's approach at least in so far as they can be proven by experiments and observations.

- Important open questions of present-day physics are resolved with a surprising ease:

  (1) *Quantum Gravity* is no longer an open point. Gravity is shown to be on the one hand a side effect of the strong force, and on the other hand, the strong force is fully covered by quantum mechanics. Thereby the presumed conflict between relativity and quantum mechanics disappears.

  (2) The *Dark Energy problem* is resolved as a result of a changing speed of light during the development of the universe. The change of c results from a physically plausible process.

  (3) The *Dark Matter problem* vanishes, as it follows from the model that every elementary particle contributes with the same amount to the gravitational field - irrespective of its mass. In the quantitative calculations, photons are

able to constitute Dark Matter particles. Furthermore, the spatial distribution of Dark Matter in the universe, which otherwise could not be explained, fits to the model.

(4) The *inertial mass* is explained including the dynamical aspects, i.e. the relativistic increase of mass and the mass-energy relation. In present physics the origin of mass is still an open point. Even if the Higgs boson should be found, which many doubt to exist, this would not constitute a complete explanation.

What has to be the next step? The basic particles have to be understood in greater detail. And primarily the behaviour of the exchange particles is an important point. We have to understand, in which way the different exchange particles, i.e. the attracting and the repelling ones, are emitted in a correlated way; further how the correlation depends on the distance, i.e. on the travelling time in a beam of such particles.

References

[1] A. Einstein, *Zur Elektrodynamik bewegter Körper*, Annalen der Physik, IV. Jg. 17, S. 891–921 (1905)

[2] H. R. Reichenbach, *The Philosophy of Space and Time*, Dover, New York (1958)

[3] H. A. Lorentz, *Electromagnetic phenomena in a system moving with any velocity smaller than that of light*, Proc. Acad. Science Amsterdam **6**: 809–831 (1904)

[4] L. de Broglie, *Radiations - Ondes et Quanta*, Comptes rendus, Vol. 177, 1923, pp. 507-510.

[5] P. A. M. Dirac, *The Quantum Theory of the Electron*. Proceedings of the Royal Society of London. Series A, Containing Papers of a Mathematical and Physical Character **117** (778): 610–624 (1928).

[6] E. Schrödinger, *Über die kräftefreie Bewegung in der relativistischen Quantenmechanik ("On the free movement in relativistic quantum mechanics")*, Berliner Ber., pp. 418-428 (1930)

[7] A. Albrecht and J. Magueijo, *A time varying speed of light as a solution to cosmological puzzles,* Phys. Rev. D 59, 043516 (1999)

[8] A. G. Riess et al, *Observational evidence from supernovae for an accelerating universe and a cosmological constant*, Astronomical J. 116:1009–38 (1998)

Dark Matter and Dark Energy

B.G. Sidharth   birlasc@gmail.com
International Institute for Applicable Mathematics & Information Sciences
Hyderabad (India) & Udine (Italy)
B.M. Birla Science Centre, Adarsh Nagar, Hyderabad - 500 063 (India)

Abstract

We consider the problem of the flattening of the velocity curves in galactic discs and the consequent postulation of dark matter from three different but converging perspectives– a change in the large scale dimensionality of space, a variation of G and the MOND approach. We argue that all astrophysical data can be satisfactorily explained invoking the dark energy cosmology which underpins the varying G approach.

1   Dark Matter or Dark Energy?

It is some seven decades now since the existence of dark, that is non luminous matter was postulated, though the identity of this dark matter has only been a matter of guess work. The reason for invoking the hypothetical dark matter is well known– the velocities of stars in galaxies should tend to zero using usual dynamics, as we approach the edge of the disc. Instead astrophysical observation has consistently shown that the velocity curves flatten out, that is the velocities tend to a constant rather than zero. So Zwicky and others postulated that there was matter other than the visible matter which gave a greater mass to the galaxies, and this in turn would explain the velocity discrepancy [1].
The question then arose, "What exactly is this dark matter?". Over the years several hypotheses have been put forth– it could be hot dark matter or it could be cold dark matter. These could range from weakly interacting massive particles (WIMPs) to cold neutrinos. Or these could be the missing monopoles, or undetectable brown dwarf stars or even black holes and so on.

To this day the question has remained unresolved. It must be mentioned however that the latest WMAP data, in a model dependent interpretation shows nearly twenty three percent dark matter. More recently there have been unconfirmed reports that there is indirect evidence of dark matter via the observation of non MOND accelerations, to be discussed below in the collision of star clusters.

There have however been alternative suggestions to explain the flat velocity curves. We will discuss three of these, two put forward by the author and the other by Milgrom, and try to find a convergence.

We would need a discussion on dark energy first. Descartes the seventeenth century French philosopher mathematician proclaimed that the so called empty space above the mercury column in a Torricelli tube, that is, what is called the Torricelli vacuum, is not a vacuum at all. Rather, he said, it was something which was neither mercury nor air, something he called aether.

The seventeenth century Dutch Physicist, Christian Huygens required such a non intrusive medium like aether, so that light waves could propagate through it, rather like the ripple waves on the surface of a pond. Hence the word luminiferous aether. In the nineteenth century the aether was reinvoked. Firstly in a very intuitive way Faraday could conceive of magnetic effects in vacuum in connection with his experiments on induction. Based on this, the aether was used for the propagation of electromagnetic waves in Maxwell's Theory of electromagnetism, which infact laid the stage for Special Relativity. This aether was a homogenous, invariable, non-intrusive, material medium which could be used as an absolute frame of reference atleast for certain chosen observers. However the experiments of Michelson and Morley towards the end of the nineteenth century, lead to its downfall, and thus was born Einstein's Special Theory of Relativity in which there is no such absolute frame of reference. The aether lay shattered once again.

Very shortly thereafter the advent of Quantum Mechanics lead to its rebirth in a new and unexpected avatar. Essentially there were two new ingredients in what is today called the Quantum vacuum. The first was a realization that Classical Physics had allowed an assumption to slip in unnoticed: In a source or charge free "vacuum", one solution of Maxwell's Equations of electromagnetic radiation is no doubt the zero solution. But there is also a more realistic non zero solution. That is, the electromagnetic radiation does not necessarily vanish in empty space.

The second ingredient was the mysterious prescription of Quantum Mechanics, the Heisenberg Uncertainty Principle, according to which it would be impossible to precisely assign momentum and energy on the one hand and spacetime location on the other. Clearly the location of a vacuum with no energy or momentum cannot be specified in spacetime.

This leads to what is called a Zero Point Field. For instance a Harmonic oscillator, a swinging pendulum for example, according to classical ideas has zero energy and momentum in its lowest position. But the Heisenberg Uncertainty endows it with a fluctuating energy. This fact was recognized by Einstein himself way back in 1913 who contrary to popular belief, retained the concept of aether though from a different perspective [2]. It also provides an understanding of the fluctuating electromagnetic field in vacuum.

From another point of view, according to classical ideas, at the absolute zero of temperature, there should not be any motion. After all the zero is when all thermodynamic motion ceases. But as Nernst, father of the third law of Thermodynamics himself noted, experimentally this is not so. There is the well known superfluidity due to Quantum Mechanical – and not thermodynamic – effects. This is the situation where supercooled Helium moves in a spooky fashion.

This mysterious Zero Point Field or Quantum vacuum energy has since been experimentally confirmed in effects like the Casimir effect which demonstrates a force between uncharged parallel plates separated by a charge free medium, the Lamb shift which demonstrates a minute oscillation of an electron orbiting the nucleus in an atom-as if it was being buffetted by the Zero Point Field-, the anomalous Quantum Mechanical gyromagnetic ratio g = 2 and so on [3]-[8],[9].

The Quantum Vacuum is a far cry however, from the passive aether of olden days. It is a violent medium in which charged particles like electrons and positrons are constantly being created and destroyed, almost instantly, infact within the limits permitted by the Heisenberg Uncertainty Principle for the violation of energy conservation. One might call the Quantum Vacuum as a new state of matter, a compromise between something and nothingness. Something which corresponds to what the Rig Veda described thousands of years ago: "Neither existence, nor non existence."

Quantum Vacuum can be considered to be the lowest state of any Quantum field, having zero momentum and zero energy. The properties of the Quantum Vacuum can under certain conditions be altered, which was not the case with the erstwhile aether. In modern Particle Physics, the Quantum Vacuum is responsible for phenomena like quark confinement, a property whereby it would be impossible to observe an independent or free quark, the spontaneous breaking of symmetry of the electroweak theory, vacuum polarization wherein charges like electrons are surrounded by a cloud of other opposite charges tending to mask the main charge and so on. There could be regions of vacuum fluctuations comparable to the domain structures of ferromagnets. In a ferromagnet, all elementary electron-magnets are aligned with their spins in a certain direction. However there could be special regions wherein the spins are aligned differently.

Such a Quantum Vacuum can be a source of cosmic repulsion, as pointed by Zeldovich and others [10, 29]. However a difficulty in this approach has been that the value of the cosmological constant turns out to be huge, far beyond what is observed. This has been called the cosmological constant problem [11].

There is another approach, sometimes called Stochastic Electrodynamics which treats the ZPF as primary and attributes to it Quantum Mechanical effects [12, 13]. It may be re-emphasized that the ZPF results in the well known experimentally verified Casimir effect [14, 15]. We would also like to point out that contrary to popular belief, the concept of aether has survived over the decades through the works of Dirac, Vigier, Prigogine, String Theorists like Wilzeck and others [16]-[21]. As pointed out it appears that even Einstein himself continued to believe in this concept [22].

We would first like to observe that the energy of the fluctuations in the background electromagnetic field could lead to the formation of elementary particles.

Indeed this was Einstein's belief. As Wilzeck put it, "Einstein was not satisfied with the dualism. He wanted to regard the fields, or ethers, as primary. In his later work, he tried to find a unified field theory, in which electrons(and of course protons, and all other particles) would emerge as solutions in which energy was especially concentrated, perhaps as singularities. But his efforts in this direction did not lead to any tangible success."

Indeed the author has argued that this dark energy or Quantum Vacuum is a dissipative medium which gives birth to inertia itself [23]. Furthermore it is the condensates of this ubiquotus dark energy at the Compton (including Planck) scale that appears as the various particles in the universe. This leads to a consistent cosmology which predicted an accelerating, expanding universe with a small cosmological constant at a time when a ruling paradigm was exactly the opposite. We will return to these considerations in Section 5 and see that there is perfect agreement with observation.

2   Less Than Three Dimensional Space?

The author (with A.D. Popova) [24] suggested that the dimensionality of space falls off asymptotically, and this would explain astrophysical observations including the dark matter problem. Indeed it had already been argued that the dimensionality of space could be expressed by a non integer number that is less than three on large scales [25, 26]. The three dimensionality of our immediate space may be necessary, for the very existence of atoms, as was pointed by Ehrenfest long ago [27]. Similarly this dimensionality may also be required for usual wave propagation [28, 29, 30]. All this is at what we may call intermediate scales. At different scales of measurement, the dimensionality could be different [31]. This fact could explain the dark matter problem, as we will now argue.

More generally, the dependence of matter on distance $M(r)$, obtained from observing $21cm$ neutral hydrogen emission of gas clouds moving around a galaxy far from its visual bounds (the continuation of a rotation curve) [32, 33] is

$$M \propto r^{1.2 \div 1.3} \tag{2.1}$$

This conclusion reflects the fact that the observed rotation velocity slightly increases at outer parts of galaxies, so the growth of M is interpreted as the presence of some dark halo besides luminous matter. Moreover, the amount of dark matter grows relatively to luminous matter when coming to larger and larger scales [34].

However, even the nonrelativistic (Newtonian) consideration of gravitational forces in spaces with lesser than three dimensions, enables us in principle to bring in correlation dynamics and "the shortage" of luminous matter. It would be very difficult to take into account the smooth fall of dimensionality because we do not know the law of such a fall. In order to make some estimates we roughly assume that on some relative distance R_0 the dimensionality changes by a leap from 3 to $n < 3$. We consider the rotation curves of disk galaxies –similar considerations apply for the dynamics of double galaxies and the dispersion of velocities in elliptical (spheroidal) galaxies. We show how one can lower the estimates of masses of these systems under the assumption that the dimensionality is less than three starting even at scales of the order of a typical galaxy's size. We also discuss the possible hierarchical change of dimensionality. We also demonstrate the possibility of lowering dynamical mass.

Certainly, we know now of a constructive physical model which can describe noninteger and nonconstant dimensionality. We outline some suggestive arguments for it. The first of possible suggestions comes from the fractal theory [35]; space itself may have a fractal-like structure. The second suggestion is that effectively, if we consider individually each object in the Universe leaving aside other objects, then we can perceive a space between us and this object as 2-dimensional because one spatial direction is fixed as a line from us to the object, and the other direction can be fixed by a vector of relative velocity of the object with respect to us. Thus, may be our space filled by distant separated objects consists of a set of (perhaps non-connected) 2-dimensional subspaces for which effectively $2 \leq n < 3$. The third suggestion comes from the existence of the large-scale structure of the Universe in distribution of galaxies, their groups, clusters, voids and superclusters. There is the tendency for matter to form oblate structures at each hierarchical level. Possibly, the structure and dimensionality of our space itself might reflect the distribution of (luminous) cosmic matter.

3 Newtonian Consideration

In 3 and n dimensions, the expressions for the gravitational forces acting between the mass M and a unit mass separated by the distance r are

$$F^{(3)} = -\frac{G^{(3)} M}{r^2} \qquad (3.2)$$

and

$$F^{(n)} = -\frac{G^{(n)} M}{r^{n-1}}, \qquad (3.3)$$

respectively, where $G^{(3)}$ and $G^{(n)}$ are relevant gravitational constants. The corresponding potentials ($\vec{F} = -\vec{\nabla}\Phi$ by definition) up to arbitrary constants $C^{(3)}$ and $C^{(n)}$ are

$$\Phi^{(3)} = -\frac{G^{(3)} M}{r} + C^{(3)}, \qquad (3.4)$$

and for $n \neq 2$

$$\Phi^{(n)} = -\frac{1}{n-2}\frac{G^{(n)}M}{r^{n-2}} + C^{(n)} \tag{3.5}$$

In the case $n = 2$,

$$\Phi^{(2)} = G^{(2)}Mlnr + C^{(2)}.$$

Now, let us assume that at some relative distance R_0 from a body the dimensionality changes by a leap from 3 to n. The condition of matching the forces (3.2) and (3.3) at R_0 gives the connection between the gravitational constants

$$G^{(n)} = G^{(3)}R_0^{n-3} \tag{3.6}$$

Thus, the improved force is (3.2) for $r \leq R_0$ and (3.3) for $r > R_0$ with (3.4):

$$F^{imp} = \begin{cases} F^{(3)}, & r \leq R_0, \\ F^{(n)}, & r > R_0. \end{cases} \tag{3.7}$$

The condition of matching the potentials (3.4) and (3.5) at R_0 is also required, and leads to the following expression for the "n-dimensional" constant ($C^{(3)} = 0$ is chosen in (3.4)),

$$C^{(n)} = \frac{3-n}{n-2}\frac{G^{(3)}M}{R_0}$$

for $n \neq 2$, and

$$C^{(2)} = -\frac{G^{(3)}M}{R_0}(1 + lnR_0)$$

for $n = 2$.

Thus, the improved potential is (3.4) for $r \leq R_0$ and (3.5) for $r > R_0$:

$$\Phi^{imp} = \begin{cases} \Phi^{(3)}, & r \leq R_0, \\ \Phi^{(n)}, & r > R_0. \end{cases} \tag{3.8}$$

Let us stress that we can only think of R_0 as a relative distance between any objects. Otherwise, first the conception of relativity of space which is an achievement of Einstein's physics, would be violated. Second, there would be troubles with the universality of gravitational attraction (Cf.ref.[24]).

Perhaps, our consideration would be less rough, if we consider the change of dimensionality which occurs by leaps several times from 3 to n_1 at R_0, from n_1 to n_2 at R_1, and so on, and from n_j to $n_{j+1} > 2$ at R_j. Then, we have the chain of relations between the gravitational constants

$$G^{(n_1)} = G^{(3)} R_0^{n_1-3}$$

$$G^{(n_2)} = G^{(n_1)} R_1^{n_2-n_1} = G^{(3)} R_0^{n_1-3} R_1^{n_2-n_1} \qquad (3.9)$$

$$G^{(n_j+1)} = G^{(n_j)} R_{j+1}^{1-n_j} = \cdots = G^{(3)} R_0^{n_1-3} R_1^{n_2-n_1} \cdots R_{j+1}^{-n_j}$$

The chain of relations between the constants in potentials is rather cumbersome; however the recurrence relation is

$$C^{(n_{j+1})} = C^{(n_j)} + \frac{n_j - n_{j+1}}{(n_j - 2)(n_{j+1} - 2)} \frac{G^{(n_j)} M}{R_j^{n_j-2}}$$

Below we present the application of the force (3.7) (the potential of (3.8)) to determinations of galactic masses.

4 Rotation Curves of Disk Galaxies

A rough calculation of rotation velocity v at the distance r far from the center of a galaxy (i.e., when all its mass is effectively concentrated near the center or spherically distributed around it) is based on the equality of the centrifugal force and gravitational force.

In the 3-dimensional space, using (3.2)

$$\frac{v^2}{r} = \frac{G^{(3)} M_g^{(3)}}{r^2} \qquad (4.10)$$

In accordance with our conception, we can write

$$M_g^{dyn} \equiv M_g^{(3)} = \frac{v^2 r}{G^{(3)}},$$

i.e., we call the dynamical galactic mass, M_g^{dyn}, a mass calculated as if our space were 3-dimensional.

In the n-dimensional space, if $r \gg R_0$ then equality (4.10) should be replaced by the following (with the use of (3.3):

$$\frac{v^2}{r} = \frac{G^{(n)} M_g^{(n)}}{r^{n-1}},$$

so that we call a mass calculated in the n-dimensional space the true galactic mass, M_g^{true}:

$$M_g^{true} \equiv M_g^{(n)} = \frac{v^2 r^{n-2}}{G^{(n)}} = \left(\frac{R_0}{r}\right)^{3-n} M_g^{dyn} \qquad (4.11)$$

where in the last equality (3.6) is used. When $2 < n < 3$ the factor at M_g^{dyn} is less than unity, therefore $M_g^{true} < M_g^{dyn}$.

The more accurate calculation of the rotation curve can be done for the case when the disk of a galaxy lies in the 2-dimensional space (or plane), and there exist no other spatial dimensions. Let the distribution of the 2-dimensional matter density in the disk satisfy the law

$$\rho = \rho_0 exp\left(\frac{r}{R_d}\right) \qquad (4.12)$$

where ρ_0 is the 2-density in the disk center, and R_d is some characteristic radius. Let $R_0 \ll R_d$. The distribution (4.12) corresponds to the observed distribution of luminous matter in [35]. The velocity square in this case is given as follows

$$v^2(r) = 2\pi\rho_0 G^{(3)} \frac{R_d^2}{R_d}\left[1 - \left(1 + \frac{r}{R_d}\right) exp\left(-\frac{r}{R_d}\right)\right] \qquad (4.13)$$

The function (4.13) monotonically increases from zero and tends to a constant value at $r \to \infty$. That is why the dynamical mass calculated with the aid of (4.13) tends to grow linearly at larger r : $M^{dyn} \propto r$. However, at any finite r we can effectively write $M_g^{dyn} \propto r^\beta$ where always $\beta > 1$. Probably, this fact could explain the dependence (2.1), meaning that our real space has the dimensionality which is very near to two at the scales of the outer parts of galaxies. This also explains the flattening of the rotational curves, without invoking dark matter.

Alternatively, we note that from the above, for $n = 2$, we get

$$v^2 = GMlnr,$$

or

$$v\frac{dv}{dr} \propto \frac{1}{r} \to 0$$

as r becomes large, so that $\frac{dv}{dr} \to 0$, because v does not $\to 0$. So, effectively, $v \to$ a constant value.

5 The Time Variation of the Gravitational Constant

We now come to the author's cosmological model which in 1997 predicted a dark energy driven accelerating universe with a small cosmological constant. It may be recalled that at that time the ruling paradigm embodied in the hot big bang standard cosmological model was exactly the opposite. This model has been discussed in detail (Cf.ref.[29, 30, 36, 37]). In this model all the so called Large Number Relations and the mysterious Weinberg formula are deducible from this theory rather than be ad hoc coincidences. We will briefly summarize this model. Our starting point is the fact that there are $N \sim 10^{80}$ elementary particles (typically pions) in the universe and further we have

$$Nm = M \qquad (5.14)$$

where M is the (luminous) mass of the Universe. A justification for (5.14), which is consistent, is that as the Universe at large is electrically neutral, the particles interact via the gravitational force, which is very weak in any case.

In the following we will use N as the sole cosmological parameter.

Equating the gravitational potential energy of the pion in a three dimensional isotropic sphere of pions of radius R, the radius of the Universe, with the rest energy of the pion, we can deduce the well known relation [38, 39, 40]

$$R \approx \frac{GM}{c^2} \qquad (5.15)$$

where M can be obtained from (5.14). (5.15) can be alternatively deduced and is consistent with observation.

We now use the fact that given N particles, the fluctuation in the particle number is of the order $\sqrt{N}$[40, 41, 36, 37, 42, 43], while a typical time interval for the fluctuations is $\sim \hbar/mc^2$, the Compton time, the fuzzy interval. Particles are created and destroyed - but the ultimate result is that $\sqrt{N}$ particles are created in this "random weak" scenario. So we have,

$$\frac{dN}{dt} = \frac{\sqrt{N}}{\tau} \qquad (5.16)$$

whence on integration we get, (remembering that we are almost in the continuum region),

$$T = \frac{\hbar}{mc^2}\sqrt{N} \qquad (5.17)$$

We can easily verify that the equation (5.17) is indeed satisfied where T is the age of the Universe. Next by differentiating (5.15) with respect to t we get

$$\frac{dR}{dt} \approx HR \tag{5.18}$$

where H in (5.18) can be identified with the Hubble constant, and using (5.15) is given by,

$$H = \frac{Gm^3c}{\hbar^2} \tag{5.19}$$

Equation (5.14), (5.15) and (5.17) show that in this formulation, the correct mass, radius, Hubble constant and age of the Universe can be deduced given N as the sole cosmological or large scale parameter. Equation (5.19) can be written as

$$m \approx \left(\frac{H\hbar^2}{Gc}\right)^{\frac{1}{3}} \tag{5.20}$$

Equation (5.20) has been empirically known as an "accidental" or "mysterious" relation. As observed by Weinberg[44], this is unexplained: it relates a single cosmological parameter H to constants from microphysics. In our formulation, equation (5.20) is no longer a mysterious coincidence but rather a consequence.

As (5.19) and (5.18) are not exact equations but rather, order of magnitude relations, it follows, on differentiating (5.18) that a small cosmological constant $\wedge$ is allowed such that

$$\wedge \leq 0(H^2)$$

This is consistent with observation and shows that $\wedge$ is very small. Indeed this has been the source of the so called cosmological constant problem. Proceeding along these lines (Cf.ref.[30]) for details, we can also deduce

$$e^2/Gm^2 \sim \sqrt{N} \approx 10^{40} \tag{5.21}$$

or without using (5.15), we get, instead, the well known so called Weyl-Eddington formula,

$$R = \sqrt{N}l \tag{5.22}$$

(5.22) is another form of (5.17) - in our case it is deduced, rather than be empirical. It can now be easily seen from the above that the gravitational constant, rather as in Dirac cosmology, has the following time dependence

$$G = \frac{\beta}{T} \tag{5.23}$$

where in our cosmology β is given in terms of the constant microphysical parameters.

Before proceeding further we would like to point out that the above scheme not only throws up the correct cosmological constant (against the ruling paradigm of that time) but it also gives another spectacular recently observed prediction namely the so called universal Milgrom acceleration $\sim 10^{-8} cms\, per\, sec^2$, which is otherwise inexplicable. This acceleration follows most simply from (5.18) - it is $H^2 R \sim c^2/R$ (as H as is well known is c/R, something which follows independently from the above). While this acceleration has been confirmed by observation (Cf.Section 7 and [45]), it appears at Hubble distances R - but then we are not at the centre of fixed space - this would be true for anywhere in the universe!

It has also been shown that arguing on these lines we get another recent observational discovery viz., irreducible energy of $10^{-33} eV$ [46] which has been equally enexplicable.

We next observe that from (5.23) it follows that

$$G = G_0 \left(1 - \frac{t}{t_0}\right) \tag{5.24}$$

where G_0 is the present value of G and t_0 is the present age of the universe and t the time elapsed from the present epoch. Similarly one could deduce that (Cf.ref.[1]),

$$r = r_0 \left(\frac{t_0}{t_0 + t}\right) \tag{5.25}$$

We next use the well known Kepler's Third law:

$$\tau = \frac{2\pi a^{3/2}}{\sqrt{GM}} \tag{5.26}$$

τ is the period of revolution, a is the orbit's semi major axis, and M is the mass of the sun. Denoting the average angular velocity of the planet by

$$\dot{\Theta} \equiv \frac{2\pi}{\tau},$$

it follows from (5.24), (5.25) and (5.26) that

$$\dot{\Theta} - \dot{\Theta}_o = \dot{\Theta}_0 \frac{t}{t_o},$$

where the subscript o refers to the present epoch,
Whence,

$$\omega(t) \equiv \Theta - \Theta_o = \frac{\pi}{\tau_o t_o} t^2 \qquad (5.27)$$

Equation (5.27) gives the average perhelion precession at time 't'. Specializing to the case of Mercury, where $\tau_o = \frac{1}{4}$ year, it follows from (5.27) that the average precession per year at time 't' is given by

$$\omega(t) = \frac{4\pi t^2}{t_0} \qquad (5.28)$$

Whence, considering $\omega(t)$ for years $t = 1, 2, \cdots, 100$, we can obtain from (5.28), the usual total perhelion precession per century as,

$$\omega = \sum_{n=1}^{100} \omega(n) \approx 43'',$$

if the age of the universe is taken to be $\approx 2 \times 10^{10}$ years.
Conversely, if we use the observed value of the precession in (5.28), we can get back the above age of the universe.
Interestingly it can be seen from (5.28), that the precession depends on the epoch.
We next demonstrate that orbiting objects will have an anamolous inward radial acceleration.
Using the well known equation for Keplarian orbits,

$$\frac{1}{r} = \frac{GMm^2}{l^2}(1 + e\cos\Theta) \qquad (5.29)$$

$$\dot{r}^2 \approx \frac{GM}{r} - \frac{l^2}{m^2 r^2} \qquad (5.30)$$

l being the orbital angular momentum constant and e the eccentricity of the orbit, we can deduce such an extra inward radial acceleration, on differentiation of (5.30) and using (5.24) and (5.25),

$$a_r = \frac{GM}{2t_o r \dot{r}} \qquad (5.31)$$

It can be easily shown from (5.29) that (on the average),

$$\dot{r} \approx \frac{eGM}{rv} \qquad (5.32)$$

For a nearly circular orbit $rv^2 \approx GM$, whence use of (5.32) in (5.31) gives,

$$a_r \approx v/2t_o e \qquad (5.33)$$

For the earth, (5.33) gives an anomalous inward radial acceleration $\sim 10^{-9} cm/sec^2$, which is known to be the case [47].

We could also deduce a progressive decrease in the eccentricity of orbits. Indeed, e in (5.29) is given by

$$e^2 = 1 + \frac{2El^2}{G^2m^3M^2} \equiv 1 + \gamma, \gamma < 0.$$

Use of (5.24) in the above and differenciation, leads to,

$$\dot{e} = \frac{\gamma}{et_o} \approx -\frac{1}{et_o} \approx -\frac{10^{-10}}{e} \text{per year},$$

if the orbit is nearly circular. (Variations of eccentricity in the usual theory have been extensively studied (cf.ref.[48] for a review). On the other hand, for open orbits, $\gamma > 0$, the eccentricity would progressively increase.

We finally consider the anomalous accelerations given in (5.31) and (5.33) in the context of space crafts leaving the solar system.

If in (5.31) we use the fact that $\dot{r} \leq v$ and approximate

$$v \approx \sqrt{\frac{GM}{r}},$$

we get,

$$a_r \geq \frac{1}{et_o}\sqrt{\frac{GM}{r}}$$

For $r \sim 10^{14} cm$, as is the case of Pioneer 10 or Pioneer 11, this gives, $a_r \geq 10^{-11} cm/sec^2$

Interestingly Anderson et al.,[49] claim to have observed an anomalous inward acceleration of $\sim 10^{-9} cm/sec^2$.

6 Other Consequences

We could also explain the correct gravitational bending of light. Infact in Newtonian theory also we obtain the bending of light, though the amount is half that predicted by General Relativity[50, 51, 52]. In the Newtonian theory we can obtain the bending from the well known orbital equations (Cf.also(5.29)),

$$\frac{1}{r} = \frac{GM}{L^2}(1 + ecos\Theta) \tag{6.34}$$

where M is the mass of the central object, L is the angular momentum per unit mass, which in our case is bc, b being the impact parameter or minimum approach distance of light to the object, and e the eccentricity of the trajectory is given by

$$e^2 = 1 + \frac{c^2 L^2}{G^2 M^2} \tag{6.35}$$

For the deflection of light α, if we substitute $r = \pm\infty$, and then use (6.35) we get

$$\alpha = \frac{2GM}{bc^2} \tag{6.36}$$

This is half the General Relativistic value.
We next note that the effect of time variation of r is given by equation (5.25)(cf.ref.[53]). Using (5.25) the well known equation for the trajectory is given by (Cf.[54, 55])

$$u" + u = \frac{GM}{L^2} + u\frac{t}{t_0} + 0\left(\frac{t}{t_0}\right)^2 \tag{6.37}$$

where $u = \frac{1}{r}$ and primes denote differenciation with respect to Θ.
The first term on the right hand side represents the Newtonian contribution while the remaining terms are the contributions due to (5.25). The solution of (6.37) is given by

$$u = \frac{GM}{L^2}\left[1 + ecos\left\{\left(1 - \frac{t}{2t_0}\right)\Theta + \omega\right\}\right] \tag{6.38}$$

where ω is a constant of integration. Corresponding to $-\infty < r < \infty$ in the Newtonian case we have in the present case, $-t_0 < t < t_0$, where t_0 is large and infinite for practical purposes. Accordingly the analogue of the reception of light for the observer, viz., $r = +\infty$ in the Newtonian case is obtained by taking $t = t_0$ in (6.38) which gives

$$u = \frac{GM}{L^2} + ecos\left(\frac{\Theta}{2} + \omega\right) \qquad (6.39)$$

Comparison of (6.39) with the Newtonian solution obtained by neglecting terms $\sim t/t_0$ in equations (5.25),(6.37) and (6.38) shows that the Newtonian Θ is replaced by $\frac{\Theta}{2}$, whence the deflection obtained by equating the left side of (6.39) to zero, is

$$cos\Theta\left(1 - \frac{t}{2t_0}\right) = -\frac{1}{e} \qquad (6.40)$$

where e is given by (6.35). The value of the deflection from (6.40) is twice the Newtonian deflection given by (6.36). That is the deflection α is now given not by (6.37) but by the formula,

$$\alpha = \frac{4GM}{bc^2}, \qquad (6.41)$$

The relation (6.41) is the correct observed value and is the same as the General Relativistic formula.

We finally come to the problem of galactic rotational curves (cf.ref.[1]). We would expect, on the basis of straightforward dynamics that the rotational velocities at the edges of galaxies would fall off according to

$$v^2 \approx \frac{GM}{r} \qquad (6.42)$$

However it is found that the velocities tend to a constant value,

$$v \sim 300km/sec \qquad (6.43)$$

This as noted had lead to the postulation of as yet undetected additional matter, the so called dark matter.(However for an alternative view point Cf.[56]. We observe that from (5.25) it can be easily deduced that[57]

$$a \equiv (\ddot{r}_o - \ddot{r}) \approx \frac{1}{t_o}(t\ddot{r}_o + 2\dot{r}_o) \approx -2\frac{r_o}{t_o^2} \qquad (6.44)$$

as we are considering infinitesimal intervals t and nearly circular orbits. Equation (6.44) shows (Cf.ref[53] also) that there is an anomalous inward

acceleration, as if there is an extra attractive force, or an additional central mass, as indeed we saw a little earlier.
So,

$$\frac{GMm}{r^2} + \frac{2mr}{t_o^2} \approx \frac{mv^2}{r} \qquad (6.45)$$

From (6.45) it follows that

$$v \approx \left(\frac{2r^2}{t_o^2} + \frac{GM}{r}\right)^{1/2} \qquad (6.46)$$

From (6.46) it is easily seen that at distances within the edge of a typical galaxy, that is $r < 10^{23} cms$ the equation (6.42) holds but as we reach the edge and beyond, that is for $r \geq 10^{24} cms$ we have $v \sim 10^7 cms$ per second, in agreement with (6.43).

Thus the time variation of G explains observation without invoking dark matter. It may also be mentioned that other effects like the Pioneer anomaly and shortening of the period of binary pulsars can be deduced [58], while new effects also are predicted.

7   The MOND Approach

Milgrom [59] approached the problem by modifying Newtonian dynamics at large distances. This approach is purely phenomenological. The idea was that perhaps standard Newtonian dynamics works at the scale of the solar system but at galactic scales involving much larger distances perhaps the situation is different. However a simple modification of the distance dependence in the gravitation law, as pointed by Milgrom would not do, even if it produced the asymptotically flat rotation curves of galaxies. Such a law would predict the wrong form of the mass velocity relation. So Milgrom suggested the following modification to Newtonian dynamics: A test particle at a distance r from a large mass M is subject to the acceleration a given by

$$a^2/a_0 = MGr^{-2}, \qquad (7.47)$$

where a_0 is an acceleration such that standard Newtonian dynamics is a good approximation only for accelerations much larger than a_0. The above equation however would be true when a is much less than a_0. Both the statements can be combined in the heuristic relation

$$\mu(a/a_0)a = MGr^{-2} \qquad (7.48)$$

In (7.48) $\mu(x) \approx 1$ when $x >> 1$, and $\mu(x) \approx x$ when $x << 1$. It must be stressed that (7.47) or (7.48) are not deduced from any theory, but rather are an ad hoc fit to explain observations. Interestingly it must be mentioned that most of the implications of MOND do not depend strongly on the exact form of μ.

It can then be shown that the problem of galactic velocities is solved [60, 61, 62, 63, 64].

8 Interrelationship

It is interesting to note that there is an interesting relationship between the varying G approach, which has a theoretical base and the purely phenomenological MOND approach. Let us write

$$\beta \frac{GM}{r} = \frac{r^2}{t_0^2} \text{ or } \beta = \frac{r^3}{GMt_0^2}$$

Whence

$$\alpha_0 = v^2/r = \frac{GM}{r^2} \quad \alpha = \frac{r}{t_0^2}$$

So that

$$\frac{\alpha}{\alpha_0} = \frac{r^3}{GMt_0^2} = \beta$$

At this stage we can see a similarity with MOND. For if $\beta << 1$ we are with the usual Newtonian dynamics and if $\beta > 1$ then we get back to the varying G case exactly as with MOND.

Furthermore, as can be seen from (3.6), when the dimensionality n gets smaller than 3, effectively G starts falling off as in the time varying case seen in sections 5 and 6. Roughly, the dimensionality falls with increasing distance, and distance increases with time as the universe expands.

9 Discussion

It is interesting to note again that the varying G approach leads to several observations that have been carried out including the precession of the perihelion of the planets, in particular Mercury, the Pioneer anomaly, the shortening of the orbital periods of binary pulsars and so on [53, 58, 29, 30]. A further interesting observation as noted is the fact observed by Milgrom that there is a curious coincidence in MOND viz.,

$$a_0 \sim H(\sim 10^{-7} cmsec^{-2})(\sim c^2/R) \tag{9.49}$$

where H is the Hubble constant. In fact this follows from the varying G theory. For, we have in this case from (5.31), (5.32), (5.33) and (6.46),

$$a_0 \sim r/t_0^2$$

Feeding the values of r, the radius of the universe $= ct_0$ and the fact that $H \sim \frac{1}{t_0}$, we get (9.49), which now shows up no longer as an ad hoc coincidence but rather as a consequence of the theory. Further, a particle obeying Newtonian dynamics, which has the acceleration (9.49) over the life time of the universe, attains the velocity of light and moreover covers a distance equalling the size of the universe.

It may be noted that the Boomerang results are in tune with MOND rather than the Dark Matter scenario, the WMAP model notwithstanding [65].

A final remark - the variation of G discussed in Sections 5 and 6, show that there is an inward acceleration in gravitationally bound systems - this would imply that such systems (galaxies included) would tend to become progressively smaller, as with binary pulsar orbits - in the absence of other dynamical considerations.

References

[1] Narlikar, J.V., "Introduction to Cosmology", Cambridge University Press, Cambridge, 1993, p.57.

[2] Wilczek, F., Physics Today, January 1999, p.11.

[3] Lee, T.D., "Particle Physics and Introduction to Field Theory", Harwood Academic, 1981, pp.391ff.

[4] De Pena, L., and A.M. Cetto, A.M., Found of Phys., Vol.12, No.10, 1982, p.1017-1037.

[5] Milonni, P.W., "The Quantum Vacuum: An Introduction to Quantum Electrodynamics", Academic Press, San Diego, 1994.

[6] Itzykson, C., and Zuber, J., "Quantum-Field Theory", Mc-Graw Hill, New York, 1980, p.139.

[7] Cole, D.C., in "Essays on the Formal Aspects of Electromagnetic Theory", Ed. Lakhtakia, A., World Scientific, Singapore, 1993, pp.501ff.

[8] Podolny, R., "Something Called Nothing", Mir Publishers, Moscow, 1983.

[9] Misner, C.W., Thorne, K.S., and Wheeler, J.A., "Gravitation", W.H. Freeman, San Francisco, 1973, pp.819ff.

[10] Zeldovich, Ya. B., JETP Lett. 6, 316, 1967.

[11] Weinberg, S., Phys.Rev.Lett., 43, 1979, p.1566.

[12] Santos, E., "Stochastic Electrodynamics and the Bell Inequalities" in "Open Questions in Quantum Physics", Ed. G. Tarozzi and A. van der Merwe, D. Reidel Publishing Company, 1985, p.283-296.

[13] De Pena, L., "Stochastic Processes applied to Physics...", Ed., B Gomez, World Scientific, Singapore, 1983.

[14] Mostepanenko, V.M., and Trunov, N.N., Sov.Phys.Usp. 31(11), November 1988, p.965-987.

[15] Lamoreauz, S.K., Phys.Rev.Lett., Vol.78, No.1, January 1997, p.5-8.

[16] Petroni, N.C., and Vigier, J.P., Foundations of Physics, Vol.13, No.2, 1983, p.253-286.

[17] Raiford, M.T., Physics Today, July 1999, p.81.

[18] Milonni, P.W., Physica Scripta. Vol. T21, 1988, p.102-109.

[19] Milonni, P.W., and Shih, M.L., Am.J.Phys., 59 (8), 1991, p.684-698.

[20] Lee, T.D., "Statistical Mechanics of Quarks and Hadrons", Ed. H. Satz, North-Holland Publishing Company, 1981, p.3ff.

[21] Hushwater, V., Am.J.Phys. 65(5), May 1997, p.381–384.

[22] Achuthan, P., et al, in "Gravitation, Quanta and the Universe", Ed. A.R. Prasanna, J.V. Narlikar, C.V. Vishveshwara, Wiley Eastern, New Delhi, 1980, p.300.

[23] Sidharth, B.G., Found.Phys.Lett., 19 (1), 2006, pp.87ff.

[24] Sidharth, B.G., and A.D. Popova, Nonlinear World, 1997, 4.

[25] Popova, A.D., Astron. and Astroph. Trans. 5 (1994), 31.

[26] Popova, A.D., Astron. and Astroph. Trans. 6 (1995), 165.

[27] Ehrenfest, P., Proc. Amsterdam Acad. 20 (1917), 200; Ann. Phys. 61 (1920), 440.

[28] Barrow, J.D., and Parsons, P., Phys.Rev.D., Vol.55, No.4, 15 February 1997, p.1906ff.

[29] Sidharth, B.G., "Chaotic Universe: From the Planck to the Hubble Scale", Nova Science Publishers, Inc., New York, 2001.

[30] Sidharth, B.G., "The Universe of Fluctuations", Springer, Dordrecht, 2005.

[31] Sidharth, B.G., Chaos, Solitons and Fractals, 12, 2001, 1369-1370.

[32] Burstein, D., and Rubin, V.S., Astroph. J., 297 (1985), 423.

[33] Kormendy, J., and Knapp, G. (eds.), "Dark Matter in the Universe", Reidel, Dordrecht, 1987.

[34] Davies, M., et al., Astroph. J. 238 (1980), L113.

[35] Turner, S., Astroph. J. 208 (1976), 304.

[36] Sidharth, B.G., Int.J.Mod.Phys.A, 13 (15), 1998, p.2599ff.

[37] Sidharth, B.G., Int.J.Th.Phys., 37 (4), 1998, p.1307ff.

[38] Sidharth, B.G., Chaos, Solitons and Fractals, 16 (4), 2003, pp.613-620.

[39] Nottale, L., "Fractal Space-Time and Microphysics: Towards a Theory of Scale Relativity",World Scientific, Singapore, 1993, p.312.

[40] Hayakawa, S., Suppl of PTP Commemmorative Issue, 1965, 532-541.

[41] Huang, K., "Statistical Mechanics", Wiley Eastern, New Delhi 1975, pp.75ff.

[42] Sidharth, B.G., in "Frontiers of Quantum Physics", Eds., Lim, S.C., et al, Springer Verlag, Singapore, 1998.

[43] Sidharth, B.G., Proc. of the Eighth Marcell Grossmann Meeting on General Relativity, Ed. T. Piran, World Scientific, Singapore, 1999, p.476-479.

[44] Weinberg, S., "Gravitation and Cosmology", John Wiley & Sons, New York, 1972, p.62.

[45] Smolin, L., "The Trouble with Physics", Houghton Miffin, New York, 2006, pp.209ff.

[46] Sidharth, B.G., Found.Phys.Lett., 19 (5), 2006, pp.499-500.

[47] Kuhne, R.W., xxx.lanl.gov/gr-qc/9809075.

[48] Berger, A.L., Astronomy and Astrophysics, 51, 1976, p.127-135.

[49] Anderson, J.L. et al., xxx.lanl.gov/gr-qc/9808081.

[50] Denman, H.H., Am.J.Phys. 51(1), 1983, 71.

[51] Silverman, M.P., Am.J.Phys. 48, 1980, 72.

[52] Brill, D.R. and Goel, D., Am.J.Phys. 67(4), 1999, 317.

[53] Sidharth, B.G., Nuovo Cimento, 115B (12) (2), 2000, pp.151ff.

[54] Bergman, P.G., "Introduction to the Theory of Relativity", Prentice-Hall (New Delhi), 1969, p248ff.

[55] Lass, H., "Vector and Tensor Analysis", McGraw-Hill Book Co., Tokyo, 1950, p295 ff.

[56] Sivaram, C., Sivaram and V. de Sabbata, Foundations of Physics Letters, 6 (6), 1993.

[57] Sidharth, B.G., Chaos, Solitons and Fractals, 12, 2001, 1101-1104.

[58] Sidharth, B.G., Found.Phys.Lett., 19 (6), 2006, 611-617.

[59] Milgrom, M., "MOND - A Pedagogical Review" Presented at the XXV International School of Theoretical Physics "Particles and Astrophysics-Standard Models and Beyond", Ustron, Poland, September 2001.

[60] Milgrom, M., APJ 270 (1983), 371.

[61] Milgrom, M., APJ 302 (1986), 617.

[62] Milgrom, M., Comm. Astrophys. 13:4 (1989), 215.

[63] Milgrom, M., Ann.Phys. 229 (1994), 384.

[64] Milgrom, M., Phys.Rev. E 56 (1997), 1148.

[65] McGauge, S.S., APJ Lett. 523 (1999), L99.

A Quantum Theory Friendly Cosmology Exact Gravitational Waves

James G. Gilson   j.g.gilson@qmul.ac.uk
School of Mathematical Sciences
Queen Mary University of London
Mile End Road London E14NS
May 1, 2008

Abstract

In this paper, it is shown that the cosmological model that was introduced in a sequence of three earlier papers under the title *A Dust Universe Solution to the Dark Energy Problem* can be expressed in a form which is quantum theory friendly. That is to say, besides not have a cosmological constant problem and also not having a coincidence problem, aspects dealt with in earlier papers and continued in the first part of this paper, it is shown that the dust universe can be expressed in a form having a close resemblance to the Schrödinger equation formalism. This resemblance cannot be seen as an identity of the two systems because the Schrödinger equation is linear and the Friedman equations are non-linear. This aspect is discussed in detail and a precise relation is shown to exist and is demonstrated to hold between cosmology theory structure and the quantum theory linear superposition of eigen-states. This relation describes cosmology's non-linearity relative to Schrödinger linearity and is called, bilinear superposition. However, in spite of not achieving an identity of structure between cosmology and quantum theory, sufficient equivalence can be shown to exist via a comparison of quantum wave motion as described by the Schrödinger equation and gravitational wave motion as described by the Friedman dust universe to suggest that a quantum theory of cosmology and gravity is likely to be possible via this route. An exact *non-linear* Schrödinger equation description for the model is obtained. In this paper's appendix, it is shown that this Schrödinger equation has an infinite multiplicity of space variable solutions that can be used to remove the usual restriction of cosmology theory to uniform space variation with dependence on epoch time only.

217

1 Introduction

The work to be described in this paper is an application of the cosmological model introduced in the papers *A Dust Universe Solution to the Dark Energy Problem* [23], *Existence of Negative Gravity Material. Identification of Dark Energy* [24] and *Thermodynamics of a Dust Universe* [32] together with applications of those papers to the cosmological constant problem, the cosmological coincidence problem and other subsidiary cosmological problems.

The conclusions arrived at in those papers was that the dark energy *substance* is physical material with a positive density, as is usual, but with a negative gravity, -G, characteristic and is twice as abundant as has usually been considered to be the case. References to equations in those papers will be prefaced with the letter A, B and C respectively. The work in A, B and C, and the application here have origins in the studies of Einstein's general relativity in the Friedman equations context to be found in references ([16],[22],[21],[20],[19],[18],[4],[23]) and similarly motivated work in references ([10],[9],[8],[7],[5]) and ([12],[13],[14],[15],[7],[25],[3]). Other useful sources of information are ([17],[3],[30],[27],[29],[28]) with the measurement essentials coming from references ([1],[2],[11]). Further references will be mentioned as necessary. The application of the cosmological model introduced in the papers A [23], B,[24] and C [32], in paper E, ([36]), is to the extensively discussed and analysed *Cosmological Coincidence Problem*. In the paper D, [34], it was shown that the quantum vacuum polarisation idea can be seen to play a central role in the Friedman dust universe model introduced by the author. In the paper, [40], it was shown that the Friedman equation structure can be converted into a *non-linear* Schrödinger equation structure. Here, this aspect is further developed by supplementing the solutions to this time only equation with a dependence on a three dimensional space position vector, $\mathbf{r}$, so that the equation remains consistent with its cosmological origin. This step then enables finding cosmological models that are not restricted to having a mass density that is certainly time dependent but otherwise remains constant over all *three dimensional* position space at every definite time.

Altogether, the objective has been to produce an alternative to the standard model which contains less paradoxical structure than does the standard model and which at the same time is hopefully adaptable to being quantized in some sense or other. The question of in *what sense is the model to be quantized*, I see to be an open question which may or may not have a unique answer and to this issue discussion will here be devoted. My strategy is to mould the cosmological model of the Friedman dust universe into a form that has a structure as near as possible to the structure of Schrödinger quantum theory by emphasising a wave motion aspect of the dust universe Friedman model. This will be explained in detail in Section **5** which is devoted to consideration of the well known essentials of Schrödinger quantum theory that need somehow to be present in the cosmological model. The basic version of this dust universe model is described by a sphere in three dimensional Euclidean space with a changing with epoch time, t, radius magnitude, r(t),

$$r(t) = b \sinh^{2/3}(\pm 3ct/(2R_\Lambda)) \tag{1.1}$$
$$b = (R_\Lambda/c)^{2/3}C^{1/3} \tag{1.2}$$
$$R_\Lambda = (3/\Lambda)^{1/2} \tag{1.3}$$
$$C = 2M_U G. \tag{1.4}$$

R_Λ is often called the de Sitter radius and C is Rindler's constant formed from twice the product of the conserved constant mass, M_U, within the total changing volume of the universe and the Newtonian gravitational constant, G. Λ is Einstein's cosmological constant. $r(t)$ is rigorously a solution to the Friedman equations and consequently also rigorously a solution to Einstein's field equation's which holds for epoch time from $t = -\infty$ to $+\infty$. The $\pm$ can usually be omitted provided the cube root of the sinh function is assumed taken after squaring so that radius, $r(t)$, is time symmetry invariant, $r(t) = r(-t)$ and no complex roots are involved. The formulae (1.1)$\rightarrow$(1.4) contain all the basic mathematical-theoretical information about the Friedman dust universe model involved in this research program. For example, the Hubble *function*, $H(t)$, the conserved mass density, $\rho(t)$, and Einstein's dark energy density, ρ_Λ, are given by

$$H(t) = \dot{r}(t)/r(t) = (c/R_\Lambda)\coth(3ct/(2R_\Lambda)) \tag{1.5}$$
$$\rho(t) = 3M_U/(4\pi r^3(t)) = \rho_\Lambda \sinh^{-2}(3ct/(2R_\Lambda)) \tag{1.6}$$
$$\rho_\Lambda = \Lambda c^2/(8\pi G). \tag{1.7}$$

The main objective of this paper is to demonstrate that the cosmological model described above can be seen to be quantum theory-friendly. That is to say it can be physically and numerically expressed so that it is not in conflict with quantum theory. The spade work for this has been done in the application papers ([34], [36]) in which it was shown that firstly the famous *cosmological constant problem* does not arise in this model and secondly the equally famous *cosmological coincidence problem* can be removed from the structure of this model, if care is taken in the use of astronomical measurements. The form of the well known cosmological coincidence problem that occurs in this model takes what I call a *critical* form because it involves the integer 2 in the result $t_0 = 2t_c$, where it was thought that t_0 should correspond to *time now* and t_c is a definitely fixed time when the universe's radial acceleration is zero. However, as time now for an observation depends when the observation is made and is in that sense variable, it is difficult to see how $t_0 = 2t_c$ can be a universal result. Thus it is more rational to define t_0 as the definite value of epoch time when the ratio of the quantity of *dark energy mass* certainly within the universe to the quantity of conserved mass which is for all time within the universe can be thought to have the value $3 = \Omega_\Lambda/\Omega_{M_U}$ as identified in terms of the Ωs by astronomical observations at time now which has also commonly been called t_0. These astronomical measurements are displayed next. The accelerating universe astronomical observational workers [1] give measured values of the three Ωs, and w_Λ to be

$$\Omega_{M,0} = 8\pi G\rho_0/(3H_0^2) = 0.25^{+0.07}_{-0.06} \tag{1.8}$$

$$\Omega_{\Lambda,0} = \Lambda c^2/(3H_0^2) = 0.75^{+0.06}_{-0.07} \tag{1.9}$$

$$\Omega_{k,0} = -kc^2/(r_0^2 H_0^2) = 0, \Rightarrow k = 0 \tag{1.10}$$

$$w_\Lambda = P_\Lambda/(c^2\rho_\Lambda) = -1\pm \approx 0.3. \tag{1.11}$$

I abandon the use of t_0 to represent time now or the rather vague time when the measurements were made and represent time now by the symbol $t^\dagger$ which still remains vague but for the purpose of theoretical discussion can taken to be the time of the present moment. The time t_0 will be used to represent the much less vague time when the universe passes through the centre value of the measurement range. I intend to re-express the second equalities above but before making that step it necessary to give a more detailed account of what the measurements above mean in their initial form in relation to the form I shall replace them by. This question of meanings and relations follows in a subsection.

1.1 Cosmological Epoch and Terrestrial Time

For clarity I now rewrite the first two equations, (1.8) and (1.9) in terms of the time, t_0,

$$\Omega_M(t_0) = 8\pi G\rho(t_0)/(3H^2(t_0)) = 0.25^{+0.07}_{-0.06} \qquad (1.12)$$
$$\Omega_\Lambda(t_0) = \Lambda c^2/(3H^2(t_0)) = 0.75^{+0.06}_{-0.07}, \qquad (1.13)$$

where the Hubble function given at (1.5) is used. The first equalities in these equations define definite $\Omega(t_0)$ functions of time, whereas the second two equalities say that known functions of t_0 lie within definite numerical ranges. These second equalities in (1.12) and (1.13) can be usefully rewritten as follows

$$\frac{3 \times 0.19}{8\pi G} < \frac{\rho(t_0)}{H^2(t_0)} < \frac{3 \times 0.32}{8\pi G} \qquad (1.14)$$

$$\frac{3 \times 0.68}{\Lambda c^2} < \frac{1}{H^2(t_0)} < \frac{3 \times 0.81}{\Lambda c^2}. \qquad (1.15)$$

Inverting these equations, we have

$$\frac{8\pi G}{3 \times 0.19} > \frac{H^2(t_0)}{\rho(t_0)} > \frac{8\pi G}{3 \times 0.32} \qquad (1.16)$$

$$\frac{\Lambda c^2}{3 \times 0.68} > H^2(t_0) > \frac{\Lambda c^2}{3 \times 0.81} \qquad (1.17)$$

and using, (1.6) and (1.7), these equations can be converted to

$$\frac{1}{0.19} > \cosh^2(3ct_0/(2R_\Lambda)) > \frac{1}{0.32} \qquad (1.18)$$

$$\frac{1}{0.68} > \coth^2(3ct_0/(2R_\Lambda)) > \frac{1}{0.81}. \qquad (1.19)$$

It follows that these two equations are saying the same thing because

$$\coth^2(3ct_0/2R_\Lambda) = \frac{\cosh^2(3ct_0/(2R_\Lambda))}{\cosh^2(3ct_0/(2R_\Lambda)) - 1}. \qquad (1.20)$$

According to the measurement (1.9) information the universe will pass through the centre of the Ω_Λ values at some time, t_0, say, given by

$$3ct_0/(2R_\Lambda) = \coth^{-1}((1/0.75)^{1/2}) \qquad (1.21)$$
$$= \coth^{-1}(2/3^{1/2}) = \cosh^{-1}(2) \qquad (1.22)$$
$$t_0 = (2R_\Lambda/3c)\cosh^{-1}(2). \qquad (1.23)$$

221

From (1.23) it is clear that we cannot find a numerical value for the special time t_0 unless we can find a numerical value for R_Λ and this is equivalent to knowing the numerical value for $\Lambda = (3/R_\Lambda)^{1/2}$. However, in the very unlikely special case of coincidence, when $t^\dagger = t_0$ we can calculate R_Λ because the relation first displayed below implies the second and then the third followed by the definite special case value for t_0 at fourth place.

$$t^\dagger = (2R_\Lambda/(3c))coth^{-1}(R_\Lambda H^\dagger/c) \tag{1.24}$$
$$= t_0 = (2R_\Lambda/(3c))cosh^{-1}(2) \tag{1.25}$$
$$R_\Lambda = 2c/(3^{1/2}H^\dagger) \approx 1.48353 \times 10^{26}. \tag{1.26}$$
$$t_0 = 4.35 \times 10^{17} \ s \approx 4.756 \times 10^{11} \ yr. \tag{1.27}$$

In fact, the value of t_0 given at (1.27) is the theoretically given value mentioned earlier for the time when $\Omega_\Lambda/\Omega_{M_U} = 3$, an event that occurs inevitably, a result independent of measurement. From the third equality above and $t_0 \neq t^\dagger$ we get the general result

$$R_\Lambda = 3ct_0 \coth^{-1}(3^{1/2}) = (3ct_0/2)\cosh^{-1}(2). \tag{1.28}$$

There is an important lesson from the general result (1.28) which is that if t_0 is determined in value then so is $R_\Lambda = (3/\Lambda)^{1/2}$ or Λ and visa versa. The time $t_{0,min}$ when the time t_0 is at the lower measurement value and the time $t_{0,max}$ when the time t_0 is at the higher measurement value are given, using (1.19) and the fact that in general R_Λ is to be determined, by

$$t_{0,min} = \left(\frac{2R_\Lambda}{3c}\right) \coth^{-1}((1/0.81)^{\frac{1}{2}}) \approx 3.8634 \times 10^{17} \ s \tag{1.29}$$

$$t_{0,max} = \left(\frac{2R_\Lambda}{3c}\right) \coth^{-1}((1/0.68)^{\frac{1}{2}}) \approx 4.8568 \times 10^{17} \ s \tag{1.30}$$

$$t_{0,mean} = (t_{0,min} + t_{0,max})/2 \approx 4.3601 \times 10^{17} \ s \tag{1.31}$$
$$t_{0,mean} - t_0 \approx 10^{15} \ s \approx 3.17 \times 10^7 \ yr \tag{1.32}$$
$$t_{0,mean}/t_0 \approx 1.00232. \tag{1.33}$$

Thus the length of the time range between which time , t_0, viewed as a variable over the measurement range, can be expected to be found is given by

$$t_{0,max} - t_{0,min} = 0.9934 \times 10^{17} \ s \approx 3.15 \times 10^9 \ yr \tag{1.34}$$
$$t_{0,max}/t_{0,min} = 4.8568/3.863 \approx 1.2572 \tag{1.35}$$
$$t_{0,max}/t_0 = 4.8568/4.756 \approx 1.02119. \tag{1.36}$$

The quantities $t_{0,min}$ and $t_{0,max}$ are here taken to be the lower and upper time limits associated with the measurements of the omegas given at equations (1.12) and (1.13). In all the evaluations of times above, R_Λ has been given the special case coincidence value. In the non-coincidence general case it would have a value different from this but this value could only be determined by some new or other experimental procedure. The coincidence value has been used just to give some idea of the various bounds of the quantities involved quantities. From these equations assumed to hold at a conceptual time, t_0, when the universe passes through the centre value of the measurement ranges, we get the exact formulae,

$$t_0 = (2R_\Lambda/(3c))\cosh^{-1}(2) \tag{1.37}$$

$$R_\Lambda = 3ct_0/(2\cosh^{-1}(2)) \tag{1.38}$$

$$t_c = (t_0\cosh^{-1}(2))\coth^{-1}(3^{1/2}) \tag{1.39}$$

$$t_0/t_c = \cosh^{-1}(2)/\coth^{-1}(3^{1/2}) = 2. \tag{1.40}$$

Having found R_Λ in terms of t_0, this value of R_Λ can be substituted into the formula for Hubble's constant, (1.41), to find the value of the *time now*, $t^\dagger$.

$$H(t^\dagger) = (c/R_\Lambda)\coth(3ct^\dagger/(2R_\Lambda)) \tag{1.41}$$

$$t^\dagger = (2R_\Lambda/(3c))\coth^{-1}(R_\Lambda H^\dagger/c) \tag{1.42}$$

$$= \left(\frac{t_0}{\cosh^{-1}(2)}\right)\left(\coth^{-1}\left(\frac{3t_0 H^\dagger}{2\cosh^{-1}(2)}\right)\right) \tag{1.43}$$

$$= \left(\frac{2t_c}{\cosh^{-1}(2)}\right)\left(\coth^{-1}\left(\frac{6t_c H^\dagger}{2\cosh^{-1}(2)}\right)\right), \tag{1.44}$$

where $H^\dagger = H(t^\dagger)$ is the present day measured value of Hubble's constant. Equations (1.43) or (1.44) is essentially the solution to the coincidence problem. If we write (1.44) in the form

$$t^\dagger/t_c = \left(\frac{2}{\cosh^{-1}(2)}\right)\left(\coth^{-1}\left(\frac{6t_c H^\dagger}{2\cosh^{-1}(2)}\right)\right) \tag{1.45}$$

$$t^\dagger/t_c = 2f(2t_c), \tag{1.46}$$

where $f(2t_c)$ gives the deviation of the ratio $t^\dagger/t_0$ from the value unity and removes the degeneracy. Expressed in another way it is the multiplicative function that breaks the coincidence at (1.40) and converts the integer 2 to a much less notable non integral value. However, we can give the formulae

223

(1.45) and (1.46) together an interpretation in terms of the uncertainties of the measurement process. This is achieved by defining the measurement *deviation* function $d_{meas}(t_0)$ as follows,

$$d_{meas}(t_0) = t^\dagger/t_0 - f(t_0) \tag{1.47}$$

$$f(t_0) = \left(\frac{1}{\cosh^{-1}(2)}\right)\left(\coth^{-1}\left(\frac{3t_0 H^\dagger}{2\cosh^{-1}(2)}\right)\right). \tag{1.48}$$

The function (1.47) is a dimensionless measure of how much the central Ω values from astronomy assumed to have occurred at t_0 differ from the time now measurement from the Hubble variable quantity $H(t^\dagger)$ taken at time now, $t^\dagger$. It is sufficient to assume that the event at t_0 is still yet to occur, $t_0 > t^\dagger$, then we see that the function d_{meas} passes through zero when the full degeneracy holds at $t_0 = t^\dagger$ and it has a maximum at $t_0 \approx 0.643 \times 10^{18} s$ when $t^\dagger$ and t_0 assume the approximate maximum deviation, 0.17. When $t_0 = 0.643 \times 10^{18}$, $t^\dagger$ can be assumed constant at the coincidence value 4.34467×10^{17} so that the maximum deviation times ratio is $t^\dagger/t_0 \approx 0.43467/0.643 \approx 0.6757$ or

$$t^\dagger = 0.6757 t_0. \tag{1.49}$$

It follows that $t^\dagger$, the time now value, can vary from t_0 down to a value of $t^\dagger \approx 0.6757 t_0 = 1.3514 t_c$. Thus the coincidence is decisively removed with $t^\dagger \neq t_0 = 2t_c$. This calculation is based on the assumption that the true physical value of the Hubble function at time $t^\dagger$, $H(t^\dagger)$, is the measured *central* value $H^\dagger$ used in the formulae (1.43) and (1.44). However, $H(t^\dagger)$ could have any value in its measurement range which according to *W. Freedman*, [37], is

$$72 \pm 8 \ Kms^{-1}Mpc^{-1} \approx (2.33 \pm 0.25) \times 10^{-18} \ s^{-1}. \tag{1.50}$$

This implies that the quantity $H^\dagger$ used in the formulae (1.43) and (1.44) could have values in inverse seconds in the range

$$H^\dagger_{min} = 2.07 \times 10^{-18} < H^\dagger < 2.58 \times 10^{-18} = H^\dagger_{max}. \tag{1.51}$$

the quantities $H^\dagger_{min}$ and $H^\dagger_{max}$ defined at equation (1.51) can then be used to produce two further versions of equations (1.47) and (1.48) referring here to the limit end points of the Hubble measurement range

2 Coincidence Deviations Diagram

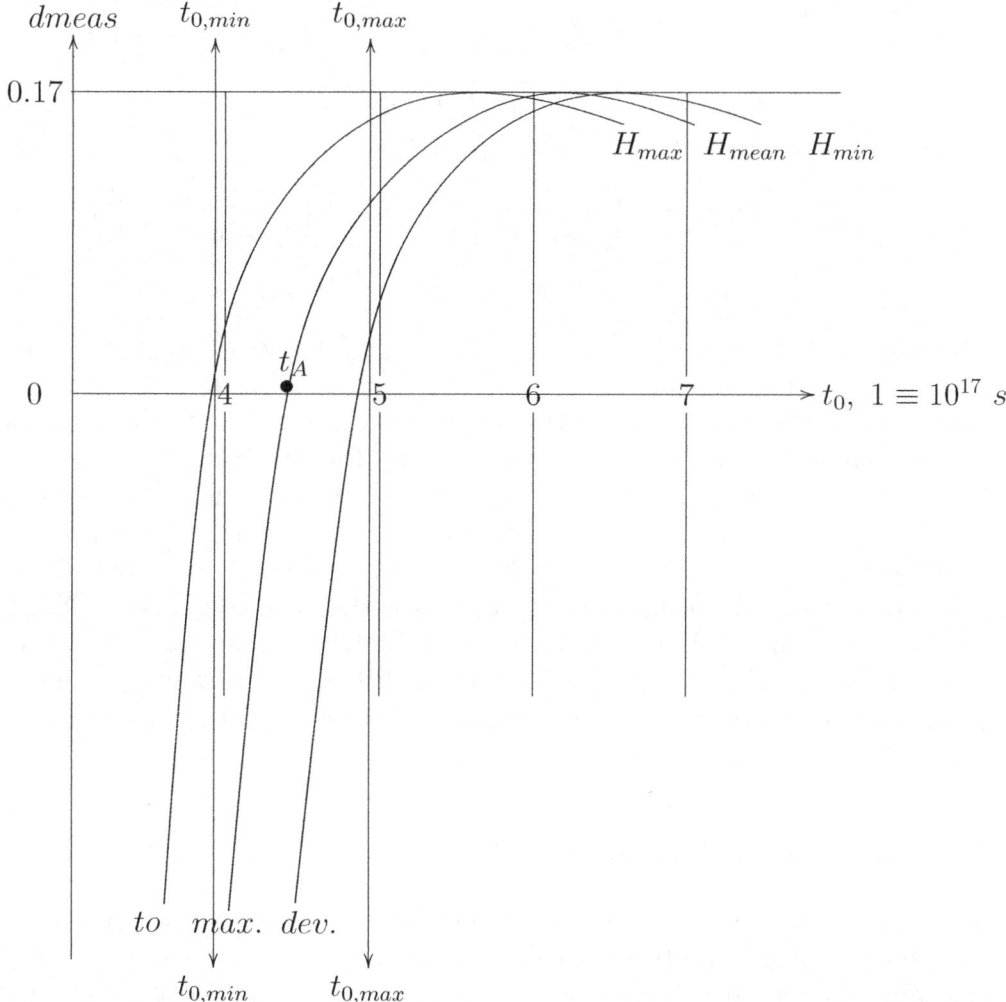

Max. neg. deviation where curve $H_{min} \to dev.$ meets abscissa $t_{0,min}$
Max. pos. deviation where curve $H_{max} \to to$ meets abscissa $t_{0,max}$
t_A astronomers' consensus universe's age, dmeas within bullet

$$d_{meas,min}(t_0) = t^\dagger_{min}/t_0 - f_{min}(t_0) \tag{2.52}$$

$$f_{min}(t_0) = \left(\frac{1}{\cosh^{-1}(2)}\right)\left(\coth^{-1}\left(\frac{3t_0 H^\dagger_{min}}{2\cosh^{-1}(2)}\right)\right) \tag{2.53}$$

$$t^\dagger_{min} = 4cosh^{-1}(2)/(3 \times 3^{1/2} H^\dagger_{min}) \tag{2.54}$$

$$d_{meas,max}(t_0) = t^\dagger_{max}/t_0 - f_{max}(t_0) \tag{2.55}$$

$$f_{max}(t_0) = \left(\frac{1}{\cosh^{-1}(2)}\right)\left(\coth^{-1}\left(\frac{3t_0 H^\dagger_{max}}{2\cosh^{-1}(2)}\right)\right) \tag{2.56}$$

$$t^\dagger_{max} = 4\cosh^{-1}(2)/(3 \times 3^{1/2} H^\dagger_{max}) \tag{2.57}$$

All three of the function $d_{meas,min}(t_0)$, $d_{meas}(t_0)$ and $d_{meas,max}(t_0)$ have a maximum deviation value a ≈ 0.17 at a value of t_0 appropriate for the function in question. However, the range of the variable t_0 in these functions is constrained by the lower and upper limits for the variable t_0, $t_{0,min}$ and $t_{0,max}$ given earlier, (1.29) and (1.30). The maximum value of the deviations with the measurements available lie outside the limits. Thus the actual positive deviations that seem to be possible under these constraints are reduced numerically to parts of the curves within the constraints. Negative deviations occur of about -0.34, and -0.51 which refer to the case when t_0 occurs before $t^\dagger$, a type of situation not discussed in previous work. The results obtained above can now be used to discuss in section (3) what I shall call the *time-now problem*.

3   Time-Now Problem

The time quantities t_0, time when the Omegas were measured, and $t^\dagger$, time now, have conventionally been taken to be the same physical quantity with the same numerical value. As I have shown above, this led to the so called *cosmological coincidence problem*. Both these time quantities have been regarded as representing *time now* and this in spite of the fact that the two separate measurements involved occurred at distinctly *different* terestrial times. Consequently, the explanation for the conceptual mistake involved in generating the cosmological coincidence problem is worth further discussion as it throws light on the nature of and reason for the coincidence difficulty that for years has seemed to be so intractable. I suggest the problem arises from the nature of *epoch time keeping* in cosmology which involves enormously large numerical values such as 10^{10} years and also from the relation between *epoch* time and *terrestrial* time. It seems that the equality of these two, very distinct time keeping systems, has been taken for granted. We can analyse these issues using the information about the measurements and the times when they occurred using the two quantities $\Omega_\Lambda(t)$ and $H(t)$. In the theoretical cosmology structure that I have been using, both of the quantities are explicit functions of the time variable t and this time variable is the epoch time that occurs in the Friedman theory and derives from Einstein's field equations. However, the astronomers refer to the closely related object pair $\Omega_{\Lambda,0}$ and H_0 as the measured values of these quantities at the time of measurement or *roughly* speaking at time now. This is indicated by the zero subscript and essentially means *time now*. The time concept that is being used by the astronomers in this context is terrestrial time, time measured by earthbound clocks of some sort or other, adjusted to some running numerical value such as Greenwich mean time. Once this competing time forms situation is recognised a whole collection of uncertainties are released into cosmological theory arena. To analyse this *conflicting times* situation between cosmological time and terrestrial time it is necessary to bring some precision into the definitions of the quantities under discussion. In particular, the idea of time now that I have previously denoted by $t^\dagger$ has always been a very vague concept. For one thing the term *time now* implies some variable like character to the value of its symbol in the sense that time now used today is numerically less than time now when used tomorrow. The reason for this vague use in cosmology is obviously due to the fact that on the cosmological scale very little will have appeared to have changed between today and tomorrow or indeed between 10 years ago or 10 years into the future and this is directly a result of the numerically large numbers associated with time passage in cosmology. This vagueness about the time now idea, I think, has been a major contributor to the time coincidence problem. For this reason I shall now abandon the use of the symbol $t^\dagger$ as representing the vague time now and firmly only use it to represent the time that is to be found from the measured central value, $H^\dagger$, of the Hubble variable from the theoretical form of that parameter, (1.5)

$$H^\dagger = H(t^\dagger) = (c/R_\Lambda)\coth(3ct^\dagger/(2R_\Lambda)) \qquad (3.1)$$
$$t^\dagger = (2R_\Lambda/(3c))\coth^{-1}(R_\Lambda H^\dagger/c). \qquad (3.2)$$

Thus mathematically nothing has changed. However, I have abandoned the terminology *time now* for the symbol $t^\dagger$ which is now to be firmly associated only with the cosmological time giving the measured value of $H(t)$. I now rename $t^\dagger$ as the 2001-Hubble measurement epoch time or briefly 1Hmeastime. This time is clearly now a fixed constant and indeed one of the very large numbers that occurs for cosmological parameters. In parallel with this new definition for $t^\dagger$, I also rename the cosmological time quantity t_0 as 2003-Ω_Λ measurement epoch time or briefly 3Ωmeastime and which is also one of the very large numbers that occurs for cosmological parameters. The time t_0 is the time when the ratio of conserved positive gravitational mass is exactly one third of the negative gravitational mass, or dark energy mass, within the universe's boundary, when $\Omega_M/\Omega_\Lambda = 1/3$. The special importance of the time t_0 is that it appears that it should be associated with the central measured value of Ω_Λ. See equations (1.29) and (1.30).

4   Coincidence Free Universe and Lambda

We have seen that the effect of using the dust universe solution in the definition for the astronomer's $\Omega_\Lambda(t)$ leads to a very effective means of eliminating the cosmological coincidence problem but results in an inability to calculate the value of Einstein's Λ because of complications due to Λ in Hubble's function $H(t)$, (3.1) and (5.19).

$$H^\dagger = H(t^\dagger) = (c/R_\Lambda)\coth(3ct^\dagger/(2R_\Lambda)) \qquad (4.1)$$
$$t^\dagger = (2R_\Lambda/(3c))\coth^{-1}(R_\Lambda H^\dagger/c) \qquad (4.2)$$
$$= (2(3/\Lambda)^{1/2}/(3c))\coth^{-1}((3/\Lambda)^{1/2}H^\dagger/c) \qquad (4.3)$$

From equation (4.3) it is clear that if $t^\dagger$ were known we could calculate Λ and vice versa. However, a measured value of $t^\dagger$ is not given by the measurements so far being used in this work so that Λ cannot be calculated. The problem of finding the numerical value of Λ rests on finding a measured value for $t^\dagger$. Unfortunately, no direct measurement of the value of $t^\dagger$ seems to be possible at this time in astronomy history. The only way out of this dilemma at the moment seems to me to be to accept the consensus result of many astronomy measurements, extrapolations and speculations that the age of the universe, t_A, is approximately 13.7×10^{10} $yrs \approx 4.320432 \times 10^{17}$ s. This value can be assigned to $t^\dagger$ as what might be called a working hypothesis, clearly not

228

correct and very approximate but perhaps the best that can be done at this time. In this spirit, I shall take this value for $t^\dagger$ together with the central measured value of $H(t^\dagger)$ in formula (4.3) to enable the finding of a *best* value for Λ. It was the identification $t^\dagger = t_0$ that lead to the coincidence problem initially and also made possible the calculation of the initially coincidence value for Λ, denoted now by the subscript C as Λ_C. This can all be seen from the following displayed equations by taking $t^\dagger = t_0$. The resulting numerical solution for Λ_C is also displayed and compared with *removed* coincidence numerical value for Λ obtained by taking $t^\dagger = t_A$ at equation (4.9).

$$t_0 = (2(3/\Lambda)^{1/2}/3c)\cosh^{-1}(2) \tag{4.4}$$
$$\approx 1.52 \times 10^8 \ yrs \tag{4.5}$$
$$t^\dagger = (2(3/\Lambda)^{1/2}/(3c))\coth^{-1}((3/\Lambda)^{1/2}H^\dagger/c) \tag{4.6}$$
$$t_A \approx 4.320432 \times 10^{17} \ s \tag{4.7}$$
$$\Lambda_C = 1.3631 \times 10^{-52} \tag{4.8}$$
$$\Lambda = 1.3536 \times 10^{-52} \tag{4.9}$$
$$t^\dagger/t_0 = \coth^{-1}((3/\Lambda)^{1/2}H^\dagger/c)/\cosh^{-1}(2) \approx 0.99 \tag{4.10}$$
$$t_0 - t^\dagger = 0.01 t_0 \approx 1.52 \times 10^6 \ yrs \tag{4.11}$$

Equation (4.10) is the ratio of $t^\dagger = t_A$ or, essentially the time-now value or age of the universe, to the time, t_0, when the amount of negatively gravitating mass to positively gravitating mass is $3/1$ and the ratio, $t^\dagger/t_0$, is less than one. This taken with equation (4.11) has the implication that we have about 1.5 million years before the $3/1$ stage in the universe's evolution is reached at time t_0. Einstein's cosmological constant plays an essential and fundamental role in the dust universe model. If Λ is put to zero within this model, it ceases to exist because most of the physical quantities involved become zero. In particular, because the vital time arguments such as $3ct/(2R_\lambda) \to 0$. Thus Λ is the essential and fundamental constant at the basis of the Friedman dust universe. I claim, ending this section, that this cosmological model, using the value for Λ at equation (4.9), is both free from the *cosmological constant problem* and the *cosmological coincidence problem*. Thus the model is of a suitable form to be used to address the problem of how to quantize cosmology. This development follows in Section 5

5 Universe Expansion as Non-Linear Wave

It is convenient here to give a *very brief* reminder of the structure of Schrödinger theory in relation to the Friedman equations. The two Friedman equations from general relativity and the Schrödinger equation from quantum theory have the following three forms,

$$8\pi G\rho r^2/3 = \dot{r}^2 + (k - \Lambda r^2/3)c^2 \tag{5.1}$$

$$-8\pi GPr/c^2 = 2\ddot{r} + \dot{r}^2/r + (k/r - \Lambda r)c^2 \tag{5.2}$$

$$i\hbar\frac{\partial \Psi(\mathbf{r}, t)}{\partial t} = -\frac{\hbar^2}{2m}\nabla^2\Psi(\mathbf{r}, t) + V(\mathbf{r})\Psi(\mathbf{r}, t) \tag{5.3}$$

$$E_n\Psi_n(\mathbf{r}, t) = i\hbar\frac{\partial \Psi_n(\mathbf{r}, t)}{\partial t} \tag{5.4}$$

$$\nabla = \mathbf{i}\partial/\partial x + \mathbf{j}\partial/\partial y + \mathbf{k}\partial/\partial z \tag{5.5}$$

$$\rho_Q(\mathbf{r}, t) = \Psi(\mathbf{r}, t)\Psi^*(\mathbf{r}, t) \tag{5.6}$$

$$\Psi(\mathbf{r}, t) = \sum_n \int c_n\Psi_n(\mathbf{r}, t). \tag{5.7}$$

At equation (5.6) the usual definition of the quantum *probability* density, $\rho_Q(\mathbf{r}, t)$, is given as the product of the wave function for the state $\Psi(\mathbf{r}, t)$ with its complex conjugate $\Psi^*(\mathbf{r}, t)$. At (5.7), is given the *crucially* important equation from quantum theory that is called the principle of linear superposition. It is here written in a somewhat symbolic form but in fact means that the solution of the schrödinger equation at (5.3) can be expressed as a discrete sum or integral or both over constants c_n times the energy eigenfunctions, given by equation (5.4), of the Schrödinger equation. In order to attempt to get at a quantum theory of cosmology or gravity it is reasonable to attempt to convert the two Friedman equations (5.1) and (5.2) into a form similar to the Schrödinger equation (5.3). The possibility of such a transformation would seem to depend on what the equations from cosmology and the equation from quantum mechanics have in common. There are two Friedman equations and in fact there are two Schrödinger equations because a real part and an imaginary part are added together to make the complex function $\Psi(\mathbf{r}, t)$ of the space and time variables. Both sets have an important density function associated with them, the mass density $\rho(t)$ in the Friedman set, and the probability density, $\rho_Q(\mathbf{r}, t)$, for the quantum set. Other than these two features they seem to have little in common and the likelihood of finding a conversion from one set to the other seems remote in the extreme. Such a conclusion is reinforced by the recognition of the well known fact that General relativity is a non-linear theory, a non-linearity it transfers to its offspring the Friedman equations, whereas the schrödinger equation is linear. However, I shall show in the following pages that this very non-linearity of the Friedman equations can be precisely evaluated and expressed relative to the linearity of the Schrödinger equation. This step leads to the formulation of a non-linear wave theory for gravitational waves in contrast with existing gravitational wave theory. The existing theory of gravitational waves usually involves a crude linearisation of general relativity describing waves that have so far not been detected. There is an exception to this by what is called an *exact* gravitational *sandwich* wave of rather unconvincing theory that has also not been detected, see Rindler, page 284, [16]. The waves that are to be described in the following pages have been detected and in fact constitute the expanding or contracting universe structure. The non-liearity of the Friedman set can be expressed relative to the linearity of the Schrödinger set, (5.3), symbolically as follows,

$$f(t) = \frac{\sum_j \int n_{f,j}(t)}{\sum_j \int d_{f,j}(t)}, \tag{5.8}$$

the space variation of the Schrödinger equation does not occur because the cosmology structure does not depend on local space variations at fixed time. The n and d functions refer to the placement of the superposed terms, either numerator or denominator. The function, $f(t)$, just represents what function is being analysed by the non-linear form. I shall call equation (5.8) the *bilinear superposition principle* for cosmology. It shows clearly that the non-linearity of cosmology involves the ratio of two linear suppositions of the quantum mechanics type. This principle will be derived in the following, while showing that it allows the contraction and expansion of the universe to be represented as a non-linear standing spherical gravitational wave of time varying radius. The next step in developing this formalism will be to analyse the main mass densities that occur in this model by showing that they can be expressed in the form of the bilinear superposition principle, (5.8). Two basic *positive everywhere and for all time* mass densities have been identified in the development of this cosmological theory. They are the mass density of positive gravitational mass $\rho(t)$ and the mass density of *negative gravitational mass* $\rho_\Lambda^\dagger$ which is also everywhere and for all time constant. A third mass density, $\rho_G = \rho(t) - \rho_\Lambda^\dagger$ is the density of positive gravitational mass relative to negative gravitational mass or alternatively it can be called the gravitational weighted mass density, $\rho_G(t) = (G_+\rho(t) + G_-\rho_\Lambda^\dagger)/G$, $G_- = -G$, $G_+ = +G$. This last mass density does become $negative$ for some epoch times. The definition for $\rho(t)$ is,

$$\rho(t) = (3/(8\pi G))(c/(R_\Lambda)^2 \sinh^{-2}(3ct/(2R_\Lambda)) \tag{5.9}$$
$$= A \sinh^{-2}(3ct/(2R_\Lambda)) \tag{5.10}$$
$$A = (3/(8\pi G))(c/R_\Lambda)^2 \tag{5.11}$$
$$R_\Lambda = (3/\Lambda)^{1/2}, \tag{5.12}$$

where the constant A is introduced as a convenient simplification at formula (5.10) and Λ is Einstein's cosmological constant. Inspection of the formula for $\rho(t)$ reveals it to be a positive function because the sinh appears squared. It is also time symmetric, $t \to -t$, leaves its value unchanged also because of the square. Thus it can be replaced by placing a modulus sign about the time variable as

$$\rho(t) = A \sinh^{-2}(3c|t|/(2R_\Lambda)), \tag{5.13}$$

and this will apply for all time, $-\infty < t < +\infty$. This means that we can use the inverse Fourier transform relation,

$$\exp(-\omega_\Lambda|t|)) = (2\omega_\Lambda/\pi) \int_0^\infty cos(\omega t)d\omega/(\omega_\Lambda^2 + \omega^2), \tag{5.14}$$

$$\exp(-\omega_\Lambda|t|)) = (2/\pi) \int_0^\infty cos(\omega_\Lambda st)ds/(1 + s^2), \tag{5.15}$$

$$s = \omega/\omega_\Lambda, \tag{5.16}$$

where ω_Λ is a constant and more conveniently in the second form where s is a none-dimensional dummy to express suitable functions of decreasing with $|t| \to \infty$ quantities as functions of integrals over the oscillatory quantity, $cos(\omega t) = cos(\omega|t|)$. $\rho(t)$ is a suitable function because

$$\rho(t) = A \sinh^{-2}(3ct/(2R_\Lambda)) \tag{5.17}$$
$$= A((\exp(\omega_\Lambda|t|) - \exp(-\omega_\Lambda|t|))/2)^{-2} \tag{5.18}$$
$$= A\left(\frac{1 - \exp(-2\omega_\Lambda|t|)}{2\exp(-\omega_\Lambda|t|)}\right)^{-2} \tag{5.19}$$

$$= A \left(\frac{1 - 2\exp(-2\omega_\Lambda |t|) + \exp(-4\omega_\Lambda |t|)}{4\exp(-2\omega_\Lambda |t|)} \right)^{-1} \tag{5.20}$$

$$= \frac{4A\exp(-2\omega_\Lambda |t|)}{1 - 2\exp(-2\omega_\Lambda |t|) + \exp(-4\omega_\Lambda |t|)} \tag{5.21}$$

$$\rho^{1/2}(t) = \frac{2A^{1/2}\exp(-\omega_\Lambda |t|)}{1 - \exp(-2\omega_\Lambda |t|)} \tag{5.22}$$

$$A = (3/(8\pi G))(c/R_\Lambda)^2 \tag{5.23}$$

$$\omega_\Lambda = 3c/(2R_\Lambda) \approx 3.0312 \times 10^{-18} \ cs^{-1}. \tag{5.24}$$

We note from equation (5.15) that by giving the constant ω_Λ the value zero the general Fourier transform result below follows

$$1 = (2/\pi) \int_0^\infty ds/(1+s^2). \tag{5.25}$$

Thus all the terms in the fraction at equation (5.21) including the zero frequency unit term in the denominator can be expressed as integrals over the dummy variable s. Both the numerator and the denominator can be expressed as a superposition of oscillatory cosines or unity, the unit term being included using equation (5.25). It should be noted that once the inverse Fourier transform involving the $\cos(n\omega_\Lambda |t|)$ functions are accepted, all such functions can be replaced with $\cos(n\omega_\Lambda t)$ because cosines are even functions anyway. The numerator and denominator of the fraction involved are as follows,

$$N_\rho(t) = 4A\exp(-2\omega_\Lambda |t|) \tag{5.26}$$

$$= \frac{8A}{\pi} \int_0^\infty \frac{\cos(2\omega_\Lambda st)}{1+s^2} ds \tag{5.27}$$

$$n_\rho(t,s) = \frac{8A}{\pi} \left(\frac{\cos(2\omega_\Lambda st)}{1+s^2} \right) \tag{5.28}$$

$$N_\rho(t) = \int_0^\infty n_\rho(t,s)ds \tag{5.29}$$

$$D_\rho(t) = 1 - 2\exp(-2\omega_\Lambda |t|) + \exp(-4\omega_\Lambda |t|). \tag{5.30}$$

$$= \frac{2}{\pi} \int_0^\infty \frac{1 - 2\cos(2\omega_\Lambda st) + \cos(4\omega_\Lambda st)}{1+s^2} ds \tag{5.31}$$

233

$$d_f(t, s) = \frac{2}{\pi} \left(\frac{1 - 2\cos(2\omega_\Lambda st) + \cos(4\omega_\Lambda st)}{1 + s^2} \right) \tag{5.32}$$

$$\rho(t) = \frac{N_\rho(t)}{D_\rho(t)} = \frac{\int_0^\infty n_\rho(t, s)ds}{\int_0^\infty d_\rho(t, s)ds}. \tag{5.33}$$

Before discussing the significance of formulae (5.27), (5.31) and (5.33), it is useful to consider the integral form for the dark energy density, $\rho_\Lambda^\dagger = (3/(4\pi G))(c/R_\Lambda)^2$. As this is a constant it can be written in the form

$$\rho_\Lambda^\dagger = \frac{4A}{\pi} \int_0^\infty \frac{ds}{1 + s^2} = 2\rho_\Lambda. \tag{5.34}$$

This can be used to find oscillatory based form for the gravitationally weighted mass density ρ_G,

$$\rho_G = \rho(t) - \rho_\Lambda^\dagger \tag{5.35}$$

$$= -2A \frac{1 - 4\exp(-2\omega_\Lambda|t|) + \exp(-4\omega_\Lambda|t|)}{(1 - \exp(-2\omega_\Lambda|t|))^2}. \tag{5.36}$$

Thus $\rho_G(t)$ can be expressed in terms of the ratio of numerator and denominator given by

$$N_G(t) = -2A(1 - 4\exp(-2\omega_\Lambda|t|) + \exp(-4\omega_\Lambda|t|)) \tag{5.37}$$

$$= -\frac{4A}{\pi} \int_0^\infty \frac{1 - 4\cos(2\omega_\Lambda st) + \cos(4\omega_\Lambda st)}{1 + s^2} ds \tag{5.38}$$

$$n_G(t, s) = -\frac{4A}{\pi} \left(\frac{1 - 4\cos(2\omega_\Lambda st) + \cos(4\omega_\Lambda st)}{1 + s^2} \right) \tag{5.39}$$

$$N_G(t) = \int_0^\infty n_G(t, s)ds \tag{5.40}$$

$$D_G(t) = (1 - \exp(-2\omega_\Lambda|t|))^2 \tag{5.41}$$

$$= \frac{2}{\pi} \int_0^\infty \frac{1 - 2\cos(2\omega_\Lambda st) + \cos(4\omega_\Lambda st)}{1 + s^2} ds \tag{5.42}$$

$$d_G(t, s) = \frac{2}{\pi} \left(\frac{1 - 2\cos(2\omega_\Lambda st) + \cos(4\omega_\Lambda st)}{1 + s^2} \right) \tag{5.43}$$

$$\rho_G(t) = \frac{N_G(t)}{D_{G,f}(t)} = \frac{\int_0^\infty n_G(t, s)ds}{\int_0^\infty d_G(t, s)ds}. \tag{5.44}$$

The dark mass dark energy time relational process ratio, $r_{\Lambda,DM}(t)$, can be expanded in oscillatory integrals as

$$r_{\Lambda,DM}(t) = \rho_\Lambda^\dagger/\rho(t) = 2\sinh^2(3ct/(2R_\Lambda)) \tag{5.45}$$

$$= 2A/\left(\frac{N_\rho(t)}{D_\rho(t)}\right) = 2A\frac{D_\rho(t)}{N_\rho(t)} = 2A\frac{\int_0^\infty d_\rho(t,s)ds}{\int_0^\infty n_\rho(t,s)ds} \tag{5.46}$$

$$= \frac{\int_0^\infty (1 - 2\cos(2\omega_\Lambda st) + \cos(4\omega_\Lambda st))(1+s^2)^{-1}ds}{2\int_0^\infty \cos(2\omega_\Lambda st)(1+s^2)^{-1}ds} \tag{5.47}$$

Hubble's function squared is a suitable function for expression as bilinear superposition

$$H^2(t) = \left(\frac{c}{R_\Lambda}\right)^2 \coth^2(\omega_\Lambda t) = \left(\left(\frac{c}{R_\Lambda}\right)\frac{1 + \exp(-2\omega_\Lambda t)}{1 - \exp(-2\omega_\Lambda t)}\right)^2 \tag{5.48}$$

$$N_H(t) = \left(\frac{c}{R_\Lambda}\right)^2 (1 + 2\exp(-2\omega_\Lambda|t|) + \exp(-4\omega_\Lambda|t|)) \tag{5.49}$$

$$= \frac{2}{\pi}\left(\frac{c}{R_\Lambda}\right)^2 \int_0^\infty \frac{1 + 2\cos(2\omega_\Lambda st) + \cos(4\omega_\Lambda st)}{1+s^2}ds \tag{5.50}$$

$$n_H(t,s) = \frac{2}{\pi}\left(\frac{c}{R_\Lambda}\right)^2 \left(\frac{1 + 2\cos(2\omega_\Lambda st) + \cos(4\omega_\Lambda st)}{1+s^2}\right) \tag{5.51}$$

$$N_H(t) = \int_0^\infty n_H(t,s)ds \tag{5.52}$$

$$D_H(t) = (1 - \exp(-2\omega_\Lambda|t|))^2 \tag{5.53}$$

$$= \frac{2}{\pi}\int_0^\infty \frac{1 - 2\cos(2\omega_\Lambda st) + \cos(4\omega_\Lambda st)}{1+s^2}ds \tag{5.54}$$

$$d_H(t,s) = \frac{2}{\pi}\left(\frac{1 - 2\cos(2\omega_\Lambda st) + \cos(4\omega_\Lambda st)}{1+s^2}\right) \tag{5.55}$$

$$H^2(t) = \frac{N_H(t)}{D_H(t)} = \frac{\int_0^\infty n_H(t,s)ds}{\int_0^\infty d_H(t,s)ds}. \tag{5.56}$$

6 Cosmological Eigen-Functions

Let us consider the bilinear form for $\rho^{1/2}(t)$, the simplest of the bilinear forms,

$$\rho^{1/2}(t) = \frac{2A^{1/2}\exp(-\omega_\Lambda|t|)}{1 - \exp(-2\omega_\Lambda|t|)} \tag{6.1}$$

$$= \frac{(4A^{1/2}/\pi)\int_0^\infty \cos(\omega_\Lambda st)ds/(1+s^2)}{(2/\pi)\int_0^\infty(1 - \cos(2\omega_\Lambda st))ds/(1+s^2)}. \tag{6.2}$$

Because $\cos(\theta) = (\exp(i\theta)+\exp(-i\theta))/2$, all the $\cos(\theta)$s in the above formula can be replaced with exponential forms with the result

$$\rho^{1/2}(t) = 2A^{1/2}\frac{\int_0^\infty(\exp(i\omega_\Lambda st) + \exp(-i\omega_\Lambda st))ds/(2+2s^2)}{\int_0^\infty(1 - (\exp(i2\omega_\Lambda st) - \exp(-i2\omega_\Lambda st))/2)ds/(1+s^2)}. \tag{6.3}$$

In the following three equations, I introduce two bilinear representations, $\Psi_{nl,\rho}(t)$ and $\Psi_{nl,\rho}^*(t)$ for the total state. They are not independent because one is the complex conjugate of the other and their normal *linear superposition* represents $\rho^{1/2}(t)$. Their use will be explained later.

$$\Psi_{nl,\rho,+}(t) = \frac{2A^{1/2}\int_0^\infty \exp(-i\omega_\Lambda st)ds/(2+2s^2)}{\int_0^\infty(1 - (\exp(i2\omega_\Lambda st) - \exp(-i2\omega_\Lambda st))/2)ds/(1+s^2)} \tag{6.4}$$

$$\Psi_{nl,\rho,-}^*(t) = \frac{2A^{1/2}\int_0^\infty \exp(i\omega_\Lambda st)ds/(2+2s^2)}{\int_0^\infty(1 - (\exp(i2\omega_\Lambda st) - \exp(-i2\omega_\Lambda st))/2)ds/(1+s^2)} \tag{6.5}$$

$$\rho^{1/2}(t) = \Psi_{nl,\rho,+}(t) + \Psi_{nl,\rho,-}^*(t) = \Psi_{nl,\rho}(t) \tag{6.6}$$

$$\rho(t) = \Psi_{nl,\rho}(t)\Psi_{nl,\rho}^*(t). \tag{6.7}$$

Thus the superposition of cos forms in these ratios *generally* can be replaced with the superposition of exp forms provided a caveat is added to the effect that an appearance of an exponential must always be accompanied with the appearance of its complex conjugate with the same coefficient. With this caveat we can take the exponential forms as representing the fundamental eigen-states of this system, the fundamental ones being taken as the continuous infinite set,

$$\Psi(s,t) = \exp(-i\omega_\Lambda st), \ 0 < s < \infty. \tag{6.8}$$

Thus if these oscillations measured by the angular frequency, $\omega_\Lambda s$, are interpreted as of quantum origin, we can define the cosmological associate energies as

$$E(s) = \hbar\omega_\Lambda s \tag{6.9}$$

and equation (6.8) can be expressed as

$$\Psi(s,t) = \exp(-iE(s)t/\hbar), \ 0 < s < \infty \tag{6.10}$$

with the consequence we get the eignen-value equation equivalent to (5.4) repeated below at (6.12) for comparison

$$i\hbar\frac{\partial\Psi(s,t)}{\partial t} = E(s)\Psi(s,t) \tag{6.11}$$

$$i\hbar\frac{\partial\Psi_n(\mathbf{r},t)}{\partial t} = E_n\Psi_n(\mathbf{r},t). \tag{6.12}$$

The local variable position vector, $\mathbf{r}$, does not occur in the cosmology version, (6.11), of this equation because all the states involved are uniformly spatially constant and only vary with time. There is also no external potential so that from the cosmology point of view the full Schrödinger equation, (5.3), *essentially* reduces to just the energy eigen-value version, equation (6.11). The continuous variable s replaces the apparently discrete subscript n but this parameter could under some circumstances also be continuous. The possible range of the dimensionless variable s given at equation (6.10) implies that the range of angular frequencies present is given by

$$0 < \omega_\Lambda s < \infty. \tag{6.13}$$

Inspection of the formulae for the various superposed quantities reveals that the second and fourth harmonics of all the range for, $n\omega_\Lambda s$, $n = 2, 4$, frequencies are also simultaneously present so that the range (6.10) or (6.13) is the complete set as it also includes all the harmonics $n\omega_\Lambda s$ and also includes the dark energy contribution, $n = 0$. It is now possible to express $\rho^{1/2}(t)$ in terms of the bi-linear superposed eigen-values as

$$\rho^{1/2}(t) = \frac{2A^{1/2}\int_0^\infty(\Psi(-s,t) + \Psi(s,t))ds/(1+s^2)}{\int_0^\infty(1 - \Psi(-2s,t) - \Psi(2s,t))ds/(1+s^2)}. \tag{6.14}$$

All the other cosmological functions mentioned above can similarly be expressed in terms of the eigen-value set (6.13) and thus it is now possible to express all the essential cosmological structure in terms of $\rho^{1/2}(t)$ and simultaneously in terms of the $\Psi(ns, t)$ by the following connections

$$\rho(t) = (\rho^{1/2}(t))^2 \tag{6.15}$$

$$\rho_\Lambda^\dagger = 2A \tag{6.16}$$

$$\rho_G(t) = \rho(t) - \rho_\Lambda^\dagger \tag{6.17}$$

$$H^2(t) = (c/R_\Lambda)^2((\rho(t)/A) + 1). \tag{6.18}$$

The wave motion followed by the dark mass dark energy time relation process seems to me to be particularly basic as it can also be identified, apart from a constant multiplier, $2A/M_U$, as a spherical volume standing wave with front, the boundary of the expanding or contracting universe. It is repeated here for convenience,

$$r_{\Lambda,DM}(t) = \rho_\Lambda^\dagger/\rho(t) = (2A/M_u)V_U(t) \tag{6.19}$$

$$= \frac{\int_0^\infty (1 - 2\cos(2\omega_\Lambda st) + \cos(4\omega_\Lambda st))(1 + s^2)^{-1}ds}{2\int_0^\infty \cos(2\omega_\Lambda st)(1 + s^2)^{-1}ds} \tag{6.20}$$

$V_U(t)$ being the volume of the universe at time t. This same type of gravitational wave motion also applies to the evolution of smaller sub-volumes of mass within the universe with a time origin different from the universe's singular epoch time, $t = 0$. [38]. This cosmology theory is thus directly derivable from the Schrödinger like eigen-equation (6.11), if the linear superposition principle of quantum theory is replaced with the *bilinear superposition principle*. The function $r_{\Lambda,DM}(t)$ is the ratio of the amount of *negatively gravitating positive mass within* the universe to the amount of *positively gravitating positive mass within* the universe as a function of epoch time, t. At equation (6.19), we see that this ratio is proportional to the volume of the universe and as a function of time it has the same shape as does the volume of the universe as a function of time. This shape is near infinite radius at $t \approx -\infty$ to zero radius at the singularity at $t = 0$ on to near infinite radius at $t \approx +\infty$. Thus the wave motion that correspond to this time shape can be thought of a converging to zero spherical wave front at negative times to a diverging from zero spherical wave front at positive times. On the other hand, the reciprocal of this ratio, the ratio of positive gravitating mass within the universe to the negative gravitating mass within the universe can be seen to be a spherical standing wave with centre fixed at $r = 0$ with time varying radius, infinite at $t = 0$ and zero at $t = \pm\infty$.

Possibly the best and most instructive image for the wave motion comes from studying the wave form of,

$$\dot{\rho_G}(t)/\rho_\Lambda^\dagger = \rho(t)/\rho_\Lambda^\dagger - 1, \qquad (6.21)$$

from which it can be inferred that the motion is *not* waves of density, but rather waves of the gravity associated with the whole body of the universe. The nodes for this motion occur at $t = \pm t_c$ with the value zero, when the acceleration changes sign through zero, and the antinodes occur at $t = 0$ and $t = \pm\infty$ with the values infinity and -1 respectively. The waves appear to be waves in the gravitationally polarised *aether* ([39]), formed from the positively gravitating universe's mass, M_U, and the negatively gravitating mass of Einstein's dark energy. However, whatever visualisation is chosen, it remains a *non-linear wave process* formed by the bilinear superposition of eigen-solutions of the Schrödinger like equation (6.11). Again it should be noted that Einstein's cosmological constant is central to the representation of gravitational waves used in this theory because all the frequencies for these waves depend essentially on it. From equation (6.13) it follows that this special frequency range would not exist if $\omega_\Lambda = (3\Lambda)^{1/2}c/2 \to 0$, as it would if $\Lambda \to 0$. Returning to the question of finding the full Schrödinger equation representation for the Friedman dust universe, an objective that can be achieved simply by operating on the non-linear cosmology state function ,$\Psi_{nl,\rho}(t)$ (6.4), with the quantum energy operator $\hbar\partial/\partial t$ without its usual imaginary unit, i. The result is

$$\hbar\partial\Psi_{nl,\rho}(t)/\partial t = (V_C(t))\Psi_{nl,\rho}(t) \qquad (6.22)$$
$$V_C(t) = -(3\hbar/2)H(t). \qquad (6.23)$$

Equation (6.22) is, not surprisingly, a non-linear Schrödinger equation with a time dependent *feedback* potential function given by (6.23) and with no dependence on a local vector position, **r**, features also not surprising. It is essentially the *quantum* description of the influence of dark energy on the conserved mass density $\rho(t)$ for this model. The missing i in the quantum energy operator in equation (6.22) can be restored by placing an i in the numerator of the $(3\hbar/2)$ factor of $V_C(t)$. There is no reason in principle why a feed back potential should not be imaginary in a quantum system involving complex wave functions. In fact, the appearance of the i in the potential function is characteristic of standing wave type solutions.

239

7 The Probability Density Issue

In order to bring the cosmological structure into line with the Schrödinger quantum structure it is necessary to decide how probabilistic concepts are to be introduced or found in relation to the cosmology eigen-states and their bilinear superposition. We can get a clue to how this might be done from the quantum equation for probability density, $\rho_Q(\mathbf{r}, t)$, (5.6) which is the Hermitean scalar product of the two amplitudes or wave functions representing this state. In the cosmology context we have seen that composite states formed from bilinear superpositions of eigen-states are needed to play the full state representation role. I have defined such a state representation for the non-linear gravitationally positive mass density, $\rho(t)$ at references (6.4) to (6.7), now repeated at (6.20) and followed by a repeat of the quantum probability density definition at (7.2).

$$\rho(t) = \Psi_{nl,\rho}(t)\Psi_{nl,\rho}^*(t) = M_U/V_U(t) \tag{7.1}$$

$$\rho_Q(\mathbf{r}, t) = \Psi(\mathbf{r}, t)\Psi^*(\mathbf{r}, t) \tag{7.2}$$

$$\rho_C(t) = \rho(t)/M_U = 1/V_U(t). \tag{7.3}$$

The function $\rho_C(t)$, the cosmological mass density for positive gravitational mass divided by the mass of the universe, M_U, is an obvious contender for representing cosmological probability density. However it is constant over all positions within the spherical volume of the universe unlike the the quantum probability density, $\rho_Q(\mathbf{r}, t)$, that is generally variable over the region under consideration, such variability described by the position vector $\mathbf{r}$. The quantum probability density can answer the following type of question. Given a spherical region, S, in which a particle is certain to be found, what is the probability for finding this particle in some, s, subregion at a fixed time? The answer can be found by integrating the density $\mathbf{r}$ over the region s. In the cosmological, situation with $\rho_C(t)$ not dependent on position the question and answer is somewhat trivial and the answer would be the value of the volume of s divided by volume of S, correct but not very interesting and clearly this is a consequence of the cosmology states being spatially uniform in most present theory, a restriction that may well be removed in the future. This type of probability question refers to a fixed time density. However, if we ask the following different question there is a more interesting answer. Given a time variable volume, $V(t)$, in which a particle is certain to be found at any time what is the probability of finding this particle in a fixed valued volume, v, at some time t. The answer to this question is, $v/V(t)$ and clearly changes with the changing volume, $V(t)$. This seem to me to be much more interesting and it fits the cosmological structure of this model when $V(t)$ is taken to be $V_U(t)$. This type of probability question refers to a changing time density. I shall assume, following the discussion above, that the density, $\rho_C(t)$, defined at equation (7.1) and (7.3) can represent a probability density of the changing time type. Thus these equations brings the cosmology structure yet closer to the quantum structure at equation (7.2).

8 Conclusions

The first four sections of this paper are devoted to an expanded discussion of the cosmological coincidence problem. Here the implications of the astronomical measurements of the Ωs and Hubble's constant are examined in more detail than in the previous paper where a solution to this problem was found. It is shown that given the present concensus of the age of the universe it is possible to derive a definite value for Einstein's cosmological constant and at the same time resolve the coincidence problem thus removing a major impediment to the quantization of cosmology. The last sections of this paper are devoted to expressing the contraction and expansion of the universe in terms of gravitational waves. This is achieved by finding the equivalent for cosmology of the quantum principle of linear superposition of states. The principle for cosmology reflects exactly the sense in which cosmology is non-linear in relation to the equivalent linear principle for quantum mechanics and is called *bilinear superposition*. The bilinear superposition is then used to express the contraction or expansion of the universe as spherical standing wave motion of varying radius. The principle is also used to describe the time dependent relation between dark energy and dark mass as a local radius varying standing wave. It is suggested that this new non-linear gravitational wave motion has been detected in the form of the expanding universe and is thus a reality unlike the usually theorised but not detected linearised wave motion of general relativity.

Acknowledgements: I am greatly indebted to Professors Clive Kilmister and Wolfgang Rindler for help, encouragement and inspiration.

9 Appendix with Abstract

Solutions of a Cosmological Schrödinger Equation for Exact Gravitational Waves based on a Friedman Dust Universe with Einstein's Lambda

In an earlier paper, it was shown that the cosmological model that was introduced in a sequence of three earlier papers under the title *A Dust Universe Solution to the Dark Energy Problem*, originally described by the Friedman equations, can be expressed as a solution to a non-linear Schrödinger equation. In this appendix, a large collection of solutions to this Schrödinger equation are found and discussed in the context of relaxing the uniform mass density condition usually employed in cosmology theory. The surprising result is obtained that this non-linear equation can have its many solutions *linearly superposed* to obtain solution of the cosmology theory problem of great generality and applicability.

The work to be described in this appendix is an application of the cosmological model introduced in the papers *A Dust Universe Solution to the Dark Energy Problem* [23], *Existence of Negative Gravity Material. Identification of Dark Energy* [24] and *Thermodynamics of a Dust Universe* [32] together with applications of those papers to the cosmological constant problem, the cosmological coincidence problem and other subsidiary cosmological problems.

The application of the cosmological model introduced in the papers *A* [23], *B*,[24] and *C* [32], in paper *E*, ([36]), is to the extensively discussed and analysed *Cosmological Coincidence Problem*. In the paper *D*, [34], it was shown that the quantum vacuum polarisation idea can be seen to play a central role in the Friedman dust universe model introduced by the author. In the paper, [40], it was shown that the Friedman equation structure can be converted into a *non-linear* Schrödinger equation structure. Here, this aspect is further developed by supplementing the solutions to this time only equation with a dependence on a three dimensional space position vector, **r**, so that the equation remains consistent with its cosmological origin. This step then enables finding cosmological models that are not restricted to having a mass density that is certainly time dependent but otherwise remains constant over all *three dimensional* position space at every definite time. It is convenient here to repeat a *very brief* reminder of the structure of Schrödinger theory in relation to the Friedman equations. The two Friedman equations from general relativity and the Schrödinger equation from quantum theory have the following three forms,

$$8\pi G\rho r^2/3 = \dot{r}^2 + (k - \Lambda r^2/3)c^2 \tag{9.4}$$

$$-8\pi GPr/c^2 = 2\ddot{r} + \dot{r}^2/r + (k/r - \Lambda r)c^2 \tag{9.5}$$

$$i\hbar\frac{\partial\Psi(\mathbf{r},t)}{\partial t} = -\frac{\hbar^2}{2m}\nabla^2\Psi(\mathbf{r},t) + V(\mathbf{r})\Psi(\mathbf{r},t) \tag{9.6}$$

$$E_n\Psi_n(\mathbf{r},t) = i\hbar\frac{\partial\Psi_n(\mathbf{r},t)}{\partial t} \tag{9.7}$$

$$\nabla = i\partial/\partial x + j\partial/\partial y + k\partial/\partial z \tag{9.8}$$

$$\rho_Q(\mathbf{r},t) = \Psi(\mathbf{r},t)\Psi^*(\mathbf{r},t) \tag{9.9}$$

$$\Psi(\mathbf{r},t) = \sum_n \int c_n\Psi_n(\mathbf{r},t). \tag{9.10}$$

The non-linear Schrödinger equation that was obtained in reference [40] has the form

$$i\hbar\partial\Psi_{nl,\rho}(t)/\partial t = (V_C(t))\Psi_{nl,\rho}(t) \tag{9.11}$$

$$V_C(t) = -(3i\hbar/2)H(t) \tag{9.12}$$

and can be compared with the general linear Schrödinger equation at (9.6). The non-linearity of the cosmological version is indicated by the feedback potential $V_C(t)$, (9.12) replacing the external potential at (9.6). The state vector $\Psi_{nl,\rho}(t)$ in the cosmology version initially has no dependence on local position denoted by the three vector, $\mathbf{r}$, as in the quantum version, (9.6). This deficiency will be rectified in the following section.

10 Position Variable Cosmology Schrödinger equation

Before starting this section, it is necessary to make some remarks about the dimensionality of the usual physical position coordinate vector, $\mathbf{r} = x\mathbf{i} + y\mathbf{j} + z\mathbf{k}$. This is often taken to have the dimension, m, physical length. The relativistic metric used in this theory is of the form

$$ds^2 = c^2 dt^2 - r^2(t)(d\dot{x}^2 + d\dot{x}^2 + d\dot{x}^2). \tag{10.1}$$

In this work up to date, I have taken the scale factor $r(t)$ to represent the *physical* radius of the universe at epoch time t so that it has the dimension

m, physical length. If as usual, c has the physical dimensions ms^{-1} and t has the physical dimension, s, then ds in the metric will have the dimension m and so the vector, $\grave{\mathbf{r}} = \grave{x}\mathbf{i} + \grave{y}\mathbf{j} + \grave{z}\mathbf{k}$, will be dimensionless and this is indicated by the above grave accent. The theory I am working with here is non-linear and attempting to use dimensioned position coordinates can lead to dimensionality chaos. Thus from now on, I shall usually work with the dimensionless position coordinates and use the grave sign to indicate this. Consistent with this policy it is useful it define the dimensionless quantities using the fundamental length R_Λ as follows and starting with a dimensionless radius for the universe, $\grave{r}(t)$,

$$\grave{r}(t) \;=\; r(t)/R_\Lambda \tag{10.2}$$
$$\grave{x} \;=\; x/R_\Lambda \tag{10.3}$$
$$\grave{y} \;=\; y/R_\Lambda \tag{10.4}$$
$$\grave{z} \;=\; z/R_\Lambda. \tag{10.5}$$

I shall also use the grave accent to indicate that *a function* is dimensionless as with $\grave{f}(r)$. My strategy in the following work is firstly, to introduce space dependence, $\grave{\mathbf{r}}$, into the cosmological Schrödinger equation (9.11) and then , secondly to show that the introduction of an $\grave{\mathbf{r}}$ dependence can be made consistent with the original Friedman equations structure without damaging their validity as a rigorous solution to Einstein's field equations. Firstly, I rewrite the purely time dependent equation (9.11) assuming an extra dependence on $\grave{\mathbf{r}}$ in the original state vector $\Psi_{nl,\rho}(t)$, while leaving the feedback term unchanged.

$$i\hbar\partial\Psi_{nl,\rho}(t,\grave{\mathbf{r}})/\partial t \;=\; (V_C(t))\Psi_{nl,\rho}(t,\grave{\mathbf{r}}) \tag{10.6}$$
$$V_C(t) \;=\; -(3i\hbar/2)H(t). \tag{10.7}$$

The first question that arises is, *can this step be done consistently?* The answer to this is in the affirmative as can be shown as follows. Rewrite (10.6) as equation (10.8) and followed by the time integration at (10.9) and then inverting the logarithm at (10.10)

$$\partial \ln \Psi_{nl,\rho}(t, \grave{\mathbf{r}})/\partial t = -(3/2)H(t) \tag{10.8}$$

$$\ln(\Psi_{nl,\rho}(t, \grave{\mathbf{r}})/\Psi_{nl,\rho}(t_0, \grave{\mathbf{r}})) = -(3/2)\int_0^t H(t')dt' \tag{10.9}$$

$$\Psi_{nl,\rho}(t, \grave{\mathbf{r}}) = \Psi_{nl,\rho}(t_0, \grave{\mathbf{r}})\exp\left(-\frac{3}{2}\int_{t_0}^t H(t')dt'\right) \tag{10.10}$$

$$\Psi_{nl,\rho}(t_0, \grave{\mathbf{r}}) = \Psi_{nl,\rho}(t_0)\grave{f}(\grave{\mathbf{r}}) \tag{10.11}$$

Thus introducing a dimensionless function, $\grave{\mathbf{f}}$, with $\grave{\mathbf{r}}$ dependence presents no problems. It means just multiplying the original time only dependent wave function, $\Psi_{nl,\rho}(t_0)$, with the purely space dependent function, $\grave{f}(\grave{\mathbf{r}})$. This also partially justifies not including any space variation in the Hubble function, $H(t)$. However, this last point will be fully justified when the affect on the purely time dependent Friedman equations, (9.4) and (9.5), is examined in the next paragraph. I should be remarked that the function $\grave{f}(\grave{\mathbf{r}})$ can be a complex valued function in the context of quantum theory wave function structure. This fact will be seen to be useful as the story unfolds.

The relation of the cosmological Schrödinger equation and the Friedman equations clearly has to be mutual consistency. A threat to this consistency is the obvious difference between the purely time dependent mass density function $\rho(t)$ in the Friedman set and the now proposed space time variability through $\grave{\mathbf{r}}$ in the Schrödinger equation wave function, (10.10). In using the original Friedman equations, (9.4) and (9.5), it has been common practice to assume that $\rho(t)$ is a purely time dependent mass density chosen as a working approximation to a correct more general time and space dependant version and so rendering difficult mathematics viable though less physically accurate. This was my starting position when I wrote the first paper, A, in this sequence of papers. However, having found the non-linear Schrödinger equation (9.11) it has become clear that the common practice position with regard to $\rho(t)$ needs some modification. My view now is that $\rho(t)$ is a correct quantity in its own right, giving information about the cosmology structure as a global entity. Its definition is repeated below,

$$\rho(t) = M_U/V_U(t) \tag{10.12}$$

$$M_U = \rho(t)V_U(t), \tag{10.13}$$

where M_U is the total conserved positively gravitational mass of the universe and $V_U(t)$ is the volume of the universe at epoch time t. If $\rho(t)$, does have a definite meaning in its own right and is not just an approximation to a better space dependent version then it can be retained with its self identity as before. This special significance of $\rho(t)$ is effectively retained by keeping it but multiplied by the space dependant contribution as in (10.11). From the existence of a possible true space and time dependent version from Schrödinger theory it can be seen that the definition for the mass, M_U, of the universe that appears in (10.12) with the space dependent density, should be

$$M_U(t) \;=\; R_\Lambda{}^3 \int \int \int_{V_U(t)} \rho(t,\grave{\mathbf{r}})d\grave{x}d\grave{y}d\grave{z} \tag{10.14}$$

$$=\; R_\Lambda{}^3 \int \int \int_{V_U(t_0)} \rho(t_0,\grave{\mathbf{r}})d\grave{x}d\grave{y}d\grave{z} \tag{10.15}$$

$$=\; M_U \;=\; a\ constant \tag{10.16}$$

$$\rho(t_0,\grave{\mathbf{r}}) \;=\; \Psi_{nl,\rho}(t_0,\grave{\mathbf{r}})\Psi^*_{nl,\rho}(t_0,\grave{\mathbf{r}}) = \Psi_{nl,\rho}(t_0)\Psi^*_{nl,\rho}(t_0)\grave{f}(\grave{\mathbf{r}})f^*(\grave{\mathbf{r}}) \tag{10.17}$$

$$=\; \rho(t_0)\grave{f}(\grave{\mathbf{r}})\grave{f}^*(\grave{\mathbf{r}}), \tag{10.18}$$

where $V_U(t)$ is the volume of the universe at time t, the time dependent spherical volume over which the integration is taken at time t and equations, (10.14), (10.15) and (10.16), holding because the total mass within the universe is a constant over time. In other words M_U is a time conserved quantity or within the universe's changing boundary, density movement should satisfy the equation of continuity which in the usual coordinates is

$$\partial\rho(t,\mathbf{r})/\partial t = -\nabla(\mathbf{v}(t,\mathbf{r})\rho(t,\mathbf{r})). \tag{10.19}$$

From equations (10.15) and (10.18), we get

$$M_U(t_0) \;=\; R_\Lambda{}^3 \int \int \int_{V_U(t_0)} \rho(t_0,\grave{\mathbf{r}})d\grave{x}d\grave{y}d\grave{z} \tag{10.20}$$

$$=\; \rho(t_0)R_\Lambda{}^3 \int \int \int_{V_U(t_0)} \grave{f}(\grave{\mathbf{r}})\grave{f}^*(\grave{\mathbf{r}})d\grave{x}d\grave{y}d\grave{z}. \tag{10.21}$$

Thus once the function, $\grave{f}(\grave{\mathbf{r}})$, is chosen, it appears that we can find the constant value of the mass of the universe, M_U. However, this appearance is deceptive because there is the complication that to get a constant valued numerical value from this equation we have to have a constant valued volume to integrate over while $V_U(t_0)$ depends on t_0 and so is in a sense time variable. It is necessary to have a value for M_U so that the value of the dimensioned length multiplier $b = (R_\lambda/c)^{2/3}(2M_U G)^{1/3}$ in the radius of the universe can be considered known,

$$r(t) = b \sinh^{2/3}(\pm 3ct/(2R_\Lambda)) \tag{10.22}$$

$$b = (R_\Lambda/c)^{2/3} C^{1/3} \tag{10.23}$$

$$R_\Lambda = (3/\Lambda)^{1/2} \tag{10.24}$$

$$C = 2M_U G. \tag{10.25}$$

Thus we seem to be left with the only options of finding the value of M_U from experiment or just accept that it is an arbitrary dimensioned constant until some alternative route to finding its value is found. The numerical value of M_U makes no difference to the theoretical structure of the theory, it only effects the numerical value of Rindler's constant, C, and any quantity in which this constant appears as a numerical multiplier which beside $r(t)$ the velocity of expansion $v(t)$ and the acceleration, $a(t)$, are involved. However, *importantly* for the non-linear Schrödinger equation, $H(t)$, does *not* involve the value of M_U,

$$H(t) = \dot{r}(t)/r(t) = (c/R_\Lambda)\coth(3ct/(2R_\Lambda)). \tag{10.26}$$

Because the integral in (10.21) is over the volume of the universe $V_U(t_0)$ which is given by

$$\frac{M_U}{V_U(t_0)} = \left(\frac{3}{8\pi G}\right)\left(\frac{c}{R_\Lambda}\right)^2 \sinh^{-2}\left(\frac{3ct_0}{2R_\Lambda}\right) = \rho(t_0)$$

$$= \left(\frac{\rho_\Lambda^\dagger}{2}\right)\sinh^{-2}\left(\frac{3ct_0}{2R_\Lambda}\right) = \rho(t_c) = \frac{M_U}{V_U(t_c)} \tag{10.27}$$

$$M_U(t_0) = \rho(t_0)R_\Lambda{}^3 \int\int\int_{V_U(t_0)} \hat{f}(\hat{\mathbf{r}})\hat{f}^*(\hat{\mathbf{r}})d\hat{x}d\hat{y}d\hat{z} \tag{10.28}$$

$$M_U(t_0) = M_U(t_0)\frac{R_\Lambda{}^3}{V_U(t_0)} \int\int\int_{V_U(t_0)} \hat{f}(\hat{\mathbf{r}})\hat{f}^*(\hat{\mathbf{r}})d\hat{x}d\hat{y}d\hat{z} \tag{10.29}$$

$$1 = \frac{R_\Lambda{}^3}{V_U(t_0)} \int\int\int_{V_U(t_0)} \hat{f}(\hat{\mathbf{r}})\hat{f}^*(\hat{\mathbf{r}})d\hat{x}d\hat{y}d\hat{z} \tag{10.30}$$

the relation, (10.30), gives a normalisation condition over physical space on a probability function density of space position variability, $\rho_{space}(\mathbf{r}) = \grave{f}(\grave{\mathbf{r}})\grave{f}^*(\grave{\mathbf{r}})$, following by cancellation of the mass of the universe M_U in the previous equation, which apparently holds from some definite time, t_0, at least. Thus the function $\rho_{space}(\mathbf{r})$ is just what is needed to describe the probability for finding mass at position $\mathbf{r}$, in the Schrödinger equation cosmology context at time, t_0. However, consistency demands that equation (10.30) holds, at least, for some specific time t_0. Thus we need to check out that such a time exists. From equation (10.27), we see much that we knew all along but, usefully, we see the value for the volume of the universe at time t_c, the time when deceleration changes to acceleration, is the obviously very constant value,

$$V_U(t_c) = \frac{M_U}{\rho_\lambda^\dagger} = \left(\frac{4\pi M_U G}{3}\right)\left(\frac{R_\Lambda}{c}\right)^2 \tag{10.31}$$

that we need to evaluate the apparently time dependent multiples integrals such as

$$1 = \frac{R_\Lambda^3}{V_U(t_c)} \int \int \int_{V_U(t_c)} \grave{f}(\grave{\mathbf{r}})\grave{f}^*(\grave{\mathbf{r}}) d\grave{x} d\grave{y} d\grave{z}. \tag{10.32}$$

Thus we seem very near a prescription for a usable cosmological Schrödinger equation. However, given a space dependent solution like (10.10) it is likely that the part $-\frac{\hbar^2}{2m}\nabla^2\Psi(r,t)$ of the quantum version at (9.6) would occur and this might render the cosmological quantum version not consistent with cosmology. It is by no means certain that such a complication would necessarily occur and not be handleable but certainly it can be avoided by playing safe and imposing the condition on this term as being zero as follows,

$$\frac{\hbar^2}{2m}\nabla^2\Psi(t,\grave{\mathbf{r}}) = 0. \tag{10.33}$$

This implies that the function $\grave{f}(\grave{\mathbf{r}})$ from equation (10.11) also satifies the Laplace equation,

$$\nabla^2\grave{f}(\grave{\mathbf{r}}) = 0. \tag{10.34}$$

The *Laplace* equation has a very large number of solutions. Thus there are many possible space dependent versions for the wave function, $\Psi(t,\grave{\mathbf{r}})$.

Furthermore, I shall show that in spite of the cosmological Schrödinger being non-linear, the many solutions of the Laplace equation can be *linearly* superposed to produce yet more solutions. Thus although the condition (10.33) reduces the number of possibilities that might be considered for the space dependent wave function it leaves us more than enough solutions to think about for a very long time. It does have another advantage that could turn out to be important concerning a possible quantum conjugate momentum, $\hat{\mathbf{p}}_C$, for the space variable $\mathbf{r}$. this can be defined as

$$\hat{\mathbf{p}}_C = \frac{\hbar \partial}{\partial \mathbf{r}} = \hbar \nabla. \tag{10.35}$$

and this momentum exists as a result of the Laplace equation (10.33) and automatically takes the form after operating on the wave function as follows

$$\hat{\mathbf{p}}_C \Psi(t, \dot{\mathbf{r}}) = \hbar \nabla \wedge \mathbf{g}(t, \dot{\mathbf{r}}), \tag{10.36}$$

where $\mathbf{g}(t, \dot{\mathbf{r}})$ is some definite vector function of t and $\dot{\mathbf{r}}$.

The wave motion followed by the dark mass dark energy time relation process can help to identify the effect that introducing position dependence has on the hyperspace vacuum. Space dependence implies the need to see this process as also space dependent. The density functions for the dark mass, dark energy and the ratio, $r_{\Lambda,DM}(t)$, of dark energy to dark mass as functions of the time only global process are respectively represented by

$$\rho(t) = (3/(8\pi G))(c/R_\Lambda)^2 \sinh^{-2}(3ct/(2R_\Lambda)) \tag{10.37}$$
$$\rho_\Lambda^\dagger = (3/(4\pi G))(c/R_\Lambda)^2 \tag{10.38}$$
$$r_{\Lambda,DM}(t) = \rho_\Lambda^\dagger/\rho(t) = 2\sinh^2(3ct/(2R_\Lambda)) \tag{10.39}$$
$$r_{\Lambda,DM}(\pm t_c) = 2\sinh^2(\pm 3ct_c/(2R_\Lambda)) = 1. \tag{10.40}$$

The *space time* dependent version for (10.37) is given simply by multiplying both sides of this equation by the space dependant contribution $\dot{f}(\dot{\mathbf{r}})\dot{f}^*(\dot{\mathbf{r}})$ giving

$$\rho(t, \mathbf{r}) = (3/(8\pi G))(c/R_\Lambda)^2 \dot{f}(\dot{\mathbf{r}})\dot{f}^*(\dot{\mathbf{r}}) \sinh^{-2}(3ct/(2R_\Lambda)) \tag{10.41}$$
$$= (\Lambda c^2/(8\pi G))\dot{f}(\dot{\mathbf{r}})\dot{f}^*(\dot{\mathbf{r}}) \sinh^{-2}(3ct/(2R_\Lambda)). \tag{10.42}$$

From equation (10.42), it follows that the cosmological constant Λ and the space dependence function can be taken together to define a *local space dependent cosmological function,* $\Lambda(\mathbf{r})$, associated with any specific solution of the Laplace equation as follows

$$\Lambda(\mathbf{r}) \;=\; \Lambda \mathring{f}(\mathring{\mathbf{r}}) \mathring{f}^*(\mathring{\mathbf{r}}) \tag{10.43}$$

$$\;=\; \Lambda \mathring{f}(\mathbf{r}/R_\Lambda) \mathring{f}^*(\mathbf{r}/R_\Lambda). \tag{10.44}$$

It follows from this definition, that the mean value of the cosmological function is equal to Λ for all solutions of the Laplace equation. In other words, the cosmological *function* is centred on Einstein's cosmological constant.

The linearity superposition of the various solutions of the Cosmological Schrödinger equation (9.11) to produce more solutions follows from (10.10) as in the following. Suppose we have two arbitrarily chosen spatially different solutions of this equation labelled with subscripts 1 and 2 as in

$$\Psi_{nl,\rho,1}(t,\mathring{\mathbf{r}}) \;=\; \Psi_{nl,\rho,1}(t_0,\mathring{\mathbf{r}}) \exp\left(-\frac{3}{2}\int_{t_0}^{t} H(t')dt'\right)$$
$$\tag{10.45}$$

$$\Psi_{nl,\rho,2}(t,\mathring{\mathbf{r}}) \;=\; \Psi_{nl,\rho,2}(t_0,\mathring{\mathbf{r}}) \exp\left(-\frac{3}{2}\int_{t_0}^{t} H(t')dt'\right)$$
$$\tag{10.46}$$

$$\Psi_{spp}(t,\mathring{\mathbf{r}}) \;=\; c_1\Psi_{nl,\rho,1}(t,\mathring{\mathbf{r}}) + c_2\Psi_{nl,\rho,2}(t,\mathring{\mathbf{r}})$$
$$\;=\; \Psi_{spp}(t_0,\mathring{\mathbf{r}}) \exp\left(-\frac{3}{2}\int_{t_0}^{t} H(t')dt'\right), \tag{10.47}$$

where c_1 and c_2 are arbitrary constants and the subscript *spp* means superposed. It follows from (10.47) that any number of solutions of the cosmological Schrödiner equation can be linearly superposed to produce yet further solutions. Thus, altogether, there is vast scope to produce solutions with almost any space form whatsoever. The common feature of all the solutions is that that they all related to the common cosmological platform defined *by and with* the same time variation structure of the space constant density function of the Friedman equations. The final prescription for finding solutions to the cosmological Schrodinger equation involve the following three steps. Find any solution, f, to the three dimensional Laplace equation and involve in this solution one initially multiplicative arbitrary constant, A_0. Form the space-time wave function for this solution, $\Psi(t,\mathbf{r})$. Find the value of A_0 by using the probability normalisation condition and integration over the Hermitian square of f over the volume of the universe at time, t_c. The wave function will then be completely determined. The probability density is also now fully determined via the definition $\rho_C(t,r) = \Psi(t,\mathbf{r})\Psi^*(t,\mathbf{r})$. The result will be a probability density function over space and time which is compatible with the Friedman equations from general relativity. The steps will be demonstrated in the next subsection for one typical case.

10.1 A Simple Example

I shall finish this paper with the simplest nontrivial example giving a universe that involves a varying space and time density. One of the simplest solutions, $f(\mathbf{\dot{r}})$, to the Laplace equation (10.34) is the sum of three variable complex numbers and just one arbitrary dimensionless constant, A_0,

$$
\begin{aligned}
f(\mathbf{\dot{r}}) &= A_0((\dot{x}+i\dot{y})+(\dot{y}+i\dot{z})+(\dot{z}+i\dot{x})) \\
&= A_0(\dot{x}+\dot{y}+\dot{z})(1+i). \qquad (10.48) \\
\dot{f}^*(\mathbf{\dot{r}}) &= A_0(\dot{x}+\dot{y}+\dot{z})(1-i) \qquad (10.49) \\
\dot{f}(\mathbf{\dot{r}})\dot{f}^*(\mathbf{\dot{r}}) &= 2A_0^2(\dot{x}+\dot{y}+\dot{z})^2 \qquad (10.50) \\
&= 2(A_0/R_\Lambda)^2(x+y+z)^2 = F(\mathbf{r}), \; say. \qquad (10.51)
\end{aligned}
$$

The definition (10.51) displays the formula in terms of the physical space coordinates, x, y, z. The normalisation condition on the probability density, (10.30), at time t_c requires the following two results

$$
1 = \frac{1}{V_U(t_c)} \int \int \int_{V_U(t_c)} \dot{f}(\mathbf{\dot{r}})\dot{f}^*(\mathbf{\dot{r}})dxdydz \qquad (10.52)
$$

$$
V_U(t_c) = \left(\frac{4\pi M_U G}{3}\right)\left(\frac{R_\Lambda}{c}\right)^2. \qquad (10.53)
$$

We need to evaluate the triple integral over the physical coordinates to find the value of the arbitrary constant A_0. This will be done in spherical polar coordinates with some condensations of notation used for the sin and cos functions,

$$x = r\sin(\theta)\cos(\phi) = rS_\theta C_\phi \tag{10.54}$$
$$y = r\sin(\theta)\sin(\phi) = rS_\theta S_\phi \tag{10.55}$$
$$z = r\cos(\theta) = rC_\theta \tag{10.56}$$
$$dxdydz = r^2dr S_\theta d\theta d\phi \tag{10.57}$$
$$0 < \theta \le \pi, \ 0 < \phi \le 2\pi, \ 0 < r \le r(t_c) \tag{10.58}$$
$$r(t_c) = (M_U G(R_\Lambda/c)^2)^{1/3} \tag{10.59}$$

Thus the function, $F(\mathbf{r}) = \hat{f}(\hat{\mathbf{r}})\hat{f}^*(\hat{\mathbf{r}})$, in the triple integral becomes

$$F(\mathbf{r}) = 2(A_0/R_\Lambda)^2(x+y+z)^2 \tag{10.60}$$
$$= 2(A_0 r/R_\Lambda)^2(S_\theta C_\phi + S_\theta S_\phi + C_\theta)^2 \tag{10.61}$$
$$= 2(A_0 r/R_\Lambda)^2((1 + 2(S_\theta C_\phi S_\theta S_\phi + S_\theta C_\phi C_\theta + S_\theta S_\phi C_\theta)) \tag{10.62}$$
$$= 2(A_0 r/R_\Lambda)^2((1 + 2(S_\theta S_\theta S_\phi C_\phi + S_\theta C_\theta C_\phi + S_\theta C_\theta S_\phi)) \tag{10.63}$$

Introducing the further notation

$$i_r = r^4 dr \qquad I_r = \int_0^{r(t_c)} i_r = r^5(t_c)/5 \tag{10.64}$$

$$i_{\theta,1} = S_\theta^3 d\theta, \qquad I_1 = \int_0^\pi i_{\theta,1} = \frac{4}{3} \tag{10.65}$$

$$i_{\phi,1} = S_\phi C_\phi d\phi, \quad I_2 = \int_0^{2\pi} i_{\phi,1} = 0 \tag{10.66}$$

$$i_{\theta,2} = S_\theta^2 C_\theta d\theta, \quad I_3 = \int_0^\pi i_{\theta,2} = 0 \tag{10.67}$$

$$i_{\phi,2} = C_\phi d\phi, \qquad I_4 = \int_0^{2\pi} i_{\phi,2} = 0 \tag{10.68}$$

$$i_{\phi,3} = S_\phi d\phi, \qquad I_5 = \int_0^{2\pi} i_{\phi,3} = 0 \tag{10.69}$$

the integral element and the integral can be expressed as

$$dI = 2(A_0 r/R_\Lambda)^2((1 + 2(S_\theta S_\theta S_\phi C_\phi + S_\theta C_\theta C_\phi + S_\theta C_\theta S_\phi))r^2 dr S_\theta d\theta d\phi$$
$$= 2(A_0/R_\Lambda)^2 i_r(d\theta d\phi + 2(i_{\theta,1}i_{\phi,1} + i_{\theta,2}i_{\phi,2} + i_{\theta,2}i_{\phi,3})) \tag{10.70}$$

$$I(t_c) = (4/5)\left(\frac{A_0\pi}{R_\Lambda}\right)^2 r^5(t_c). \tag{10.71}$$

252

The last expression for $I(t_c)$ is all that is left after integration. The normalisation condition at time t_c using (10.31) can now be used to find the numerical value of A_0 by

$$1 = I/V_U(t_c) = (4/5)\left(\frac{A_0\pi}{R_\Lambda^2}\right)^2 r^5(t_c)\left(\frac{3c^2}{4\pi M_U G}\right) \qquad (10.72)$$

$$= \frac{3A_0^2\pi}{5}\left(\frac{M_U G}{c^2 R_\Lambda}\right)^{2/3} \qquad (10.73)$$

$$A_0 = \left(\frac{5}{3\pi}\right)^{1/2}\left(\frac{c^2 R_\Lambda}{M_U G}\right)^{1/3} \qquad (10.74)$$

Thus the full solution for the wave function, the probability density and all the constants involved is as follows:

$$\Psi_{nl,\rho}(t,\mathbf{\grave{r}}) = \Psi_{nl,\rho}(t_c,\mathbf{\grave{r}})\exp\left(-\frac{3}{2}\int_{t_c}^{t} H(t')dt'\right) \qquad (10.75)$$

$$\Psi_{nl,\rho}(t_c,\mathbf{\grave{r}}) = \Psi_{nl,\rho}(t_c)\grave{f}(\mathbf{\grave{r}}) \qquad (10.76)$$

$$\Psi_{nl,\rho}(t_c) = (\rho_\Lambda^\dagger)^{1/2} \qquad (10.77)$$

$$\grave{f}(\mathbf{\grave{r}}) = (A_0/R_\Lambda)(x+y+z)(1+i) \qquad (10.78)$$

$$\rho(t,\mathbf{r}) = 2(A_0/R_\Lambda)^2\rho_\Lambda^\dagger(x+y+z)^2\exp\left(-3\int_{t_c}^{t} H(t')dt'\right) \qquad (10.79)$$

$$\rho_\Lambda^\dagger = \frac{\Lambda c^2}{4\pi G} \qquad (10.80)$$

$$A_0 = \left(\frac{5}{3\pi}\right)^{1/2}\left(\frac{c^2 R_\Lambda}{M_U G}\right)^{1/3}. \qquad (10.81)$$

11   Appendix Conclusions

In an earlier paper, it was shown that a non-linear Shrödinger equation can be obtained from the Friedman cosmology equations which is entirely consistent with those equations. Here, the time evolution of this Schrödinger equation is examined in relation to conservation of the universe's total positive gravitational mass. This leads to the identification of a wave function for cosmology states with a definite time evolution and consequently also to a probability density for cosmology. This cosmological probability density can depend on spatial variability in addition to just the time variability of the Friedman equation structure. Consistency of the new Schrödinger equation with its originating Friedman set is achieved by restricting solutions to the condition that they satisfy the Laplace equation in hyperspace. It becomes clear that, even with this restriction, a multiple infinity of solutions remain available and applicable. The structure of this theory seems to confirm the view often expressed about the *quantum vacuum* that it is a bubbling cauldron of activity in the form of random quantum transitions, such as pair production and annihilation, between short lived virtual states of fundamental particles. The expansion of the universe can be explained in such terms as a spherical advancing and evolving wave of quantum *before and after measurement type conditions* in reverse through the expanding boundary. Just outside the expanding boundary, the vacuum chaotic states as described by the *wave function*, resourced by the multiplicity of solutions of the Laplace equation, are progressively converted from chaos to a definite gravitational form sufficient to describe the mass density that has taken up residence within the expanded boundary. The universe expansion colonises surrounding hyperspace so as to accommodate within its boundary its *conserved positive* gravitational mass with more territory and in a quantum form that can hold non-transient positive gravitational mass. Outside the universe the solution holds but remains a linear superposition of many varied chaotic transient states with mass density value centred on the value of twice Einstein's dark energy mass density $\rho_\Lambda^\dagger$.

References

[1] R. A. Knop et al. arxiv.org/abs/astro-ph/0309368
New Constraints on Ω_M, Ω_Λ and ω from
an independent Set (Hubble) of Eleven High-Redshift
Supernovae, Observed with HST

[2] Adam G. Riess et al xxx.lanl.gov/abs/astro-ph/0402512
Type 1a Supernovae Discoveries at $z > 1$
From The Hubble Space Telescope: Evidence for Past
Deceleration and constraints on Dark energy Evolution

[3] Berry 1978, Principles of cosmology and gravitation, CUP

[4] Gilson, J.G. 1991, Oscillations of a Polarizable Vacuum, Journal of Applied Mathematics and Stochastic Analysis, **4**, 11, 95–110.

[5] Gilson, J.G. 1994, Vacuum Polarisation and The Fine Structure Constant, Speculations in Science and Technology , **17**, 3 , 201-204.

[6] Gilson, J.G. 1996, Calculating the fine structure constant, Physics Essays, **9** , 2 June, 342-353.

[7] Eddington, A.S. 1946, Fundamental Theory, Cambridge University Press.

[8] Kilmister, C.W. 1992, Philosophica, **50**, 55.

[9] Bastin, T., Kilmister, C. W. 1995, Combinatorial Physics World Scientific Ltd.

[10] Kilmister, C. W. 1994 , Eddington's search for a Fundamental Theory, CUP.

[11] Peter, J. Mohr, Barry, N. Taylor, 1998, Recommended Values of the fundamental Physical Constants, Journal of Physical and Chemical Reference Data, AIP

[12] Gilson, J. G. 1997, Relativistic Wave Packing and Quantization, Speculations in Science and Technology, **20** Number 1, March, 21-31

[13] Dirac, P. A. M. 1931, Proc. R. Soc. London, **A133**, 60.

[14] Gilson, J.G. 2007, www.fine-structure-constant.org The fine structure constant

[15] McPherson R., Stoney Scale and Large Number Coincidences, Apeiron, Vol. 14, No. 3, July, 2007

[16] Rindler, W. 2006, Relativity: Special, General and Cosmological, Second Edition, Oxford University Press

[17] Misner, C. W.; Thorne, K. S.; and Wheeler, J. A. 1973, Gravitation, Boston, San Francisco, CA: W. H. Freeman

[18] J. G. Gilson, 2004, Physical Interpretations of
Relativity Theory Conference IX
London, Imperial College, September, 2004
Mach's Principle II

[19] J. G. Gilson, A Sketch for a Quantum Theory of Gravity:
Rest Mass Induced by Graviton Motion, May/June 2006,
Vol. 17, No. 3, Galilean Electrodynamics

[20] J. G. Gilson, arxiv.org/PS_cache/physics/pdf/0411/0411085v2.pdf
A Sketch for a Quantum Theory of Gravity:
Rest Mass Induced by Graviton Motion

[21] J. G. Gilson, arxiv.org/PS_cache/physics/pdf/0504/0504106v1.pdf
Dirac's Large Number Hypothesis
and Quantized Friedman Cosmologies

[22] Narlikar, J. V., 1993, Introduction to Cosmology, CUP

[23] Gilson, J.G. 2005, A Dust Universe Solution to the Dark Energy
Problem, Vol. 1, *Aether, Spacetime and Cosmology*,
PIRT publications, 2007,
arxiv.org/PS_cache/physics/pdf/0512/0512166v2.pdf

[24] Gilson, PIRT Conference 2006, Existence of Negative Gravity
Material, Identification of Dark Energy,
arxiv.org/abs/physics/0603226

[25] G. Lemaître, Ann. Soc. Sci. de Bruxelles
Vol. A47, 49, 1927

[26] Ronald J. Adler, James D. Bjorken and James M. Overduin 2005,
Finite cosmology and a CMB cold spot, SLAC-PUB-11778

[27] Mandl, F., 1980, Statistical Physics, John Wiley

[28] Rizvi 2005, Lecture 25, PHY-302,
http://hepwww.ph.qmw.ac.uk/~rizvi/npa/NPA-25.pdf

[29] Nicolay J. Hammer, 2006
www.mpa-garching.mpg.de/lectures/ADSEM/SS06_Hammer.pdf

[30] E. M. Purcell, R. V. Pound, 1951, Phys. Rev.,**81, 279**

[31] Gilson J. G., 2006, www.maths.qmul.ac.uk/~ jgg/darkenergy.pdf
Presentation to PIRT Conference 2006

[32] Gilson J. G., 2007, Thermodynamics of a Dust Universe,
Energy density, Temperature, Pressure and Entropy
for Cosmic Microwave Background http://arxiv.org/abs/0704.2998

[33] Beck, C., Mackey, M. C. http://xxx.arxiv.org/abs/astro-ph/0406504

[34] Gilson J. G., 2007, Reconciliation of Zero-Point and Dark Energies
in a Friedman Dust Universe with
Einstein's Lambda, http://arxiv.org/abs/0704.2998

[35] Rudnick L. et al, 2007, WMP Cold Spot, Apj in press

[36] Gilson J. G., 2007, Cosmological Coincidence Problem in
an Einstein Universe and in a Friedman Dust Universe with
Einstein's Lambda, Vol. 2, *Aether, Spacetime and Cosmology*,
PIRT publications, 2008

[37] Freedman W. L. and Turner N. S., 2008, Observatories of the Carnegie
Institute Washington, Measuring and Understanding the Universe

[38] Gilson J. G., 2007, Expanding Boundary Pressure Process.
All pervading Dark Energy Aether in a Friedman Dust Universe
with Einstein's Lambda, Vol. 2, *Aether, Spacetime and Cosmology*,
PIRT publications, 2008

[39] Gilson J. G., 2007, Fundamental Dark Mass, Dark Energy Time
Relation in a Friedman Dust Universe and in a Newtonian Universe
with Einstein's Lambda, Vol. 2, *Aether, Spacetime and Cosmology*,
PIRT publications, 2008

[40] Gilson J. G., 2008, . A quantum Theory Friendly Cosmology
Exact Gravitational Waves in a Friedman Dust Universe
with Einstein's Lambda, PIRT Conference, 2008

The Physical Origins of Dark Matter and Dark Energy: Exploring Shadows on the Cave Wall

James E. Beichler
P.O. 624
Belpre, Ohio 45714
Jebco1st@aol.com

Abstract

Abstract: Nearly three decades ago, a physical anomaly that should have shaken physics to its very foundations was observed: The galaxy rotation problem. The new concept of Cold Dark Matter was invoked to explain this anomaly. Alternative explanations such as Modified Newtonian Dynamics and models utilizing specific variations in Newton's Universal Gravitational Constant have been proposed, but altering the basis of Newtonian physics by adding fudge factors to the laws of motion and/or gravity are aesthetically displeasing as well as completely unnecessary. The problem has been further complicated by the recent discovery of Dark Energy, which must be related to Dark Matter at some fundamental level. Both the Dark Matter halo that surrounds galaxies and the Dark Energy that seems to be propelling expansion in the universe at an increasing pace can be easily explained if scientists are willing to accept the physical reality of a fourth dimension of space, which amounts to a fifth dimension of space-time. Although space-like, this new dimension is uniquely different from the normal three dimensions of space. This solution may seem radical, but there is ample evidence in other areas of physics to support the existence of a fourth spatial dimension. What is commonly called Dark Matter is no more nor less than an extrinsic curvature of common space-time that is not directly associated with local matter, but is instead a local variation in the total or global curvature of the universe induced by nearby matter.

Keywords: galactic rotation problem, halo, Dark Matter, CDM, HDM, Dark Energy, General Relativity, Kaluza, five-dimensional, fourth dimension, MOND, Modified Gravity, Randall-Sundrum, Branes, space-time, curvature

Introduction

The scientific community has now been faced with a major crisis for more than three decades, the galactic rotation problem, and that crisis has been further complicated by an even greater crisis over the last decade with the detection of an increase in the expansion rate of the universe. Stars in the rims of spiral galaxies move at constant speeds as if large quantities of invisible matter surround the galaxies in 'halos'. Vera Rubin and Kent Ford discovered these constant speeds nearly three decades ago (Rubin and Ford, 1970; Rubin, et.al., 1985), although Fritz Zwicky and Sinclair Smith had originally predicted Dark Matter (DM) during the 1930s (Zwicky, 1933; Smith, 1936; Zwicky, 1937a; Zwicky, 1937b). The existence of these halos was further confirmed when clusters of galaxies were observed to exhibit motions that could require as much as ten times the material content of the visible portions of the galaxies that makeup the clusters (Oort, 1940). Scientists assume that the halos are made of Cold Dark Matter (CDM) since no apparent source of these gravitational attractions is visible. The 'matter' in the halo is assumed cold because it has no discernible (or very low) kinetic energy, *i.e.*, it is devoid of motion whatever it is. It is dark because it neither emits nor reflects visible light nor other electromagnetic waves. The gravitational source is assumed to be material simply because science knows of nothing other than matter that can act gravitationally.

The intimate physical relationship between matter and gravity is expressed in all of its beauty and elegance by General Relativity (GR), which equates matter (as the energy-stress tensor $\mathbf{T}_{\mu\nu}$) directly to space-time curvature (as the metric tensor $\mathbf{G}_{\mu\nu}$):

$$\mathbf{G}_{\mu\nu} = \kappa\mathbf{T}_{\mu\nu},$$

where the constant κ is equal to $8\pi G/c^4$ and G is Newton's universal gravity constant. In the purest mathematical interpretation of general relativity, neither side of the relation between matter and curvature should be more nor less real than the other side. However, science is materially biased to believe that matter is the reality and curvature is just a product of that matter. For example, John Archibald Wheeler often described our real world of matter as the cause of curvature, while the curvature moved matter: "Matter tells space how to curve. Space tells matter how to move". Two and a half millennia ago, Plato created a metaphor to explain a similar situation within the context of his own knowledge of physical reality. He described what humans sense and perceive as reality as a shadow on a cave wall, while the true reality was that which cast the shadow. The same can be said for the relationship between matter and curvature, but new questions arise when this metaphor is applied to the modern situation. Is matter the true reality? Is curvature the true reality? Or is the true reality something of a hybrid between the two? If curvature is the reality, then is matter merely how humans and other living beings perceive curvature?

Given the present level of theoretical research and scientific observation, it would be difficult to answer the question whether matter is the shadow or matter is casting the shadow, so which is the reality? Yet this question does not even seem to exist in modern science. Einstein's equation does not reveal whether matter or curvature, shadow or shadow source, represents the ultimate physical reality of our world, so we do not know which is which. In other words, there is nothing in physics that would restrict space-time curvature to being the product of matter and have no other reality except that which is attached to it by matter, such that curvature could only exist due to the local presence of matter. In the end, there is no logical or theoretical reason why curvature could not exist independent of local matter and still influence nearby matter gravitationally, even though our five normal senses predispose us to accept matter rather than curvature as the reality. In other words, if we assume the existence of a real fourth space-like dimension (the fifth dimension of space-time) that is macroscopically extended, then space-time curvature could account for the CDM content of the galactic halo as well as Dark Energy (DE), even lacking local matter as a direct source for the curvature. The DM and DE problems can then be reduced to simple geometrical solutions within a suitably extended interpretation of General Relativity.

The evolving crisis

Unfortunately, science has failed to recognize either the simplicity of the solution or the extremely critical nature of the galactic rotation problem, preferring instead to believe that the present paradigms of science will eventually solve the DM crisis. Quantum answers to the problem of explaining the anomalies in galactic speeds range from hypothetical WIMPS (Weakly Interacting Massive Particles) to hypothetical non-baryonic particles and super-symmetric particles, but the existence of these particles has never been verified. Physics almost seems driven to invent new particles every time an anomaly arises, but the growing list of particles only presents new problems whose solutions science must put off until a later day and a higher level of knowledge. Quite clearly, the invention of new particles cannot answer the question of CDM and the galactic halo. Nor does the practice seem adequate to account for DE.

Mordechai Milgrom's Modified Newtonian Dynamics, or MOND theory, (Milgrom, 1983) presents a more classical solution to the problem. Milgrom proposed that Newton's second law of motion, F = ma, be rewritten by adding a correction factor of $\mu(x)$, which is a function of distance and can be altered at will to fit any given gravitational discrepancy. However, modifying Newton's basic laws of motion by adding a correction factor without any other evidence of a problem with Newton's laws of motion does not exactly represent the best of science. Milgrom's addition of an *ad hoc* factor to Newton's second law would almost seem to be more of a 'fudge factor' without additional independent corroboration or at least a need for the new factor from another unrelated source. It is extremely easy in physics to just add a mathematical factor to explain away problems rather than to do the physics and solve the problems correctly without corroborating support or necessity. Under these circumstances, MOND seems more a method to explain away the galactic halo phenomenon than to get at the fundamental physics that is necessary to explain CDM and the observed galactic halo.

Jacob Bekenstein recently developed (Bekenstein, 2004; Bekenstein, 2007) a more advanced version of MOND, or rather a more complete relativistic theory that reduces to MOND in the Newtonian limit at large distances. It is called TeVeS (Tensor-Vector-Scalar theory) because the space-time metric of GR is supplemented by vector and scalar terms. However, Bekenstein's theory refers more to DE than it does to DM. Without such a modification, MOND alone suffers from having no association with relativity and does not address relativistic limits, even though GR forms the basis of modern cosmology and astrophysics, just the field of study to which MOND applies. TeVeS clarifies this issue. In so doing, TeVeS obviates the need for DM. TeVeS reduces to MOND in the Newtonian limit, but Lorentz Invariance is spontaneously violated producing a major problem for the theory. So, even this extended version of MOND is not completely adequate.

Both quantum solutions and Newtonian solutions seem somewhat artificial since the galactic rotation problem is a gravitational problem in cosmology, which is defined by GR, rather than a problem for the fundamental laws of motion. In recognition of this fact, some scientists have suggested that the gravitational constant G may be different inside galaxies and outside of galaxies (Drummond, 2000). In other words, the value of G conveniently differs on the large and small scales as studied in cosmology. If there is nothing 'Dark' and there is no galactic halo, then gravity must differ from the common value of G that science measures in the laboratory on the small scale. This suggestion represents only one of a group of models that are known as modified gravity theories (Bludman, 2006), or just Modified Gravity (MG). These theories can either alter the present equations that are used to model space-time or retain the present equations and alter the value of G. In the latter case, the value of G may depend upon the location of gravitational phenomena within the greater universe rather than be a true single-valued constant for the whole universe. Given only the existence of the DM halos and nothing else, there is no logical reason to accept such a suggestion. Nor is there any observational or experimental evidence to corroborate the possibility that G is multi-valued depending on the relative location of any particular gravitating bodies that are being considered.

Yet it is clearly understood that any variation in either the Newtonian or relativistic theories of gravity will also modify Newton's second law, $\mathbf{F} = m\mathbf{a}$, so MG theories have a great deal more to overcome than is immediately evident. Modifying either G or $\mathbf{F} = m\mathbf{a}$ would have unintended consequences elsewhere in our normal physical laws, so the idea is to minimize the consequences, as best as possible, in all but galactic-size distances. Any such variations in F = ma would hopefully be limited to extremely small values of acceleration, which are extremely difficult to measure or they would affect other physical systems. As it now stands, G has been measured accurately in the laboratory down to an acceleration value of 5×10^{-14} m/sec^2 (Gundlach, et.al., 2007). Yet even this accurate a value does not disprove the MG theories. In order to do so, the experimental value of G must be measured outside of the perturbative effects of nearby gravitational fields, which has never been done. Otherwise, this accurate a measure at such a small scale still tends to render the MG models less likely. It certainly does not help them. Instead, it pushes them out of the DM arena and into the K-essence or quintessence arena, which again raises questions that deal more with the DE issue than the DM mystery.

A more recent theory, the Randall-Sundrum (1999a; 1999b) braneworld model could also account for some of the dim objects thought to fill the halo as CDM. Charles R. Keeton and A.O. Petters have used the Randall-Sundrum model to predict the existence of microscopic black holes throughout the universe (Keeton and Petters, 2006). This prediction is in strict disagreement with general relativistic predictions, according to which primordial black holes would have developed earlier in the life of the universe (Hawking, 1974; Carr and Hawking, 1974; Carr, 1975), but would have evaporated by now due to Hawking radiation (Page, 1976a; Page, 1976b). NASA plans to launch the Laser Interferometry Space Antennae (LISA), a group of three satellites that will form an extended five kilometer array deep in space, by 2018. LISA will seek to detect gravity waves and their sources in the 0.1 to 100 mHz range of electromagnetic waves, which cannot be detected on Earth due to atmospheric interference. Proponents of the model expect that this array should be able to detect microscopic black holes if they exist in our solar system, thus verifying the Randall-Sundrum braneworld model. Should such black holes be discovered, they might account for some of the CDM in the galactic halos, but this partial solution of the missing matter problem is a long shot at best and the Randall-Sundrum model has other problems, as do all superstring and brane models.

Enter Dark Energy

The simplest solution to the DM mystery is that it only seems at present that no single solution exists that fits all of the observed parameters and relates to all of the observed phenomena, which raises other suggestions and new questions. Instead of following a piecemeal road to solving the 'dark' mysteries of matter and energy, perhaps science should look for a single and complete solution to all of the problems together. Science may only be addressing and treating superficial phenomena and not looking directly at the underlying problems from a fundamental enough level. The CDM and HDM models dominate each of their respective areas of application and observation, but the nature of either type of DM has yet to be determined. Science should be asking the more fundamental question, what is matter? Under these conditions, the DE issue enters the fray. In many cases scientists seem to use the terms DE and DM interchangeably, further complicating the issues. For example, some scientists claim that DE shapes galaxies while others claim that DM shapes the galaxies. On the other hand, some scientists claim that HDM fills all of space, others that DE fills all of space and still others that both occupy the 'empty' vacuum of space.

In either case, the DM problem faced by the scientific community deepened, quite radically, and took a new twist in 1998 when teams headed by Saul Perlmutter and Adam Riess detected an increase in the expansion rate of the universe, which could be explained as the result of a new form of energy with negative pressure (Perlmutter, et al, 1999; Riess, et al, 1998) called DE. Both groups were investigating redshifts exhibited by Type Ia supernovae and noticed that their redshifts were dimmer by a small amount from that expected. The values thus obtained could only be explained by assuming that the expansion rate of the universe is increasing. This finding was completely unexpected since the standard model of cosmology posits that the expansion should be slowly decreasing due to gravitational attraction. The only thing that could counteract gravitational attraction would be a small negative pressure and the concept of DE was born. Type Ia supernovae are ideal for the precise measure of stellar redshift because their life cycles are very well defined, such that they provide a standard candle for consistency in measuring the absolute magnitude among different Type Ia supernovae.

The actual existence of DE has been questioned and is still open to debate, but a great deal more evidence for the existence of an increasing expansion rate has accumulated in the past few years and DE is still the leading hypothesis to explain the rate increase (Frampton, 2004). The evidence now comes from a variety of different astronomical sources, confirming the findings of Perlmutter and others measuring the redshift of Type Ia supernovae. Other astronomers have confirmed the original findings and attempts have been made to use other types of supernovae for corroborating measurement (Knop, et.al., 2003), but Type Ia supernovae still offer the most reliable and accurate results. Measurements of an anisotropy in the cosmic microwave background (CMB) have also been used as evidence for the existence of DE (Melchiorri, et.al., 1999; Hu and Dodelson, 2003) as well as gravitational lensing (Mellier, 1999) and the large-scale structure of the universe (Tegmark, et.al., 2003; Linder, 2003). These can be further supplemented by baryon acoustic oscillations (Linder, 2006) and galaxy cluster observations (Bahcall, 2000; Allen, et.al., 2002). All of the evidence that has been gathered so far confirms that the universe is expanding at an increasing rate rather than slowing as expected by the standard cosmological model based on classical GR. New methods of gathering data through astronomical observation and satellite technology are in the planning and implementation stages (Frampton, 2005; Bean, et.al., 2005; DETF, 2006). These measurements are not just being gathered to further confirm the increasing rate of expansion, but also to gain more information that might lead to distinguishing between the different models and hypotheses, if not to lead to new models and theories.

DE is not very dense, measuring about 10^{-26} kg/m^3 throughout empty space, yet it does seem to account for approximately 70% of the mass-energy content of the universe. It has certainly been established that our universe is not normal or Baryonic matter dominated, no matter what else has been accomplished, since astronomers have only managed to observe and measure a small fraction of the amount of normal matter needed to collapse the universe. Normal matter seems to constitute about 5% of that needed to close the universe and DM about 25% (Spergel, et al, 2006). DE is a more dynamical physical quantity than the CDM within the galactic halo, so its existence offers a more practical road to developing new models and theories than DM in the eyes of many scientists. A flock of different models have been suggested (Lima, 2004). In fact, a complete list of the suggested models would easily fill a whole page, even a page with very small type. However, most models are related to each other at some level, in recognition of the fundamental nature of the mystery. The various models thus seem to fall within three broad categories, although these categories do not directly represent all of the suggested solutions: Adding a cosmological constant, developing a type of quintessence or dynamical fluid and the breakdown of GR and the standard model of cosmology, which would at least require that substantial changes or additions be made to classical GR.

It is generally assumed that the conservation of DE could indicate that either the standard model of cosmology and/or the general theory of relativity upon which it is based are incorrect or incomplete. The failure to find gravitons and a particulate solution to the DM problem certainly indicate that quantum fields do not offer an effective alternative and this bodes poorly for an explanation of DE based on quantum considerations. However, the evidence could well indicate that the universe is be dominated by some as yet unidentified particle (which grows more and more unlikely as evidence accumulates), some type of quantum field that has negative pressure or GR must include a non-zero cosmological constant. One such particulate model posits new particles called 'accelerons' that only interact with neutrinos as if there were a stretched rubber band between them (Kaplan, et.al., 2004). The last choice of a non-zero cosmological constant is usually referred to as the Lambda-CDM model, where Lambda represents the energy density of vacuum space. If the cosmological constant Lambda is particle dominated, which would indicate a quantum solution to the problem, then new problems arise. In particular, the quantum vacuum energy of quantum field theory is greater than the amount of DE measured by astronomical observations by a factor of 10^{120} (Kilbinger and Hetterscheidt, 2006; Carroll, 2003). So once again, the answer to DE seems unrelated to the quantum theory, if not beyond the quantum theory. Yet any of these models would require fundamental alterations to physics as it is now accepted in science. All of the evidence gathered so far seems to confirm the Lambda-CDM model. Even though the Lambda-CDM model is the leading contender to explain DE, the model is far from confirmed and the ultimate nature and source of DE is far from being understood.

Otherwise, scientists can either modify gravity or accept the presence of DE. On the one hand, keeping gravitational theory as it is and accepting the reality of DE requires an explanation of DE, which has not exactly been forthcoming. However, the second possibility of modifying gravity and thus obviating the need for DE would mean either adding a new mathematical term to the left of Einstein's equation or adding one on the right side of the equation. From the physical point-of-view, these modifications would mean introducing new fields or extending known fields; at least that is the common interpretation. In this scenario, the DE problem changes to a problem of finding the proper mathematical modification and then explaining the physics that might support that modification. The new fields obtained in this manner may be independently dynamical, as is the case with DM, or they could result from ordinary matter, as with some MG models. The new dynamical fields could otherwise be occupied by some truly exotic form of matter such as quintessence or K-essence (Caldwell, et.al., 1998). But whatever the case may be, the answers offered so far seem extremely complex and far more complicated than may be necessary.

As of yet, there is no single experiment or known observation that can distinguish between the two main theories, quintessence and Lambda-CDM, let alone eliminate all of the other theories and models. Nor is there any way to even distinguish between MG and DE (Bludman, 2006), while the theoretical side of the DM and DE questions is still wide open and nearly any new suggestion for a possible solution must be taken seriously. However, attempts are being made to measure anisotropic stress, which can be observed with precise measurements of weak gravitational lensing (Kochanek, 2004) and these attempts should eliminate some of the models and physical possibilities from the field of contenders.

The problem is so important that a Dark Energy Task Force (DETF) was convened in 2006, under the auspices of the Department of Energy, the National Science Foundation, the National Aeronautics and Space Administration and other agencies. The short amount of time from the initial discovery of DE to the convening of a government sponsored Task Force emphasizes the significance of the problem for basic physics and science. The DETF concluded that

> Dark energy appears to be the dominant component of the physical Universe, yet there is no persuasive theoretical explanation. The acceleration of the Universe is, along with dark matter, the observed phenomenon which most directly demonstrates that our fundamental theories of particles and gravity are either incorrect or incomplete. Most experts believe that nothing short of a revolution in our understanding of fundamental physics will be required to achieve a full understanding of the cosmic acceleration. For these reasons, the nature of dark energy ranks among the very most compelling of all outstanding problems in physical science. These circumstances demand an ambitious observational program to determine the dark energy properties as well as possible (DETF, 2006).

It was certainly not easy for members of the DETF to conclude that our theories of matter are either 'incorrect or incomplete'. This conclusion flies directly in the face of modern trends in physics to place all known phenomena under the auspices of a single theory, usually based upon the quantum paradigm. If the presently accepted paradigms are either 'incorrect or incomplete', then a new revolution in science must be at hand and the Task Force came to that exact conclusion.

The DETF was composed of a distinguished group of scientists who voluntarily contributed white papers on the subject, without reward except for the knowledge that they could be contributing to the eventual solution of one of the greatest questions of science. The revolutionary nature of the existence of DE was clearly cited by the Task Force and the historical significance of the concept was clearly established. Quite frankly, the DM and DE problems that beset science strike directly at the very nature of matter and thus at the very heart of physics and science. Although this simple fact was not stated in the DETF report, it was certainly implied.

Although scientists are attacking the DM and DE problems separately, because

267

they represent two distinctly different sets of cosmological problems, most will admit that both problems must ultimately be solved together. DM and DE must be related at some fundamental level, just as normal matter and energy are related. Logically speaking, E = mc² for normal matter and there is no reason to believe that the same or a similar relation will not hold for DM and DE, whatever explanation is accepted as correct in the end. So it is safe to assume that solving one problem, say the nature of the CDM in the halo, will lead to a solution of the other problem, as many scientists suspect. Yet when push comes to shove, all the suggestions made to date either fall short, introduce new problems or otherwise fall victim to unintended and unwanted consequences. In reality, one small change in basic physics solves all of the DM and DE problems, yet that change is quite revolutionary in spite of its fundamental simplicity.

The DM problem

While many scientists are seeking a solution to the DE question first because it is a more dynamical physical quantity, it is actually simpler to find a geometrical solution to the CDM problem first and then use geometry to solve the mystery of DE. A typical plot of rotational speeds of stars around a galactic core as a function of radial distances shows that the speeds become approximately constant outside of the core; even though Newtonian gravity theory predicts that the speeds must fall off as the radial distance increases according to the inverse square law. Newtonian normal gravity is normally used to explain stellar motions in a spiral galaxy because it is generally believed that there are no relativistic effects at large distances from gravitating bodies, such that space-time curvature is a local effect of material bodies. A simple graph of the speeds for the Andromeda galaxy offers a case in point.

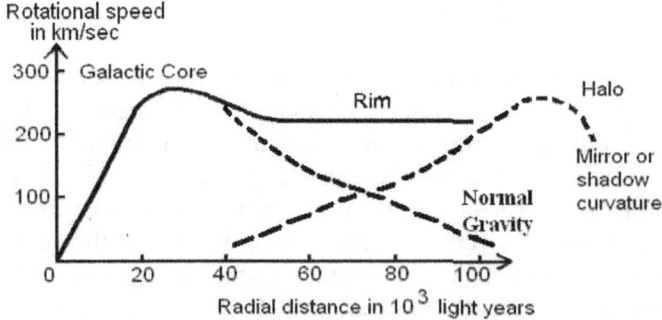

Fig 1: A suspected source for Andromeda's rotational speeds

These speeds could only be equalized by a gravitational source surrounding the galaxy that is strong enough to add a velocity component to the objects along this radius at least as large as that contributed by the core itself. This line of reasoning suggests that a mirror or shadow space-time curvature would have to extend completely around the galaxy in the form of a halo. This idea is well understood in cosmology and astrophysics, but it means that the amount of normal dim matter that many scientists would like to believe causes the curvature of the halo would need to be several times greater than the total amount of matter in the galactic core. Early findings for the case of galactic clusters indicate that the amount of CDM could be as high as ten times the normal visible matter that inhabits galaxies, yielding about 90% of the total amount of matter in galaxies. In this particular model the dim matter would also be spread out evenly throughout the halo yet concentrically around the galactic center beyond the rim, which is highly unlikely for different reasons.

Filling the galactic halo with dim bodies of baryonic matter such as black holes, brown dwarfs, large dark errant planets, non-luminous interstellar gases and other MACHOs (MAssive Compact Halo Objects) to account for the gravitational anomalies seems grossly inadequate. Normal forms of dim matter could not possibly make up the vast amount of halo material that would be necessary to account for the observed equalizing of rotational speeds in the galactic rims. Even if there were far more dim objects in the halo than in the galaxy proper, science would need to explain why. The presence of so much dim matter in the halo could raise even more questions than it answers. For example, if normal baryonic dim matter could be dispersed in such a halo pattern, why would dim matter alone fill the halo without the presence of any type of normal baryonic visible matter? It would be hard to even imagine a physical scenario whereby dim matter would be so highly concentrated in a region of space in this manner. So the existence of normal dim matter in such large concentrations is highly unlikely, let alone in such an odd geometrical distribution. Otherwise, the dim matter in the halo would seem to exhibit some very strange gravitational characteristics.

In other words, there are specific logical problems with a dim matter model of the halo. Baryonic matter tends to clump together in lumps due to gravitational attractions, but the extremely large amount of dim matter in the halo would very nearly form a very dense and stable material string or ring around galaxies rather than clump together at a central location. This structure could well indicate a non-traditional form of non-clumping gravitational attraction. Otherwise, if the halo were found to be formed from baryonic dim matter or some strange unknown form of matter acting during the original building phase of the galaxy, then scientists need to ask why the dim matter did not accrete with a greater concentration more toward the core as is normal. Yet the string or ring of matter forming the halo of dim matter would still need to attract material bodies in the galactic rim according to normal concepts of gravitational attraction to account for the uniform speeds. It would therefore seem unlikely that any type of normal matter, including dim matter, could ever account for the necessary space-time curvature exhibited by the halos. In other words, space-time is definitely curved in the galactic halo, but there is no known material source or process that could account for space-time curvature in the halo.

A 'single' (but radical) solution to multiple problems

It should be obvious, under these conditions, that a new approach to this 'Dark' problem is necessary, while the only viable solution to the problem is the first solution offered: A halo of CDM must surround spiral galaxies even though the nature of that CDM is neither defined nor identified as material, at least in any form with which scientists are presently familiar. Under these circumstances, science needs a fundamental new conceptual understanding of matter itself, as well as its relation to 'curvature', a fact that many scientists recognize. The existence of CDM clearly lacks a theoretical basis, a problem that is completely unnecessary. A theoretical basis for both CDM and Hot Dark Matter (HDM), as well as DE, can be found in one simple yet fundamental change in general relativity and our fundamental view of physical reality: The addition of a physically real fourth spatial dimension.

Theodor Kaluza added a fifth dimension to GR in 1921 (Kaluza, 1921). He was then able to derive both Maxwell's and Einstein's equations from a 'single field' model of the universe. But Kaluza's fifth dimension was merely a mathematical device to which he gave no physical meaning or reality. Kaluza obtained the proper mathematical formulations for gravity and electromagnetism in spite of this self-imposed restriction to his theory, but this restriction left his model with no predictive capabilities. It merely duplicated the formulas and equations that were already known, so Kaluza's theory gained very little support within the scientific community. In other words, Kaluza's five-dimensional model was not falsifiable as it was originally conceived.

However, endowing Kaluza's fifth dimension with a physical reality guarantees the falsifiability and explanatory power of the five-dimensional model while retaining the unification of general relativity and electromagnetism that Kaluza obtained. All that is necessary is to determine the physical properties of the fifth physical dimension, which could be difficult because it is generally assumed that any higher embedding dimension

270

has no direct sensible or measurable effect on physical phenomena. This belief, albeit false, renders the simple solution of a fourth dimension of space quite radical because it requires such an assumption to explain 'why' the fourth dimension does not influence normal phenomena. In fact, the influence of the embedding higher dimension is quite common across the spectrum of phenomena investigated by science, if one knows what to look for.

The fifth dimension of the space-time continuum is space-like rather than temporal, but differs radically from the ordinary three dimensions of space. Matter is clearly three dimensional as are the electric field and potential, yet the magnetic field is three dimensional with a scalar potential element that the mathematical formulas indicate is perpendicular to the three dimensions of normal space. So a fourth dimension of space is implied by Maxwell's electromagnetic theory, although this spatial dimension would be unique in its differences with the other three dimensions of space. The physical space constituting our universe could therefore be considered a four-dimensional space for the purposes of scientific analysis under proper time-independent physical conditions (Beichler, 2007). A single field that is the precursor to the known normal physical fields as well as matter itself, all of which are specific field-density structures within the single field, occupies the five-dimensional space-time indicated by Kaluza's mathematical model. Although this fourth dimension of space is macroscopically extended, it remains closed with respect to the three normal dimensions of space, such that it does not alter Kaluza's unification in any manner (Einstein and Bergmann, 1938; Einstein, Bergmann and Bargmann, 1941; Beichler, 2007). Within this context, the four-dimensional spatial model reduces CDM and the galactic halo to a simple geometric problem.

The universe as a whole can be pictured as either a three-dimensional Riemannian positively curved surface in a four-dimensional embedding space or a curved four-dimensional space-time embedded in a fifth spatial dimension, both of which are completely compatible with GR (Misner, et.al., 1973). Under normal conditions, three-dimensional matter and thus material objects are confined to the three-dimensional surface of the Riemannian sphere (or similar positively curved surface) that forms our material universe. The radius of the surface, the Einstein Radius R_E, is so large that there would be no discrepancies between normal matter and small-sized material objects with respect to the overall global curvature of space or space-time. The global curvature of space-time in the small local scale varies so little that space-time would seem to be flat locally. In other words, individual material objects, be they small or large, would only seem or 'appear' to conform to the overall large-scale global curvature of space-time. They would 'appear' to bend with the three-dimensional surface relative to the higher

271

fourth dimension of space because their own contribution to the curvature due to their own gravitational fields would be so much greater than the local bend in global curvature, which is imperceptibly small on a scale relative to the rest of the universe. Furthermore, the local curvature due to massive objects must also be taken into account, rendering the total curvature of the actual three-dimensional surface 'lumpy' on the local scale, which makes it even less likely to notice the effect of the overall positive global curvature of space on the local scale. However, a material structure as large as a spiral galaxy would not necessarily conform to the global Riemannian curvature of the universe. The local curvature due to the central core material of the galaxy would extend a great distance although is effect relative to the global curvature would diminish rapidly according to the inverse square law, allowing for an ever greater relative effect of the global curvature as distance from the galactic center increases. So a galactic core, as massive as it is, would only mask global curvature near the center of a galaxy and not throughout the whole galaxy.

If we only consider the galactic plane of a common spiral galaxy as representing the material content of the whole galaxy and place it on a positively curved Riemannian surface such as a sphere, the fact that a galaxy extends beyond the Riemannian curvature of the universe becomes evident. As a material object, the galaxy is too large to conform to the overall global curvature of space-time in the higher embedding spatial dimension.

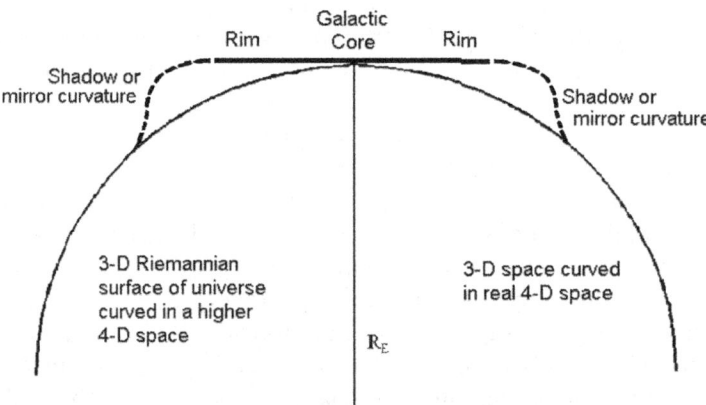

Fig 2: 4-dimensional geometry of the CDM halo

The galaxy must remain three-dimensional within its own material self, according to its

own internal forces and fields, but it must also appear to conform to the three-dimensional global curvature of the universe, thus producing a physical discrepancy or what could be called a dimensional gap. For smaller objects, this gap between the three-dimensionality of the object and that of space poses no problem, but for extremely large astronomical objects it does pose a problem. This discrepancy would only be noticed for large astronomical objects such as galaxies whose radial size extends beyond the local gravitational curvature of their central cores and the amount of the discrepancy (the size of the gap) would depend directly on the radius of the galaxy. This discrepancy can only be overcome if the global curvature of the universe bends toward the outer edge of the galactic rim as shown above, thus maintaining the continuity of the field. This bending of space-time curvature would appear as a shadow or mirror image of the curvature of the same galactic core that produces the normal three-dimensional gravitational forces that bind the galaxy together as an individual three-dimensional material object.

Having said this, we must admit that our common space of sensation and perception is three-dimensional so that all material bodies and objects such as galaxies must appear to conform to the three-dimensionality of our common space. For all intents and purposes, the shadow or mirror curvature would 'appear to pull' the dimensionally displaced galactic rim down to conform to the overall global curvature of space-time, forming the CDM halo that has been observed to surround galaxies and closing the dimensional gap. However, the idea of the universe pulling the galactic rim down to its surface is far too anthropomorphic a description.

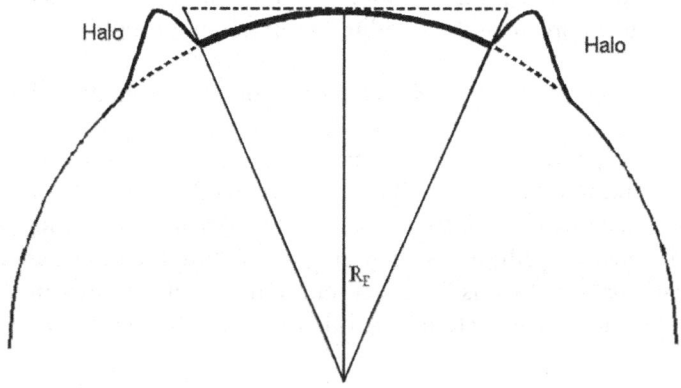

Fig 3: The CDM halo as interpreted as a 3-dimensional object

Astronomers should detect an unspecified and normally unsuspected 'halo' of space-time curvature around the observed three-dimensional galaxy even though that 'halo' curvature is not directly related to the gravitational curvature set by the core's matter. From our relative position within the three-dimensional material surface, astronomers could only view the halo and the galaxy as they appear within the same three-dimensional surface that provides our perceived material reality, even though the galaxy is actually extended beyond the global curvature of the surface that is the three-dimensional universe. In other words, galaxies larger than a certain minimum radius, as defined by the material content in their cores, could only exist in three-dimensions if the halos exist as described. Otherwise, those galaxies would extend out into the four-dimensional space at large and they would be invisible to astronomers and observers, except for their central cores.

In reality, the halo evolves a small amount at a time as the core and spiral arms of the galaxy emerge from primordial gaseous clouds. As the core forms and matter accretes to the core over time to form a galaxy, from the center outward, the gravitational forces of the clumping matter act in only three dimensions. So the galaxy forms outwardly along a three-dimensional tangent line that is perpendicular to the radius of the four-dimensional positively curved surface of the universe. According to simple relativity theory, matter curves space-time rather than the reverse, so the very slight overall global curvature of space-time is not strong enough to 'pull' the accreting galactic plane of matter 'down' to the surface of the positively curved universe as the galaxy forms. As the material galaxy grows by the accumulation or accretion of material bodies, the halo also grows in direct proportion to the radius of the galactic body, until the building (or material accumulation) phase of the galaxy is complete and the radius stabilizes at a roughly constant value.

A far more accurate view of the phenomenon would include the local curvature due to the matter within the galactic core. The core matter provides the gravitational forces that direct and guide the various star systems in their orbits around the galactic axis. The orbiting bodies actually follow geodesics along the 'extrinsically' curved surface of four-dimensional space-time due to the presence of matter at the core, but appear as three-dimensional elliptically curved paths within the three-dimensional surface of the Riemannian surface. In this case, the curvature would represent the gravitational field potential that orbiting star systems and other material bodies follow.

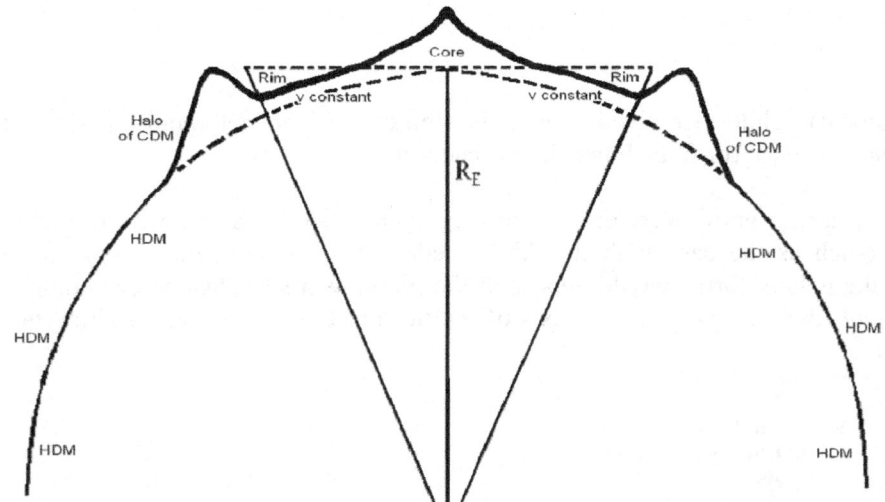

Fig 4: The internal curvature of a galaxy relative to curvature of the universe

The rim of the galaxy, which conforms to the overall global curvature of the space-time continuum between the galactic core and the halo, would form a gravitational equipotential surface. In other words, the speeds of all objects lying in the galactic rim would move at approximately equal speeds, except for local variations, if they travel along the equipotential surface.

The equipotential surface is as much a product of the mirror curvature in the halo (or the rest of the matter in the universe) as it is a result of the core material. They both contribute to the gravitational potential, adding non-gravitational potential to the orbiting star systems that occupy the galactic rim. Since it comes from the CDM in the halo, this additional amount of energy could be called DE and forms the basis of the existence of DE as exhibited by the increasing expansion rate of the universe.

A similar effect is quite commonly known in normal physics under other circumstances. For example, all material objects orbiting a planet at the same altitude do so with equal speeds because they all move along the same gravitational equipotential surface, a geodesic path in curved space-time. Similarly, all stars, star systems and other material bodies that lie beyond the immediate local curvature of the galactic core would move at the same speed because they all follow paths across the same equipotential surface relative to the universe as a whole, accounting for the constant speeds in the galactic rim. So the galactic rotational problem has been easily and simply solved by the simple recognition that our common space is four-dimensional (space-time is five-dimensional) while perceived matter is limited to our commonly perceived three-dimensional space (or four-dimensional space-time).

If a new graph were drawn showing galactic speeds as a function of the galactic radius (such as the case with the Andromeda galaxy above), the tangent plane along which the galaxy forms would appear in the graph as a straight line extending from the zero point (the galactic center or axis of rotation) to the furthest star orbiting the core in the rim.

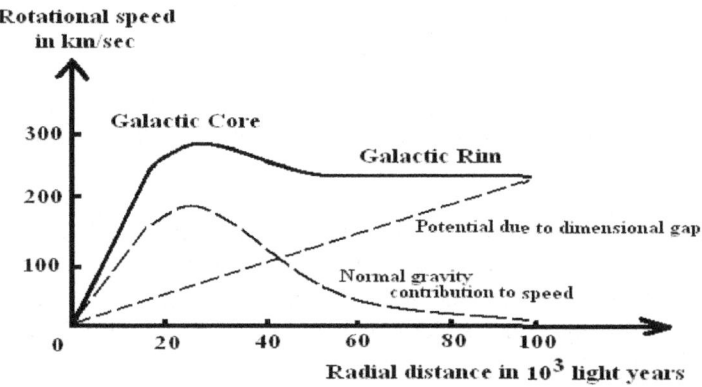

Fig 5: True energy contributions to galactic rotational speeds

Of course, that line would be amended according to the gravitational potential of the core itself. The remaining line 'exactly' duplicates real graphs for different galaxies. The dotted line indicates the amount of potential energy necessary to maintain constant speeds in normal galaxies, as graphed. In other words, this model exactly duplicates the observed physical characteristics of the CDM halo.

The simplicity of the four-dimensional explanation of CDM is not just limited to the physical model. Many calculations dealing with this new view of curvature can be reduced to a matter of simple geometry, simplifying the mathematics involved. For example, the Pythagorean Theorem can be used to determine the extent or amount of curvature, and thus the amount of CDM, at any point along the galactic plane.

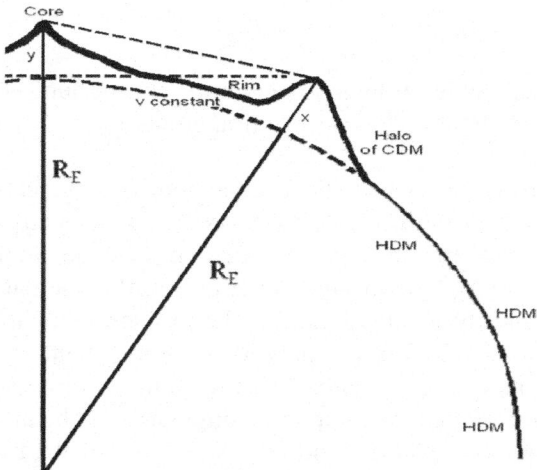

Fig 6: Pythagorean Theorem can be used to calculate the halo curvature

In this instance, the amount of curvature that constitutes the CDM halo surrounding the Andromeda galaxy is easy to calculate. According to the Pythagorean Theorem, we have

$$R_E^2 + R_{Andromeda}^2 - (R_E + x)^2 \,,$$

where x is the extent of curvature at the boundary between the galaxy and the halo. Using values of 10^{10} light years (Butterworth, 2005; Blair, 1996; Cornish, et.al., 2003) for the Einstein Radius and 10^5 light years for the radius of Andromeda, the extent of the curvature would be about 0.5 light years or 4.7 million kilometers.

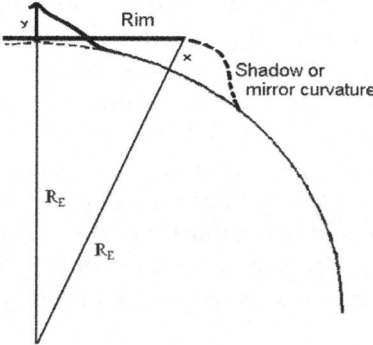

Fig 7: Pythagorean Theorem used to calculate dimensional gap distances in 4th-D of space

Given the calculated value of the halo curvature, the amount of curvature at the center of the galaxy could then be found by simple triangulation.

A problem could arise with this method in those cases where a singularity or black hole is present at the center of a galaxy. A mathematical singularity has infinite curvature. In this case, the shadow curvature in the halo might be distorted because it could not perfectly mirror the infinite curvature of a mathematical singularity. However, an interesting possibility for a solution to the problem also arises from this model. The halo curvature would need to compensate in some manner for the net effect of the singularity within its own curvature. Yet how could a finite shadow of curvature result from the infinitely extended curvature of a singularity? The answer is simple. A physical singularity such as a black hole is not necessarily equivalent to its mathematical model since the real fourth dimension of space is macroscopically extended but closed with respect to the normal three dimensions of space, as specified in the modified Kaluza model. In other words, a physical singularity cannot have an infinite curvature as specified by the mathematics, although the real curvature does approach a zero value, but never reaches it, for the diameter (the singularity limit) in three-dimensional space at an extremely large but finite extension in the fourth dimension. This bit of information should lead to a new view of black holes that could result in a better understanding of their physical nature and internal structure.

Dark Energy revisited

Given this explanation of DM and the CDM halos, the DE mystery has already begun to unravel and solve itself: The four-dimensional gap between the three-dimensional flatness of an individual galaxy and the three-dimensional curved surface of the universe supplies an extra amount of gravitational potential energy to the star systems in that galaxy's rim. This gravitational potential energy is the source of DE in the galaxies, raising the possibility that a similar source of DE must exist throughout the normal space-time continuum. This turn of events clearly indicates that DE is either a direct measure or at least directly proportional to a 'thickness' of the three-dimensional surface in the fourth direction of space. Even though the vacuum of space is empty of all matter it is still occupied by field, so it must also have some 'thickness' in the fourth direction of space due to the presence of the 'single' field occupying the higher dimension. The four-dimensional 'thickness' constitutes the DE content of the vacuum in three-dimensional space.

In both classical and normal space-time physics, there are only two fundamental forms of energy: kinetic and potential. Potential energy is energy that is derived from the relative position of matter within various physical fields. On the other hand, kinetic energy is derived from the change of position in space-time relative to the same external fields. The higher fifth dimension of space-time is occupied by a 'single field' that is the precursor to our normal gravitational, electrical and magnetic fields as well as the fundamental particles of matter that we perceive in three-dimensional space-time. This single field was the basis upon which Einstein sought his unified field theory and developed the original Einstein-Kaluza model. From the perspective of a higher-dimensional model of reality, kinetic energy disappears and all energies become relative positions within the higher-dimensional space-time relative to their position in the 'single field'. In the higher-dimensional models, there is no change of position, but only different positions relative to the space-time continuum. In other words, all energies, including three-dimensional kinetic energy, are just potential energies of relative position within the single field that occupies five-dimensional space-time. Kinetic energy in three-dimensional space is reduced to potential energy in four-dimensional space-time while potential energy in four-dimensional space-time is reduced to field energy potential in five-dimensional space-time, just as velocities in normal space-time are geodesic paths in a higher dimension of space-time. So potential energy in four-dimensional space-time is no more than relative position in the fourth direction of space, at least in the five-dimensional model of space-time.

Therefore, the apparent or 'minimum thickness' of our common three-dimensional space continuum in the fourth direction of space would give each point in common three-dimensional space an 'effective length' and thus an 'effective position' in the fourth direction of space relative to an infinitesimally thin median position, or density limit, in the center of the three-dimensional 'sheet' that constitutes our common relative space and material reality. The 'effective width' of four-dimensional space-time in the fifth direction would thus constitute a special form of energy potential, rather than a potential energy, that could be called DE. Space-time curvature is already related to potential within GR, so this description of DE follows directly from the five-dimensional extension of relativity theory. DE, like HDM, fills all of three-dimensional space between material bodies because there is a component of curvature at each and every geometrical point in the three-dimensional space of physical reality, as defined by the space-time metric. In the past few decades, evidence that HDM and DE fill all of space has been accumulating (Bergh, 1999; Bergh, 2000; Tuttle, 2006), supporting this prediction of the five-dimensional model of space-time. Given the three-dimensional surface 'area' of the closed Riemannian surface that is the perceivable three-dimensional space of our universe, an approximate value for the total amount of dark energy in the universe could be calculated as the 'volume' of the three-dimensional curved surface or 'sheet' in the fourth direction of space.

Three additional predictions from the geometrical model

Any physical model or theory is only as good as its usefulness for calculations and its ability to make verifiable predictions. This model is no different. However, unlike other models and explanations of DM and DE, this model makes several testable and easily verifiable predictions. Three of these predictions are of immediate interest. In particular, this model predicts that the expansion rate of the universe is undergoing a period of increase. The increase occurs during the mature and old age phases of a galaxy's lifetime. Conversely, the expansion was decreasing or slowing at an unprecedented rate during an earlier period of the history of the universe corresponding to the galaxy-building era or phase.

Quite beyond any questions whether the expansion will ultimately stop and reverse, continue forever, or stabilize, the standard model assumes that the rate of expansion is roughly constant. Within this context, the Einstein radius is increasing and the universe is moving toward a flatter curvature over time.

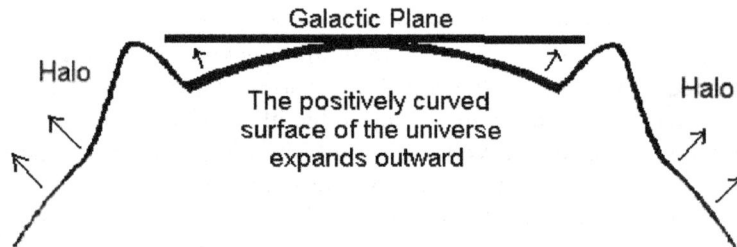

Fig 8: Geometry of an expanding universe relative to a galactic plane

As the universe expands, the dimensional gap between the positively curved surface of the universe and any given spiral galaxy is decreasing or closing. The gravitational potential energy that is normally derived from this dimensional gap is slowly leaking away, but it must return to the universe as a whole. This means that the DE and/or DM in and around the galaxies, whichever concept any particular scientist chooses to link to this phenomenon, is decreasing. According to the conservation of energy, this energy or curvature must return to the universe as a whole and thus increases the expansion rate of the universe. Given the number of galaxies in the universe, their average radius and mass distribution, the amount of DE returning to the universe could easily be calculated and compared to the observed increase in expansion rate.

The same methodology could also be applied to calculate the DE content in the spatial vacuum between stars, galaxies and other material bodies. Since the Pythagorean theorem can be used to calculate the 'height' of the dimensional gap between the galactic plane and the global curvature of the universe at any point in space or space-time, this calculated discrepancy could be expressed in energy units when compared to the amount of gravitational potential energy supplied by the discrepancy. The potential energy supplied to star systems in the galactic rims would be directly proportional to the linear measure of discrepancy in curvature and thus to the amount of DE present. Then, given the 'effective width' of the vacuum or free space in the fourth spatial direction, a simple proportion would yield the DE content of vacuum space, a predicted quantity from this model. This value could then be compared to the accepted value of 10^{-26} kg/ meter^3.

And finally, the diameter or radius of any given galaxy should be slightly less than that expected when only gravitational sources for galactic rotation speeds are taken into account.

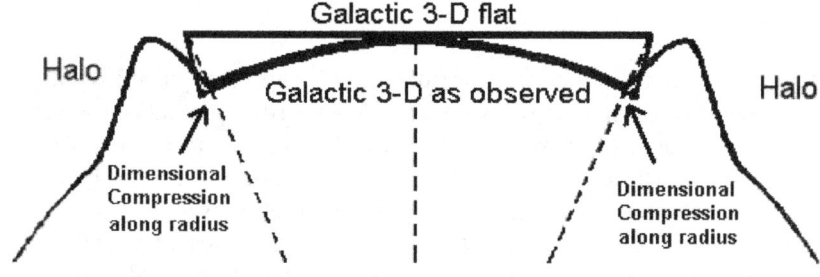

Fig 9: Galactic radius compression as predicted from the geometry

In other words, as the CDM halo 'appears' to 'pull' the three-dimensional flat galaxy 'down' to the three-dimensional curved surface of the universe, a small foreshortening of the galaxy along the radius or diameter would appear from the three-dimensional perspective of observers constrained to the three-dimensional surface. Once again, the actual amount of this visual field compression of the galactic diameter can be calculated for any given galaxy using the Pythagorean Theorem and simple geometrical considerations and then compared to observation. These three predictions offer enough new physics to render this model easily 'falsifiable'. All that is needed to make the calculations are the correct values for the necessary variables from astronomical observations.

The mathematical model

A mathematical model that matches the four-dimensional geometry of DM and DE is not that evident, so it presents a problem for the theory. However, such a model can be derived from considering that the source of DM could be particulate. The very notion that the CDM halo could consist of separate undetected particles of matter within the halo is itself untenable. All particles of matter, at least that science has been able to identify thus far, exhibit gravitational attraction. If the so-called particles invented to explain the halo followed that same prescription, they would have been attracted to the core of the galaxy during the accretion phase and would thus form part of the normal galaxy itself. Otherwise, perhaps they could exhibit anti-gravity or levity. Yet in this case they would have been repelled during the accretion phase and since dissipated out into the vastness of empty space between the galaxies rather than accumulate around the galaxies to form halos. The only possibility would be if these unknown particles exhibited a tangential form of gravitational force.

In a well-known common experiment, two rings are placed on a long rod and attached together by a spring or rubber band. As the rod spins about a center along its length between the rings, the rings move outward along the rod, away from the axis of rotation in spite of the attractive force afforded by the spring or rubber band. CDM seems to act in this same manner. It acts gravitationally by an attractive force along the radius of the galaxies, but it also gathers to the outside of galaxies and that fact implies a previously unknown and undetected tangential component to normal gravity. Or rather, it seems that gravity exhibits a tangential component in association with its normal radially directed component of attraction. So it would seem that the simple law of gravity must be extended to account for the CDM halo, although the source of the halo would then be found in the normal material particles that constitute the galaxy rather than independent, undetected or unknown quantum particles such as MACHOs and axions.

If gravity does exhibit a tangentially directed component in conjunction with its normal radial component, then it must act something like electromagnetism. The total electromagnetic force has two components: an electrical component and a magnetic component. Electric force is directed radially between attracting and repelling particles, but the magnetic force derives from a magnetic field forming 'around' a moving electrical charge that interacts with an external magnetic field filling surrounding space, just as DM seems to form around a spiraling galaxy composed of moving material bodies in the form of stars and star systems. The combined or total electromagnetic force is represented by the Lorentz equation:

$$F_{EM} = qE + qv \otimes B .$$

So, if gravity also has a tangential component, like magnetism, then the total gravitational force would be represented by the equation

$$F_{Gr} = mg + mv \otimes \Gamma .$$

The term **mg** merely represents the normal weight or gravitational attraction for a material object, **g** being the gravitational field strength at the point in space near a

gravitating mass where the object is located. The second term, $mv \otimes \Gamma$, represents the

new tangential component of gravity and offers a complete description of both DM and DE. In this case, the quantity v is the relative speed of the material object and Γ represents the external gravitational field, *i.e.*, the total or collective gravitational effect of the whole universe at that point in space. This formula is simply pregnant with new physics.

The formula itself is not new to physics. It has been proposed by other scientists, although it was derived independently, without any prior knowledge of earlier attempts to use the formula, in this case. Other scientists have sought a formal relationship between electromagnetism and gravity so they merely wrote the equation down and then tried to derive new physics from it. They were looking for a torsional effect in gravity. However, the interpretation of the formula in this case is completely different from any offered before, especially since it refers directly to a higher-dimensional space in both the electromagnetic and gravitational applications.

The first task is to identify the new term $m\nu \otimes \Gamma$. Since $m\nu$ is the momentum or

moving inertia of a material body (as in Newton's first law of motion) relative to Γ, the collective external gravitational field representing the total amount of matter in the

universe, then the term $m\nu \otimes \Gamma$ is no more nor less than a mathematical representation of

Mach's Principle: The inertia of a body is derived relative to the rest of the matter in the universe. Or, in other words, the relative velocity ν could never be zero since that would constitute an absolute reference frame, *i.e.*, a frame at rest with respect to all matter in the universe, which is not possible in a completely relative space. And indeed, the velocity v could never be zero because any object would always be moving at some speed v relative to at least one other object in the universe. Since the term $m\nu$ represents a momentum in three-dimensional space, the cross product would mathematically yield a four-dimensional conclusion, which means that inertia must be a four-dimensional property of three-dimensional matter or material bodies. That is why inertia has never been precisely defined or directly measured and that is why curvature, extrinsic to three-dimensional space, can be equated to the mass or amount of matter in an object.

The units of Γ are also of interest. Its units are second $^{-1}$, which renders Γ a frequency. In other words, the collective gravitational attraction of the universe interacts with individual material particles or objects as a wave with the frequency of Γ. Since general relativity predicts the existence of gravity waves, this fact fits well with general relativity, however it carries other implications. Since gravitational attraction is collective in nature, in that it never has a negative or repulsive component and cannot be blocked or shielded, then the value of Γ should also be constant. A constant Γ would seem to be necessary to guarantee a constant inertia or mass for any given material particle or object. However, if material objects interact gravitationally through gravity waves, then the CDM halo surrounding galaxies could also be interpreted as a standing wave in the fourth dimension of space due to the interaction of the matter in the galaxy and the total matter of the universe as a whole. This description fits the above geometrical model quite well.

In fact, expressing the new gravitational force equation in terms of energy instead of a momentum, we have the equation

$$F_G = mg + 2(KE/v)_{rel}\, \hat{r} \otimes \Gamma\ .$$

Since the kinetic energy KE is due to the gravitational attraction of the core material in the galaxy, the velocity v is also due to the core material and thus relative to the core material. The velocity v decreases as the distance from the core, along the radius of the galaxy, which increases. This relationship is linear and would graph as a straight line from the zero point (galactic center) to the outer radius from the galaxy in speed and radius graph of a galaxy. This linearity exactly fits the model and graphs shown above. So the new equations that include a tangential component of the classical gravity force precisely explain DM and the CDM halo, mimicking the geometrical model described above.

The new tangential component of gravity could be called 'gravnetism' to complete the parallel or analogy between gravity and electromagnetism. The two formulas

$$F_{EM} = qE + qv \otimes B \quad \text{and}$$

$$F_{Gr} = mg + mv \otimes \Gamma$$

offer a complete picture or model of the dynamics of the universe. In both cases, the terms to the left of the 'plus' sign represent three-dimensional spaces and the terms to the right of the 'plus' signs represent a four-dimensional embedding space. These formulas imply a five-dimensional space-time structure such as that offered in the Einstein-Kaluza theory. However, the dimensionality of this new space structure could be better represented with other formulas.

285

In standard electromagnetic theory, curl $\mathbf{A} = \mathbf{B}$ and div $\mathbf{A} = 0$, where the quantity $\mathbf{A}$ represents the electromagnetic vector or magnetic potential and $\mathbf{B}$ is the external magnetic field strength. Although $\mathbf{A}$ is a standard part of electromagnetic theory, it has never been directly detected or measured. Its existence has only been verified through implication by experiment. Electrical or scalar potential has been verified and is measured as common volts, but there is no such thing as a 'magnetic volt' even though theory implies that it must exist. The failure of science to either measure or detect the electromagnetic vector potential derives from the fact that the vector emanates from a point in three-dimensional space, but has a length in the fourth dimension of space. In other words, the electromagnetic vector potential in three-dimensional space is the same thing as an extended magnetic potential in four-dimensional space. It is just a matter of perspective. Since both the curl and div mathematical operations use the del function, which is a partial derivative taken simultaneously along all three directions in three-dimensional space, the product of the curl or cross multiplication must be orthogonal or mathematically normal to the three dimensions of common space, *i.e.*, a fourth dimension of space. The fact that the div operation yields zero confirms this fact. So magnetism is four dimensional.

A similar group of formulas would also exist for gravity and the 'gravnetic' field, such that

curl $\mathbf{W} = \mathbf{\Gamma}$ and   div $\mathbf{W} = 0$.

In this case, $\mathbf{W}$ represents a gravitational vector potential and $\mathbf{\Gamma}$ the 'gravnetic' field strength. The same mathematical argument would apply, demonstrating that $\mathbf{W}$ and $\mathbf{\Gamma}$ are both four dimensional quantities. However, the gravitational vector potential $\mathbf{W}$ could further be identified as the source of DE itself, completing this mathematical model and its description of DM and DE. So $\mathbf{W}$ is an extended vector in four-space, emanating from each and every point in three-dimensional space. When the potential $\mathbf{W}$ interacts with matter it is DE, *i.e.*, energy that cannot be accounted for by any classical means yet known.

And this model still yields more new physics. Since $\mathbf{\Gamma}$ represents the frequency of a gravitational wave, it is also expedient to render the momentum in terms of a wave using deBroglie's concept of a matter wave. Since $\lambda_{matter} = h/mv$ and $v = f\lambda$, the momentum $mv = hf/v$, where h is Planck's constant. Therefore yielding the equation

$$F_{Gr} = mg + (hf/v) \otimes \Gamma \,.$$

This equation clearly demonstrates that gravity acts by the interaction or superposition of waves. Yet it also represents a new representation of quantum gravity, a well known concept in physics that has never been successfully explained. This model alone indicates how a complete unification of gravity with the other forces of nature should proceed as well as the unification of relativity and quantum theory.

Additional considerations

From all of our 'normal' experience and observations of the past, science has only been able to conclude that matter is three-dimensional at both the microscopic and macroscopic levels of reality. Science normally determines the three-dimensionality of space from the relative positions of three-dimensional bodies of matter. However, this neither guarantees nor necessitates the three-dimensionality of space itself. Space could have any number of dimensions so long as matter itself, as far as we perceive it, is confined to just three of those dimensions. All normal matter must be confined to the same three-dimensional space. So our three-dimensional material reality would amount to either a 'slice' across the four-dimensional space that is occupied by material bodies or a 'sheet' of three-dimensional curvature that is embedded (or extended) in the four-dimensional space. In either case, the concept of a macroscopically extended fourth dimension becomes questionable since there are no material bodies outside of the 'slice' or 'sheet' to determine relative distance in the higher dimension. The term macroscopic was only used in this model to denote the fact that space would have been expanding in the fourth dimension for the same amount of time since the big bang as have the other three commonly perceived dimensions of space. However, the concept of 'compactification' of any higher dimension, or even dimensions, is meaningless, unnecessary and irrelevant under these circumstances.

In reality, we only 'perceive' the outward surfaces of material particles and bodies and that perception tells us nothing about their inward nature or physical properties. In other words, the absolute outward three-dimensionality of elementary material particles and extended material bodies is sufficient to explain the relative positions of matter, but it is not enough to guarantee that space is limited to only three dimensions. So the perceived existence of three-dimensional matter neither necessitates nor guarantees the corresponding three-dimensionality of space. Nor does the perceived outward three-dimensionality of matter require that all characteristics and properties of matter be three-dimensional, *i.e.*, intrinsic to three-dimensional space. The only property of matter that need be intrinsic to three-dimensional space is the outward appearance of material particles by which their position relative to other material bodies is determined. So matter or material particles could be inwardly higher dimensional while showing an outward 'face' of three-dimensionality. Under these conditions, matter could have other-dimensional (extrinsic to three-dimensional space) properties that are independent of the normal properties of matter in three-dimensional space.

In reality, the internal shape of an elementary particle such as a proton has nothing to do with its outward appearance or dimensionality. So it would be quite possible that a proton inwardly resembles an exponential curve in a higher (or fourth) dimension of space that could be modeled or expressed by a formula similar to that modeling Yukawa potential and field (Beichler, 2007), which would 'seem' to correspond to a singularity relative to three-dimensional space. The Yukawa potential is given by the formula

$$Y = -g^2 \frac{e^{-kr}}{r} \, ,$$

where g is a coupling factor and k is a function of the mass of the binding particle field particle (Yukawa, 1935). The single factor of 'r' in the denominator refers to the spherical configuration and nature of the potential in the normal three dimensions of space. Both normal gravity and electricity have potentials that are proportional to 1/r, which is normal for a three-dimensional space. However, the Yukawa potential has another factor of 'r' in the numerator, which would imply some type of 'inward' spatial curvature in a dimension other than the normal three dimensions of space. In fact, given the form of the higher-dimensional component of the Yukawa potential as an exponential expression in the numerator, it is easy to conclude that the potential conforms to an exponential curve in a higher fourth dimension of space, even though it would 'seem' that a singularity exists at the proton's center relative to three-dimensional space.

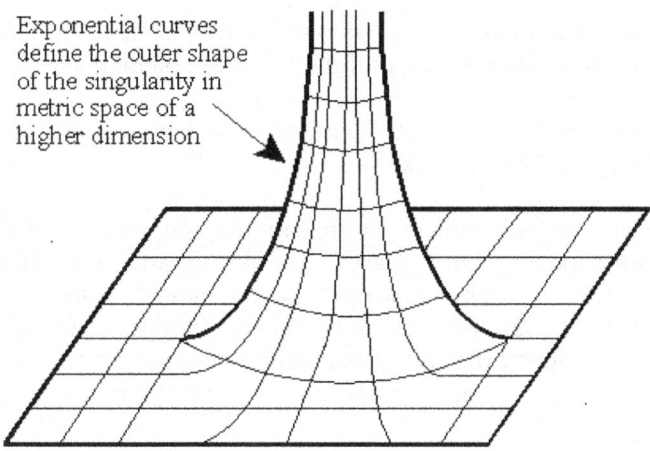

Fig 10: The Yukawa potential mimics particle surface curvature in a higher dimension

This simplistic interpretation flies directly in the face of the normal interpretation of the Yukawa potential as indicating the existence of a meson-field that acts to combine nucleons in an atomic nucleus. However, there is a relativistic precedent for adopting this novel interpretation of the Yukawa potential that incorporates space-time curvature.

Vu B. Ho has come to a similar conclusion within a completely different context. He has shown that the space-time curvature of GR can be used to derive the Yukawa potential. According to Ho,

> The result has shown that within the short range of strong force, the field equations of general relativity admit a line element that takes the form of Yukawa potential for strong interaction. This leads to the conclusion that if there is no other form of matter, besides the mass and charge, that characterizes strong interaction, then it would be possible to consider strong interaction also a manifestation of general relativity at short range (Ho, 1995).

In the case studied by Ho, the strong nuclear force associated with the Yukawa potential became an artifact of the singularity that characterizes the interior curvature of a material

289

particle. If Ho's derivation is placed within the context of an 'extrinsically' curved four-dimensional space-time, requiring an embedding fifth dimension, then his results would correspond to the above interpretation of the Yukawa potential and field.

In yet another case, Lisa Randall seems to have come to a similar conclusion within a completely different context.

> The space-time is "warped"; it appears that the strength of the apparent four-dimensional gravity decreases exponentially with distance away from the "Planck brane" that traps the graviton. Another way of stating this is that although the strength of the gravitational coupling is the same everywhere, physical mass scales decrease exponentially with distance from the brane, so that gravity far from the brane is weak (Randall, 2002).

In this case, Randall argues that the gravitational force is weak within three-dimensional space and rapidly goes to zero in the fourth direction, given the fact that our common four-dimensional space-time is a 'brane' curved in a five-dimensional 'bulk'. But the effect is the same whether one is talking about gravitons or the actual internal shape of a material particle as it curves into the higher fourth dimension of space. Rather than assuming the brane and bulk as separate objects characterized by different dimensions, as braneworld theorists do, evidence shows that the true picture is a single continuous five-dimensional space-time, whereby our common space-time is an embedded four-dimensional surface or 'sheet'. Randall and Sundrum's 'bulk' is just the fourth dimension of space as expressed in this model. The 'bulk' in the braneworld models is actually the continuous 'single field' that occupies five-dimensional space-time. The weak gravitons traveling between branes in the Randall-Sundrum model are actually the point-to-point continuous A-lines extending from one side of the 'sheet' to the other that are required to represent closure in the higher dimension in Kaluza's original model, the same closure condition that Randall and Sundrum seem to have ignored in their model.

According to the five-dimensional single field model, normal gravity is a field stress placed on three-dimensional space surrounding material particles by the internal extension of normal space-time into the higher embedding dimension. A three-dimensional material particle is constrained to the three-dimensional surface even though its curvature has been 'stretched' and is 'stretching' further into the higher embedding dimension within its interior. Newton's universal constant of gravitation should therefore be viewed as a stress factor on the three-dimensional curved surface of our observable three-dimensional space. The stress would be spread across the circumference of a closed loop extending from the top of the three-dimensional space-time surface to the point where the loop returns below the surface in the higher dimension, according to Kaluza's model, yielding a value of $2\pi R_E$, where R_E is the Einstein radius of the Riemannian surface. However, the exponential curvature that represents the higher-dimensional curved interior of the elementary particle has a negative radius of curvature relative to the three-dimensional center of mass of the particle, so the Einstein radius would have a negative exponent, yielding a preliminary factor of $2\pi R_E^{-1}$, relative to the three-dimensional surface of normal space.

This factor would act through the radius of a fundamental particle such as a proton in each of the normal three dimensions of space, yielding a complete stress factor of

$$. G = \frac{2\pi R_E^{-1}}{r_{proton}}$$

Using a value of 10^{10} light years for the Einstein Radius (Blair, 2004; Butterworth, 2007; Cornish, et.al., 2004) and 10^{-15} meters for the diameter of a proton, this formula yields a value of approximately 6.6×10^{-11} m^{-2} for G. This example yields the correct units of meter^{-2}, which corresponds to the four-dimensional equivalent of a pressure.

The value thus obtained for G as a stress factor is very close to the normally accepted value for Newton's gravitational constant except for the units, but the units in four-dimensional space are different from those for three-dimensional space so the problem of units is forgivable. Granted, only approximate values were used for the calculations because exact values are not known, but the similarity of this value to the actual value of Newton's gravitational constant is none-the-less significant. At present, the most accurate values for the radius of the proton range from 0.805 ± 0.011 to 0.862 ± 0.012 femtometers (Stein, 1995) and the Einstein Radius could range from 10 billion light years to a maximum of about 13.8 billion light years if the universe had been constantly expanding at the speed of light. But exact values are not available, so the importance of these calculations is not so much in their small inaccuracy, but rather in the fact that this model yields so close a value to the known value of G, thereby lending more credence to the model. The closeness of this value validates the four-dimensional approach to a small degree and offers a preliminary confirmation of this five-dimensional model with a macroscopically extended embedding dimension. Gravity is undoubtedly a stress in three-dimensional space caused by the discrepancy between four-dimensional physical reality and the three-dimensional materiality of particles.

More realistic values for G as a stress factor could be found using more exact values for the radius of the universe, if such values were available. It is, however, possible to use this model of gravity as a hyper-dimensional stress in normal three-dimensional space to calculate or 'predict' an actual value of the Einstein Radius, at least in the fourth spatial dimension. Given that more exact values of G and the radius of a proton exist, 6.67×10^{-11} and 0.8×10^{-15} meters, the Einstein Radius is calculated to be 12.44 billion light-years. This value is less than the maximum value of about 13.8 billion light years, but is greater than the minimum of 10 billion light years. However, it was calculated for the fourth dimension of space and there is no reason why the Einstein Radius for the fourth dimension should be exactly the same as that for the three normal dimensions of space, given the differences between their physical characteristics and properties. In fact, the fourth dimension is not isotropic with the ordinary three dimensions, even though the normal three dimensions of space are isotropic with respect to each other. The fourth dimension is unique and physically distinct from the other three by its varying density, among other things. So obtaining this accurate a value is significant and a bit remarkable.

Carrying this model further, we can consider gravity a 'surface tension' along the three-dimensional surface of the universe curved in the fourth spatial dimension, literally a three-dimensional spatial or four-dimensional space-time 'sheet', extrinsically warped in the next higher spatial dimension. Electricity then would amount to a pressure or stress acting through a three-dimensional 'cut' across the 'sheet'. This description conforms exactly to Kaluza's theory by which a four-transformation yields the gravitational equations and the cut transformation results in an equation analogous to the electromagnetic formulas. So electrical forces act through the 'sheet' while gravitational forces act 'across' the 'sheet' surface, at least from the four-dimensional perspective. Yet both would act spherically through the surface of the actual three-dimensional material particles from the three-dimensional perspective.

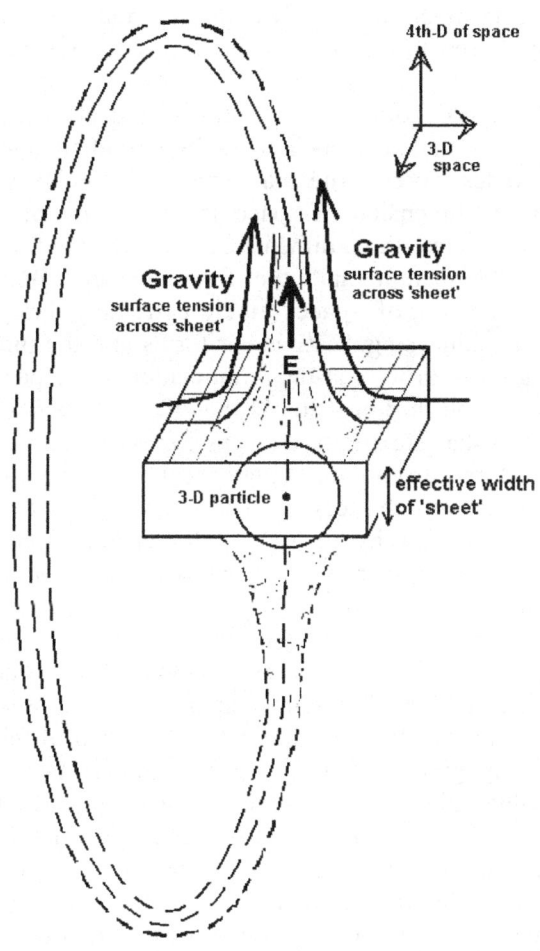

Fig 11: The geometrical source of **E**- and **g**-fields as stresses in a higher-D space

Gravity is always attractive and always acts toward the three-dimensional center of gravity in material particles, which corresponds to the physical singularity at any fundamental particle's center, while electricity acts from the spherical surface of charged particles. Within this context, the 'force' of gravity would necessarily be far weaker that the electrical 'forces' associated with the same particle, just as the surface tension across

293

a water surface is much weaker than the internal cohesive forces within the body of water, which conforms to scientific observations of physical reality.

Under these conditions, gravitational and electrical forces would both seem to merge together as a single center-directed force of elementary particles if the center of the elementary particles were a true mathematical singularity. Even as the particle wraps around the fourth dimension of space from the top of the 'sheet' to the bottom, it maintains a minimum three-dimensional width in four-dimensional space that is characteristic of the quantum and other field constants. But the interior particle extension in the fourth direction of space, albeit extremely large, is not infinite, so a true mathematical or 'point' singularity never forms and the three-dimensional density of the particle never goes to infinity at the point center of the particle. The extended portion of the particle in the fourth dimension is thus like a 'tube' with a separate interior portion and surface, such that electricity and gravitation do not merge into a single force, but remain separate forces even though both act through the three-dimensional surface of the material particle toward this center. That is why gravity is always attractive and acts toward the center of gravity of a three-dimensional material body or object, leaving electricity as a stress in either direction, radially into or out of a particle through its three-dimensional surface.

While G is reduced to a stress factor in three-dimensional space relative to the four-dimensional field, a three-dimensional material object must also be defined by similar field constraints. It is not the material content of the particles that determines the materiality of any given material object as much as it is the fields that define the materiality of physical objects. A common object such as a table or chair is mostly empty space, relative to the actual amount of matter from which the table or chair (as measured by the mass of the elementary particles that constitute the atoms and molecules in the object) is normally identified as material. All common material objects are mostly empty space, far more devoid of matter than not, given the total volume or spatial extension of the object. What we sense as material objects are actually defined by the electromagnetic and gravitational fields that hold the true material particles of the object together and allow us to identify them as individual material objects for physical considerations. A spiral galaxy is a typical although extremely large three-dimensional material object. So the gravitational, electrical and magnetic forces that hold it together define a spiral galaxy, not the curved space-time through which we 'sense' or observe the galaxy from afar.

The three-dimensionality of a galaxy is not defined by the three-dimensionality of the universe, as represented by the surface curvature of space or space-time. Nor is it defined by the actual material mass of the galaxy. Three-dimensionality is a property of the object as a whole, not external to the object and thus not determined by the three-dimensionality of external relative space. So a three-dimensional object, if it is large enough, need not follow nor be restricted by the three-dimensional surface of the universe as it curves in the higher fourth dimension. This logical argument further confirms the notion that a three-dimensional galaxy is independent of the three-dimensionality of the external spatial curvature of space-time that surrounds it. Thus it forms a halo of curvature as described above.

The other shoe drops

Since the fourth dimension of space is real and curvature is an extrinsic property of four-dimensional space-time in the higher dimension, rather than an intrinsic property of space-time as commonly believed, the overall or global curvature of space-time also becomes the source of the observed HDM. The global curvature of space-time in a three-dimensional Riemannian sphere embedded in a four-dimensional space results from the total material content of the universe. This curvature extends throughout the normally empty space (at least empty of normal matter) that exists between all material bodies in the universe in roughly equal proportions, just as HDM is believed to fill all of space between material bodies. Therefore, HDM is just the average and normal space-time curvature of the universe in a Riemannian surface in the higher dimension. DM itself is nothing but space-time curvature that is not normally associated with local matter, but is instead either associated with, or the product of, the total (global) material content of the universe.

Yet within this context, a new problem arises. If space is four-dimensional and the fourth dimension is macroscopically extended, why then are material bodies limited to a three-dimensional existence such that all material objects are restricted to a single common three-dimensional surface? In other words, what distinguishes the unique measure of the three-dimensional surface in the fourth direction of space? In particular, if a large three-dimensional object like a galaxy can be three-dimensional within itself while it also extends beyond the three-dimensional curvature of the universe as a whole in the fourth direction, what then determines the separate three-dimensionality of the galaxy as opposed to the rest of the material objects in space? Or, why is matter three dimensional at the microscopic level such that all matter appears to exist within the same three-dimensional continuous 'sheet' even though space is actually four dimensional and continuous in that higher dimension? What defines and limits the three-dimensional surface that is curved in and continuous with an embedding fourth dimension as unique and at the same time allows large objects such as spiral galaxies to extend beyond the limits of that three-dimensional surface in the fourth dimension? These questions, regarding the uniqueness or discreteness of three-dimensional space in the fourth direction, which is just the curved 'sheet', need to be answered before a four-dimensional model of space can be understood and appreciated.

In pure geometry, a line is one-dimensional and a surface is two-dimensional. However, physics and physical reality diverge from mathematics at this point. A one-dimensional physical line is different from its mathematical model or counterpart because it must have some measurable thickness in a higher dimension, just as a two-dimensional physical surface must have a real measurable thickness in the third-dimension even though it can be modeled mathematically as a purely two-dimensional object. The same would be true for a three-dimensional surface embedded and curved in a higher four-dimensional space. Physical reality is modeled by mathematics, but the mathematical model is not the real physical 'thing' that it is modeling, so there are subtle differences between the two that are inherent in their perceived reality. This is not just a polemical argument that has no meaning. These concepts need to be understood, or rather need not to be misunderstood, before relativity, representing the continuous field, and quantum theory, representing discrete particles, can be unified. These concepts also need to be clarified so that the mathematical model of GR that uses an extrinsic curvature in the higher space can be derived and understood.

By the same token, the three-dimensional matter/field surface of the universe must have some small but finite thickness in the fourth direction of space because it is a physical reality not just a mathematical model. This minimum thickness of the three-dimensional surface determines the materiality or the material integrity and inertial reality of all objects and bodies that exist in three-dimensional space. In a sense, this is just the geometric equivalent to Wheeler's statement that 'matter curves space-time and space-time curvature moves matter'. Yet the gravitational field as well as the surface must be continuous in all dimensions in order to form the halo as described above, even while the three-dimensional surface and the galaxy are independent three-dimensional objects occupying different regions of the four-dimensional extension of three-dimensional space. The physical displacement or dimensional gap between the three-dimensional galaxy and the perceived three-dimensionality of space that leads to the existence of the halo and HDM is simultaneously quite large at the outer edge of the galaxy, guaranteeing that the overall extension of space in the fourth direction of space is macroscopic.

This simple fact poses a serious problem, at the very least, for any theory that depends on the compactification of higher dimensions. Under these circumstances, compactification is neither justified nor required from either the theoretical or a purely philosophical perspective. Kaluza's original five-dimensional model depended upon two stated mathematical conditions, one assumed condition and one suggestion. Kaluza required that (1) the higher dimension be closed with respect to four-dimensional space-time and (2) the A-lines that connected points in normal four-dimensional space-time continuum through the higher dimension are of equal length. Kaluza assumed continuity in the higher dimension, just as continuity exists in the normal dimensions. Otherwise, he only suggested that extension in the fifth direction must be very small because we cannot sense or detect that higher dimension.

Oskar Klein took advantage of Kaluza's suggestion and related the periodicity in the higher dimension, what has been called the 'cylindrical condition', to the quantum (Klein, 1926a; Klein, 1926b; Klein, 1927; Klein, 1939; Klein, 1947). Yet Klein did not go beyond Kaluza's mathematical conditions, nor did he contradict them in any way. In this manner, Klein only followed one of the possible legitimate paths that could be used to extend Kaluza's theory. Klein was trying to unify relativity and the quantum, but his efforts failed, by his own admission. In the 1980s, the string theorists were in a quandary on how to develop their own quantum model further when they rediscovered and then adopted Klein's extension of Kaluza's model, but they expanded Klein's extension of Kaluza's model to include several more dimensions without considering how doing so might affect the original unification that was accomplished using only a five-dimensional model. The Superstring theorists, as they have become known, have neither defied nor broken Kaluza's original mathematical conditions for unification, but they have stretched the notion of 'curvature' within GR to a conceptual breaking point. However, the braneworld and similar models have completely violated Kaluza's original mathematical conditions, even though their theoretical models are ultimately based upon Kaluza's unification of gravity and electromagnetism. In other words, the latest braneworld models have 'de-unified' the Kaluza model of GR and electromagnetism upon which they are based.

Yet Klein's model does not represent the only possible way to interpret and extend Kaluza's theoretical model. The density of space in the higher dimension can be varied such that our material world only inhabits a thin three-dimensional 'sheet' of extremely high field density, extending through a slice of the four-dimensional space or five-dimensional space-time, while the overall extension in the fourth direction of space is macroscopic. The width of the three-dimensional 'sheet' could be extremely small rather than infinitesimally small and still define the quantum, while the width of the 'sheet' defines our material world of senses and perceptions, corresponding to Newtonian physics. Einstein and his colleagues followed a second path of application similar to this model in the late 1930s, but applied it unsuccessfully because they never completely understood the versatility and finer geometrical points of the five-dimensional approach.

If the extension of our three-dimensional space in the fourth direction is macroscopic, then a small but finite 'minimum' width of three-dimensional space that is necessary to create the three-dimensionality of matter as observed must exist. On the other hand, there is absolutely no reason to assume that this minimum thickness in the fourth direction is as small as the Planck length. Describing the three-dimensional 'thickness' or 'effective width' of the 'sheet' as a 'minimum' value better conforms to the quantum perspective of material reality and is, in fact, proportional to the quantum. Thus the discrete quantum enters the continuous field model through the concept of a 'minimum' or 'effective width' of our three-dimensional space in the fourth direction.

Still, the thickness of three-dimensional space must be extremely small because we do not normally perceive the higher dimension of space, as Kaluza noted, but the total extension in the fourth direction of space must also be macroscopically large at the same time. So an extremely small but definable 'minimum' thickness of three-dimensional space in the fourth dimension becomes necessary to establish the material content of the universe. Otherwise, the fourth dimension of our common space must be continuous as are the other three dimensions of space as well as time, so three-dimensional space must therefore have an extremely small 'effective width' in the fourth dimension rather than a true width as a common two-dimensional sheet of paper would have. So the three-dimensional surface of materiality that is our common space could not be a 'brane' curved in a separate 'bulk'. There is no separation between the three-dimensional 'sheet' with its minimum 'effective width' and the actual macroscopic extension of four-dimensional space-time in the higher dimension, as would be the case between a 'brane' and its embedding 'bulk'.

The 'effective width' defines the common quantum of action as expressed in the quantum theory. This 'effective width' of three-dimensional space would be equal to the product of the fine structure constant and proton width, or approximately $1/137 \times 10^{-15}$ meters, leaving the 'sheet' continuous with the macroscopic extension in the fourth direction. To further maintain continuity in the fourth direction, we could picture successive layers of three-dimensional space-time, each having the same 'effective width', stacked on top of each other in the fourth direction like the pages of a book or onion skins. Each successive layer, defined by equal 'effective widths', would correspond to the principle quantum numbers used in quantum mechanics, while the density of the single field would fall off by a factor of $1/r^4$ as distance from the center of the three-dimensional 'sheet' increases in the fourth direction of space.

Protons and electrons would amount to folds and bumps in the 'sheet', respectively. The smallest or minimum amount of local curvature possible, representing the smallest possible fundamental material particle, the neutrino, should equal the 'effective width' of three-dimensional space and four-dimensional space-time in the higher embedding spatial dimension. In other words, neutrinos correspond to the smallest possible local puckers or burble in the three-dimensional 'sheet' that is the three-dimensional surface of curvature. Therefore, the width of a neutrino should be about 1/137 femtometers. Coincidentally, the best present particulate candidate for explaining HDM is the neutrino, at least from the quantum point-of-view whereby particles are needed to explain everything. However, HDM is a relativistic effect and not a quantum phenomenon even though the quantum and relativistic explanations are thus afforded a unique correspondence with respect to the size of the neutrino. In other words, one quantum of 'effective width' of the 'sheet' has the same numerical value as the diameter of the neutrino. This correspondence between the size of the neutrino and the 'effective width' of the space-time continuum is therefore logically justified, at least as much as anything in science can be 'justified'. Science normally depends on verification, not justification. The rest mass of the neutrino could also be calculated or predicted from this model, rendering the model falsifiable in yet another sense.

Conclusion

The confirmed existence of DM in both the galactic halo and more generally throughout the emptiness of space between material bodies offers clear observational evidence of the existence of a macroscopically extended fourth dimension of space. This evidence is irrefutable, although alternative hypotheses to explain DM will surely persist far into the future because the implications of a real fourth dimension of space are extremely radical compared to the present worldview of science and culture. As an added bonus, this model also explains and identifies the source of DE as well as unifying the quantum and relativity in a single theory. DM is nothing more than the real extrinsic curvature of the three-dimensional space-time continuum that cannot be associated with local matter and is thus a consequence of the non-local or global (total) material content of the universe, while DE is just the field 'thickness' of three-dimensional space in the fourth dimension. DE results from the new concept of gravitational vector potential.

299

These explanations of DM and DE are extremely simple, but it will be difficult for many scientists and scholars to accept them on s number of different intellectual levels. Simple ideas are quite often the hardest to accept and the easiest to overlook when proposed as solutions to the most difficult and pernicious problems in science.

Scientists have sought evidence of a higher dimension of space, as exhibited by curvature, since the days of Friedrich Gauss (Scholz, 2005), but the search has never before been a priority because there has never been more than an esoteric need to discover the next higher dimension of space. William K. Clifford attempted to develop a model of electromagnetism based upon a three-dimensional space curved in a single-polar elliptical fourth dimension as early as 1869 and thereafter, however most of his theoretical work has been lost to both science and history. Instead, he is remembered for stating that matter is nothing but curvature and the motion of matter nothing other than variations in that curvature. In this regard, he is considered a pre-cursor to Einstein's general theory of relativity, even though his first concern was magnetism rather than gravity. Yet his work did foster interest in the possibility that space could actually have more than three dimensions in the late nineteenth century and a few forward thinking scientists actually used parallax measurements of stars to look for the suspected space curvature. That possibility shook the very foundations of the scientific community, although not everyone noticed. The French mathematician Henri Poincaré even stated that he would rather change the basic laws of optics than accept the possibility that space might be curved (Poincaré, 1892). Poincaré's statement clearly reflected the existence of a three-dimensional bias in science and that bias is still prevalent today.

In spite of Clifford's failure to fully develop a working higher-dimensional theory of matter, due more to his early death from consumption than rejection by the scientific community, the idea never really caught on. It was later overshadowed by Einstein's assumption of time as the fourth dimension and the rise of strict positivistic attitudes in science. Then Kaluza's five-dimensional extension of relativity theory entered the scene in 1921 and Klein's 1926 adaptation to the quantum. These developments are well known although there is still a lot of historical mythology surrounding the efforts. However, it is not so well known that others tried to develop their own versions of the five-dimensional space-time structure. Sir Arthur Eddington, heavily influenced by Clifford, tried to develop a model using a three-dimensional space with two dimensions of time, dubbed the Fundamental Theory.

However, the most extensive efforts came from Henry T. Flint who published more than thirty articles and a book on the subject over a period spanning the 1920s to the

1960s. He conducted most of his theoretical research alone, but was aided at various times by other scientists such as William Wilson. In the end, Flint unified gravity, electromagnetism and quantum theory into a single theory, publishing his last contribution in the form of a book in 1966, *The Quantum Equation and the Theory of Fields*. Then his work seems to have completely disappeared from science, thereby becoming a lasting monument to the general bias of the scientific community against the basic concept of continuity and the reality of higher dimensions of space. Many details of Flint's model are similar to this model and his final work forms an extremely important contribution to the general body of five-dimensional unified field theories. And finally, Paul S. Wesson (1998, 2006) has also developed a five-dimensional theory that he calls the Space-Time-Matter Theory. His theory also shares some common features with this new model of five-dimensional space-time. But Wesson seems more eager to demonstrate that his theory is equivalent to the five-dimensional brane and superstring models. All in all, a large enough amount of mathematical work has been completed on the various five-dimensional theories to keep science busy for decades applying the theories and understanding the concepts, if only the scientific community would look seriously at what has been accomplished so far.

The scientific bias against the possibilities of a real extended higher dimension of space and the fundamental nature of continuity is still alive and well. It is symptomatic of a deeply rooted positivistic attitude that has dominated science since the early 1900s and poses a real threat to solving the DM and DE crises, even today. Yet a need for using higher-dimensional spaces in physics has developed in the past few decades and hyper-dimensional theories have become commonplace in theoretical physics to explain some of the stranger and more exotic phenomena observed in nature. Unfortunately, this newer and more recent version f hyper-dimensionality has only been accepted because of the supposed compactification of the higher dimensions that 'saves the phenomena' described within the present paradigm of physics that rely on discreteness: Compactification is no more than a positivistic compromise to maintain three-dimensionality even while utilizing higher-dimensional spaces as extra degrees of freedom. With compactification, there is no need to explain why the other dimensions are not detectable, beyond that they are just too small to worry about. Ignoring reality because it is beyond human sensations and detection, or because it is inconvenient for one's *a priori* theories of reality, is a positivistic trick. So the use of compactification has no real physical justification beyond the fact that it can be used. In fact, compactification is irrelevant to higher dimensions because distance is a strictly three-dimensional property of the material world.

Scientists who are presently attempting to develop 'compactified' hyper-dimensional theories in physics seem never to have studied the mathematical properties and consequences of the next higher dimension (the fourth dimension of space) or they would have realized that only a single higher dimension is necessary to unify physics and account for the observed physical properties of reality. Instead, they have adopted as many as six or seven extra dimensions, if not more, to account for the extra degrees of freedom that they need to unify different physical theories, which is a bad practice. They are merely offering a purely mathematical and completely non-physical compromise between physical reality and the quantum paradigm. The fourth dimension of space alone has a rich enough group of properties to render still higher dimensions of space unnecessary at this time: The fourth dimension of space differs from the ordinary three by far more than just another degree of freedom for physicists to play with. It has unique physical properties and special characteristics that will impact all of science and physics, so ordinary physics will eventually need to be rewritten to cope with this new higher-dimensional reality. Many of the physical properties of a higher-dimensional space were derived in the late nineteenth century and thereafter forgotten as positivistic attitudes won over science. However, scientists now seem uninterested in these properties because ideas expressed in the late nineteenth century are considered classical and tainted rather than modern by their tainted reckoning.

Unfortunately, those scientists and scholars who have adopted them will not easily give up their hyper-dimensional compromises or the alternative theories that they have constructed on that hypothesis and accept the simplicity of a single continuous higher dimension. So the legitimate five-dimensional theories are stuck between a rock and a hard spot: Nearly all scientists either believe in too few dimensions (only a four-dimensional space-time) or too many dimensions (ten, eleven, twenty-six or more). Too many scientists are too enamored with the quantum and their own mathematical prowess and dexterity to seriously consider the simpler physics and mathematics of continuity in the next higher dimension. Accepting the reality of a fourth dimension of space will be neither simple nor easy for science even though the idea is directly implied in both classical and modern physics and it simplifies the modern physics, but science will eventually be forced toward acceptance by the evidence provided by nature.

Whatever five-dimensional theory is finally established as the most accurate because it corresponds best to nature, we can be assured that revolutionary changes in science will occur as the fourth dimension of space or the fifth dimension of space-time is accepted by the scientific community. The idea that revolutionary changes in science will accompany any solution to the DM and DE crises is certainly not new, as stated in the concluding statement of the DETF. Yet no one seems to recognize the fact that the simplest ideas in science often lead to the greatest advances because nature is simple and simple changes in our understanding of nature usually represent the most profound theoretical consequences for science. Science even follows a rule of thumb to cover such situations when the need arises, called Ockham's razor. According to the modern interpretation of William of Ockham, all things being equal between two competing theories, the simpler theory will usually be correct. Yet the basic idea of the simplicity of nature goes much further than Ockham's razor would indicate: The simplicity is inherent in nature itself. Unfortunately, science sometimes overlooks the inherent simplicity of nature and physical reality while Ockham's razor only considers competing theories, not the simple understanding of nature that scientific theories are supposed to exemplify. Neither the simple structure of our theories nor the simple nature of reality is even considered in many of the more complex theories of science.

In advanced theoretical physics, it almost seems that the more mathematically complex models have become the preferred norm of science even though developing more complex models in order to simplify the physics is a contradiction. The mathematical models created by scientists would be as simple as nature itself, if science were to completely understand the true nature of reality. The closer scientific theories come to the true nature of reality, the simpler they should become. So the overly complicated theories that physicists have recently developed clearly demonstrate how far science has transgressed in its misunderstanding of the true nature of reality. Many modern theorists are committing a fundamental Cartesian error by developing ever more complex mathematical theories simply because they expect to eventually reach a point where nature agrees with their mathematics, rather than exploring physical reality and developing the simple theories required by nature from the beginning.

Nature does not follow the philosophical and mathematical whims of humankind or fabrications of the scientific establishment. Nature guides humankind to its understanding of reality, but nature does so on her own terms, not the terms set by science or culture. Such is the case with adopting complex mathematical hyper-dimensional models rather than utilizing the simple physics of a four-dimensional space to represent physical reality. The concept of a real four-dimensional space is simple, but it will have far-reaching consequences in both science and culture. It is easier in the short run for science to reject the simple answer and instead compromise with invented mathematical realities that prop up the status quo of the present paradigms, but this course of action unnecessarily complicates the long run of the scientific search for truth and often misleads science in the final analysis of phenomena.

303

It is a clear and simple fact that past scientific revolutions have resulted from extremely simple but profoundly fundamental changes in scientific thought as expressed in the occasional crises that science has had to overcome. Scholars have previously concluded that the Second Scientific Revolution started when simple solutions were found to account for the failure to detect the luminiferous aether and explain blackbody radiation. Neither scientists nor scholars considered these failures crises during the pre-revolutionary period of the late 1800s. They were merely seen as problems that scientists would eventually solve by applying the accepted theories and paradigms of the era, much as present-day scientists believe that the DE and DM problems will eventually be resolved within the present paradigms of physics. Nor did anyone even suspect the radical changes that science would experience by finding novel new solutions to the late nineteenth century crises. The same situation has once again emerged with regard to the existence of DM and DE. DM and DE are playing the same historical role in the present development of science that the Michelson-Morley experiment and blackbody radiation played before the last scientific revolution. Physics has always been about the nature of matter itself and the 'motion of matter', ever since Aristotle first wrote his book on the subject, but the true nature of matter has never been determined. Finding the true nature of matter has been circumvented by modern quantum theory and the Heisenberg Uncertainty Principle. Now, both the concept of matter (in so far as it now exists) and its true physical nature have been thrown into turmoil by the discoveries of DM and DE, forcing science to consider the possibility that its time-honored concepts of matter are at least 'incomplete' if not 'incorrect'. Science can no longer afford to talk about matter without defining it.

The first Scientific Revolution resulted from natural philosophers' attempts to fulfill Aristotle's program of explaining nature in terms of 'matter in motion' and culminated in Newton's laws that govern 'matter in motion'. The Second Scientific Revolution merely refined those laws of 'matter in motion' in extreme cases (relativistic speeds, astronomical distances, exceptionally large masses and the world of the extremely small), but did not replace them. So physics is still based upon the fundamental concept of 'matter in motion'. However, real solutions to the DM problem (rather than stopgap compromises) will both challenge and change our basic and fundamental concepts of matter, and possibly of motion itself, which will be a truly revolutionary accomplishment in the continuing evolution of human thought. This simple fact was implied in the final report of the DETF. For this reason, if no other, scientists are suspicious of new theories, especially theories that directly challenge their most fundamental and most cherished concepts (or beliefs, although scientists might object to this word). However, that does not mean that these radical ideas are untrue.

It would seem from the historical record that crises in science can only be defined within the context of the revolutions that result from solving them correctly, so crises are seldom, if ever, identified as such before revolutions begin. Yet science is already beginning to accept the existence of new 'crises' in the discoveries of both DM and DE, so it would seem that science is slowly recognizing the beginnings of a new third scientific revolution, if only begrudgingly. Science now needs to recognize the true nature of that revolution by the adoption of a five-dimensional theory of matter and reality. The theoretical adoption of a four-dimensional space has the potential to solve many of the problems in physics and related sciences, but it also represents the beginning of a new scientific revolution that will lead to a recognition that the fourth dimension of space is a reality, not just a 'working hypothesis' developed to 'save the phenomena' (or the present paradigm) or save a mathematical model of reality that has already run its course. The fourth dimension of space is neither a mathematical artifice nor a scientific convenience invoked to find a fast and easy partial solution, but rather a direct observation of physical reality. For that reason alone, the acceptance of the concept will not come easily or quickly although its acceptance is inevitable before science can progress and take its next big leap forward.

Bibliography

S.W. Allen, R.W. Schmidt and A.C. Fabian. (2002) 'Cosmological constraints from the X-ray gas mass fraction in relaxed lensing clusters observed with Chandra". *Monthly Notices of the Royal Astronomical Society* 334, L11. Eprint at arXiv: astro-ph/0205007v1.

N. Bahcall. (2000) "Clusters and Cosmology". *Physics Reports* 333: 233-244.

Rachel Bean, Sean Carroll and Mark Trodden. (2005) "Insights into Dark Energy: Interplay Between Theory and Observation". Eprint at arXiv: astro-ph/0510059v1.

James E. Beichler. (2007) "Three Logical Proofs: The five-dimensional reality of space-time". *Journal of Scientific Exploration*, to be published.

Jacob D. Bekenstein. (2004) "Relativistic Gravitation Theory for the MOND Paradigm". *Physical Review* D70: 083509. Eprint at arXiv: astro-ph/0403694.

Jacob D. Bekenstein. (2005) "An alternative to the dark matter paradigm: relativistic MOND gravitation". Eprint at arXiv: asto-ph/0412652v3

Sidney van den Bergh. (1999) "The Early History of Dark Matter". *Publications of the Astronomical Society of the Pacific* 111: 657-660.

Sidney van den Bergh. (2000) "A Short History of the Missing Mass and Dark Energy Paradigms". Eprint at arXiv: astro-ph/0005314.

William P. Blair. (2004) "Size Scales in Astronomy", v1.3. Johns Hopkins University. At <fuse.pha.jhu.edu/~wpb/scale.html>.

Sidney Bludman (2006) "Cosmological Acceleration: Dark Energy or Modified Gravity?" Eprint at arXiv: astro-ph/0605198.

Paul Butterworth. (2004) "Measuring the Size of the Universe". At <imagine.gsfc.nasa.gov/docs/ask_astro/answers/971124x.html>

R.R. Caldwell, Dave Rahul and Paul J. Steinhardt. (1998) "Cosmological Imprint of an Energy Component with General Equation of State". *Physical Review Letters* 80: 1582-1585. Eprint at arXiv: astro-ph/9708069v2.

Bernard J. Carr and Stephen W. Hawking. (1974) "Black holes in the early universe". *Monthly Notices of the Royal Astronomical Society* 168: 399-416.

Bernard J. Carr. (1975) "The primordial black hole mass spectrum". *Astrophysical Journal* 201: 1-19.

Sean M. Carroll (2003) "Why is the Universe Accelerating?" Eprint at arXiv: astro-ph/0310342v2.

Neil J. Cornish, et al. (2004) "Constraining the Topology of the Universe". *Physical Review Letters* 92: 201302. Eprint at arXiv: astro-ph/0310233v1.

Dark Energy Task Force. (2006) "Report". Eprint at arXiv: astro-ph/0690591

I.T. Drummond (2000) "Biometric gravity and "Dark Matter"". *Physical Review* D63, 043503. Eprint at arXiv: astro-ph/0008234.

Albert Einstein and Peter Bergmann (1938). "On a Generalization of Kaluza's Theory of Electricity". *Annals of Mathematics* 39 (3): 693-701.

Albert Einstein, Peter G. Bergmann and Valentine Bargmann (1941). "On the Five-Dimensional Representation of Gravitation and Electricity". *Theodor von Karman Anniversary Volume*. Pasadena: California Institute of Technology: 212-225.

Henry T. Flint. (1966) *The Quantum Equation and the Theory of Fields*. London: Methuen.

Paul H. Frampton. (2004) "Dark Energy – a Pedagogical Review". Eprint at arXiv: astro-ph/0409166v3.

Paul H. Frampton. (2005) "Introduction to Dark Energy and Dark Matter". Eprint at arXiv: astro-ph/0506676v1.

Gundlach, et al. (2007) " Laboratory Test of Newton's Second Law for Small Accelerations". *Physical Review Letters* 98.

Stephen W. Hawking. (1974) "Black Hole Explosions". *Nature* 248: 30.

Stephen W. Hawking. (1975) "Particle creation by black holes". *Commun. Math. Phys.* 199-220.

Vu B. Ho. (1995) "A metric of Yukawa potential as an exact solution to the field equations of general relativity". Eprint at arXiv: hep-th/9506154: 2.

Wayne Hu and Scott Dodelson. (2002) "Cosmic Microwave Background Anisotropies". *Annual Review of Astronomy and Astrophysics* 40: 171-216. Eprint at arXiv: astro-ph/0110414v1.

Theodor Kaluza. (1921) "Zur Unitätsproblem der Physik". *Sitzungsberichte der Preussischen Akademie der Wissenschaften* 54: 966-972.

David B. Kaplan, Ann E. Nelson and Ann Weiner. (2004) "Neutrino Oscillations as Probe of Dark Energy". *Physical Review Letters* 93: 091801.

Charles R. Keeton and Arlie O. Petters. (2006) "Formalism for testing theories of gravity using lensing by compact objects. III. Braneworld geometry". Eprint at ArXiv: gr-qe/06033061. Accepted in *Physical Review D*.

Martin Kilbinger and Marco Hetterscheidt. (2006) "Dark Energy Dominates the Universe". Argelander Institut für Astronomie. At <www.astro.uni-bonn.de/~kilbinge/de/index.html>

Oskar Klein. (1926a) "Quantentheorie und fünfdimensionale Relativitätstheorie". *Zeitschrift fur Physik* 37: 895-906.

Oskar Klein. (1926b) "The Atomicity of Electricity as a Quantum Theory Law". *Nature* 118: 516.

Oskar Klein. (1927) "Zur fünfdimensionale Darstellung der Relativitätstheories". *Zeitschrift fur Physik* 46: 188-208.

Oskar Klein. (1939) "On the Theory of Charged Fields". *New Theories in Physics*. Paris: International Institute of Intellectual Cooperation: 77-93.

Oskar Klein. (1947) "Meson Fields and Nuclear Interaction". *Arkiv for Mathematik, Astronomi och Fysik* 34A: 1-19.

R.A. Knop, et al. (2003) "New Constraints on Omega_M, Omega_Lambda, and w from an Independent Set of Eleven High-Redshift Supernovae Observed with HST". *Astrophysical Journal* 598: 102. Eprint at arXiv: astro-ph/0309368v1.

Chris Kochanek. (2004) "Where does the Dark Matter Begin?" Eprint at arXiv: astro-ph/0412089v1. To be published in *The Impact of Gravitational Lensing on Cosmology*. Eds. Y. Mellier and G. Meylan.

Eric V. Linder (2003) "Cosmic Structure Growth and Dark Energy". *Monthly Notices of the Royal Astronomical Society* 346: 573. Eprint at arXiv: asro-ph/0305286v2.

Eric V. Linder. (2006) "Theory Challenges of the Accelerating Universe". Eprint at arXiv: astro-ph/0610173v2.

J.A.S. Lima. (2004) "Alternative Dark Energy Models: An Overview". *Brazilian Journal of Physics* 34: 194-200. Eprint at arXiv: astro-ph/0402109v1.

Y. Mellier. (1999) "Probing the Universe with Weak Lensing". *Annual Review of Astronomy and Astrophysics* 37: 127-189. Eprint at arXiv: astro-ph/9812172.

Mordechai Milgrom. (1983) "A modification of the Newtonian Dynamics as a possible alternative to the hidden mass hypothesis". *Astrophysical Journal* 270: 362-370.

Alessandro Melchiorri, Paul Bode, Neta A. Bahcall and Joseph Silk. (2003) "Cosmological Restraints from a Combined Analysis of the Cluster Mass Function and Microwave Background Anisotropies". *Astrophysical Journal* 586: L1-L3.

Charles Misner, Kip Thorne and John A. Wheeler. (1973) *Gravitation*. San Francisco: Freeman: 417-428.

Jan Hendrik Oort. (1940) "Some Problems Concerning the Structure and Dynamics of the Galactic System and the Elliptical Nebulae NGC 3115 and 4944". *Astrophysical Journal* 91: 273-306.

Don N. Page (1976a) "Particle emission rates from a black hole: Massless particles from an uncharged, rotating hole". *Physical Review* D13: 198-206.

Don N. Page (1976b) "Particle emission rates from a black hole. II. Massless particles from a rotating hole". *Physical Review* D14: 3260-3273.

Saul Perlmutter, et.al. (1999) "Measurements of Omega and Lambda from 42 High Redshift Supernovae". *The Astrophysical Journal* 517: 565-586. Eprint at arXiv: astro-ph/9812133.

Plato (c.375 BCE) *The Republic*. Book VII: 514A-520A.

Henri Poincaré. (1892) "Non-Euclidean Geometry", Translated by W.J.L. *Nature* 45: 407.

Lisa Randall and Raman Sundrum. (1999a) "A Large Mass Hierarchy from a Small Extra Dimension". *Physical Review Letters* 83: 3370-3373. Eprint at ArXiv: hep-ph/9905221.

Lisa Randall and Raman Sundrum. (1999b) "An alternative to compactification". *Physical Review Letters* 83: 4690-4693. Eprint at ArXiv:hep-th/9906064.

Lisa Randall. (2002) "Extra Dimensions and Warped Geometries". *Science* 296: 1423-1424.

Adam G. Riess, et.al. (1998) "Observational Evidence from Supernovae for an Accelerating Universe and a Cosmological Constant". *The Astronomical Journal* 116: 1009-1038. Eprint at arXiv: astro-ph/9805201v1.

Vera Rubin and W. Kent Ford. (1970) "Rotation of the Andromeda Nebula from a Spectroscopic Survey of Emission Regions". *Astrophysical Journal* 159: 379.

Vera Rubin, W. Kent Ford, D. Burstein and N. Thonnard. (1985) "Rotation Velocities of 16 Sa Galaxies and a Comparison of Sa, Sb, and Sc Rotation Properties," *Astrophysical Journal* 289: 81.

Erhard Scholz. (2005) "Curved spaces: Mathematics and Empirical evidence, ca. 1830 1923". Preprint at Wuppertal. At <www.mathg.uni-wuppertal.de/~scholz/preprints/ES_OW2005.pdf>.Shorter version to appear at Oberwalfach Reports.

Sinclair Smith. (1936) "The Mass of the Virgo Cluster". *Astrophysical Journal* 83: 23.

D.N. Spergel, et.al. (2006) "Wilkinson Microwave Anisotropy Probe (WMAP) three year results: Implications for cosmology". *American Journal of Physics* in press. Eprint arXiv: astro-ph/0603449.

B.P. Stein. (1995) "Physics update". *Physics Today* 48: 9.

M. Tegmark, et al. (2003) "Cosmological Parameters from SDSS and WMAP". *Physical Review* D69: 103501. Eprint at arXiv: astro-ph/0310723.

Helen Tuttle. (2006) "Dark Matter Observed". *SLAC Today*, 22 August. At <today.Slac.Stanford.edu/feature/darkmatter.asp>

Paul S. Wesson. (1998) *Space-Time-Matter: Modern Kaluza-Klein Theory*. World Scientific.

Paul S. Wesson. (2006) *Five-dimensional Physics: Classical and Modern Consequences of Kaluza-Klein Cosmology*. World Scientific.

Hideki Yukawa. (1935) "On the interaction of elementary particles". *Proceedings of the Physics and Mathematics Society, Japan* 17.

Fritz Zwicky. (1933) "Die Rotverscheibung von extragalaktischen Nebeln". *Helvetica Physica Acta* 6: 110-127.

Fritz Zwicky. (1937a) "Nebulae as Gravitational Lenses". *Physical Review* 51: 290.

Fritz Zwicky. (1937b) "On the Masses of Nebulae and of Clusters of Nebulae". *Astrophysical Journal* 86: 217-246.

Different Algebras for one Space-time Reality

José B. Almeida

Universidade do Minho, Physics Department

Braga, Portugal, email: bda@fisica.uminho.pt

Abstract

The most familiar formalism for the description of geometry applicable to space-time physics comprises operations among 4-component vectors and complex real numbers; few people realize that this formalism has indeed 32 degrees of freedom and can thus be called 32-dimensional. We will revise this formalism and we will briefly show that it is best accommodated in the Clifford or geometric algebra $\mathcal{G}_{1,3} \times \mathbb{C}$, the algebra of 4-dimensional spacetime over the complex field.

We will then explore other algebras isomorphic to that one, namely $\mathcal{G}_{2,3}$, $\mathcal{G}_{4,1}$ and $\mathbb{Q} \times \mathbb{Q} \times \mathbb{C}$, all of which have been used in the past by various authors to formulate their respective approaches to physics. $\mathcal{G}_{2,3}$ is the algebra of 3-space with two time dimensions, which Carroll used implicitly in his formulation of electromagnetism in $3 + 3$ spacetime, $\mathcal{G}_{4,1}$ was and it still is used by myself in a tentative to unify the formulation of physics and $\mathbb{Q} \times \mathbb{Q} \times \mathbb{C}$ is the choice of Rowlands for his nilpotent formulation of quantum mechanics. We will show how the equations can be converted among isomorphic algebras and we also examine how the monogenic functions that I use are equivalent in many ways to Rowlands nilpotent entities.

PACS numbers: 04.50.-h; 02.40.-k.

1 Introduction

We see Physics as a discipline that creates mathematical models of physical reality. In practice, we write mathematical equations whose solutions allow us to predict the outcome of experiments and observations. One physical model is just as good as the predictions it allows and the most successful models become known as physical theories. Every model makes use of a limited set of independent variables, which can be operated among themselves; we say that the model uses an underlying algebra. The model must also give physical meaning to the independent variables and algebraic operations performed among them, so that everybody can then translate into reality the results of operations performed within the model.

In view of what was said above, one sees that an algebra is an intrinsic component of any physical model, but it happens quite often that several algebras are only apparently different and can be shown to be isomorphic to each other. When this happens, models incorporating such algebras are frequently equivalent, although the insight one has over problems addressed with two equivalent models may be entirely different. In the following sections we will discuss the algebras associated with models proposed by various authors, showing that they are in many cases isomorphic. We will also show how to convert equations between isomorphic algebras. In the case of a model proposed by Carroll [1], considering a space with 3 spatial and 3 temporal dimensions, the associated algebra is a superalgebra of several 5-dimensional algebras, so, the isomorphisms that can be found apply only to a subalgebra of the one proposed by that author.

2 Algebras most frequently used in space-time physics

Both Newtonian mechanics and Maxwell's equations are models based on 4 independent variables, 3 space coordinates and 1 scalar time variable; it must be pointed out, however, that the latter is relativistically invariant while the former is not. The algebras used to operate with those variables are the algebras of real and complex numbers complemented with vector algebra, but it is easy to see that this system lacks consistency. For instance, two vectors $\mathbf{a}$ and $\mathbf{b}$ determine a parallelogram with area given by $|\mathbf{a} \times \mathbf{b}| = |\mathbf{b} \times \mathbf{a}|$. We make use of an operation among two vectors and then define the area as a scalar quantity. It makes more sense to define a new product whose outcome is an oriented area, called *outer product* and denoted $\mathbf{a} \wedge \mathbf{b}$. The outcome of the outer product is precisely the area of the parallelogram defined by the two vectors, with a sign defined by the direction of movement from one vector to the other.

Clifford algebras are based on the *geometric product* or simply the product of vectors, incorporating both the inner and outer products. For any two vectors it is

$$\mathbf{a}\mathbf{b} = \mathbf{a} \cdot \mathbf{b} + \mathbf{a} \wedge \mathbf{b}. \tag{2.1}$$

The geometric product is associative and so it is possible to have products of 3 vectors, leading to a grade-3 element of the type $\mathbf{a} \wedge \mathbf{b} \wedge \mathbf{c}$ which, if not zero, represents an oriented volume. We can thus say that the algebra associated with the spatial part of Newtonian mechanics and Maxwell's equations is Clifford algebra of dimension 3, also known as *geometric algebra* of dimension 3 and denoted $\mathcal{G}_3$ or $\mathcal{G}_{3,0}$. The volume elements of this algebra, as well as the highest grade elements of any Clifford algebra, are called *pseudoscalars*. For an extensive treatment of geometric algebras see [2, 3].

Further problems with the Newtonian and Maxwellian models reside in the fact that time is treated as scalar but it has to be differentiated from the scalar coefficients of vectors. This is solved by special relativity, because it proposes that time is to be treated as a dimension of space-time, thus increasing the dimensionality of the associated geometric algebra to 4; the highest grade element is now a 4-dimensional hypervolume. There are two possible algebras, $\mathcal{G}_{1,3}$ and $\mathcal{G}_{3,1}$, associated with one positive and 3 negative norm frame vectors or one negative and 3 positive norm frame vectors, respectively. In more common terms, what this means is that the quadratic form can be expressed either as

$$(\mathrm{d}s)^2 = (\mathrm{d}x_0)^2 - (\mathrm{d}x_1)^2 - (\mathrm{d}x_2)^2 - (\mathrm{d}x_3)^2 \tag{2.2}$$

or

$$(\mathrm{d}s)^2 = -(\mathrm{d}x_0)^2 + (\mathrm{d}x_1)^2 + (\mathrm{d}x_2)^2 + (\mathrm{d}x_3)^2. \tag{2.3}$$

The former is the most common choice and it allows the formulation of most physics equations, including quantum mechanics, as done by Hestenes [4]. In order to fully accommodate quantum mechanics in Dirac's approach one must, however, allow for complex coefficients, a possibility not considered in [4].

Starting with the work by Theodor Kaluza, who proposed a 5-dimensional unification of electromagnetism with general relativity [5], some authors have used higher dimensional spaces to try and unify the equations of physics. My own work makes use of 5-dimensional space-time and bears a strong relation to Kaluza's [6, 7]. The geometric algebra associated with 5-dimensional space-time in this formulation is $\mathcal{G}_{4,1}$ but other authors have used the opposite signature $\mathcal{G}_{1,4}$ [8]. How different and how similar are all these approaches?

In order to answer the question we start by examining the overall dimensionality of the different algebras, starting with the algebra of physical space, $\mathcal{G}_{3,0}$. We realize that the elements of the algebra can be classified into 4 grades: scalars, vectors, areas and volumes, or better, grades 0, 1, 2 and 3. While both scalars and volumes have no associated orientation besides positive and negative, vectors and areas have 3 possible orientations, so, the total number of degrees of

Table 1: Matrix representation of Geometric Algebras $\mathcal{G}(p,q)$, with p positive and q negative norm frame vectors. The notation $\mathbb{F}(n)$ is used for the n-dimensional matrix algebra over the field $\mathbb{F}$; $\mathbb{R}$ stands for real numbers, $\mathbb{C}$ for complex numbers and $\mathbb{Q}$ for quaternions. The notation $^2\mathbb{F}(n)$ identifies the sum $\mathbb{F}(n) \bigoplus \mathbb{F}(n)$, meaning that two matrices of dimension n are needed in the isomorphism.

q \ p	0	1	2	3	4	5	6	7
0	$\mathbb{R}$	$\mathbb{C}$	$\mathbb{Q}$	$^2\mathbb{Q}$	$\mathbb{Q}(2)$	$\mathbb{C}(4)$	$\mathbb{R}(8)$	$^2\mathbb{R}(8)$
1	$^2\mathbb{R}$	$\mathbb{R}(2)$	$\mathbb{C}(2)$	$\mathbb{Q}(2)$	$^2\mathbb{Q}(2)$	$\mathbb{Q}(4)$	$\mathbb{C}(8)$	$\mathbb{R}(16)$
2	$\mathbb{R}(2)$	$^2\mathbb{R}(2)$	$\mathbb{R}(4)$	$\mathbb{C}(4)$	$\mathbb{Q}(4)$	$^2\mathbb{Q}(4)$	$\mathbb{Q}(8)$	$\mathbb{C}(16)$
3	$\mathbb{C}(2)$	$\mathbb{R}(4)$	$^2\mathbb{R}(4)$	$\mathbb{R}(8)$	$\mathbb{C}(8)$	$\mathbb{Q}(8)$	$^2\mathbb{Q}(8)$	$\mathbb{Q}(16)$
4	$\mathbb{Q}(2)$	$\mathbb{C}(4)$	$\mathbb{R}(8)$	$^2\mathbb{R}(8)$	$\mathbb{R}(16)$	$\mathbb{C}(16)$	$\mathbb{Q}(16)$	$^2\mathbb{Q}(16)$
5	$^2\mathbb{Q}(2)$	$\mathbb{Q}(4)$	$\mathbb{C}(8)$	$\mathbb{R}(16)$	$^2\mathbb{R}(16)$	$\mathbb{R}(32)$	$\mathbb{C}(32)$	$\mathbb{Q}(32)$
6	$\mathbb{Q}(4)$	$^2\mathbb{Q}(4)$	$\mathbb{Q}(8)$	$\mathbb{C}(16)$	$\mathbb{R}(32)$	$^2\mathbb{R}(32)$	$\mathbb{R}(64)$	$\mathbb{C}(64)$
7	$\mathbb{C}(8)$	$\mathbb{Q}(8)$	$^2\mathbb{Q}(8)$	$\mathbb{Q}(16)$	$\mathbb{C}(32)$	$\mathbb{R}(64)$	$^2\mathbb{R}(64)$	$\mathbb{R}(128)$

freedom is 8 and we say that total dimensionality is 8. In a similar way, the total dimensionality of a general geometric algebra, $\mathcal{G}_{p,q}$ is 2^{p+q}, if only real coefficients are allowed for all grades. If complex coefficients are allowed the total dimensionality is either doubled or remains unaltered relative to the real coefficient version. Some algebras can be classified as *complex algebras*, because their pseudoscalar elements have negative square and commute with all other elements. In complex algebras the unit pseudoscalar doubles as the complex imaginary, so, introducing complex coefficients does not bring in any extra dimensions. In non-complex algebras the introduction of complex coefficients doubles the degrees of freedom, doubling the total dimensionality.

All geometric algebras are isomorphic to one particular matrix algebra, over one particular field that provides the coefficients. What this means is that all operations performed in a particular geometric algebra have equivalent operations in the isomorphic matrix algebra. The use of matrix algebra isomorphism is useful for classification purposes, but it is usually not recommended for performing operations since all the links with geometry are lost. Table 1 shows the matrix algebras isomorphic to the lowest order geometric algebras. The entries in the table are of the type $\mathbb{F}(n)$, which stands for algebra of n-dimensional matrices with coefficients in the field $\mathbb{F}$. The coefficients' field can be real numbers ($\mathbb{R}$), complex numbers ($\mathbb{C}$) or quaternions ($\mathbb{Q}$). A few algebras are non-simple and are denoted $^2\mathbb{F}(n)$; this means that two matrices of dimension n are needed in the isomorphism. For instance, while the algebra $\mathcal{G}_{1,0}$, associated with the quadratic form $(\mathrm{d}s)^2 = (\mathrm{d}x)^2$, is isomorphic to a pair of real numbers $^2\mathbb{R}$, the algebra $\mathcal{G}_{0,1}$, associated with the quadratic form $(\mathrm{d}s)^2 = -(\mathrm{d}x)^2$, is isomorphic to complex numbers $\mathbb{C}$.

Looking up the table for the matrix representation of physical space algebra, $\mathcal{G}_{3,0}$, we see that we must use 2-dimensional matrices with complex coefficients. Usually we associate the frame vectors $\{\sigma_m\}$ to the Pauli matrices $\{\hat{\sigma}_m\}$, as follows:

$$\sigma_1 \Leftrightarrow \hat{\sigma}_1 = \begin{pmatrix} 0 & 1 \\ 1 & 0 \end{pmatrix}, \quad \sigma_2 \Leftrightarrow \hat{\sigma}_2 = \begin{pmatrix} 0 & -\mathrm{i} \\ \mathrm{i} & 0 \end{pmatrix}, \quad \sigma_3 \Leftrightarrow \hat{\sigma}_3 = \begin{pmatrix} 1 & 0 \\ 0 & -1 \end{pmatrix}, \quad (2.4)$$

where we introduced the circumflex over a symbol to denote the matrix representation of the entity represented by that symbol; for instance, if a designates a vector, $\hat{a}$ is the matrix representation of that vector. Under matrix isomorphism scalars are represented by the product of a real number by the identity matrix, vectors by linear combinations of matrices $\hat{\sigma}_m$, areas by linear combinations of two Pauli matrix products and volumes by the product of a real number by $\hat{\sigma}_1\hat{\sigma}_2\hat{\sigma}_3$. Since the product of the three Pauli matrices is the identity matrix multiplied by the complex imaginary, we see that the unit pseudoscalar of the algebra actually doubles as imaginary.

Minkowski space-time is most frequently associated with $\mathcal{G}_{1,3}$ algebra, although several authors prefer the $\mathcal{G}_{3,1}$ alternative; Hestenes [3, 4] and Doran and Lasenby [2] follow the former approach. No physical significance is attributed to the choice of signature, but one sees from Table 1 that the corresponding algebras are not isomorphic; there is probably some deep meaning in this choice that has escaped physicists so far. For the matrix representation of $\mathcal{G}_{3,1}$, the most direct route starts with Majorana gamma matrices, which have only imaginary elements, proceeding to assign the four frame vectors from the algebra by the equation

$$\sigma_\mu \Leftrightarrow \mathrm{i}\hat{\gamma}_\mu; \quad (2.5)$$

the circumflex indicates that $\hat{\gamma}_\mu$ are matrices and the product by the imaginary ensures that the frame vectors are represented by real matrices. For the $\mathcal{G}_{1,3}$ algebra we should, in principle, select Pauli matrices $\hat{\sigma}_1$ and $\hat{\sigma}_3$ over the quaternion field. There is a workaround that avoids the discomfort of quaternions, which consists on allowing for 4-dimensional matrices with complex elements and restricting the matrix coefficients to real numbers. There are several possible alternatives for the assignment of basis vectors to matrices, the most common being derived from Dirac-Pauli representation; this is

$$\gamma_0 \Leftrightarrow \begin{pmatrix} I & 0 \\ 0 & -I \end{pmatrix}, \quad \gamma_m \Leftrightarrow \begin{pmatrix} 0 & \hat{\sigma}_m \\ -\hat{\sigma}_m & 0 \end{pmatrix}. \tag{2.6}$$

These matrices have both real and imaginary elements, but used with real coefficients they still provide a basis representation for $\mathcal{G}_{1,3}$, avoiding the use of quaternions.

In 5-dimensional space-time the representation is much easier with $\mathcal{G}_{4,1}$ then with $\mathcal{G}_{1,4}$, because the latter not only needs quaternions but it is also a non-simple algebra; we will not pay much attention to this case. With $\mathcal{G}_{4,1}$ we have a beautiful scenario; we can use 4-dimensional matrices with complex elements and complex coefficients. Among the various possible assignments we propose the following one, which is derived from the Dirac-Pauli representation, as we shall see below:

$$e_0 \Leftrightarrow \begin{pmatrix} 0 & 0 & 1 & 0 \\ 0 & 0 & 0 & 1 \\ -1 & 0 & 0 & 0 \\ 0 & -1 & 0 & 0 \end{pmatrix}, \quad e_1 \Leftrightarrow \begin{pmatrix} 0 & 1 & 0 & 0 \\ 1 & 0 & 0 & 0 \\ 0 & 0 & 0 & -1 \\ 0 & 0 & -1 & 0 \end{pmatrix}, \quad e_2 \Leftrightarrow \begin{pmatrix} 0 & -i & 0 & 0 \\ i & 0 & 0 & 0 \\ 0 & 0 & 0 & i \\ 0 & 0 & -i & 0 \end{pmatrix}$$

$$e_3 \Leftrightarrow \begin{pmatrix} 1 & 0 & 0 & 0 \\ 0 & -1 & 0 & 0 \\ 0 & 0 & -1 & 0 \\ 0 & 0 & 0 & 1 \end{pmatrix}, \quad e_4 \Leftrightarrow \begin{pmatrix} 0 & 0 & -1 & 0 \\ 0 & 0 & 0 & -1 \\ -1 & 0 & 0 & 0 \\ 0 & -1 & 0 & 0 \end{pmatrix}.$$

$$\tag{2.7}$$

It can be verified that this matrix assignment produces the desired signature, with

$$(e_0)^2 = -1, \qquad (e_i)^2 = 1, \qquad i = 1, \ldots, 4. \tag{2.8}$$

For the remainder of this chapter we will use the symbols $\{e_\alpha\}$, defined above, as the basis for 5-dimensional space-time with signature $(-, +, +, +, +)$.

We have not covered in this section the matrix representations for Carroll's $\mathcal{G}_{3,3}$ [1] or Rowlands' $\mathbb{Q} \times \mathbb{Q} \times \mathbb{C}$ [9, 10], although the former can be looked up in the table. We will consider these algebras in the next section.

3 Converting equations among algebras

We have seen in the previous section that there are isomorphisms between the algebras of different spaces, which means that it is feasible to translate all equations from one algebra to any of its isomorphic algebras. Although the equations can be translated, the geometric connection varies substantially an so does the insight one has over the equations. As an example, take the Dirac equation, which appears formulated as a matrix equation in every textbook. The standard formulation does not allow any geometrical interpretation, because matrices have no connection to geometry whatsoever. The fact that the Dirac equation can be translated into geometric algebra provides the necessary link to geometry and the solutions can be interpreted geometrically [7].

If all we are interested in is the formulation of general relativity, 4-dimensional space-time is the adequate choice, which has a total dimensionality of 16. Physics equations, however, involve the use of complex numbers, at least for quantum mechanics. The total dimensionality implied by the set of physics equations for general relativity and quantum mechanics is then 32 and our task is then to translate equations among algebras with this total dimensionality. We start with Dirac-Pauli matrices, as defined in Eq. (2.6), and we follow the usual procedure for the definition of matrix $\hat{\gamma}_5$:

$$\hat{\gamma}_5 = i\hat{\gamma}_0\hat{\gamma}_1\hat{\gamma}_2\hat{\gamma}_3. \tag{3.1}$$

The translation between Dirac algebra and the algebra of 5-dimensional space-time, $\mathcal{G}_{4,1}$, is made directly by the following relations

$$e_\mu \Leftrightarrow \hat{\gamma}_\mu\hat{\gamma}_5, \quad e_4 \Leftrightarrow -\hat{\gamma}_5. \tag{3.2}$$

This equation can be interpreted both as a matrix or a geometric algebra equation. Indeed, if the γ_μ represent the frame vectors of Minkowski space-time, the equation can be read as a geometric algebra equation and allows the transposition from Minkowski into 5-dimensional space-time. The inverse transposition follows the rules:

$$\hat{\gamma}_\mu \Leftrightarrow e_4e_\mu, \quad \hat{\gamma}_5 \Leftrightarrow -e_4. \tag{3.3}$$

We turn our attention now to Rowlands' algebra, whose elements are sets of two quaternions and one complex number. For convenience we shall represent a general element of this algebra with the notation **q**qc; boldface and sanserif characters represent two independent quaternions and a normal character represents both real and complex imaginary. The elements in the set can be commuted, so,

319

the total dimensionality of the algebra is $4 \times 4 \times 2 = 32$, just as Dirac's algebra. The basis for Rowlands' algebra is given by the sets

$$\{1, \mathbf{i}, \mathbf{j}, \mathbf{k}\},$$
$$\{1, i, j, k\},$$
$$\{1, i\};$$

the real part of each set is always represented with a normal character. The first quaternion basis verifies the relations

$$\mathbf{i}^2 = \mathbf{j}^2 = \mathbf{k}^2 = -1,$$
$$\mathbf{i}\mathbf{j} = -\mathbf{j}\mathbf{i} = \mathbf{k}; \tag{3.4}$$

similar relations hold for the other quaternion. In order to set up the conversion relations for $\mathcal{G}_{4,1}$ we start by defining 3 anticommuting elements that can be associated with the 3 physical space dimensions; for this we set

$$e_1 \Leftrightarrow \mathbf{i}i, \quad e_2 \Leftrightarrow \mathbf{j}i, \quad e_3 \Leftrightarrow \mathbf{k}i. \tag{3.5}$$

We note that the unit volume is now

$$e_1 e_2 e_3 \Leftrightarrow \mathbf{i}\mathbf{j}\mathbf{k}\, i = -i. \tag{3.6}$$

Now we need to find an element that anticommutes with the former ones, with negative square, for e_0, and a second one, squaring to unity, for e_4. A possible choice is

$$e_0 \Leftrightarrow \mathbf{j}, \quad e_4 \Leftrightarrow i\mathbf{k}. \tag{3.7}$$

We need to check that the unit pseudoscalar coincides with the complex imaginary, that is, it squares to -1 and commutes with all elements of the algebra. For this, we do

$$e_0 e_1 e_2 e_3 e_4 \Leftrightarrow i\,\mathbf{j}i\mathbf{k} = i. \tag{3.8}$$

The inverse relations are very easy to establish. With the help of the above conversion relations it becomes a feasible task to convert all equations between Dirac's, Rowlands' and my own notations but, if physics is the same in all notations, the insight and comprehension one has over the problems at hand can gain a lot from different approaches.

The best equation to demonstrate the conversion relations is arguably the Dirac equation; this is written, in terms of matrices, as

$$\hat{\gamma}^\mu \partial_\mu \psi + im\psi = 0. \tag{3.9}$$

Upper indices are used here and elsewhere to denote a change of sign, with respect to the corresponding lower indices, for those elements that square to -1 ($-I$ in the matrix case). Multiplying on the left by $\hat{\gamma}^5$ and using conversion relations from Eq. (3.3), the Dirac equation becomes

$$e^\mu \partial_\mu \psi + im\psi = 0. \tag{3.10}$$

We now establish that $im\psi = \partial_4 \psi$, that is, we establish that the wavefunction dependence on x^4 is harmonic and is governed by the particle's mass; this is very similar to a compactification of coordinate x^4. The Dirac equation acquires a new form:

$$e^\alpha \partial_\alpha \psi = \nabla \psi = 0. \tag{3.11}$$

The index α runs from 0 to 4 and the symbol ∇ represents what is known as the vector derivative of the algebra. Any function ψ that is a solution of Eq. (3.11) is called monogenic. There are plane wave solutions for this equation, with the general form

$$\psi = \psi_0 e^{i(p_\alpha x^\alpha + \theta)}. \tag{3.12}$$

The monogenic equation implies that $e^\alpha p_\alpha \psi_0 = 0$, which can only be true if $(e^\alpha p_\alpha)^2 = 0$ and ψ_0 includes a factor $e^\alpha p_\alpha$. We say that the vector $p - e^\alpha p_\alpha$ is a nilpotent vector. In the above cited works, Rowlands uses the nilpotency condition as first principle, but we see here how this can be derived from the monogenic condition. If one establishes monogeneity as first principle, then the nilpotency condition is implied.

In its matrix version, Dirac's equation accepts column matrix solutions, which are called Dirac spinors. In order to find the geometric equivalent of these we define 4 orthogonal idempotent elements by the relations

$$\begin{aligned}
f_1 &= \frac{1}{4}(1 + e_3)(1 + ie_1e_2), \\
f_2 &= \frac{1}{4}(1 - e_3)(1 + ie_1e_2), \\
f_3 &= \frac{1}{4}(1 - e_3)(1 - ie_1e_2), \\
f_4 &= \frac{1}{4}(1 + e_3)(1 - ie_1e_2).
\end{aligned} \tag{3.13}$$

These elements are called idempotents because their powers are always equal to the element itself. They are orthogonal because the product of any two different idempotents returns zero; they also add to unity. We can then split the original monogenic function into four components as in

$$\psi = \sum_{i=1}^{4} \psi f_i = \sum_{i=1}^{4} \psi_i. \tag{3.14}$$

Each of the terms ψ_i is still a monogenic function and it is the geometric version of a Dirac spinor. Rowlands' nilpotents have 4 components and they are also another form of spinors.

Now, the case of Carroll's $\mathcal{G}_{3,3}$ algebra [1] does not readily fall into the algebras we have discussed above because, being 6-dimensional, it has a total dimensionality of 64, doubling the dimensionality of those algebras. However, Carroll argues that there is one special time dimension, which corresponds to ordinary time, and two orthogonal time dimensions, which must be treated differently. Carroll's proposed wavefunction is the solution of the second order equation

$$- (\partial_{s1}^2 + \partial_{s2}^2 + \partial_{s3}^2)\psi + m^2\psi + \partial_{t3}\psi = 0; \tag{3.15}$$

where

$$m^2\psi = (\partial_{t1}^2 + \partial_{t2}^2)\psi. \tag{3.16}$$

For the purpose of this equation we can define a combined time coordinate, using $t1$ and $t2$, by

$$tc = \frac{1}{2}(t1 + t2). \tag{3.17}$$

Equation (3.16) is then a $\mathcal{G}_{2,3}$ algebra equation and we see from Table 1 that this algebra is isomorphic to $\mathcal{G}_{4,1}$. In order to convert between the two algebras we define the vectors for $\mathcal{G}_{2,3}$ by

$$\begin{aligned} e_{sm} &= ie_m, \\ e_{t3} &= ie_0, \\ e_{tc} &= e_4. \end{aligned} \tag{3.18}$$

With this conversion it is easy to verify that Eq. (3.16) is indeed a second order version of Eq. (3.11). We don't discuss here other implications of Carroll's 6-dimensional approach, the purpose of this discussion being only to show that there is an implied 5-dimensional algebra isomorphic to the other ones presented above.

322

4 Conclusion

Many authors resort to different algebras for the exposition of their own approaches to fundamental physical equations, such as Maxwell's, Dirac's and Einstein's. Quite frequently authors propose their own versions of those equations, highlighting the virtues of their approaches. The task of comparing results is difficult because the form and not content is dependent on the particular algebra that the author has chosen. We have shown that the algebra used by Dirac has an overall dimensionality of 32, the same as several 5-dimensional algebras proposed by different authors. The tensor algebra that most people use for general relativity is indeed a 16-dimensional sub-algebra of the Dirac algebra, so, it does not need to be addressed specifically. Some authors, like Hestenes [4], have proposed a version of the Dirac equation in this algebra.

Particular examples of algebras isomorphic to the Dirac algebra are those used by Rowlands [9, 10] and myself [6, 7]. We have shown how to convert between those two algebras and the Dirac algebra. A slightly different case occurs with the algebra used by Carroll [1], because this has an overall dimensionality of 64. Here we have shown that some of the proposed equations can be set in an algebra isomorphic to the previous ones and we presented the means for converting equations between Carroll's algebra and the remaining ones.

The choice of a particular algebra is irrelevant from the point of view of the mathematical validity of equations, but it may make a significant difference to the perception and comprehension of the physics behind the equations. Quite often, no single choice of an algebra offers the definitive approach to an equation. Looking at a particular problem from different angles usually broadens our perspective over that problem, so, it makes sense to have equivalent equations written in varied algebras. However, we need to be able to convert among algebras in order to unify the various approaches.

References

[1] J. E. Carroll, *Electromagnetic fields and charges in 3+1 spacetime derived from symmetry in 3+3 spacetime*, 2004, arXiv: `math-ph/0404033`.

[2] C. Doran and A. Lasenby, *Geometric Algebra for Physicists* (Cambridge University Press, Cambridge, U.K., 2003).

[3] D. Hestenes and G. Sobczyk, *Clifford Algebras to Geometric Calculus. A*

Unified Language for Mathematics and Physics, Fundamental Theories of Physics (Reidel, Dordrecht, 1989).

[4] D. Hestenes, *Spacetime physics with geometric algebra*, Am. J. Phys. **71**, 691, 2003.

[5] T. Kaluza, *On the problem of unity in physics*, Sitzungsber. Preuss. Akad. Wiss. Berlin. (Math. Klasse) pp. 966–972, 1921.

[6] J. B. Almeida, *The null subspace of G(4,1) as source of the main physical theories*, in *Physical Interpretations of Relativity Theory – IX* (London, 2004), arXiv: `physics/0410035`.

[7] J. B. Almeida, *Hidden geometric character of relativistic quantum mechanics*, J. Math. Phys. **49**, 012301, 2007, arXiv: `quant-ph/0606123`.

[8] S. S. Seahra and P. S. Wesson, *Null geodesics in five dimensional manifolds*, Gen. Rel. Grav. **33**, 1731, 2001, arXiv: `gr-qc/0105041`.

[9] P. Rowlands, *The nilpotent Dirac equation and its applications in particle physics*, 2003, arXiv: `quant-ph/0301071`.

[10] P. Rowlands, *Zero to Infinity: The Foundations of Physics*, vol. 41 of *Series on Knots and Everything* (World Scientific, Singapore, 2007).

Dynamic Universe and the Conception of Reality

Tarja Kallio-Tamminen

Physics Foundations Society, Finland
www.physicsfoundations.org

Abstract

The article gives an overview of the structure and ontological implications of the Dynamic Universe theory. It inquires whether the theory should be taken as a serious candidate for a new theoretical framework, comparable to Newtonian approach in its own time. Classical physics justified the particle-mechanistic outlook and the world was seen as a huge clockwork until the beginning of 20th century when quantum physics and relativity theory challenged the previous metaphysical presuppositions and obscured the common world view. Dynamic Universe decisively breaks out from the particle-mechanistic context and provides an alternative perspective to reality – a new holistic framework with new set of assumptions and legitimate presuppositions applicable to all natural events and things. If we have a theory whose scope and profound simplicity surpasses the capacity of present theories – a theory which manages to handle correctly all the known physical and cosmological phenomena with minimum amount of postulates and with no fitting parameters – it should be carefully studied with all its features and consequences.

Since antiquity the growth in physical understanding of reality has characteristically been related to achieving a proper perspective and further knowledge on the interrelations between such concepts as mass, energy, motion, space and time. Dynamic Universe once more successfully changes the perspective and provides a new connection between these perennial concepts. By introducing a hidden motion into the 4th dimension it reveals an unexpected sight to the internal structure of the universe.

1. Introduction: Revising the conception of reality

Each culture is fundamentally overshadowed by its understanding of reality. What is the ultimate nature of all the different things we encounter and how are they related together? This cultural inheritance defines the general orientation of individuals who seldom question the view they are adapted to. The main features of an accustomed paradigm are difficult to overcome but nevertheless the common world-view in western culture has changed radically in antiquity and at the turn of the modern era. At these turbulent times also the deepest metaphysical questions are reflected on and understanding about the basic nature of reality may move on.

Modern picture of reality was decisively triggered by physics at the turn of the modern era. Newtonian framework stimulated all fields of research with its objective method suitable for particle-mechanistic world. Ideas of atomism, reductionism and determinism were supposed to be applicable to all relevant problems. Also philosophers had to take seriously the new concept of matter even if it was insufficient to enlighten mental phenomena. Descartes introduced the idea of substance dualism which set the stage for subsequent discussions. Advocates of idealism and materialism emerged and the schism between the hard science and humanistic concerns still continues. It can hardly be solved if the general mechanistic-deterministic outlook of the world does not change.

The advancement in understanding the nature of reality has historically happened as an interplay between science and philosophy. New empirical findings may affect the underlying metaphysical assumptions, the deepest ontological and epistemological views concerning the foundations of reality and knowledge. These assumptions invariably coordinate and motivate the research activities whether they are explicitly recognised or not. Yet they may change when natural science finds new invariances and interconnections in nature. More sophisticated theories give support for or against certain metaphysical views. Modern physics started from the particle-mechanistic basis which led to many successes culminating in the ability to manipulate individual atoms in quantum- and nanotechnology. Further examination related to the constitution of matter gradually revealed also more and more unsuspected events which show that everything is not reducible to individual building blocks moving in spacetime. Complex holistic phenomena observed in quantum physics or self organising systems give evidence for such internal relations and emergent phenomena which cannot be understood within the particle-mechanistic framework, the clockwork universe alleged by Newtonian physics.

The common metaphysical and methodological principles related to the Newtonian approach cannot any more provide further insight into the constitution of reality. The interpretation of quantum mechanics has remained unsettled almost for a century as the new phenomena cannot be explained on the previous basis. Yet the complex phenomena encountered in physics are facts concerning reality. They are interesting new findings whose relevancy is sadly dismissed by the instrumentalists who refrain from using theories as windows into reality. They lose the historic opportunity to achieve a better understanding of reality which might also allow a preferable view on how humans are related into the larger scheme of things.

There is a genuine call for a new conception of reality, for a new paradigm comparable to Newtonian approach in its coherent structure and exact general method applicable to all natural events and things. The advent of such a comprehensive framework is admittedly improbable and it may be greeted with disbelief. Very few would ever vote for a profound paradigm change to take place and many dismiss the very idea of getting further understanding concerning the nature of reality. All the trials to unify quantum mechanics and relativity theory lead to complicated and highly abstract mathematics which is impossible to conceive in common terms. Nevertheless, it is known that planetary mechanics was facing mathematical complications and increasing number of epicycles before Copernicus made his breakthrough. Since antiquity the growth in physical understanding of reality has characteristically been related to achieving a proper perspective and further knowledge on the interrelations between the concepts of mass, motion, space and time. Copernicus' findings overruled the Aristotelian view on the universal lawfulness, but the final hierarchy of these perennial concepts has not been decided yet.

Tuomo Suntola in his Dynamic Universe –theory (DU)[i] once more successfully changes the perspective by introducing a hidden motion into the 4th dimension. The theory is able to handle correctly all the known physical and cosmological phenomena with minimum amount of postulates and with no fitting parameters. The precision and scope of the theory seems to surpass the capacity of present ones and it also gives a plausible model of reality. Thus the Dynamic Universe should be carefully studied as a serious candidate for a new theoretical framework. The powerful basic principles of the theory provide a possibility to grasp amazingly profound questions and thus the theory is pregnant with fundamental metaphysical implications whose explication might provide a leap into a better understanding of reality.

According to Thomas Kuhn a proper paradigm shift typically means a complete gestalt shift; a new paradigm is incompatible with the old one and assumes the world as being made of different kinds of entities. The form of relevant questions may change as things are approached from different perspective. In the DU framework many of the annoying features in present theories do not exist any more. We do not need to worry about the uncommensurability or puzzles of quantum and relativity theories as they do not contribute to the new description. Yet the situation does not lead to relativism. DU certainly is more than just another instrument for organising observations. Its holistic structure and new interconnections undoubtedly reveal some real features in nature which permit the handling of a wider variety of phenomena under one and same mathematical formalism.

From the philosophical point of view perhaps the most significant and unexpected result is the solution DU provides for the age-old controversy concerning the relation between parts and the whole. In antiquity, the general outcome in natural philosophy was that reality was an organic whole capable of regulating itself. The outlook was overturned at the turn of the modern era when everything was supposed to be reducible to countless tiny particles that obeyed strict deterministic laws. In the DU, the schism between atomism and holism can be reconciled because the new concept of mass allows the elementary units to be seen as a result of an energetic diversification of the whole.

2. The historic controversy between atomism and holism

Extended debate on natural philosophy started in antiquity. The early natural philosophers discarded mythical explanations and concentrated on universal principles, natural causes and invariances. The Milesians looked for a basic substance whereas Heraclitus asked for a more abstract basis, *logos* to take care of the continuos change of everything. Pythagoras trusted in mathematical harmony. Numbers, quantities and their relations would be the basic operators behind all the appearances. But on the basis of strict logic Parmenides then questioned whether anything that is eternal and invariable could ever produce any change. The relation between being and becoming, one and many, temporal and eternal become central for subsequent natural philosophy. How can the continuously changing sense world be explained on the basis of something that is eternal and changeless? The answer of Empedocles and the Atomists was that change was caused by eternal atoms out of which all the things were composed. The change was brought in as the atoms were continuously rearranging themselves, joining together and disassembling. For them mechanical materialism was enough to understand reality. Soul and the gods could be explained by consisting of especially long lasting arrangements.

At the bloom of ancient thinking Plato and Aristotle discarded the particle-mechanistic conception defended by the Atomists. They did not believe that "dead" matter could ever explain organic phenomena or the activity of humans. In addition to matter some kind of organising form or essence was also needed. *Materia prima* as such was something indefinite. It was not detectable until it was connected to an organising form. Plato believed that the eternal forms inhabited an upper world of ideas and they were reflected in the sense-world objects whereas Aristotle argued that the forms were hidden potentialities that were immanent in the actualised objects. Aristotle was a great scientist and systematizer who refuted the claim of Parmenides that change is unintelligible. In the fourth century B.C he studied the relations between time, space and motion and the close connection between weight and movement of elemental substances. One of his concerns was the study of lighter and heavier in terms of the speed of falling bodies.[ii]

Aristotle's ideas on physics and natural philosophy became an integral part of the medieval way of thinking. His dealing with causes and the motion of bodies was considered superior to that of the Atomists, and his ideas could not be overcome in a thousand year. It was only in the beginning of the modern era in the 16th century when natural philosophy once again reached the level of discussion experienced in Antiquity. Copernicus challenged the Ptolemian astronomy and found a new cosmological order and locus for the earth. Kepler highlighted the mathematical approach defended by Pythagoras and Galileo united mathematical laws with the atomistic and materialistic outlook originating from Democritus and Epicurus. Newton systematized everything into a uniform axiomatic form and his mechanics was applicable to all matter in motion. It was generally believed that Newton had revealed the true structure of the world. It was comparable to a clockwork, mechanical, quantitative and without any purpose.[iii]

Thus reality was no more considered as an organism. Classical physics abandoned the Aristotelian concept of matter and returned to the idea of "dead matter" defended by the Atomists. Matter consisted of indestructible tiny particles that obeyed strict deterministic laws. Any kind of form, abstract essence and activity inherent in matter became irrelevant and mind, reason and all subjective phenomena were stripped from the objective reality. Descartes divided reality sharply into two parts: *res extensa* and *res cogitans*. He proclaimed that "neither God nor any rational soul present in the world will ever disturb the ordinary course of nature in any way". Physics began to study the measurable world of extended objects, whereas subjective phenomena were excluded.

The world was seen as a huge clockwork until the beginning of the 20th century when quantum physics and relativity theory were born. These theories shattered the metaphysical presuppositions related to Newtonian physics. Theory of relativity connected time to space. The time intervals and spatial separations became frame-dependent and the idea that body's velocity, length or kinetic energy are invariant, real properties was obscured. Many quantum phenomena like wave-particle dualism, entanglement or statistical predictions are still more difficult to conceive within the particle mechanistic framework. Non-local phenomena simply cannot be explained with the idea of reducing everything into separate material particles. Quantum physics and the theory of relativity certainly provided further knowledge and refined handling of things but they also obscured the common world view. On the basis of these abstract mathematical theories we do not really understand in what kind of reality we are living. Against all evidence many of us still keep on thinking the world as a huge clockwork – that is even if modern physics in many ways overstepped the limitations of classical physics, classical metaphysics still very much guides our thinking and imagination. The conventional ontological and epistemological ideas concerning objects, their characteristics and their relationships, as well as our own relationship to them, are well suited to the particle-mechanistic framework. As theories do not fit into the picture they are often considered to be just usable instruments which are not meant to provide any further understanding concerning reality; we should not worry if we are not able to understand the encountered new phenomena and locate them into a comprehensive and all-embracing scheme of things.

Yet, against all instrumentalists' bias, modern physics actually collapsed the world view created in the beginning of modern era. The foundational presuppositions of classical physics turned out to be just half-truths which are suitable in the macroscopic world. The particle-mechanistic picture of reality defended by the ancient Atomists was falsified but otherwise the perennial problems concerning the nature of reality are still very much the same. According to Thomas Kuhn deep metaphysical questions come into the foreground only when there is a quest for a profound paradigm change.[iv] For the change actually to happen an alternative view is needed. In these circumstances the potential of Dynamic Universe theory should be studied carefully; whether it actually contains insights mighty enough to move the understanding of reality a major leap forward.

The next disposition is fragmentary and tentative. It is just meant to encourage further studies by giving hints of the marvellous ingredients contained in the theory.

3. The Dynamic universe

The Dynamic universe provides a holistic framework covering all phenomena from micro structures to cosmology. It is a precise highly structured mathematical formalism with well established physical assumptions and strict predictions. The theory provokes a new vision to the internal structure of the universe by revealing an unexpected connection between the concepts of mass, energy, motion and space. In the new hierarchy the totality of mass is a fundamental invariant which links everything together into a holistic composition. Through energy excitation, mass determines the motion, volume and time development of space, and regulates all the local structures in space. The theory re-establishes a universal frame of reference. By dealing with the common coordinate quantities which locate the exact position and time of all incidents, DU manages to give a comprehensible model of the constitution and evolution of the universe.

The Dynamic universe theory was created by a distinguished Finnish scientist Tuomo Suntola who has done a remarkable career in industrial technology. The breakthrough in the development of the DU occurred in 1995 and ever since Suntola has documented the subsequent expansion of the theory in series of monographs. The latest "The Dynamic Universe – Toward a Unified Picture of Physical Reality" came out in 2009. The proponents praise the wholesomeness of the approach. "The

Dynamic Universe describes physical nature from a minimum amount of postulates. It accurately explains observed physical and cosmological phenomena without any fitting parameters. It develops the ideas of Einstein and Feynman into a complete theory."[v] The theory has aroused interest in conferences but until now quite a few professionals have taken the trouble to actually examine the theory in detail. Even if the new theory could explain and predict all the observed phenomena, it goes against the accepted paradigm by creating an alternative perspective into physical examination. The utility and rationale of the new set of postulates and procedures may be difficult to understand from the context of prevailing theories which it aims to replace. Yet DU is difficult to dismiss. Experiments cannot decide between the standard theories and the Dynamic universe as the predictions for local physical phenomena are essentially the same. At extremes – at cosmological distances and in the vicinity of local singularities – differences arise and DU seems to work better. Even if, from the instrumentalist point of view, physics might do without DU, people with realist inclination should notice the bonus it provides by widening the horizons of the present world view.

The main features of the theory

Dynamic Universe illustrates space as a 3-dimensional surface of a four sphere. This is actually quite an old idea pondered in the 19th century e.g. by Bernhard Riemann and Ernst Mach. It was also Einstein's original cosmological view[vi] which he had to give up when looking for static solution and introducing time as the fourth dimension. Mixing up space and time demolished the readily understandable picture of reality but the spacetime concept was considered unavoidable. DU, however, succeeds in separating the space and time coordinates when carrying thorough the idea of the universe as a dynamic four sphere. Space, the surface of the metric four space, appears as a spherical pendulum tediously oscillating in the fourth dimension. It is presently in phase of expanding at the velocity of light along the 4-radius. This kind of precise dynamical geometry allows a parameter-free derivation of the primary cosmological quantities, which gives strong evidence for the factuality of DU's depiction. Ideas like dark energy or accelerating expansion of space are not needed. The motion of the whole space in an unseen direction may cause some doubt, but in Copernicus' days most people could not take seriously the idea that earth was moving around the sun as the proposal was in such contrast to everyday observation.

DU thus differs from the theory of relativity in rejecting the spacetime concept as well as the postulate of the constancy of the velocity of light. The spacetime appears to be just an unwarranted mathematical trick. In local environments the adjustment of the coordinate quantities may disclose the observations correctly but renouncing the idea of fixed measures of time and distance, which are central for human conception, obstructs the overall view of what is actually going on in reality. Neither does it promote further physical understanding; for example it does not tell why the velocity of light is observed as constant and why it is the maximum velocity in space. The dilemma gets an obvious solution in the framework of DU because of the linkage between the velocity of light and the expansion of space along the 4-radius. DU does not need the postulates of Lorenz transformation or the equivalence and relativity principles either, but the enriched framework elucidates the background and contents of these twisted ideas. In addition to the precise geometry and the postulate of an overall zero energy principle, DU just needs to fix the total amount of mass to get a universe going. Like in a pendulum the sum of the energies of motion and gravitation are equal throughout the cosmic expansion-contraction process and the total energy is conserved in all interactions in space. These postulates are very concrete and corporeal when compared to the mathematical approach characteristic to the theory of relativity.

According to the DU theory, most of the universe is outside space, i.e. the observed 3-dimensional reality whose volume and internal hierarchy are orderly expanding. The geometrical 4th dimension is inaccessible even if it constantly contributes to everything that is happening. The evolution is fuelled internally because of the active energy process in which the motion in the fourth dimension is always balanced by gravitation. The counteracting process determines the form, volume and time development of space. It also regulates the emergence of all local structures in space as during the expansion process part of the primary energy converts into material, electromagnetic etc. phenomena in space. Thus everything we observe in space springs from the abstract, immaterial realm of being. It may be non-empirical but nevertheless it is very real and definite.

The DU theory clarifies the concept of mass and its relation to matter and energy. The totality of mass is the fundamental invariant which links everything together into a holistic composition. Mass as such is extremely abstract and inconceivable, devoid of dimensional extension or form. It is the persistent "core of being" that is immanent in all events and things. A basic substance, if you like, the unseen, eternal basis of being itself. Thus mass is not an amount of some impenetrable solid stuff, neither is it a property of matter or fields but rather their cause. Mass in the DU resembles the abstract *materia prima* defended by Plato and Aristotle. It cannot be conceived as such, since it obtains reality only when activated by energy. Thus energy turns out to be reminiscent of the idea of form proposed by the great philosophers in antiquity, an inherent organising factor, eternal pattern or potentiality in things. Nothing is just matter, but it cannot be without matter either, as already Aristotle stated. The concept of zero-energy principle also incorporates the ideas of Anaximander and Heraclitus who hold that all separated elements came into being as opposite pairs, out of one undifferentiated and indivisible basis. For Anaximander it was an abstract substance called *apeiron* whereas Heraclitus defined it more quantitatively as *logos,* a governing principle which preserved a balance between various dynamic opposites, out of whose struggle everything was being born.

Everything in the changing phenomenal world, the becoming of all physical structures and their interactions, emanates from mass which is excited by energy. Mass is the most abstract and at the same time the most concrete factor behind everything – always measured in kilograms. In space, the energy of motion mass possesses in the 4th dimension is observed as the rest energy of matter. This is a natural consequence of the fact that on the surface of the expanding sphere we are at rest related to the motion. The basic form for energised mass is wavelike dark matter which in suitable conditions converts into radiation, electricity or material particles. DU relates an explicitly defined mass equivalence to all different things and phenomena. There is no question of wave-particle dualism as basically there is no conventional corpuscular matter. Phenomena like EPR-paradox or Young's two-slit experiment can be elegantly explained. Many obstacles vanish already when DU unravels c (the velocity of light) from the traditional Planck constant. An even more crucial fact disclosed by the theory is the universal system of cascaded energetic frames whose inherent holism explains the apparent non-locality of diverse phenomena.

The cascaded energetic frames

In the DU, reality is a dynamic, structured whole which links mass, energy, space and motion inseparably together. More precisely, all the different parts of the universe are internally related to the whole via a system of nested energetic frames. The primary motion of mass with the expansion of space, balanced by global gravitation arising from the total mass in space, generates the ultimate frame, the homogenous space, whose evolution naturally gives rise to further relatively autonomous subsystems or frames, whose configuration always conserves the local as well as the overall energy balance. The frame provides both local and universal state of rest. It controls the behaviour of its parts which may contain sub-frames behaving accordingly. Thus atomic reductionism is left behind. When starting from the overall mass and zero-energy balance, the multiplicity of the units is a result of the diversification of the whole. Reality is not based on discrete bodies, the "dead" particles of the Atomists. The cause and dynamics for matter in motion has been decisively clarified.

Starting from the homogenous space, all the minor frames, which range from galaxy groups and solar systems to earth and all the individual material objects are formed at the cost of reduced local rest energy, (which means a dent in space and reduced velocity of light). There is no definite answer to what breaks the ideal symmetry of homogeneous space and leads to the accumulation of mass into subsequent sub-structures in which energy finds a localised balance and mass receives an extended form. Nevertheless, the dynamics of this unmistakable structuring is precisely presented, including an elaborated handling of local singularities and black holes. The concept of energetic frame also helps to clarify the idea of an object. It fulfils all the criteria for being called an object. Naturally, all objects are real extended structures which host a specific amount of mass and energy. Composite objects may contain various befitting sub-structures or frames but elementary particles are described as plain structures which capture the mass wave into a closed pattern whose shape determines the energy states available for it.

All the different frames or objects are relatively autonomous. They are localised and closed structures, which have a certain locus and degrees of freedom within the whole. The specific amount of mass and energy which each frame hosts is a strict share of the total energy of the universe. Thus the energy state of a frame defines its relation to the whole as well as it determines the energy states available for its sub structures. Local velocities are related to the velocity of space in the fourth dimension, and local gravitation is related to the total gravitational energy in space. The holistic approach in the DU means relativity of local to the whole. It is not relativity between observer and an object but a consequence of the limited amount of energy in space – relativity is a measure of the locally available share of the total energy.

In the DU, conservation of total energy links local interactions to the rest of the space – providing a solid theoretical basis and a quantitative expression to Mach's principle and a natural explanation for the relativity of observations. Because of global gravitation all the relatively autonomous local structures, the material objects in space are united to the whole. The structured rest energy of matter (caused by the energy of motion in 4th dimension) is always locally distributed in space whereas the counterbalancing energy of gravitation is due to all the remaining mass in space. Thus there exists a complementarity between the local and the whole for objects. They are located in a specific position but because of global gravitation they are also omnipresent. The frames interpenetrate and sub-frames are throughout in inherent immediate contact with the parent frame, ultimately the hypothetical homogeneous space. In a way, the parts are "aware" of the whole universe through global gravitation.

The complementarity allows reconciling of the old schism between atomism and holism. Because of global gravitation there exists a holistic aspect in reality, which allows it to be considered an organic whole capable of regulating itself. Yet the whole divides itself into autonomous parts whose behaviour is strictly regulated and follows exact mathematical rules. Basically, the trick is to reveal the proper distinction between matter in motion, and mass and understand the twofold energetic diversification of the latter. The mass that manifests itself in corporeal objects is always related to energy. Balanced excitation of energy creates space, motion and all the conceivable structures which consistently remain internally related to the whole.

4. Conclusion: Implications for the world view

The conception of reality is a result of human deliberation. The previous assumptions and presuppositions can be questioned or altered on the basis of new findings and better theories, typically offered by natural science. DU seems to exceed present theories in its capacity to deal with physical phenomena, and its structure appears to be coherent and well-founded on sound postulates. Confirming the correctness of the theory with detailed examination would be essential – worthwhile both for physics and philosophy of science. A concurrent theoretical framework of this size provides an excellent opportunity to test and employ the tools philosophy of science has produced for evaluation and comparison of theories. The study of DU could also enlighten the topical discussion concerning the relation of theories to reality; in what sense do theories create reality or correspond to it? And naturally the situation provides a proper chance to examine the revising picture of reality and the mechanism of paradigm changes.

A general approximate interpretation of the DU formalism is in many ways obvious like was the case with Newtonian mechanics. Everybody could easily understand the basic features of the theory when the world was described as a huge clockwork. It is no more difficult to conceive a cosmic pendulum with contraction and expanding phases. The fact that space is described as a 3-dimensional surface of a pulsating 4-sphere provides room for additional relations which are impossible to locate into mechanistic-deterministic reality. The details of the theory provide indispensable material for accurate philosophical explication, which may lead to new metaphysical hypotheses and better understanding of reality. Especially the clarification of the concepts of mass and energy gives a real chance to continue the age-old philosophical discussion concerning the proper character of this evolving corporeal reality we observe and live in.

DU starts from the idea of a complete unbroken whole, which is contrary to the present tendency to extrapolate global phenomena from local rules and observations. Particle-mechanistic view is based on the idea that there exist merely external relations between individual particles. The whole emerges as a sum of the elementary units and no immediate internal connections are acknowledged. In the DU, elementary units result from the diversification of the whole, and local conditions are determined by the large-scale structure of the universe. The dynamic composition of the theory gives an opportunity to bring back the idea of form or essence cherished in antiquity. Because of energy excitation, there exist an inherent activity within matter bearing similarities to the ideas of Plato and Aristotle. Reality, macrocosm, once again contains a more refined layer or aspect, which can be reflected in the constitution of humans, the microcosms. Cartesian dualism is overstepped. Mental phenomena need not be stripped from the objective reality as was thought in the beginning of modern era when the idea of mechanical "dead" matter was adopted.

During the last centuries there has been a strong faith in atomism and mechanical materialism, but in physics a more holistic trend has strengthened since the study of electromagnetic phenomena and the concept of field, long before the idea of quantum holism. Yet the holistic tendency in DU is much more coherent and penetrating providing a new comprehensive perspective to material phenomena and reality. A new framework which breaks out from the particle-mechanistic world can once again, like Newtonian framework in its time, stimulate both physics and philosophy. The new principles can be applied to innumerable practical problems, and their explication provides room for new metaphysical ideas and philosophical discussion. In addition to Plato and Aristotle, many philosophers like Spinoza, Leibniz or Hegel have been in search of form and unity. They have tried to justify the view that passive mechanical matter is not enough to explain all the phenomena we encounter, and thus something that is active, rational and conscious should also have a locus in reality.

This kind of profound discussion, typical for the great system builder philosophers, has not been in fashion in philosophy in recent times. The task of clarifying the concept of reality and its most important attributes is mainly left to physics, whose empirical method may be limited. Nevertheless, physics has managed to specify the contents of the terms mass and energy. Newton connected mass to inertia and distinguished inertial mass from gravitational mass, even if he equalled them. Newton was not able to give inertia any clear cause, but Ernst Mach in the end of the 19th century proposed that inertia of a body would arise from its relation to the totality of all other bodies in the universe. Albert Einstein arrived at the famous equation $E = mc^2$, which states that the energy E of a physical system is numerically equal to the product of its mass m and the speed of light c squared. Einstein proclaimed the formula to be "the most important upshot of the special theory of relativity" and it certainly is very useful in many cases, but what does it actually mean? The equivalence of mass and energy has caused a philosophical controversy on its real meaning and ontological consequences. Can mass and energy really be conceived as the same property of physical systems and in what sense mass is "converted" into energy?[vii]

On the basis of present theories we are not able to answer these ontological questions. We do not actually understand where the relation between mass and energy comes from, no more than we know whether the fields should be considered real entities, or what is the actual shape and destiny of the universe. Especially, the ontological status of energy is blurred. Is it some kind of stuff, or property of things like velocity, or is it just a useful theoretical fiction, a bookkeeping device.

In classical physics energy describes accumulation of force, the work that has been done into a system. Its reality can be questioned as we are not able to define an absolute value or distribution for energy. Conservation laws are applicable only for closed systems and thus, on the basis of present theories, we are not able to identify what kind of thing energy is.[viii] DU considers reality to be a closed system, the three dimensional surface of a 4-sphere, containing a universal frame of reference, which allows the amounts of mass and energy be compared in an absolute manner. Thus energy can be a real property of objects, whose locality can also be maintained. Energy belongs to the objects in a way it exhibits the essence of them. Mass and energy are clearly separated. Energy, in a manner, triggers or forms the abstract indefinite mass giving rise to extended detectable things: matter, radiation, electromagnetic phenomena, and all possible forms of closed vibrations.

In the DU, the famous equation of Einstein gains a more precise form taking into account the frame where phenomena are studied. DU also gives a clear vision of the long-term development of space. During the expanding phase the energy of motion, the rest energy of matter, diminishes paying back the debt it owes to global energy of gravitation. The process defines circumstances in which new structures or frames are born. The frame establishes the amount of energy and degrees of freedom available for its parts and it also provides a local and the universal state of rest.

This kind of rich tapestry of relations including the idea of balanced activity and inherent unity of everything would certainly be helpful in understanding humans and their relation to the larger scheme of things. Also humans naturally consist of many energetic frames, and belong to various kinds of them. We are able to host specific states of energy and, like all ontological structures, we are in a complementary manner connected to reality. Our internal aspect is ultimately connected to global gravitation which unites all the relatively autonomous local structures or material objects in space. In the DU the origin of gravitation is no more a mystery. By being the attracting energetic aspect of excited mass, it gets an understandable locus and role in totality.

The concept of global gravitation might be useful in explaining consciousness. It makes more understandable old philosophical ideas, like Leibniz's *monads*, which are able, in variable grades, directly reflect reality or the *world spirit* of Hegel. When related to gravitation the spirit might be coming more conscious about its own nature, as in the present energetic phase the global energy of motion is converting into gravitation causing increased structuring in space. DU certainly provides interesting analogies which might be used in modelling human psychology and mental phenomena. Humans with all their characteristics are influential parts in the evolving whole. They naturally "feel" the impact of totality in their bodies, and gradually they may understand its structures better. In addition to external frames there may be numerous mental frames, internal bonds which are more difficult to notice than the external objects conceived from outside. Discerning all the frames, however, is essential for getting control over them. Only, when finding out a local state of rest, one is able to "tune" into a more fundamental and energetic frame.

– In a particle-mechanistic world there are no internal relations, whereas DU gives an option for an access from local and relative world to a more fundamental reality. In a sense it restores the concept of aether which in Greek mythology originally was the personification of the "upper sky", space and heaven.

337

Acknowledgement: I express my gratitude to Dr. Tuomo Suntola for many inspiring discussions when figuring out the structure and postulates of the Dynamic Universe theory.

---

[i] Suntola Tuomo, *The Dynamic Universe, Toward a Unified Picture of Physical Reality*. ISBN 978-952-67236-0-0. Physics Foundations Society, Espoo, Finland (2009).

[ii] Hahm David E., *Weight and Lightness in Aristotle and His Predecessors* in Motion and Time, Space and Matter. Interrelations in the History of Philosophy and Science. Ed. Machamer Peter K. and Turnbull Robert G. p.56-82. Ohio state university press, (1976).

[iii] Kallio-Tamminen Tarja, *Reality Revisited, From a Clockwork to an Evolving Quantum World*. p. 16-66. VDM Verlag Dr Muller, Saarbrucken, (2008).

[iv] Kuhn Thomas, *The Structure of Scientific Revolutions*. University of Chicago Press, (1970).

[v] Quotations from the back cover of Tuomo Suntola's book *The Dynamic Universe, Toward a unified picture of physical reality*, signed by prof. Ari Lehto and dr. Heikki Sipilä. Physics Foundations Society, Espoo (2009).

[vi] Einstein Albert, *Kosmologische Betrachtungen zur allegemeinen Relativitätstheorie*, Sitzungsberichte der Preussischen Akad.d.Wissenschaften (1917).

[vii] Flores Francisco, *The Equivalence of Mass and Energy*. http://www.plato.stanford.edu. (2007).

[viii] Lange Marc, An Introduction to the Philosophy of Physics. s. 124-130. Blackwell Publishers, Oxford (2002).

Back to the Ether

Erik Trell,
Faculty of Health Sciences, University of Linköping,
Se-581 83, Linköping, Sweden
E-mail: erik.trell@gmail.com

Abstract

In spite of the success of the standard Bohr model, the Positron/Electron has defied precise charting of its own detailed properties in an orbital representation. Evidence from modern high-resolution nanotechnology and mineralogy supports the original Ether concept that the Universe has a regular solid structure where the tetrahedron and octahedron cyclically expand their motifs in flat Euclidean space. Since they are all made of the unit straight line, the Universe is thereby digital. In the present investigation of the classical Euclidean space and law-bound Lie algebra realizations therein updated to the new information, a primary electron module (with positron inverse) occurs in the first, $(2^3)^3$, spherical phase transition root vector lattice expansion in a Cartesian segment diagonally skewed from the cubical 2^3 ground space Neighbourhood of the Nucleon. Over the Nucleon surface the so charged t isospin root vectors of unit length connect to a circuit of length $2\pi \times 2^{1/2}$ and mass number $938.28/(2\pi \times 2^{1/2})$ MeV = 105.59 MeV, identical to the Muon. However, in the extra-Nucleon continuation of the lattice they step by step outline the sides of the polygonal mesh by triple-coil nodes consisting of two tetrahedra and one octahedron of sum margin length 12. In sheetwise, net 180° axially twisted continuous alignment, 152 such nodes distribute into the $(2^3)^3$ region by a Bohr spectrum layered, Hydrogen-matching space-filling truncated octahedron of mass $938.28/12 \times 152 = 0.514$ MeV in comparison with actual Electron mass 0.511 MeV. In turn, these modules may fuse to honeycombs with proportionally increased pivotal inertia and continuing self-similar exponentially doubling volume expansion cycles to Atoms fitting the periodical system and able to join to molecules with tandem spin and other attributes as in reality as well. Also the organic realm is tessellated by the same bricks with added pentagonal arrangement, too, in conductive aqueous milieu. The Ether is not an alien medium for analogue transmission but the digital hologram processor and display per se.

Introduction

Back to the Ether – back to the future. It is remarkable how the ancient Ether, whose "beautiful forms unfold from unity", is returning at the forefront of modern natural sciences from Nanotechnology to Cosmology by its "building blocks of three-dimensional space, central to architecture, chemistry and atomic physics" as well as "cornerstones of mathematical and artistic inquiry since antiquity" (Sutton [1998]). They are the regular solids and what they can tessellate by their only vector constituent, the unit straight line; the single, digital element of the plenum thereby entirely filled by itself: the self-made Ether (Greek αιθήρ, *Aither or Aether*). In Greek philosophy its original both divine and mechanistic designation was *Protogenos*, like the Vedic *Akasha* the first-born "substance of light" (http://www.theol.com/Protogenos/Aither.html); the categorical state of coming into being from Nothing-Whatsoever; from absolute Absence to the equally obligate contrast of Presence or, for that matter, of Big Bang through every coordinate; coming and being at once and forever by its real possibility. And not as a quiescent dissociated interstitial medium as relegated to in the middle ages, but the "soul of the world from which all life emanates" and springs up in "the bright, glowing upper air of heaven" (Ib.) as the very flames of it. This is the tie-up and identity of the Ether with the regular solids; the tetrahedron(s) of fire mingling with the octahedron(s) of air in the cube(s) of the Earth and, as passed on from the geocentric reference, Euclidean space alike.

Methods

These three are the simplest of the regular solids, like the icosahedron and dodecahedron step by step contoured by the unit straight line, and hence naturally digital. But only the tetrahedron, octahedron and cube are by themselves able to span all from the infinitesimal level a dichotomous space-filling lattice system quite akin to modern Nanotechnology, where polygonal structures "folded from planar substrates" (Whitesides and Grzybowski [2002]) similarly "self-organize…even forming hierarchies" (Ikkala and ten Brinke [2002]) "from the smallest to the largest scale" (Guth and Kaiser [2005]). And they were likewise atomistic and mechanistic, because the ancient Greek who developed the system over many centuries from before Pythagoras (569-475 BC) to Diophantus (200-284 AC) and beyond delegated the ontological questions to the "mythical cosmogonies" (http://www.theol.com/Protogenos/Aither.html) while themselves setting out to pragmatically test the feasible building models.

In Plato's *Timaeus*, one version is described, where triangular planes (Fraser [2001]) "played the role of 'the subatomic particles' in his theory of everything" and "in turn, these triangular particles consisted of the three legs (which we might liken to quarks)" of their sides, and hence infinitesimal straight line bits deepest down built up the solids which "Plato regarded…as the 'atoms' or corpuscles of the various forms of substance" (http://www.mathpages.com/HOME/kmath096.htm).

In the delightful book *Platonic and Archimedean Solids. The Geometry of Space.* (Sutton [1998]), a plausible aboriginal scenario is suggested of the primordial mind pondering over the riddle of the ready-made world and one's part in it and the prototype likely to have come out of it: "Imagine you are on a desert island, there are sticks and sheets of bark. If you start experimenting with making three-dimensional structures you may well discover five 'perfect' shapes" possible to iterate over space just from and as they are; and simplest of them the tetrahedron, the octahedron and the cube.

The rational logical corollary is that it not only suffices with but demands just one uniform element to expand ground-up in such way without being obstructed by incongruent ingredients, and that this element is the equilateral stick, i.e., a line bit of common unit length that at the threshold minimal, *limes* scale where it enters the world, automatically by its equivalence class division is as short as its equally infinitesimal thickness – so that the whole world really consists solely of it – so that it is self-templating and self-assembling to that infinitely ascending end in which all realizations must also be non-overcrossing, that is, since every locus consists only of itself, there cannot be more than the one own occupancy of it – for the moment as well as for ever.

Results
1. The Cube – Element of Euclidean and Diophantine Space
That in many cultures, behind an often deified mythological symbolism was thought to work a primary structural execution of the world with homogenous volume realization of its elements can be traced at least from Neolithic time (Ib.) over to Hellenic age and even further. This likewise applies to the cubical space where everything – including the space – directly takes place by itself. Notably, "the geometry that Euclid learnt from his Ionian teachers was originally based on watching how people built", and "the measurement of volume by the number of cubes with sides of standard length required to fill a solid space was probably first used by the Sumerians, who built with bricks" (Hogben [1937]).

Towards Euclid's era (325-265 BC) it was widely accepted that space whether firmament or vacuum is rectilinear and its shape cubical, viz., the same as the fabric of the Earth and hence the ground solid and as such *per se* inert and interactionless but, as shown by the quotation before, possible to build with. In its utter simplicity it still resolves the dilemma of the "leading candidate for a 'theory of everything'…(which)…suffers from a fundamental weakness…the strings move in a spacetime whose shape has been chosen from the beginning, as if they were actors on a previously constructed stage. A truly fundamental theory of gravity, everyone agrees, would build the stage itself", to which end "a few physicists have…concocted a theory that precisely describes spacetime on the smallest length and time scales…just as matter is made of atoms and elementary particles, space consists of tiny indivisible bits" where the "warping of the very fabric" (Cho [2002]) forms the template for "untangling the Universe….into the equations of string theory" (Trefil [2004]).

However, there are "no strings attached" (Cho [2002]) in the 'canvas' (Kamionkowski [2002]) of the ordinary flat space here but as first shown by Ptolemy it still "constructs

341

itself" (Cho [2002]) holographically in maximally three dimensions (Fig. 1a) with profound Diophantine equation as well as Cartesian coordinate, Lie algebra (fig 1b), and up-to-date Nilpotent Universal Computer Rewrite System (NUCRS) (Rowlands [2006]) mathematical strength.

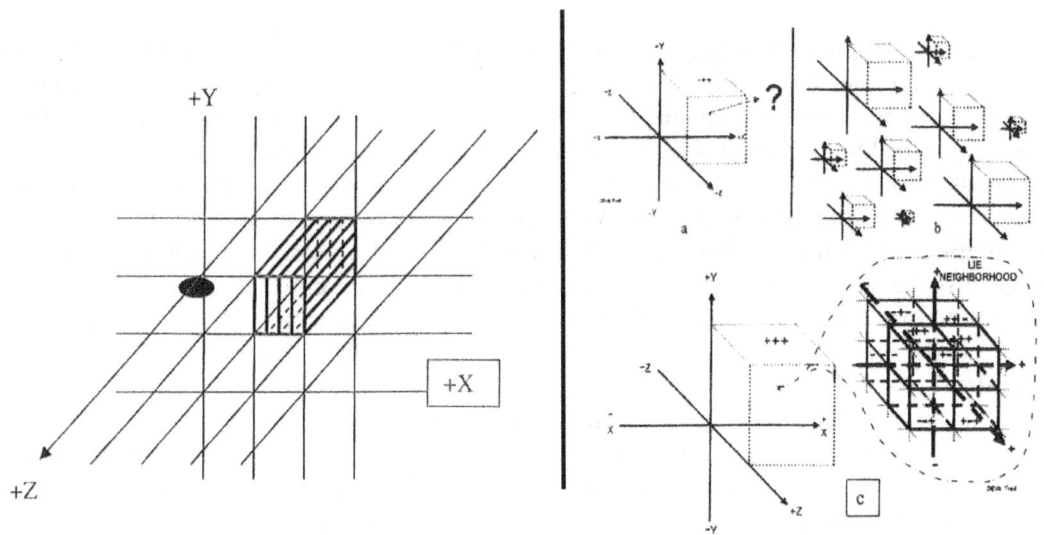

Fig 1a *Eigenspace and Eigenelement*   ***b*** *Smallest space point keeps Car-*
of flat Euclidean space (after Ptolemy)   *tesian order in Lie Neighbourhood*

In Fig. 1a it is envisaged how Ptolemy in his book *On distance* (150 AC) established the maximal three-dimensionality of rectilinear space and the equivalence class sifting out of its (here slightly distorted) infinitesimal cubical eigen-element (or "pixel") by the global spanning of the straight line by itself in a thereby 'Gödel-immunized' *sensu strictu* (and hence also 'Popper-ratified') falsification of its own conditions alone: "Draw three mutually perpendicular lines. Try to draw another line perpendicular to all of these lines. It is impossible. The fourth perpendicular line is entirely without measure" (http://scholar.uwinnipeg.ca/courses/38/4500.6001/Cosmology/dimensionality.htm).

Thus at every lattice intersection is established a primary Cartesian coordinate system which persists at all scales (Fig. 1b) and at the infinitesimal threshold assumes the works of a Lie algebra Neighbourhood, that is, the *limes* region where mathematical operations may be germinated then to be expanded into continuous geodesics by exponentiation or other iterative procedure.

This is because at that smallest level the walls of the Cartesian segment cubes are contracted to those of the spatial ground eigen-element, or "CuBit", and hence made up of the likewise infinitesimally small partial derivatives *dx, dy* and *dz* at the origin of differential equations so that they are the instant unit space numbers and Diophantine

342

integer digits alike (Fig. 2) with deep connections also to the NUCRS algebraical grammar and taxonomy (Rowlands [2003, 2006, 2008]) as well as Duffy's fundamental re-geometrization of the Ether (2004).

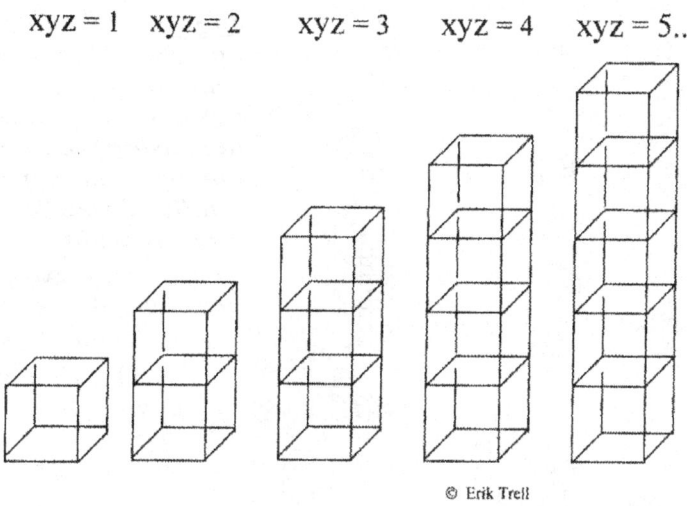

$$xyz = 1 \quad xyz = 2 \quad xyz = 3 \quad xyz = 4 \quad xyz = 5...$$

© Erik Trell

Fig. 2 *Three-dimensional Diophantine whole-number cells (or, after Penrose [1995], polyominoes), one-dimensionally joined together in the vertical direction to infinite series of integers of the first degree by the same discrete amount of the ground unit cubicle, or 'cuBit'.*

Based upon earlier ideas, there were "the ancient Greek" who first more systematically formulated the "completely brilliant idea...to use spatial images to represent numbers" in this manner (Noel [1985]). And "for the Pythagoreans and through the sixteenth century, one was seen as the root of every number" (Fraser [2003]), and was in three dimensions since time immemorial in ground form represented as a unit cube possible to apportion into this original whole-number bit system (Fig. 2), by which the genuine Euclidean space as well as the Diophantine equations and the operations and constellations therein can be directly brick-laid. Regarding next question - how did the building proceed? - there are at least two main continuous alternatives, one of which has been brought to the fore again both theoretically by e.g. Penrose [1995] and in recent nanotechnological "layer-by-layer" material self-aggregation and -organisation (Velikov et al [2002]). It can be described as a stepwise eccentric winding over the surface of the expanding box and has been used to literally underpin a previous proof of Fermat's Last Theorem (FLT) (Trell [1997, 1998a, 2002, 2003c,d]).

The other, and most straightforward at the bottom level is to first pave the floor, starting by a row from a corner along the side, after that turning for the next row, and so on till the ground square or rectangle is filled. Then, with unbroken succession in reverse order in the next tier, and so on, till the box is filled in a hence really analytical way, too, i.e. continuous, spacefilling and non-overcrossing (Fig. 3). Although this mode would

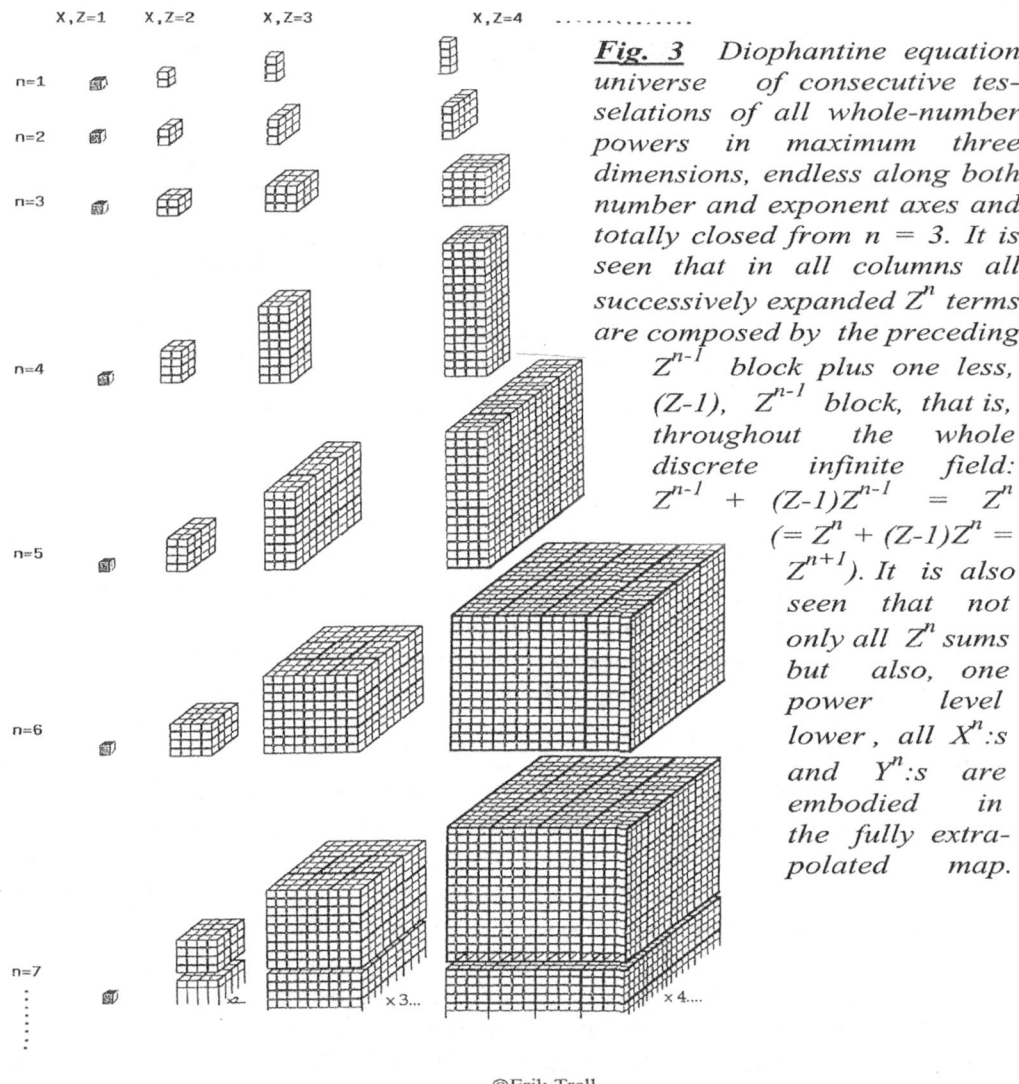

Fig. 3 _Diophantine equation universe of consecutive tesselations of all whole-number powers in maximum three dimensions, endless along both number and exponent axes and totally closed from n = 3. It is seen that in all columns all successively expanded Z^n terms are composed by the preceding Z^{n-1} block plus one less, $(Z-1)$, Z^{n-1} block, that is, throughout the whole discrete infinite field: $Z^{n-1} + (Z-1)Z^{n-1} = Z^n$ $(= Z^n + (Z-1)Z^n = Z^{n+1})$. It is also seen that not only all Z^n sums but also, one power level lower, all X^n:s and Y^n:s are embodied in the fully extrapolated map._

probably be closest at hand for Diophantus as well as for Pierre de Fermat, both may be used optionally. For it is important, that the comparatively late Diophantus himself "stated the traditional definition of numbers to be a collection of units" when in his equations they "were simply put down without the use of a symbol" (Heath [1964], Zerhusen [1999]). The effective quantum leap in relation to modern linear functions is of course the integer and spatial instead of point and imaginary nature of the numerical unit.

And pointless, too, would be to make this a heuristic controversy since it is all about reality: reality for the founders, reality of means and ends; reality of the very facts and findings of the case, i.e., that when ancient mathematicians well up to Cardano calibrated joint numerical and physical space they used what during thousands of years between the Sumerian bricks and Roman *tessellas** was the most refined of manufactured self-assembling forms: the cube, the irreducible (but in its products rationally divisible) whole-number bit; cubicle, kaba, cuBit©, "nanocube" (Murphy [2002]) of arbitrary unit side, providing the atomic set of a myriad literal dice not alone for God to throw but for themselves to stow by cumulative fulfilment (Noel [1985], Sutton [2002]) of their own, "nanobox" (Murphy [2002]) properties.

In order to reconstruct the original procedure, it may be reminded that computation in those days was much like surveying (Noel [1985]). For the first degree, *positio* alignment, the unit number cells (Figs.s 2, 3) then automatically deliver the measuring-rod by longitudinal plus or minus stacking like in the contemporary *abacus* (and the Inca *Quipu* threads) over a single axis; here, referring to digits, chosen as the vertical.

However, as outlined in Fig. 3, the added, in a double sense manifold value of the direct spatial realisation of whole numbers does not become apparent until with Diophantus formalising their exponentiations and subsequent equations. As mentioned, the natural procedure that offers for a serial power expansion is a sideways instead of length-wise multiplication of the digit by itself, producing at the second degree stage a square tile, step-by-step like the Sumerians did till the quadrate or rectangle is continuously and non-overcrossingly tessellated (Fig. 3). This mode is also documented by Aristotle when, criticising its application to the passage of time, he wrote that "the movement of the units (μοναδα) will be lines" and "a moving line will be a plane" (McGinnis [2003]).

Then, in the same fashion, next layer is filled, and next, and next, till the resulting first-order third degree 'hypercube' is also analytically completed (Fig. 3). In turn, that 'hypercube of the first order' in same periodic progression re-multiplied by the base number yields a 4th power in the shape of a quasi-one-dimensional 'hyper-rod of the second order', which in forthcoming multiplications generates a 5th degree second order hypersquare, then 6th degree hypercube, then 7th degree hyperrod, 8th degree hypersquare etc. in an endless cyclical "self-assembly at all scales" (Whitesides and Grzybowski [2002]) that eventually contains all whole-number (and fractional) powers that there at all are (Fig. 3). In both this and the previously mentioned eccentric mode which may be geared in for physical realizations within and from any such hierarchical level, it is important to re-emphasise that the build is successive also within each sheet by the respectively winding and zigzag alignement of the individual tessellas so that they never

---

* Oxford Concise Etymological Dictionary of the English Language: *Tessella* is Latin for little cube, diminutive of *tessera* = a die (to play with), a small cube. *Tile, tiling* stem from the Latin word, *tegula.*

clash. The entire Diophantine equation Block Universe is thus generated by a recursive, perpendicularly revolving algorithm in a maximum of three dimensions, thereby reproducing the hierarchically retarded, non-overcrossing, i.e. analytical space-filling of consecutively larger constellations, imaginable up to the size and twist of galaxies, no matter if taking place during actual time or an instantaneous phase transition in the sufficient ordinary Cartesian co-ordinate frame. In such geometrized mathematical iteration, a stepwise continuous "rod-coil-rod…self-assembly of phase-segregated crystal structures" (Kato [2002]) – which "in turn form assemblies or self-organize, possibly even forming hierarchies" (Ikkala and ten Brinke [2002]) – precipitates in a completely saturating, consecutively substrate-consuming way, displacing other stepwise cumulative syntheses (Fig. 3). This is of utmost relevance, since, with bearing to and like Fermat´s Last Theorem (FLT), "far from being some unimportant curiosity in number theory it is in fact related to fundamental properties of space" (www.groups.dcs.stand.ac.uk/~history/HistTopicsFermat'slast_theorem.hyml) as well as of integers (www.coe.uncc.edu/cas/flt,html). And the uniformity, that all whole-number powers from n = 3 and onwards are realised in sufficiently three dimensions as saturated regular parallelepipeds which per definition are composed by integer blocks alone, is of equal cardinal importance for the demonstrations *ad modum* Cardano to be exposed in the continuation.

That the (Western) situation was essentially the same up to the days of Cardano, and hence also current for Fermat, is namely another undeniable mathematical and philosophical fact, as most clearly demonstrated by the former in his Ars Magna [1545]. Quoted from Parshall [1988]; "For quadratic equations, Cardano, like his ancestors, built squares, but for third degree equations, he constructed cubes". He concluded "that only those problems which described some aspect of three-dimensional space were real and true. In his words: *"For as positio [the first power of the unknown] refers to a line, quadratum [the square of the unknown] to a surface, and cubum [the unknown cubed] to a solid body it would be very foolish for us to go beyond this point. Nature does not permit it"'* (Ib.). That indeed Nature does not allow a truly analytic (that is, continuous, space-filling and non-overcrossing) simultaneous physical distribution over more than three linearly independent dimensions had been shown already by Aristotle, and was the state of the art also for Fermat in 1637, when in the exclaimed (but unexplained) *demonstrationem mirabilem* of his last theorem he manipulated plain *"cubos"* in equal *en bloc* manner without the use of algebraic symbols (www.coe.uncc.edu/cas/flt,html).

But whereas Cardano "was unable to conceive of….a four-dimensional figure" geometrically (Parshall [1988]), this, and its continuation may well have been that instant flash of insight for the one century younger Fermat mind: just perpetuating the identified row-rectangle-octagon cycle to ensuing powers by the same undulating iteration and reiteration of the ground unit cube which comprised the genuine whole-number atom of the still prevailing protagonist era. The consequences would have been immediately recognised, too, for Fermat, but why he did not pass on the veritable blockbuster remains an enigma. Perhaps he did not want to destroy future number theory fun, or it was just an

346

act of that cryptic jeopardy game which seems to have been going on in the esoteric circles when mathematics was often a jealously protected secrecy.

His plausible modus operandi will here be expressed by the simplest – but undeniable – 'schoolboy mathematics' formulas – by which yet his FLT (Fermat's Last Theorem) as well as the latter-day progeny called Beal's Conjecture (BC) can be proved. Expressed in the forefather FLT designation, BC states that all possible whole-number power, $X^n + Y^m = Z^p$, additions must share an irreducible prime factor in all its terms (Mauldin [1997-], Mackenzie [1997]). By extrapolation from Fig. 3, it can be observed that all manifold blocks grow from the preceding one in the same column by adding upon this one less of the same than its base number:

$$X^n + (X-1)X^n = X^{n+1},$$

which has an all-power solution when $(X-1)$ is of n.th power, e.g. $9^3 + (9-1)9^3 = 9^4 = 9^3 + (8)9^3 = 9^4 = 9^3 + (2 \times 9)^3 = 9^4$. This borders to trivial but has profound bearings and consequences, notably in regard of the prevailing X = integer requisite. First, it is a universal relation; All X^n.s are represented, both by the first summand term and by the sum one step up (or gradually higher by the relations $X^n + (X^2-1)X^n = X^{n+2}$ (as in $3^3 + (3^2-1)3^3 = 3^5$) and, with non-integer roots of the multiplicative coefficient, $X^n + (X^3-1)X^n = X^{n+3}$, $X^n + (X^4-1)X^n = X^{n+4}$ etc. ad infinitum, according to the general formula, $X^n + (X^p-1)X^n = X^{n+p}$. And with rare exceptions like in $3^3 + (3^2-1)3^3 = 3^5$, only (X^1-1) can have a whole-number n:th root of power $n \geq 3$, and $(X-2,3,4...)$ is too small to raise the sum to higher power.

However, using X^n as coefficient in the second term generates a FLT/BC equation where all terms are integer powers and thus emptying the whole X^n set:

$$(X^n + 1)^n + X^n(X^n + 1)^n = (X^n + 1)^{n+1},$$

giving one solution alone to each X^n. It is easy to exemplify for any X^n, e.g. $(12345^{6789}+1)^{6789}$:

$$(12345^{6789}+1)^{6789} + (12345^{6789}) \times (12345^{6789} + 1)^{6789} = (12345^{6789}+1)^{6790} =$$

$$(12345^{6789}+1)^{6789} + [(12345)(12345^{6789} + 1)]^{6789} = (12345^{6789}+1)^{6790}$$

And so it goes on, for every consecutive X and every consecutive n, and hence, for every whole-number X^n introjected in the second term there is but one pure FLT/BC equation where all terms are ground whole-number powers, i.e., in the irreducible form with all external coefficients = 1, screening off other solutions. Since the equation thus drains the whole space of binary additions of integer powers it also proves both FLT and BC because (stated in most general form) $(X^n+1)^n + X^n(X^n+1)]^n = (X^n+1)^{n+1}$

347

excludes n.th power sums (FLT), and the mutual (X^n+1) shares least prime factor (BC). The total occupation of the FLT/BC space becomes even clearer when X^n and Y^n are entered together in the equation according to the formula:

$$X^n(X^n+Y^n)^n + Y^n(X^n+Y^n)^n = (X^n+Y^n)^{n+1} =$$

$$[X(X^n+Y^n)]^n + [Y(X^n+Y^n)]^n = (X^n+Y^n)^{n+1},$$

which likewise gives an infinity of integer solutions (like $16 \times 97^4 + 81 \times 97^4 = (2 \times 97)^4 + (3 \times 97)^4 = 194^4 + 291^4 = 97^5$). Clearly, and also when $X^n = Y^n$, the two first terms are thus permutatively engaged by every possible X^n and Y^n whole-number power pair and giving in the third term a sum whole-number power sharing prime factor but in a higher degree. Inserting each successively larger X^n and Y^n thus proves both BC (by all X and Y permutations) and FLT (by all n powers) *en bloc* by the ascending differential "layer-by-layer…complete close-packed" (Velikov et al. [2002]) sequential iteration gradually sweeping over and so covering the overall Diophantine equation space.

In conclusion, what has been done here is a "brute force" "infinite machine" (Davies [2001]) exposition that every discrete X, Y and Z power can be explicitly retrieved by the smallest = only possible solution from the first term, the second term, the first and second term, the sum term and all terms in the FLT/BC equation by a simple but universal numerical formula. The ascending addition acts infinitely and successively ties every second term specifically to the complementary first term, hence making also the sum sharing the mutual least prime factor. So, with BC in tow, the proper spelling-out of the FLT acronym should now righteously be Fermat's Last Triumph. Contrary to the assertion that "the problem may require a brand-new approach that would not only re-prove the Fermat theorem but a whole lot more" (Mackenzie [1997]), the brand-old directions yield even better.

There are indeed a lot of further interesting and significant things one can do with the Euclidean/Diophantine space numbers and equations, like natural 'self-parameterization' and 'self-encapsulation' likewise solving FLT and BC; three-dimensional rendering of prime numbers and prime number products etc. (Trell [1997, 1998a, 2003b, 2004 a-d, 2005 a-c, 2006 a-c, 2008]). However, the main aim here has been the direct and irrefutable demonstration of FLT and BC and showing that also an endless rigid cube can be of fundamental complementary relevance and interest for the true scientist.

2. Straight-to-Round phase transition Ether dynamics
In Eastern science, and also with traces from the Egyptian civilisation (Pavlov [2006]) the notion of all-pervading, i.e. identical one-by-one local and aggregated global reciprocity between space and matter, and that they commutually engender and sustain each other has remained vivid. In recent Kolkata PIRT conferences reference was made to the ancient Vedas (Trell [2003c, 2004a, 2005 b,c, 2006c]), probably the

oldest written texts on our planet and supposed to have been passed through oral tradition for over 10,000 years before written down in their mnemonic Sutra form between 6,000 to 4,000 years ago. As here quoted from Haselhurst [2003], especially the Rigveda epically envisions how "the Universe reveals itself in two fundamental properties: as motion and as that in which motions take place, namely Space...This space is called Akasa and is that through which things step into visible appearance, i.e. through which they possess extension and corporeality...Akasa is derived from the root Kas, "to radiate, to shine" and therefore direct part of the "breath of life...that underlies and makes possible all the multiplicity...that forms the cosmic rhythm...through the vibrations, and the action and interactions of vibrations produce all the phenomena."

A wave/string constitution of matter is thus realised, spun between the curved loop and the straight cap and spokes of that "wheelwork of the Universe" (Ib.) bipolar generator which is symbolised in the original, regrettably later so abused, Veda "mill of motion" (Fig. 4).

 ***Fig 4*** *The Veda wheel of procreation and movement, generated between the inter-facing straight frame and round perimeter extremities with the step gradient outlined by the divergence and potential gap between equal length of them.*

These ideas of procreation by rebounding contrast were further developed during many centuries by the ancient Greek philosophers and notably among them Aristotle (384 – 322 BC), from whose *Physics* (~350 BC) is cited (http [a]): "All thinkers then agree in making the contraries principles...Democritus also, with his plenum and void...he speaks of differences in position, shape, and order, and these are...contraries, namely, of position, above and below, before and behind; of shape, angular and angle-less, straight and round". Aristotle reduced the superficial and secondary and paradoxical of such contraries, for instance, that "the not-musical man becomes a musical man" and arrived at the following specification of "the principles of natural objects which are subject to generation:...it is clear that there must be a substratum for the contraries, and that the contraries must be two" (Ib.). And of such principle substrata, *primum movens*: the world-line of movement, where, although he somewhat unconvincingly questioned it, Aristotle conceded upon the prevailing conception that "if there is a contrary to circular motion, motion in a straight line must be recognized as having the best claim to that name" (Stocks [1922]).

Due to its ontological consistency, the self-generated dynamic Ether was bound to reincarnate in the scientific evolution, for instance, under the evocative designation "vortex sponge" as strongly advanced but in the end rejected by, above all, William Thomson, later Lord Kelvin, around the turn of the 19th century (Lindley [2004]). With forerunners in the moving particle Ether concept of both Descartes and Newton, it was designated from heuristic and mechanistic considerations to resolve "the Electromagnetic

349

World View" and wave equations not only as "a reference frame…through which all action was transmitted" but simultaneously the inherent generator and, in current terminology, "event-particles" and vehicle per se of its own "dynamical ether…activity" (Duffy [2004]). Such a hybrid necessitates an internally and externally coherent "two-way reaction" and interaction (Ib.), calling to mind analogies like a conduction plate or an engine's piston and cylinder block. Thomson proposed "a fine mixture of rotating and non-rotating elements", which 40 years later, as Lord Kelvin failing "to pin down…the exact nature of the little rotating element in his sponge ether…permissible under Newtonian mechanics", he denounced so radically that direct pursuit on that instant classical track became and has remained virtually barred since then (Lindley [2004]). His "tragedy" (Ib.) was largely related to the vortex sponge debacle and determined by his living before of the 20th century's elementary particle discoveries (Ib.). These would almost certainly have enabled him to verify and explain his intuition of strikingly simple ordinary geometrical patterns and transformations, fulfilling the stipulated wave-mechanistic criteria of rotational symmetry and torque of the vortex (or 'piston') component in the non-rotational real space rectilinear sponge wall (or 'block') moiety together making up the virtual dualistic dynamo that each vacuole in the mesh so constitutes and yields with obligate continuity from centre to collective periphery.

Whether of primarily substantial or immaterial stuff is of no consequence when everything is part of the same binary give and take. One might take for granted a random building kit strewn out in ready-made distance space and immediately cogging in (and not wonder why the pieces are all kin). Or one might feel the profound need of a more systematic organization; the imperative potential, spark and twist between no more – or less – than two contrasting and yet infinitely approximating principal philosophical (cum mathematical cum physical) categories always kept in juxtaposition and confrontation with each other. The present consensus goes in the latter direction whether the complex-forming agonists are designated "string and loop quantum gravity" (Cho [2002]), or reciprocal 'mortise and tenon' (MacKenzie [2002]) moieties, or "eternal-universal Branes" (Seife [2002]), or Yin-Yang, or the dual lattice phase transition (Trell [2005 a-c]) "between the curved and the straight…at the heart of Greek geometry and indeed of geometry in general" (Netz [2002]). With the triumph of Victorian engineering and the industrial revolution at large, it was natural that "the Maxwellians", in that golden age of Science unsurpassed since the Renaissance "believed that we are immersed in a medium in intense spinning motion, the equal counterpart of matter…a complex system of strains and vortex motions in the ether, that tenuous but all-pervading medium" (Coey [2004]). And "to deepen their understanding and get ideas of the working of the ether the Maxwellians turned to their models", the finest of which was George Francis FitzGerald's now regrettably lost "array of brass wheels mounted in a large array on a mahogany base and connected by indiarubber bands which were strained as the wheels turned" (Ib.).

It displayed several functions illustrating the "real electromagnetic phenomena" (Ib.) but the missing links of the replication were those of the interior of the spinning wheels and of the ether "base", or actually encasement that they were co-acting in and with, and of which later elementary particle spectroscopy findings as well as still dormant Lie group and algebra neighbourhood geodesics would doubtlessly have provided sufficient clues for synthesis. Which were the inner springs, the "standing waves" (Duffy [2004]) that the outer ones were the harmonic continuations and iterations and resultants of? And what was the conformation of the coalescing resonating cavity rather than inert base plate of the oscillations? For mere lack of information the black box became sealed for forthright further exploration.

Only recently has casually imitated "quantum foam" replaced vortex sponge with "strings" instead of springs and with a disjoint and heteromorphic instead of interrelated and harmonic "spin network" (Cho [2002]); and on such an incredible, not to say absurd scale - billions of times smaller than the electron and yet inflatable to that of the Universe - and hypercomplicated constitution - eleven dimensions wrapped up into themselves - that for this reason alone a revisit to the more tangible and verifiable prototype would seem highly profitable. The moratorium that initial failure and subsequent quantum mechanics and misunderstood general relativity laid on the pioneering contrivance has had its day so it is high time to open up the promising corridor anew. Some striking results of this venture will be the aim of the present discourse together with a brief recollection of associated philosophical, mathematical and physical merits and utility. These findings are noteworthy and convincing as such, but it is hoped that still more outcome-oriented research will be stimulated, above all on the electron and associated second- and ensuing generation external properties, eventually allowing also an animation of the "Protein Universe" (Service [2005 a,b]).

3. Geometrization of numbers and Arithmetic

It is important that such a veritable Ether incubator is fully compatible with special as well as general relativity. Einstein himself, in his famous Leyden lecture [1920] was quite "in favour of the ether hypothesis. To deny the ether is ultimately to assume that empty space has no physical qualities whatever. The fundamental facts of mechanics do not harmonize with this view...the ether of relativity...helps to determine mechanical (and electromagnetic) events...at every place determined by connections with the matter...which are amenable to law in the form of differential equations...We know that it determines the metrical relations in the space-time continuum, e.g. the configurative possibilities of solid bodies as well as gravitational fields but we do not know whether it had an essential share in the structure of the electrical particles constituting matter...there can be no space nor any part of space without gravitational potentials for these confer upon space its metrical qualities, without which it cannot be imagined at all...two realities completely separated from each other conceptually, although connected causality, namely, gravitational ether and electromagnetic field, or as they might also be called space and matter...the elementary particles of matter are also, in their essence, nothing else than condensations of the elctromagnetical field...Of course it would be a great advance if we could succeed in comprehending the gravitational field and the electromagnetical field together as one unified conformation...Recapitulating we may

351

say that according to the general theory of relativity space is endowed with physical qualities, in this sense, therefore, there exists an ether...space without ether is unthinkable, for in such a space there not only would be no propagation of light but also no possibility of existence for standards of space and time (measuring rods and clocks), nor therefore any space-time intervals in the physical sense."

The extensive quotations may be justified by their canonical nature in support of a necessary ever-present distribution also of the Cartesian co-ordinate system, i.e., of commensurable cubical space segments, in any localized mathematical or physical realization. Or, as expressed in the colloquial present-day idiom of string and loop quantum gravity, where "area-conveying links connect little chunks of space...: a recipe for transporting direction-indicating vectors through space-time...in order to tell you which chunks of space that talk to each other" (Cho [2002]). But it is still just the static part of the synopsis. How enters dynamics into the Ether block? In this dilemma the original concepts paired with modern theories again give direction. In fact, there are recent developments both in hypercomplex numbers and in re-geometrization topology (Duffy [2004], Pavlov [2006]) to that end. However, here will be focussed upon the new 'nilpotent vacuum', Rowlands/Diaz holographic universal computation rewrite system [2003, 2008] and Santilli's iso-, geno- and hypermathematics [2001]. There is full rapport with both because, adjusting for the alternate modality at hand, an equivalent dualistic potential builds up, commencing between the antipodal Zero (=nothing) versus All (= anything) state vector set and eventually bootstrapping a three-dimensional volume-preserving orthogonal Lie algebra by the system's infinitesimal generators, which in both cases form a 3×3 diagonal matrix with Det = 1 (Trell [1991]).

In the real, recently reconfirmed perfectly flat (Bahcall et al. [1999], Kamionkowski [2002]) Euclidean space, the immediate make-up is infinite extension (Fig. 1), whose converse is not a hollow void but just no extension at all. At this principal level, the null hypothesis of absolute non-being thus shrinks to: absolutely nothing, not even a point, and also without any charge or other property. It is total Absence; the bare Zero (annulled both if existing or non-existing) and in relation to which the equally radical and obligate, antipodal state of something at all has a monovalent and unlimited positive category exposition, here endless extension whose beam is length with immediate realization the infinite straight line (Fig. 1 a,b).

However, a thought experiment consideration of any line in the figure, say, the arbitrary Z, shows the further expansion of an O(3) distribution space (Trell [1991]) as the full obligate contrast of the impossible non-being alternative. Everywhere along the line, the virtual observers may look forward in any other direction maximally covering a circumference of 360° and so setting up a dense radial array together filling the space, demonstrating it to be three-dimensional and rotationally symmetric and furthermore enclosing its infinitesimal cubical iso-vectors inside itself (Fig. 1a).

Thus, textured space just is there, ready-made, though so far coordinate- and orientationless. However, coordinates and orientation come automatically, too, because of the local observer-centred condition which brings in the SO(3) group and so joins up with Rowland's nilpotent vacuum conditions [2003] "that physics has its origins in a symmetrical structure which preserves its conceptually zero content...a principle of duality...for instance...between + and − applied to the unspecified entities which are generically described as the reals". Putting the observer-site in the origin sets up the dichotomy of Fig. 1 a,b, i.e., that there is a relational plus and minus side of the lines. In the whole space they still stretch out in every angle, but when ordered by means of the infinitesimal generators created by the 90° and 180° direction sines, they form the Lie algebra SO(3) around each observer site, all of these in turn co-ordinated by their global interaction necessarily likewise aligned to remain in the algebra (Trell [1991]).

While the nothing/everything, so called *categorical duality* refers to monistic fundamental category and is mutually exclusive (if nothingness exists, anything does not exist and game is off, if anything exists nothingness doesn't and anything is just the obligate offset), the +/− opposition may be called *bipolar duality*, as in pure mathematical terms also advanced in Santilli's pioneering hadronic mechanics [2001], which, like the universal rewrite system comprises an extra, +1 dimensionality for dynamics.

However, before revealing it in the present version, too, there is much more to say about perpetual Euclidean space as such, most of it remarkably overlooked/forgotten in the correspondingly detached modern postulations. Nonetheless, recent observations confirm that real Euclidean space as originally conceived is indeed flat (Bahcall et al. [1999]) and thereby unambiguously fulfils also its pure logical and axiomatic principle in three dimensions as an endless cubical Cartesian coordinate system (Trell [1997, 1998a, 2003b, 2004 a-d, 2005 a-c, 2006 a-c, 2008]), or "canvas" (Bahcall et al. [1999]), Kamionkowski [2002]) for everything appearing in it (Fig. 1b).

Hence, modern Cosmology's equally mathematical as physical space in its total span from the infinitesimal to the infinite is by nature as well as definition not curved itself but, again, of "zero curvature" (Mackenzie [2004]), thus forming to actual curves a mutual complement and virtual blackboard. 'Curved space' is therefore a contradictio in adjecto; light-rays in space may be bent but were their screen bent, too, both would appear parallel. Only in their own reciprocal capacity the straight and round categories are equally much part of the full 'mortise and tenon' (Ib.) world panorama where "just as matter is made of atoms and elementary particles, space consists of tiny individual bits" (Cho [2002]), which in the faithful eigen-vector reduction come out as cubical (Trell [1997, 1998a, 2003b, 2004 a-d, 2005 a-c, 2006 a-c, 2008]) (Fig.s 1-3).

It is important to emphasise that Euclid himself in the formulation of his geometry did not even consider a fourth dimension, and that, as previously quoted, Ptolemy in his book *'On Distance'* gave a proof (AD 150) that the fourth dimension, in Euclidean space, as defined, is antithetical: "Draw three mutually perpendicular lines, he suggested. Try to draw another line perpendicular to all of these lines. It is impossible. The fourth perpendicular line is entirely without measure and without definition" (http [b]). The same appears from the following Aristotelian quotations: "A magnitude if divisible one way is a line, if two ways a surface, and if three a body. Beyond these there is no other magnitude, because the three dimensions...three directions...are all that there are...bodies which are classed as parts of the whole are each complete according to our formula, since each possesses every dimension...either straight or circular or a combination of these two...as body found its completion in three dimensions...its movement completes itself...the reasoning which applies to the whole applies also to the part" (http [c]).

The perfect Euclidean space is entirely spanned by three coordinate axes, and hence three coordinate axes are all that simultaneously exist in the perfect Euclidean space. It is not a circular but truly peripatetic argument coming back to the identity. And when as constituent parts applying to the whole, the Euclidean eigenvector "bits" (Cho [2002]) according to both Aristotle, current nano-technological self-similarity (Ikkala and ten Brinke, Kato, Velikov et al., Whitesides and Grzybowski [all 2002], Aizenberg et al. [2003]) and isomorphic differential reduction (Trell [1997, 1998a, 2003b, 2004 a-d, 2005 a-c, 2006 a-c, 2008]) come out as cubes, too, they instantly provide the primary building blocks and numeric units of the one simultaneous physical/mathematical space.

However, since ancient time, a profound insight in relation to the straight and round forms is that they are both absolutely endless, yet radically distinct and irreconcilable over a gap of *limes* (the last decimal of) π. They thus present a *facultative duality* with an eternal *'primum movens'* subsidence potential fall between their maximally dilated versus maximally contracted form of infinity, respectively. One may here quote Aristotle: "everything that comes to be comes into being from its contrary and in some substrate, and passes away likewise in a substrate by the action of the contrary into the contrary" and (again) "if there is a contrary to circular....a straight line must be recognized as having the best claim to that name" (Stocks [1922], http [d]).

The still valid corollary of the above is that dynamics dualistically occurs between the pair-wise juxtaposed curvature extremes, as may of course also inert projections like parameterization, while upon and within each of them the non-inert processes, like travelling and transformations and assembly, are negentropic.

A similar awareness is now also re-entering Physics and Cosmology. Two decades after its inception, it has recently been concluded that "string theory, humanity's best attempts at the ultimate explanation of matter and energy, space and time...has yet to pass...fundamental scientific tests...especially in particle physics, in order to maintain the theory's credibility" (Cho [2004]). A critical obstacle thereby, as also previously quoted is that this "leading candidate for a 'theory of everything'....suffers from a fundamental weakness...the strings move in a spacetime whose shape has been chosen from the beginning, as if they were actors on a previously constructed stage. A truly fundamental theory of gravity, everyone agrees, would build the stage itself", to which end "a few physicists have…concocted a theory that precisely describes spacetime on the smallest length and time scales…just as matter is made of atoms and elementary particles, space consists of tiny indivisible bits" where the "warping of the very fabric" (Cho [2002]) forms the template for "untangling the Universe….into the equations of string theory" (Trefil [2004]).

In other words, a reciprocal 'mortise and tenon' (Mackenzie [2004]) relation with a "hope that the two approaches will merge someday" is envisaged in the mingled "string and loop quantum gravity theories" (Cho [2002]) in question, where, however, the mortise "moduli" conjectured in elementary particle physics, like the "versions of vacuum" in cosmology are virtually legion; "works on moduli stabilization suggest that there are a whopping 10^{300} different stable vacua, and theorists have no way to choose among them" (Cho [2004]). In consequence, there is a "feel that there's some missing idea or some very difficult mathematics that needs to be done" (Ib.).

A prominent candidate in that regard is "Inflatory Cosmology", aiming at "exploring the universe from the smallest to the largest scales" whereby likewise the "understanding…depends critically on insights about the smallest units" (Guth and Kaiser [2005]). At considerable credit costs in terms of suitably parameterizing and otherwise setting the stage, the theory manages to re-create flat and homogeneous global space, and also ripples in it, by the endless expansion of the surface of the inflating "pocket…hopping up the wall of potential energy rather than down".

But it fails at smaller scale and, again, at the remaining question: "What, then determined the vacuum state for our observable universe?.....the authors hope that some principle can be found….it must have had a past boundary, before which some alternative description must have applied. One possibility would be the creation of the universe by some kind of quantum process" (Ib.).

So, it is back to the outset. What is around – "before", and for ever? Where are the so far missing "insights about the smallest units" of matter (Ib.) as well as of space (Cho [2002]) that would hold the key to resolution? When today nanotechnology goes deeper and deeper in scale and everywhere reveals a coherent physical constitution from the infinitely large to the infinitesimally small (Ikkala and ten Brinke, Velikov et al., Whitesides and Grzybowski [all 2002], Aizenberg et al. [2003]), also the mathematical and chemical formulas and signs are taken back to their common origin as real composition and shape. Even string theory now feels the need to "get real" and "must find a way to account for experimental observations….especially in particle physics" (Cho [2004]).

To that end, there is in fact no "missing idea", and although it is essentially mathematics that provides the way it is not "very difficult" (Ib.) at all. On the contrary, as often

355

happens even in advanced science, the overlooked string of evidence may not be the very complicated, but the very simple one – where *one* has a double meaning and embodiment of real consequence in the given context.

While the ordinary spatial characteristics are primarily retrieved in rectilinear geometry, the elementary particle symmetries are spherical, with the 'mortise and tenon' (Mackenzie [2004]) relation directly between them (Lie [1871], Trell and Santilli [1998]) and genuine dynamics of particulate matter therefore first occurring in injective and surjective mappings (Adhikari & Adhikari [2003]) "between the curved and the straight…at the heart of Greek geometry and indeed of geometry in general" (Netz [2002]).

One discerns that olden "wheelwork of the universe" (Haselhurst [2003]) as epitomized in the Veda mill of motion (Fig. 4) and Yin Yang metaphors; the tenseless juxtaposition, potential gap, spark and flow between polar, yet approximating infinite categories, one of the best verbal expressions of which to my knowledge is Mark Kurlansky's in his very interesting book on salt [2003]: "Nature seeks completion…salt was a microcosm for one of the oldest concepts of nature and the order of the universe. From the fourth-century B.C. Chinese belief in the forces of yin and yang, to most of the world's religions, to modern science, to the basic principles of cooking, there has always been a belief that two opposing forces find completion, one receiving a missing part and the other shedding an extra one".

When the Aristotelian equally logical as real differential substrate and dynamo for what must after all be identified as Cosmic coming into being and passing away are the two alone ultimately irreducible endless forms that Straight and Round comprise, and, as he also noted, the latter one is the most condense to which straight thus strives but also may be hyperbolically reflected from (Mackenzie [2004]), we may here contemplate the eternal theme of *Tao*, *Bhagadvita*, the elder *Edda* and other mythologies (Trell [2003a]).

We may perhaps also envisage the more fantastic outline of a binary phase motor with flat block and curved cylinder innumerably multiplied and overall dispersed between simultaneous observer-centred Big Bang and Crunch apertures. In any case we would then, as mentioned, see something quite akin to the *Vortex Sponge* contrivance and even machine models of "a complex system of strains and vortex motions in the ether" brought to near fulfilment by "the Maxwellians" including William Thomson, later Lord Kelvin, around the previous turn of the century but withering away in lack of solid fine-scale observations just a few years before these started to arrive (Coey [2004], Trell [2004b,c]).

And now having them at hand, including a faithful English translation (Trell and Santilli [1998]) of Marius Sophus Lie's Ph. D. thesis *Over en Classe Geometriske Transformationer* [1871], the evidence strengthens even more that the elementary particle generation is in the form of a homomorphism between the straight and the spherical curvature categories. The virtual cathode is the Euclidean space, whose constitutional "indivisible bits" (Cho [2002]), or "cuBits" (Trell [2003 b,c, 2004 a-d, 2005]), are necessarily cubical themselves and at their first interaction level gather into miniature Cartesian coordinate systems of 2^3 space segments according to the Aristotelian prescription (http [a-c]) that what "applies to the whole applies also to the part" (Fig 1b).

We here have the Det = 1 Santilli isounits of individual and aggregated infinitesimal systems [2001] echoed in their parameter frames, where the first operative order with full Cartesian plus/minus range is the 2^3 "cubicule" (in analogy with molecule) outlined in Fig 1. Thereby, paraphrasing Dr. Watson, "all the elements are falling to place". When, in the words of a more recent observer, "untangling...the fabric of Cosmos...without recourse to anything more esoteric than words and pictures", the first choice is henceforth not such "abstruse things" and "fantastic...quantum fluctuations" as the "six-dimensional Calabi-Yau space" and similar "implausible ideas" of fanciful computer complexity "floating around these days" (Trefil [2004]), but again that simultaneous unit phase transition "between the curved and the straight" (Netz [2002] in a classical "geometrized vortex sponge" (Duffy [2004]) twosome.

As illustrated in Fig 5, it is possible to envision a mutual interplay and intermorphing

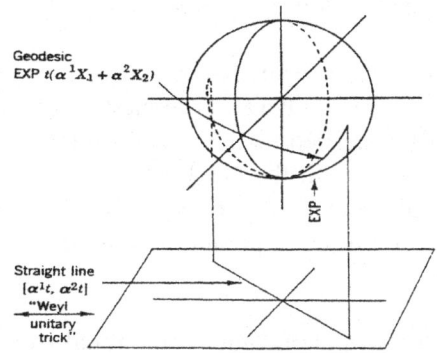

Geodesic
EXP $t(\alpha^1 X_1 + \alpha^2 X_2)$

EXP

Straight line
$|\alpha^1 t, \alpha^2 t|$
"Weyl
unitary
trick"

Fig. 5 *Graphical exposition of Lie´s projection of the surface "fundamental relation that takes place between the Plücker line-geometry and a geometry whose elements are the space's sphere" (after Gilmore [1974])*

between the rectilinear Cartesian matrix of the Euclidean "quantum foam" and the curved "superstring" geodesics of the interior and interstitial particulate symmetries by that SU(3) = SO(3) × O(5) group and algebra describing the elementary particle symmetries (Trell [1982, 1983, 1990, 1991, 1992, 1998c, 1999, 2000, 2003b,c, 2004a-d, 2005a-c, 2006a-c, 2008]) as well as Lie´s original, "on philosophical reflections upon the nature of Cartesian geometry" based, "transformations by which surfaces that touch each other are turned into similar surfaces...between the Plücker line geometry and a geometry whose elements are the space's spheres" (Lie [1871], translated in Trell and Santilli [1998]).

But it is significant of the present-day alienation of theoretical physics from the prime Lie groups and algebras that it has been stated that these are only "mystically fit to describe mathematically" the elementary particles and their patterns and behaviour (Jaffe [1977]). On the opposite, and on the proper infinitesimal physical neighbourhood plane, by their direct geometrical, nowadays labelled SO(3) × O(5) decomposition of SU(3) "we find between the corresponding transformations of R: all movements (translation-movement, rotation-movement and the helicoidal movement), semblability-transformation, transformation by reciprocal radii, parallel transformation...etc." (Lie [1871]).

4. The Baryons
Bringing this in phase with the vortex sponge dual motor is then a hybridization operation of centrally accommodating maximally contracted endless Round, viz. the surface of the sphere here epitomizing the Nucleon, within the commensurate portion of maximally extended endless Straight, viz. the open-ended rectilinear space grid of the eight cuBits of the surrounding positive and negative Cartesian co-ordinate quadrants of the same unit scale (Fig. 6).

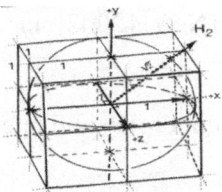

Fig 6 *The hybridization of the unit sphere within the cubical Lie neighbourhood sets up an interstice, in the universal iteration of which the basic, tetra- and octa-hedral regular solid phase transition is immanent*

That the sphere as the inner subsidence forms the matter heart (or 'clockwork orange') member is just as perceived also in modern quark, "Bag", QCD and related confinement theories (Trell [1993, 1998c]) where it is explicitly specified that "the hadron must be an extended, geometrical object" of spheroidal symmetry with the Nucleon as the spherical "preferred (ground) state of the system", whose all other "properties are attributable to this non-perturbative ground state" trough "a semi-classic approach similar in spirit to Bohr´s treatment of the hydrogen atom" (Jaffe [1977]).

The elementary particle states are governed by the complementary orthogonal subspace O(5) coset of the "canonical real form involutive automorphism" of SU(3) (Gilmore [1974], Trell [1991]). In this the unit radii of the enclosed sphere are statically extended (e.g. in a Neutron star) along the ordinary x, y and z axes as the neutral t isospin root vectors directly ingrained in the junction of flanking neighbourhood cuBit segments (Fig. 6) (Trell [1982, 1983, 1990, 1991, 1992, 1998c, 1999, 2000, 2003b,c, 2004a-d, 2005a-c, 2006a-c, 2008]). In addition there are over the walls of these the diagonal H_1 and H_2 vectors (Ib.) of length $2^{1/2}$, bisecting the 60° inclined t isospin vectors over which the charged states are displayed (Fig.s 7, 8), and providing tracks for a rolling medial motion along any of the Cartesian space axes. There is in consequence no deviation from them, so that this spherical mode is that of the neutron. Two coordinate systems are discerned, where the orthogonal SO(3) is that of the previously derived Euclidean space lattice, and O(5) the diagonal generative and accommodation matrix of the neutral and charged elementary particles, which are by step and/or angle divergently phased into the universal unit gauge so that their still three and ordinary physical extensions have the aberrant signature of quarks.

Fig. 7 a-d summarises how the infinitesimal spectroscopic events necessarily originate

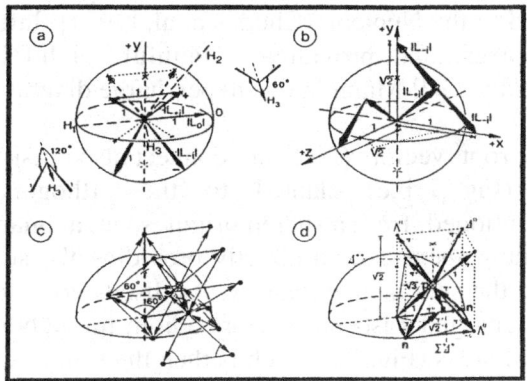

Fig. 7 (a) SO(3) x O(5) mapping of duplicated Lie algebra A_2 root space diagram into unit sphere with (b-c) three-dimensional transition lattice by unit length pion and lepton vectors there, and (d) some basic super-multiplet transitions produced by eightfold way in either of the eight Cartesian space segments as space-filling regular tetra- and octahedrons, none of which can fill the space separately due to different side lengths in lateral and bottom planes.

in the individual cubits, of which there are two in each geodetic neighbourhood half-plane so that SO(3) like the A_2 diagrams is duplicated resulting in sum 2×3 + 5 = 11 dimensions (degrees of e.g. combinatorial freedom) of the canonical coset decomposition of SU(3). In consequence, it is, like the original Lie algebras (Lie [1871], Trell & Santilli [1998]), a transformation process between surfaces, a phase transition between the alternative 90-180° and 60-120° lattice orientations of the common unit length root vector element. The following exposition will focus on the infinitesimal morphogenetic outcome of this phase transition between Straight and Round in the Lie neighbourhood which twists the unit cubical vectors of this into the spherical root space (Fig. 8). This is composed of two flat A_2 SU(3) commutation diagrams accommodated in the unit sphere, bringing the realisation from the complex to the parent three-dimensional ordinary space according to the canonical coset decomposition SO(3) x O(5) of SU(3).

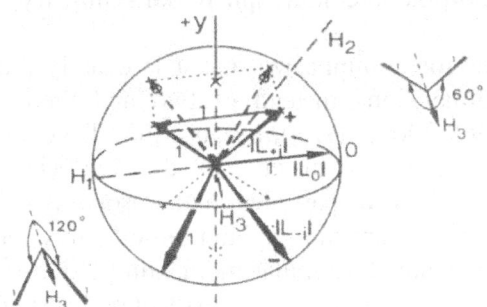

Fig 8 Real form three-dimensional spherical Lie algebra neighbourhood with duplicated A_2 root space diagrams

359

It is seen that the continuation of the cubic sides and internal diagonals sets up an infinite space lattice around any observer origin (Fig.s 1, 6), and likewise it is clear that the spherical A_2 root vectors (Fig. 7,8) from the mutual centre connect to a global lattice that is 60° skewed to the horizontal and vertical planes, and non-commutating with the latter (Fig. 7d). The correspondence to the quark three-dimensionality in the observed elementary particle spectroscopy is apparent and the close coincidences persist with all attributes of this into a virtual 'binary phase motor' transformation system set up by the A_2 axes of the unit sphere domain assigned to the Nucleon (Chodos et al. [1974], Jaffe [1997]) in relation to the straight space axes, and providing a faithful "eightfold eightfold" three-dimensional version of the plane Gell-Mann lattice hypercharge diagram.

The diagonal A_2, so called charged t isospin root vectors (Fig. 7 a), connect also outside the sphere to an endless hexagonal lattice (Fig. 7 b,c) slanted to the orthogonal Euclidean co-ordinate axes and thus, as mentioned, from a shared origin span a quark space matrix aberrant to the cubical arrangement and so directly providing the still rectilinear phase transition of this turned to the spherical symmetry. As discussed more in detail later it is quite significant that the charged t isospin vectors adjoin throughout space into an infinite continuous mesh of unit sides (Fig. 9), which is then the runner for the one-dimensional electron step distribution into atoms and larger coherent portions in

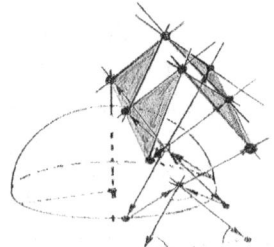

Fig 9 *A dissection of a small portion in the first and second extranucleon layer of the charged t isospin root vector lattice, showing that it connects as octahedron sides without involvement of any neutral root vector elements*

a cyclically cumulative course with all the apparition of the likewise stepwisely sequential orbital model including quantum indeterminacy and related evasive behaviour under any in comparison vastly broader, currently feasible probing. Regarding the impenetrable Nucleon, this infinitesimal, i.e. absolutely smallest sphere can not be further shrunk itself, nor can its complementary form of endlessness be effectively changed. But when impacted it can, by its mere unattainability, be shape transformed and then by necessity preserving both volume and, isomorphic to colour, spheroidal symmetry.

Since described in detail earlier, it suffices to re-emphasize that it is exactly the Gell-Mann eightfold way in the real three dimensions instead of two and therefore an "eightfold eightfold way" (Trell [1998, 1999, 2000, 2008]), because the (diagonally into anti-versions mirrored) transformations may occur in any of the Cartesian space segments. Considering that all observed Baryon particles and resonances in the Λ, Σ, Δ, N, Ξ, Ω and also full charmed series (Ib.) are directly and reproducibly retrieved with just and no more than the actual states, channels, angular momentums, charge levels and precise mass numbers, and moreover in a faithful three-dimensional realization of the accepted eightfold way according to the original Lie prescriptions, the results are true and lasting and it is remarkable, too, that they are projected over the regular solid space axes and sides (Fig. 10).

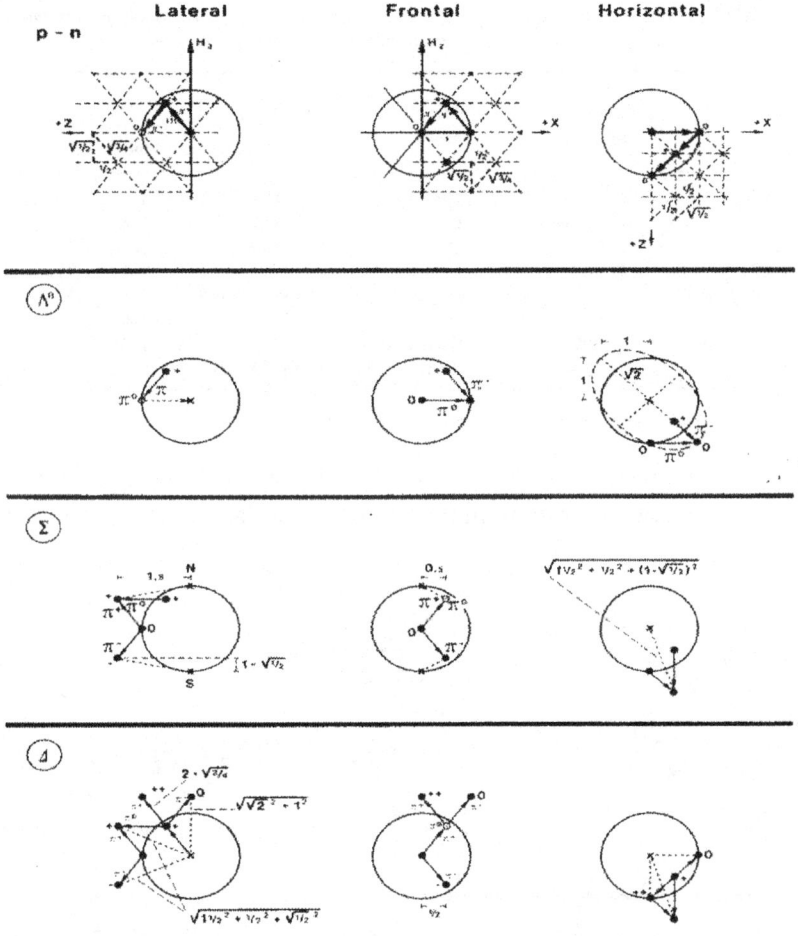

Fig 10 *The Λ^0, $\Sigma^{+,0,-}$ and $\Delta^{++,+,0,-}$ transformations*

To reach the transformation, the same root vector steps as in the Gell-Mann supermultiplet diagrams and the observed spectroscopy alike are taken, leading to new endpoints for an ellipsoidal reconfiguration of the parent state, whereby the masses (given in MeV) according to the quark pressure formula, $\Delta p = \hbar/\Delta x$, come out reciprocally to the proton mass by the minor semiaxis length. In fig. 10, the plane graphs show the channels and the major semiaxis endpoints arrived at in the $\Lambda^{0,-}$, $\Sigma^{+,0,-}$ and $\Delta^{++,+,0,-}$ Hyperons with lengths to the origin given by the root expression, and further that also the charge levels are retrieved exactly and exhaustively as in reality. The global, quark-skewed hexagonal spherical root space lattice is indicated in the p-n transposition and (the equatorial plane of) the volume-preserving ellipsoidal reconfiguration in the Λ^0

361

state. Table 1 shows the exact correspondences obtained by this proven method in all the basic Baryon states and which persist throughout their full respective series:

Table 1. *Lambda, sigma, delta, xi, sigma{1385}, lambda{1405}, xi{1530} and omega hyperons.*
Masses calculated according to formula: 938.28 · 1/minor semiaxis

	Major semiaxis	Minor semiaxis	Mass	
			Calculated	Observed
Λ^0	$\sqrt{2}$	$\sqrt[4]{\frac{1}{2}}$	1115.8	1115.6
$\Sigma^{+,0,-}$	1.60804	0.788591	1189.8	1189.4 — 1197
$\Delta^{+,+,+,0,-}$	$\sqrt{3}$	$\sqrt[4]{\frac{1}{3}}$	1234.8	1230 — 1236
$\Xi^{0,-}$	1.975	0.7116	1318.5	1314.9 — 1321.3
$\Sigma(1385)^{+,0,-}$	$\sqrt{4.71} - \sqrt{4.75}$	$0.679 - 0.678$	$1382.2 - 1385$	1383 — 1386
$\Lambda(1405)^0$	$\sqrt{5}$	$\sqrt[4]{\frac{1}{5}}$	1403	1405 ±5
$\Xi(1530)^{0,-}$	$\sqrt{7.06}$	0.6134778	1529.5	1528 — 1534
Ω^-	$2.505 - 2.51$	$0.561 - 0.560$*	$1673.5 - 1677$	1672 — 1674

* Minor semiaxis changed in the transformation (c).

5. The Mesons

The Mesons likewise appear directly in the regular solid lattice (Fig. 11) as differentials

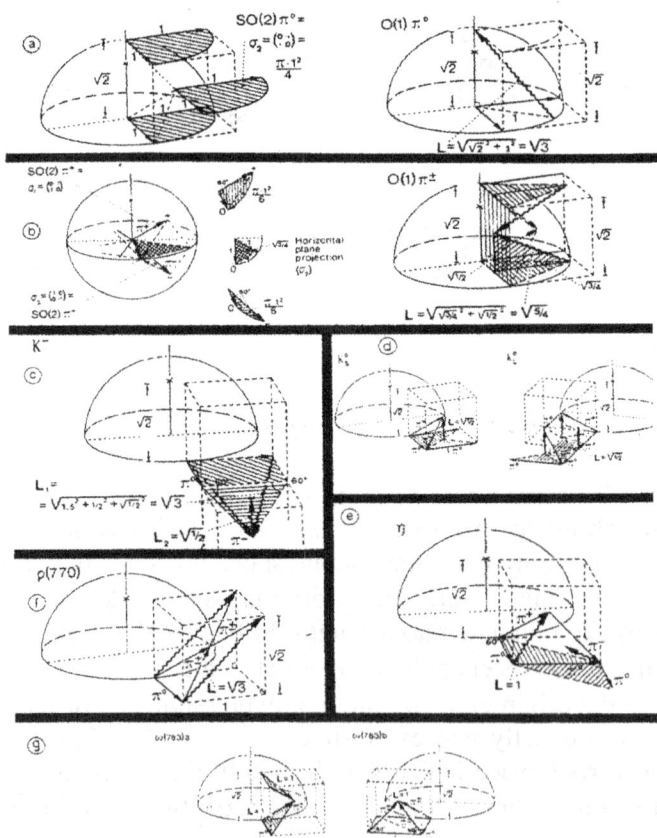

Fig 11 *The basic Mesons. (Same order as in Table 2) Mass given by the O(2) plane times the inverse of the O(1) angular momentum difference vector in relation to Proton mass. By this theoretically prescribed method in fact all Mesons, also the Charm and D and B states can be directly and exactly reproduced as differentials between successive Hadrons (including also themselves)*

362

between Hadron states , and in their spatial shape, in the geometric form of the established (symmetric) SU(2) x U(1) (antisymmetric) product group of the weak force (Trell [1982, 1983, 1990, 1991, 1992, 1998c, 1999, 2000, 2003b,c, 2004a-d, 2005a-c, 2006a-c, 2008]), they come out as polyhedrons, too, albeit not equilateral in all their extensions and therefore, like the Hyperons, unsustainable in the universal symmetry. All other Mesons, too, including charmed, D and B states and by the total vector collection in a Proton-Antiproton pair also the Gauge Vector Bosons (Trell [1998a, 2000, 2003b]) are equally fully and exactly retrieved. Again it is striking and convincing that polyhedral root space elements, both differential and equilateral, are so in double sense straightforwardly involved and that it is possible to exactly and exhaustively match the observed elementary particle spectroscopy by classical regular solid metamorphoses.

Table 2 exemplifies the mass number calculations, in the charged states involving projections of the hexagonal A_2 planes to the Euclidean observational reference, yielding unprecedented close matches in all quantum number and channel respects. It is remarkable and convincing that this continues in the entire legion of mesonic differentials and transformations between all the towering hadrons up to charmed and D and B and even T flavours (Ib.).

Table 2. Basic Mesons calculated and observed mass numbers (MeV)

π^0	$1/4 \times 938.27 \times 1/3^{-2}$	135.4	135.0
$\pi^\pm$	$1/6 \times 938.27 \times 1/(5/4)^{-2}$	139.9	139.6
$K^\pm$	$938.27/4 \times 1/3^{-2} + 938.27/4 \times 1/3^{-2} + 938.27/6 \times 1/2^{-2}$	492.0	492.7
K^0_S	$938.27/(4 \times 1/2^{-2}) + 938.27/(8 \times 1/2^{-2})$	497.6	497.67
K^0_L	$938.27/(8 \times 1/2^{-2}) + 938.27/(8 \times 1/2^{-2}) + 938.27/(8 \times 1/2^{-2})$	497.6	497.67
η	$938.27/6 + 938.27/6 + 938.27/4$	547.33	548.8 ±0.6
$\rho(770)$	$(938.27/1/2^{-2})/(3/4)^{-2}$ or $(938.27 \times 2^{-2})/3^{-2}$	766.1	768.3 ±0.5
$\omega(783)$	$938.27/4 + 938.27/4 + 938.27/6 + 938.27/6$	781.9	781.95±0.14

6. The Leptons

Paradoxically, despite their plain one-dimensionality and limited number of states, the Leptons stand forth as the perhaps most elusive of the elementary particles. Their antisymmetric Lie algebra is U(1), whose geometric isomorphism is the ordinary real line, the composed length of which may accordingly vary. However, already in the existing wave model it is at the *limes* level put together by infinitesimal derivatives which are straight unit bits meaning that, innermost, the Lepton scalar "world function" (Garasco [2007]) emerges as digital. So is likewise the case in the regular solid lattice which offers complete correspondences with the actual spectroscopy in this respect as well.

The infinitesimal straight line digit, or 'pixel' is immediately embodied in the uniform, sole ingredient unit root vector element of either neutral or charged inclination, whose iteration is everything that constitutes the lattice and the hence eigen-spacefilling

363

geodesics there. That close matches with the Leptons are indeed manifest in it is therefore not surprising in regard of the regular solids' (slightly oxymoronic) 'unique universality', but nonetheless truly remarkable. In fact, the Leptons weave the extra-nucleon world, and with such extreme simplicity that is has been overlooked for that very reason.

Starting with the particulate Leptons, there are two principal ways of connecting the needle-like sharp charged root vectors of same sign, here exemplified by the positive Muon and the Positron, namely, in the first case, by 90-180° turns (Fig 12 a-c), and, in the second case, 60-120° turns (Fig 12 d,e).

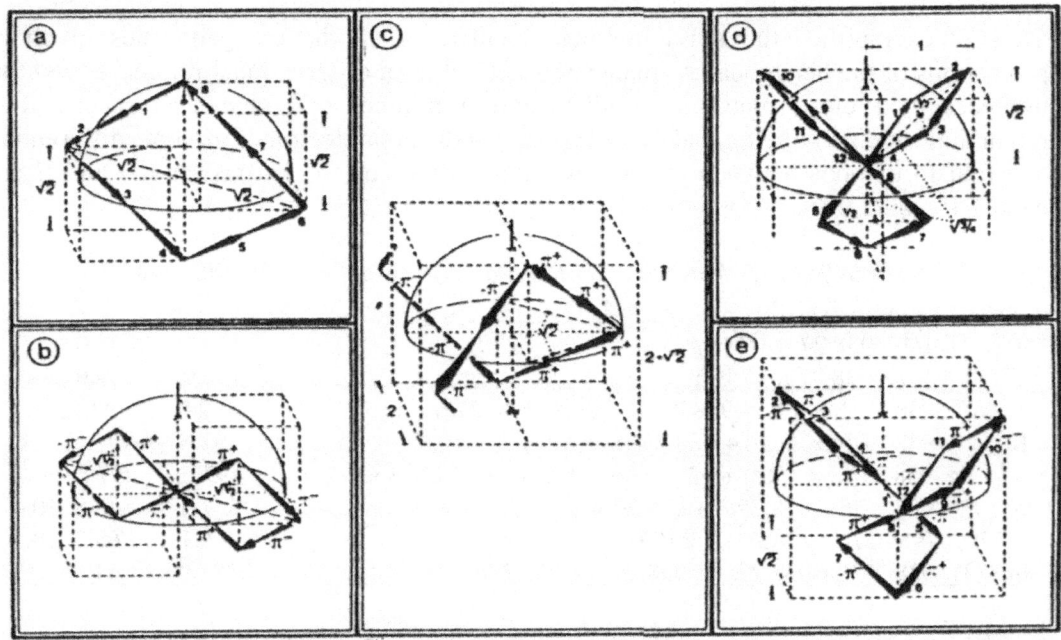

Fig 12 *Cores of (in this case positively charged)*
Lepton geodesics over the Nucleon surface

The first alternative forms plane or helical orbits from, over and outside of the Nucleon surface with a unit scale length in all varieties of $(2\pi \times 2^{-2})$ or $(2\pi \times 2 \times 1/2^{-2})$ and resulting mass number $1/(2\pi \times 2^{-2}) \times 938.27 = 1/(2\pi \times 2 \times 1/2^{-2}) \times 938.27 = 105.59$ MeV in comparison with the measured Muon$^{\pm}$ mass of 105.66 MeV.

In the second alternative, a three-winged orbit can be tied together (Fig. 12 d) and leads out of the Nucleon surface, so that it is natural to associate it with the Positron/Electron trajectory. The circular orbital length of the ground rosette is easy to calculate as $3 \times (2\pi \times 1/2^{-2})$ in unit gauge, but it is well known that one has to multiply with the fine structure constant, 137.035986.., to obtain the first, in this case 'Mercedes star' three-pronged circumference, so that the ground state Positron/(mirror)Electron mass number comes out as $1/(137.035986 \times 3 \times 2\pi \times 1/2^{-2}) \times 938.27 = 0,514$ MeV in comparison with the recorded 0,511 MeV. However, there are problems with the orbital model, for instance, in terms of the then alien, empty region under and between its rings, so that a truly spacefilling distribution is wanted. Being a sequence of unit steps there would be no difference in principle in relation to the orbital model, which, at the *limes→0* threshold, is also composed of iterated infinitesimal straight line intervals. And there exists such a possibility which can be patched together to larger structures in a hierarchically periodic fashion just as in modern nanotechnological self-assembly (Trell [2006 a-c, 2008]).

This is the truncated octahedron which is a composite space-filling Archimedean solid that already Kepler [1619] saw as fundamentally engaged in the cosmographical architecture. The truncated octahedron distribution of a full Positron/Electron turn follows from the only space-filling sequence of the charged root vector lattice (Fig.s 7, 12), namely (Fig. 13), a twelve-step, two-tetrahedrons/one octahedron triple coil node, or 'rosette' (that may at extremes also squeeze into the likewise twelve sides of the unit cube from the hexagonal symmetry, so that there is, for instance, at both extremes of the temperature range, an inverse of either group operation), which returns towards

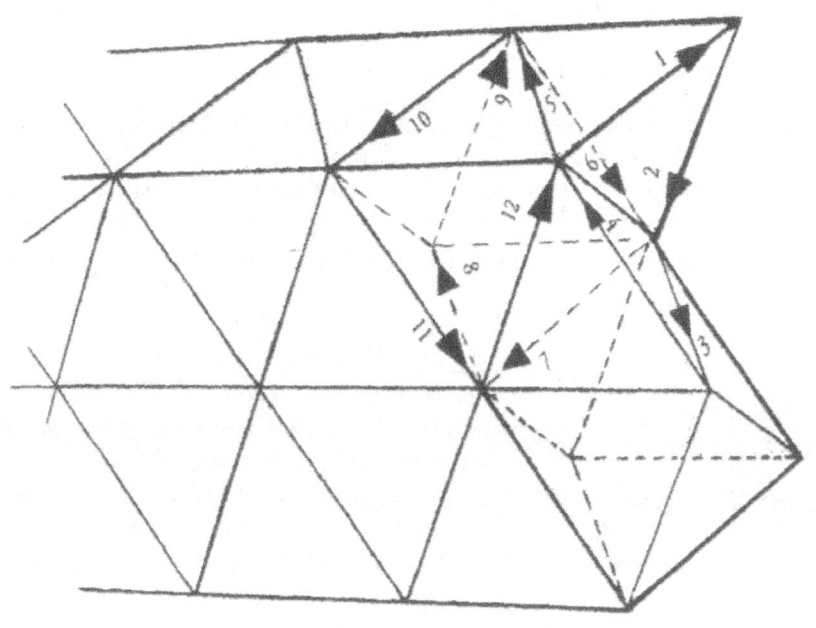

Fig 13 *The Electron (and Positron) can directly form the subunit of the transition in all universe*

its origin to proceed with the ensuing nodes which so eventually occupy the next serially expanded Cartesian spacefilling solid portion of continuous dynamic iso-space

(Fig. 14). It is seen that the resulting truncated octahedron is filled by 152 such 12-step Electron rosettes, in all 1852 charged root vector steps, so that the total inertia/mass is $1/1852 \times 938.27 = 0.514$ in comparison with measured 0.511 MeV.

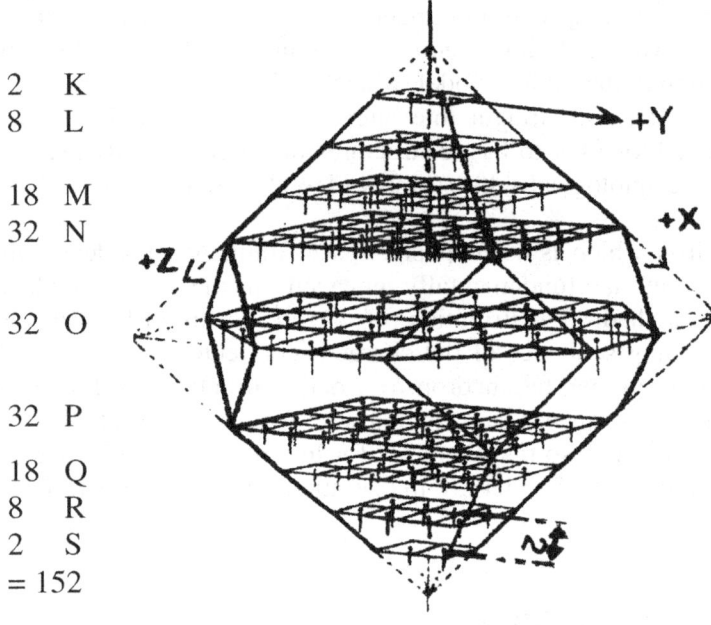

2 K
8 L

18 M
32 N

32 O

32 P

18 Q
8 R
2 S
= 152

Fig 14 Truncated octa-Hedron distribution of 152 electron twelve-step nodes (their triple coils just indicated as rods) in two vertically joined Cartesian segments. The corresponding Bohr orbital shells are shown to the left.

Before coming back to this projected primary Electron cloud of the Hydrogen ion and its onward atomic and Periodic Table expansions, the remaining Leptons; the Photon and the Muon and Electron Neutrinos and Antineutrinos (Tau is not included here) will be briefly considered. They occur as one-dimensional residues, e.g., when a larger differential envelope such as the neutral Pion decays like an imploding bubble into two γ.s (Fig.15)

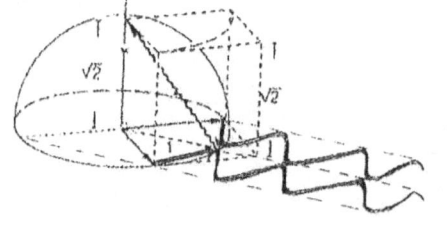

Fig 15 Example, in the rotating top differential neutral Pion, how photons are generated when root vectors in e.g. particle decay or Brehmsstrahlung bendings within same charge plane snap back to their space axes setting up a zig-zag ripple between them of infinite length; thus zero mass; and amplitude/frequency also determinable

or when an orthogonally inclined Muon trajectory is destined to bend into the prevailing spheroidal symmetry extranuclear Electron geodesics leaving a Muon Neutrino and

Electron Antineutrino pulse in the field (Fig 16). All of them have endless trajectories with mass expression $1/\infty \times 938.27 = 0$ MeV as in confirmed reality, too. Table 3 summarizes the results in the Leptons accounted for here.

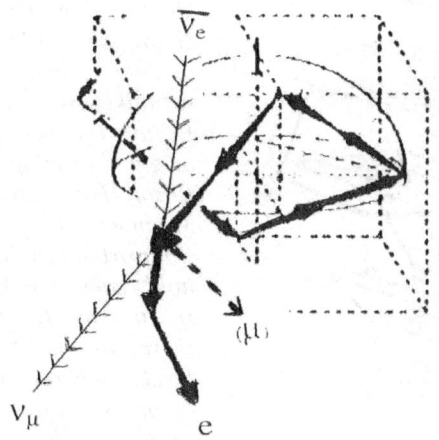

Fig 16 *Example (compare Fig 12 c) how a transition of, in this case, the Muon into the Electron geodesics leaves straight forward Bra/backward Ket direction vectors corresponding to Muon Neutrino and Electron Antineutrino, respectively, both endlessly extending over successive space lattice intervals and hence of $1/\infty$ = zero mass and nil amplitude*

Table 3. Basic Lepton calculated and observed mass numbers (MeV)

$\mu^{\pm}$	$1/(2\pi \times 2^{-2}) \times 938.27$ or $1/(2\pi \times 2 \times 1/2^{-2}) \times 938.27$	105.59	105.66
$\epsilon^{\pm}_{orbital}$	$1/(137.035986 \times 6\pi \times 1/2^{-2}) \times 938.27$	0.514	0.511
$\epsilon^{\pm}_{solid}$	$1/(152 \times 12) \times 938.27$	0.514	0.511
γ	$1/\infty \times 938.27$	0	$0\ (<3\times10^{-33})$
$\nu_{\mu,\epsilon...}$	$1/\infty \times 938.27$	0	$0\ (<17 - 35)$

7. Archimedean Atom Honeycombs

With the Leptons all elementary particles are reproduced by an instantaneous principal phase transition where the electron cloud comes out as a spacefilling mesh segment of defined, second-order regular solid form. This truncated octahedron module can be seen as a diagonal cube (Fig. 17) possible to tessellate into different shapes which, in turn, may self-template into cyclically larger portions of same or modified form to go on filling space in a three-dimensional Tetris way, and, at any such stage, to combine with each other in various full-packing conformations.

There is nothing different from the orbital model in that regard, under one crucial provision: that the continuous transition lattice can also be continuously delineated. This is indeed the case under a Fermion half-spin rotation around the forward diagonal axis (Fig. 18) bringing the end of the line one charged root vector step and one or two (or at lowest quantum, Bose-Einstein Condensate state zero) neutral space axis steps from

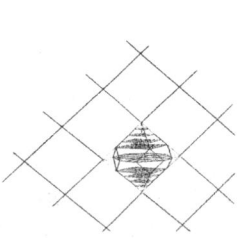

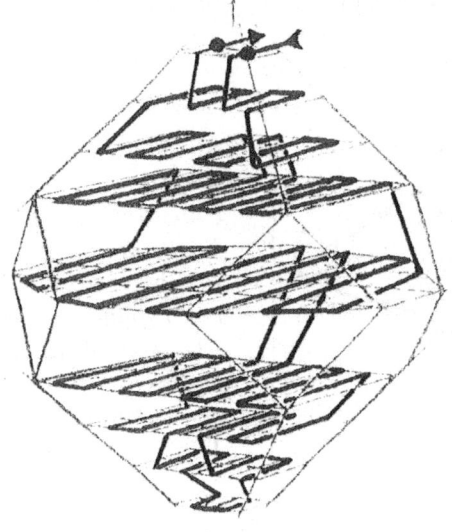

Fig 18 One possible, net Fermion continuous sequence of singlet rosettes in Hydrogen Electron module connecting with likewise Fermion Proton root vector at origin, forming net Boson Hydrogen Atom. Module tiers match the Bohr orbital shells and hold same number of rosettes as Electrons there; also in P to S levels where higher amounts have not been seen in reality

Fig 17 The Electron module is surrounded by other modules in second-generation global lattice, and hence doubly bound to segmental shape

the origin which, not taking part in the Electron formation, appears as the reciprocal pivot, each point of which is 1852 times longer lasting than the Electron with proportionately higher inertia and consequential mass number: $1852 \times 0.514 = 938.27$ MeV. The advantage is that the distribution solid can be used as structural bricks, and this double cast of Electrons as "wave functions or transition matrix elements" is in line with recent Hydrogen ground state research (Martin et al. [2007]) and the instant material "modular building block" (De Weck et al. [2005]) nature of the Electron is pending in modern nanotechnology, molecular biology etc.

Fig. 19 a illustrates the complex of one electron module linking with the Proton in an

a

b

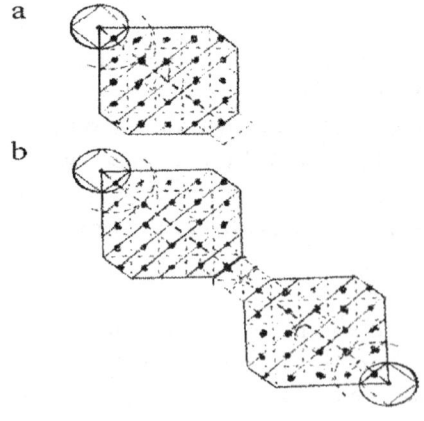

Fig 19 a) Horizontal plane projection of single extra-nucleon module with open end and so realizing H. b) When two H ions are linked end-to-end (or side) the H_2 molecule is formed

upper Cartesian segment and so matching the Hydrogen atom. The opposite end of the complex is free to bind with another open-ended ion, here a second H into the H_2 molecule (Fig. 19 b). It is a variety of "nested polyhedra...which can in turn be put together in spatial arrangements", e.g. "helicoidal progression" (Huybers [2007]; in the present case creating the Bohr orbit signature of the singlet nodes in the forward plane. And when instead under strong pressure two Hydrogen ions will fuse so that one is pushed a step upwards, still rooting with the upper Proton pole in the Nucleon and the other with the under and thereby also the in-between Neutrons' space axis points are involved, a two-module truncated octahedron honeycomb is generated (Fig. 20), closing the ground (K) sheet of lattice intersections and thus stable so as to faithfully realize the Helium atom.

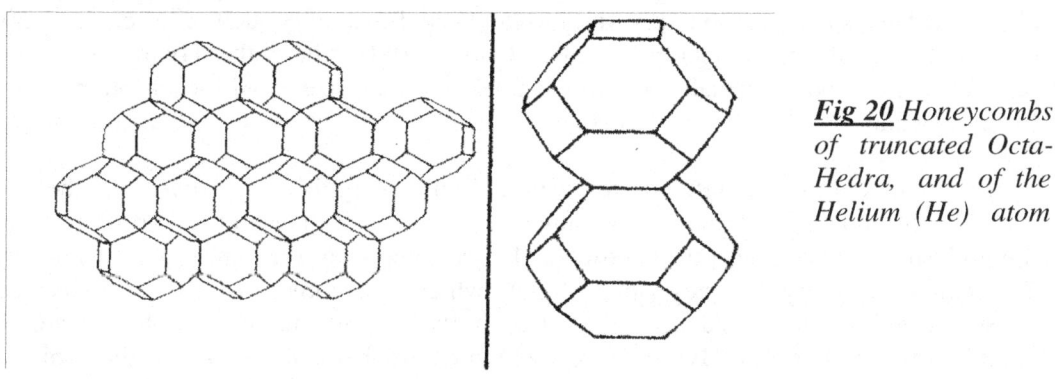

Fig 20 Honeycombs of truncated Octa-Hedra, and of the Helium (He) atom

In that way the singlet sites can be dragged in under an expanding central boundary as Nucleon centres of consecutively larger honeycombs, which thereby are templated in steps and constellations of the periodical system and onwards to further self-similar spacefilling, for instance, of crystalline lattices, deposits, rocks, planets etc. Exemplifying the mechanism only in the first three atoms from the next (L) sheet (Fig. 21), the Lithium honeycomb is variably triangular with one free end for molecular coupling whereas the square or rhombic Beryllium can combine with two atoms/ions/complexes

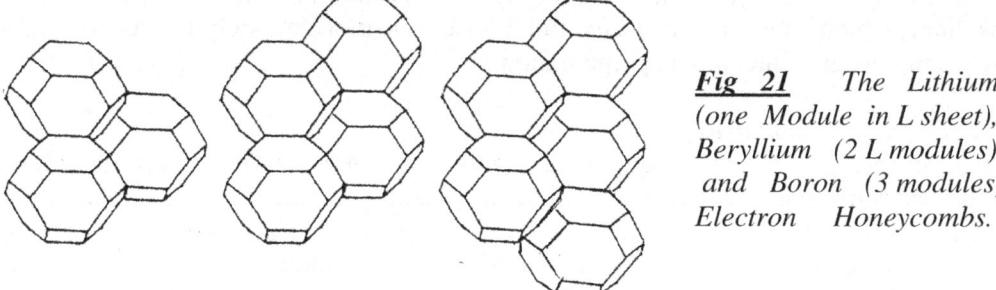

Fig 21 The Lithium (one Module in L sheet), Beryllium (2 L modules) and Boron (3 modules) Electron Honeycombs.

and Boron with three. Not illustrated, Carbon can permutatively couple/chain with four including itself, whereas Nitrogen holds five of the L positions and so has three to offer; Oxygen then two, and Fluorine very strongly one; and when the L shell is filled a new saturated and hence stable atom, Neon, is established. And so it continues and the correspondences are so extensive, that there can be little doubt that it is how matter will be ultimately reconstructed in forthcoming nanotechnology.

The paradoxical aspect is that even in the heavier elements there is only one root vector string fulfilling the respective honeycomb by an extended and convoluted figure-of-eight course so that head-on intercepting by an observation instrument with that individual electron arrow is a matter of quantum indeterminacy and probability statistics. And so far the reproductions only comprise the first extranucleon level, or quantum state of the elements. This is the situation prevailing in the Bose-Einstein condensation but also at the other extreme of the temperature scale (Cho [2008]) where the Hydrogen electron modules are extended virtually to straight lines and may merge with each other into the fusion cascade.

Therefore, it is group-theoretically relevant and interesting that as an inverse at the other, zero Kelvin end of the temperature range, the charged root vector suit runs back into 'empty' space, that is, into the rectilinear lattice moiety by the same route it appeared. This can only be by a honeycomb singlet, which can come so close whereas larger constellations are too distanced. But also in the singlets there is a difference, first exhibited by the Fermion Hydrogen ion which comes back ½ step along the projection planes away from the origin so that retrograde access to the Euclidean space is blocked as it were, whereas the larger Boson Helium will come at +1 and so hitting the entrance; which Hydrogen can do by pairing (Ib.).

Similarly, two Fermion Lithium modules settle at $1½ + 1½ = 3$, while Boson Beryllium reaches 2 at once…and so on in same pattern as recorded throughout the Periodical System (Ib.). Apart from supporting the faithfulness and pertinence of the regular solid scheme it also illustrates that perfect precision, which after all characterises "distilled" physical reality (Schmidt and Lipson [2009]), can only run through all magnitudes of this if generated already at the outset by the not only perfect but categorical precision of an absolutely unique and universal quantum phase transition; the only that exists per se by itself and thus guaranteeing its perpetuation.

8. Stacking the Atom hives

When proceeding from the singlet honeycombs to the atoms there is a cyclical expansion of the respective basic motif which can be described as a stacking of exponentially larger boxes in a Tetris-like manner, each generation templated by the previous along the beaten track of the preceding ones so that the single electron geodesics remains unbroken. The number of generations is then, as cause and effect, temperature-dependent. From just one at both the hot and cold extremes, it increases towards the logarithmic mean which for many reasons would be around where water flows and life is formed, and where, as also in other quantum levels, the equal Avogadro pressure of equally many atoms (in gaseous state) reflects the different number of root vector steps in their completion. Fig. 22 shows the principle in a horizontal plane projection towards three cycles in the Hydrogen atom stacking, and it is seen that it does not take many cycles before the cross-section is increased ten-thousand-fold as is also the ratio between the actual elementary particles and Atoms.

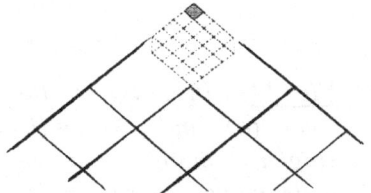

__Fig 22__ Schematic equatorial plane projection of first three self-similar cycles of Electron module in the Hydrogen atom as well as of the vertically doubled (Fig.17) Helium Atom

Single as well as fused in honeycomb and molecular aggregates, the modules heap up the joint structural architecture as veritable Lego pieces, patching together already at the infinitesimal level every three-dimensional real shape from their consecutive own and composed combinations. This does not mean that they are some static wire bundles, but the second-generation, $2^3)^3$ periodical partition of the continuous space-filling charged root vector lattice into the next sclf-similar segment of the global transition matrix. Its outline may be distended in, for instance, accelerations, but then the surrounding modules, whether occupied or empty at the moment, will, too, and the apportioned volume share remains preserved.

One possible sequential ordering of the electron singlet subunits (Fig. 18) runs through the (here) upper Cartesian segment from its origin and returns half a turn twisted beside this and so deplaces it correspondingly many charged and neutral rot vector steps and changes the module progression to the opposite direction, describing a virtual cross-section rotation with Bohr orbital signature, then to continue like the pattern of a knitting. And (continuing here the actually quite relevant knitting analogy) when in larger atoms their fabric is formed by the iteration of equally much as their atomic number reduplicated figure-of-eight segmental loops, the pivotal inertia of the interstitial charged and neutral root vector content of the casting-on stitch coming back to the needle as it were will manifest as the reciprocal amount of Protons and Neutrons.

For instance, since the electron geodesic is wrapped throughout the entire atom it matches the "quantum superposition…qualitative picture of all possible electron paths conspiring together" (Ambjørn et al. [2008]) with proportionately low probability of hitting it in a particular infinitesimal interaction cone. And the propagation of the atoms themselves when they occupy their consecutively inflated domains would be determined by their template form so that highly symmetrical shapes, like the noble gases, would proceed in one-dimensional curves and accordingly be gaseous while sharply bent honeycomb modules, like Lithium, regardless of its low weight would go into dense, net two- or three-dimensional convolutions so as to be solid (until heated/excited so that it starts to boil into orbit). And since the offset 'caps' that the honeycombs' collective truncation leaves at the top contain the abandoned central isospin vectors, there will be a reciprocal nucleus, always with as many charged, Proton roots as the atomic number, while the Neutrons can be more numerous reflecting the lateral displacements possible under acceleration, e.g. in ^{11}Li (Sherill [2008]). Apart from Hydrogen, only Helium, Neon and Argon are illustrated here (Fig. 22, 23 a,b) by an (utterly schematic) horizontal plane projection.

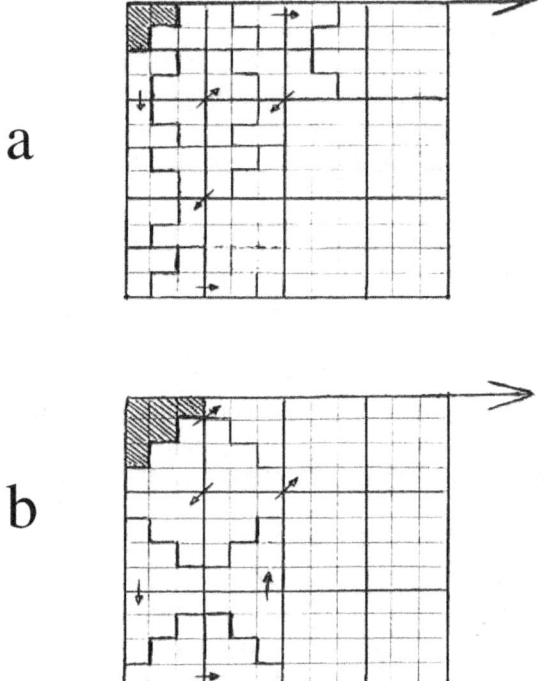

a

b

Fig 23 *Iteration of the Neon (a) and Argon (b) singlet motifs can be distributed as rectangles or squares over the horizontal plane. Extrapolating from this and figure 19 the outward widening of the Cartesian segment, it is seen that equally many further steps out there as the core atom module is provided locus and mould for same volume of motif expansion or, between unsaturated atoms, molecular formation in each of the pyramid corners*

Helium's dispersion there is equal to the Hydrogen atom (Fig. 19), while both Neon and Argon go into ordinary rectangular or quadratic planes (Fig. 20). Since all the Noble gases wholly occupy their vertical columns they are adapted to the Euclidean space also in that direction. Obviously, these rough sketches as well as all other Atoms leaves a lot of further, but rewarding, work to be done. Nonetheless it can be discerned that they represent a general procedure from which the full inorganic realm and its likewise regularly polygonal macroscopic minerals and crystals can be reconstructed with all the characteristics of the periodical system.

At this stage some renewed qualification might be appropriate of the investigation going on, because it is not the triumphant exclamation of the one and exclusive and overturning Truth, but a rational research into the classical Ether and its reaches by the pragmatic scientific method of consequential testing and outcome. Is it indeed feasible to manufacture a homogeneous and coherent, perpetually sliding manifold by the categorical antipode to structural nothingness, namely, the straight line via its convolution to equilateral planes and the iteration of these to spatial structurers?

When the horizontal spread of an equilateral triangle triplet, quadruplet and pentaplet, respectively, is raised like a tent from the ground centre (which is the mutual meeting point of their apices), the tetra-, octa- and icosahedron vaults will form, whereas the hexaplet completely fills the horizontal plane (Sutton [1998]) and therefore cannot stretch out above it. But a triplet of pentagons has a narrow open slit at the bottom and so inflates (a small) volume, too, and in iterations of this the dodecahedron. The problem is to make the pentagon, which the 90-180 and 60-120 angles employed in the rectilinear to spherical root vector lattice phase transition cannot do so that the ground elementary particle and atomic realization is solely a matter of the cube and the amalgamated tetrahedron and octahedron solids.

A self-directive building block set is thus provided, which thoroughly paves a regular polyhedral matrix of the genuine ancient Ether; But hence not the whole of it. The obtained latticework is rigid and edgy and stereotypical, and its zero to positive curvature (Mackenzie [2004]) linearity allows little infrastructural variation beyond repetitive crystal or grainy agglomerations. One recognizes that it is the inorganic realm that is minutely reconstructed, and this may be good enough – but lacking vitality.

Life ether
So far, attempts to replicate the soft texture and forms of living things by spheriodal rendering of polyhedral ground elements have been unsuccessful. There have been initiatives to apply an "expanding-cube design" and algorithms for wiring, "morphing" and "navigating that world" (Mackenzie [2003a,b], which offer a basic self-similar nanotechnology blue-print whose continuous and space-filling modus operandi through individual topological "body plan" (Ib.b) adaptations has close counterparts in the symmetry co-ordination and cubical-rectangular parallelepiped modulations of the infinitesimal cuBit relatives here.

That there will be a tendency to aggregation is also natural, as is collective dextro- or levorotation from equal corner occupancy, and hence to helicity and innumerable other modes of folding due to neighbours who "compete for the same space" (Ib.a,b). When the turns of the "self-folding delivery boxes" (Service [2006]) are 90 degrees the resulting form will be cubical, too. But with more intricate and heterogeneous ingredients and neighbours one might anticipate phenomena like when "in a lattice robot…stacks of lattice modules can reshuffle themselves into a nearly limitless variety of shapes…also attach end to end and form wheels" (Mackenzie [2003a]).

Thereby inspired to "animal metaphors, such as snakes, spiders, and centipedes" (Ib.) one approaches a new principal division line of twofold Nature: that between the inorganic and organic realms. The name organic chemistry was proposed in the 19th century by the

373

Swedish chemist Jöns Jacob Berzelius, and related to the ancient protagonists by invoking a vital force in the generation. Now the designation is carbon chemistry, which to some extent overlooks its quite remarkable characteristics; so fragile and yet covering some 95 per cent of all known compounds, to exhaustion combining the lightest of elements, and filling such narrow a slot of admissible physical habitat…that precisely therefore in an endless Ether it is bound to iterate.

Searching for the distinctive morphogenetic properties of Carbon that enables this 'hyperbolic jungle' (Mackenzie [2004]) of life, one notes that it may itself enter into both inorganic and organic compounds the tetra- and octahedral way.

But living matter's "three-dimensional meshwork with hydrogel-like properties" (Frey et al. [2006]) cannot be outlined by such rigid etchings but need a softer, aquarelle brush, which, continuing the parables, delivers the same terminal "quantized electronic structure" (Walldén [2006]) pigment grains on a pentagon palette where the dodecahedron is the quintessential element and water supplies the lacking linear medium and solvent and butress for the "protein universe" (Service [2005a]).

It is true, that carbon rings and Fullerenes and nanotubules besides hexagons also forge pentagons and larger, but still inert and without life, for which here on Earth and everywhere water with its complementary set of unique properties and ingredients is a *conditio sine qua non*. Firstly, its own molecular configuration is a dynamic domed triangle that may asymmetrically complete a skewed icosahedron-generating pentaplet. And secondly, it is a temperature-dependent "two-structure liquid" matrix of both inorganic and organic disposition by its "quartz-like structure based on tetrahedral units" and "dodecahedral (chlatrate) model associated with special hydrogen-bond rings" (Marjan et al. [1999]), respectively.

Under the latter symmetry, add Nitrogen, and the basic construction kit of proteins (name likewise coined by Berzelius from Greek *protos*, first, *proteios*, primary) is provided by the lightest of ubiquitous elements which in rare, yet all-over standard circumstances permutatively and exhaustively will law-bound "choreograph life's most fundamental processes" (Service [2005a]) from the thinnest, universally determined K and L layer "standing-electron-wave state" (Walldén [2006 assemblies. That is, a self-replicating crystallization of defined pieces (Crystal comes from Greek *crustallos*, meaning clear ice, and is defined as a structure of small repeating geometrical patterns). The crystal side of the "Protein Universe" (Service [2005a]) is further echoed by the X-ray crystallography means of charting it, only that it is again an infinitely descending reconstruction via back-scattering diffraction instead of direct stacking.

Water was originally identified with the icosahedron, and there are close coincidences between the icosahedral and dodecahedral arrangements, for instance, in relation to planetary orbits, the universal constants, the golden section and the Fibonacci series (Martineau [2001], Johansen [2005 a,b, 2008]), and their intersections will merge into the fourth and last of the spacefilling Platonic/Archimedean bodies, i.e. the rhombic dodecahedron (Sutton [2002]).

It is clear that there is indeed a vibrant dual "aether, the life force, here enveloping lively Earth" (Martineau [2001]) and that in spite of its rarity it is so fully defined and exhaustive that it is bound to be prolifically reproduced under the same conditions at copious places, not only in principle but in its exact executions, like human beings (or perfected dinosaurs where the asteroid roulette ball did not happen to extinct them), and homologous derivatives, like the cognitive process (Gregori [2006]), syntax and language (Johansen [2005 a,b, 2008], Marcer [2006]).

In other words, the 'Protein Universe' is as fixed as the mineral and possible to Lego style build up by its amino acid and other bits and pieces and their form and frame provisions. In the assembly the Platonic solids are again engaged, albeit in comparatively colossal composite casts built up to precise form by the many-generation atomic and molecular module lattices.

Pentagonal symmetry can enter in many ways in this biological conveyor-belt manufacture, including the (composite) dodecahedron of which Plato stated: "God used this solid for the whole universe, embroidering figures on it" (http://www.mathpages.com/HOME/kmath096.htm) – and by it. Given exact infra-structure and shape of every part the reproduction and exploration of Proteins can then be made with the same precision and detail as the explosion diagram of a car. Fundamental morphological work in this field is under way above all by Hill and Rowlands (2008).

Eventually, it will be better understood by such "topology from the bottom up…how proteins interact" and "cluster into four structural classes…as four elongated arms emerging from a common centre (whereas) much of the protein structure space is empty because proteins with certain shapes are unstable…the global protein landscape is a bit like the cosmos, where galaxies cluster" (Service [2005 b]).

Discussion
This has been a somewhat wordy survey, still focussed on the reproducible descriptive results. It stops at the atom stage, but can be extended over molecules and larger compounds to the cosmological scale (Seife [2003], Battener and Florido [1998]). Like many other current models, it is a lattice system, however, almost embarrassingly simple in comparison. Therefore, the verbal report tends to assume a slightly surrealistic ring so that is has been aimed at a complementary pictorial account hopefully catching some of the ideas behind.

Yet, it has been said that there is a crisis in today's mostly analogous Physics (Cho [2008]) so that a concrete digital alternative may be welcome. Composed as in a computer just of I, the one straight line; one, unit, binary digit; that can be angled into curves, it is an immediate structure but also the irreducible vector element of pure existence, of anything at all and as abstract as real, as mathematical as material.

A more antrophic argument for Straight is that we and our perceptions are directly parts of and resonating with the actual world of ours all from the quantum level. In other words, we should pay much attention to testimonies like the following (noticed at Tate Modern): "Piet Mondrian (1872-1944) believed that all complex forms could be reduced to a 'plurality of straight lines in rectangular opposition'…his paintings…also represent a physiological reality about the brain…the cells of the visual brain are responsive to straight lines of specific orientation and the field of view to which they respond is rectangular in shape".

And this applies to our binary branching thought processes as well, i.e. intelligence and logic (Marcer [2006], Johansen [2008]) where the straight line bit and its Platonic concatenations and expansions constitute a faithful morphogenetic ground modality of the Nilpotent Universal Computer Rewrite System (NUCRS) (Rowlands [2003, 2006, 2008]) and likewise are engaged in the three-dimensional orthogonal twist "processes of Encryption/Decryption" utilized in "Quantum Holography, defined by means of the Heisenberg nilpotent Lie Group" and "applied at Bletchley Park in World War Two using various machines including the Turing Bombes and Colossus" (Marcer [2008]) as well as more recently in Magnetic Resonance Imaging (Schempp [1997]).

However, this is beyond the present scope, which focuses upon the ancient Ether, and whose enduring contribution is the re-introduction of the modular make-up of it. The identification of the spacefilling electron building brick is in double sense particularly relevant thereby, also because it complies well with newer evidence in other fields that the electron at "boundary conditions" is a "quantized structure" of "discrete character" which "varies in steps given by the thickness of an atomic layer" (Walldén [2006]). Precisely this is the case with its truncated octahedron 'quantum box', which space-fillingly tessellates the extra-nuclear interstitium. The inertially and formatively differential piling of the isomorphic pieces is likewise determined by their "standing wave thickness" (Ib.) from the nodes where the modules' straight Plücker line complex strands start unravelling and recombining their gauge-invariant share of space in unprecedented compliance with the scientific records.

It is all the way the same electron node, but flexible and pliable in its atomic number presentation with corresponding shape accommodation, where even pores and gaps may occur in simultaneous aggregates due to locally influenced geodetical pre- or postponement. A further advantage is the harmonic co-habitation of the electron's wave and quantum double-nature in the common box.

There is therefore no doubt in my by considerable rational scientific experience (searchable in Pubmed: www.pubmed.com) well trained objective judgement that the modular electron model is true and that the truncated Octahedron is its attractor and actuator and will enable also in the organic realm the working out of real bottom-up density cage diagrams and structural rendering not just approximately weighted around anonymous atomic dots but as their intricate flowering filigree.

Also, there is in principle nothing different between a continuous orbital distribution of the electron or a likewise twisting polygonal mesh outlining of it where individual singlet loops at the outer boundaries can be sequestered in their figure-of-eight terminal connection leaving the rest of the motif correspondingly reduced but still in operation.

However, there remains enormously much to be done, but it is creative and productive work. And when the findings so far are judged by their reproducibility, extent and precision, they are of lasting value. Also, as earlier summarized (Trell [2002, 2003b,c, 2004a-d, 2005a-c, 2006 a-c, 2008]) they comply very well with existing paradigms and theories, for example, the nilpotent vacuum (Rowlands [2003, 2006, 2008]), philosophical "infinite machines" (Davies [2001]), the world liquid crystal model (Kleinert [2008]), Wittgenstein matter-of-factness (Hossack [2000]), Hilbert's formalism and so endorsed "Euclidean geometry games" (Devlin [2002]), Scientific Realism (Kukla [1998]), Kant's teleological *als ob* archetypes (Laubichler [2003]), the Turing computer (Earman and Norton [1996]), the Klein bottle (Johansen, Purcell [both 2006]), quantum holography (Marcer [2006]), deployable mathematical software (Petti [1995]), nanotechnological self-assembly (Whitesides and Grzybowski, Velikov et al. [both 2002]), fluctuations in reinstituted flat universe (Bachall et al. [1999], Rees [2000]), eternal-universal brane inflation (Seife [2002], Guth and Kaiser [2005]), and, not the least, the "bright bio-inspired future" (Douglas [2003]) where "crystals with exquisite micro-ornamentation directly develop within preorganized frameworks" (Aizenberg et al. [2003]).

Further, the employed Lego brick and other parables are no more profane than the "tables and chairs" of Scientific Realism (Kukla [1998]), and the exposition at large no more jargon and verbose than many a trendy cosmological text (Cho [2002]), and no more outlandish than dramatic Eternal-Universal Branes (Seife [2002], Guth and Kaiser [2005]).

On the contrary, when the 'Occam razor' rational scientific criteria are applied that among "alternative interpretations…the one which finds the most widespread acceptance is the one which provides the most comprehensive, simple and accurate interpretation of phenomena, and which solves outstanding problems without introducing complex *ad hoc* conceptual or methodological devices" (Duffy [2004]), then the present updating of classical "geometrized vortex sponge World-Ether" (Ib.) and its reproducible results deserves serious consideration.

There is in effect no ontological conflict with other, and more sophisticated models (Ib.), but the contribution lies on the descriptive and quantitative plane and with the appropriate adaptations fully complies with them. It is hard to see a more congenial alternative or come around it, when deeper and deeper ever-sharper nanotechnology confirms the fundamental self-similarity of the physical world all from its very entry.

When Quantum Chromodynamical (QCD) "tamed equations of quark theory" (Seife [2004]) have been awarded the Nobel Prize and together with Quantum Electrodynamics (QED) found to be "the final solution to relativistically invariant quantum mechanics" (Royal Swedish Academy of Sciences [2004]), their real form representation and animation should be useful and valuable. Extremely simplified, both QED and QCD are local gauge symmetries, where "in order to compute a physical quantity we must constrain the field, i.e. specify a gauge" (Ib.). Thereby, in a geometrical/geodetical rendition, QED is volume-preserving and QCD (spheroidally) symmetry-preserving, and concretely realised by the canonical coset decomposition, SO(3) x O(5) of SU(3), yielding both precise and exhaustive reproduction of all elementary particles and their properties (Trell [1981, 1982, 1983, 1990, 1991, 1992, 1998c, 1999, 2000, 2003b,c, 2004a-d, 2005a-c, 2006a-c, 2008]) and also reintroducing the respective Lie groups and algebras in their original geodetic neighbourhood form [1871]. Therefore, this is basically an orthodox exposition and serves to bridge the unnecessary split between Philosophy, Mathematics and Physics and in the end reconciling relativity and quantum mechanics.

Starting with the latter, the very synonym of a quantum is a unit. And the customary conceived reference/rendition of a quantum in an implicitly rectilinear quantum cavity 'incubator' stages strikingly equal equation circumstances as the direct geometrical projections here. In fact, the corresponding local quantum mechanics has since the late sixties been extensively developed in the iso-, geno- and hypermathematics of Santilli and associates [2001]. Taking into account that "the most fundamental quantities…of physical and chemical theories are not abstract mathematical notions (but) the basic units", a rich spectrum of "Lie-admissible" formulas by the respective double-valued, Det = 1 isodual operators and operations transfer and connect global quantum mechanics to the internal elementary particle fields and systems, hence, like the present graphical reproduction, actually bringing further support and homage to the acknowledged standard model (Ib.). Also relativity is maintained because, as noted already by Aristotle, "each part is determined relatively to that part which is next to it by contact" (http [a]).

Moreover, at any single moment the realized spheroidal matter/wave-front by the same mechanism will dispense the quantum indeterminacy and randomness in individual interactions and scatterings that further leads on to the "fantastically filigreed hyperbolic jungle" typically occurring as iterated "cylindrical (or convex)…mirror hall…reflections" along diagonal instead of orthogonal directions (Mackenzie [2004]).

That like Carl Sagan and others have theorized, other physical environment might allow other cybernetic circuits in deeper Electron shells does not change the uniform infinitesimal conditions. And we live in the minutely set K-L table and this is satisfied per se as the singular composed space of our vibrant Nature where the prevailing "three-dimensional topology" (Ib.) still does not hold an independent time extension. However, while its principal absoluteness puts stringent restrictions upon whatever co-existing 'multiverse', it allows forking alternative realizations from any of its own statutory three-dimensional co-ordinates. But really empty gaps are excluded, which is of relevance when coming to the dodecahedra and by them the organic realm of restless life. They are not spacefilling, which means that their hyperbolic conformations either have to be supported by force in an interstitial medium – notably water – or encapsulated, or will fall back into the lower energy. For the flat Euclidean/Cartesian 'canvas' the situation is analogous; The cube is self-replicatively filling its own space in all endlessness, so endlessness is no complication per se – endless Euclidean space just is endless Euclidean space.

The projective deduction here is the spatial arrangement of the Electron (Fig.s 14, 17-19, 22),   It sets out by a complex of one central Octahedron and two laterally flanking Tetraedra. The space-filling packing of the electron module proceeds diagonally over a Cartesian segment in the Euclidean space in which it carries on, that is, paired over the forward axis in thereby doubled, quadratically enlarging frontal plane sheets (Fig 14). There are convincing data, that the serial magnification of successive three-dimensional structural realization lots is "obtained from the Planck scale by period doubling" (Lehto [2006]). With the Nucleon occupying the 2^3 neighbourhood (Fig. 6), that of the Electron would thus accrue towards $(2^3)^3$ (Fig. 17)); next self-similar orders being $(2^3)^3)^3$, $(2^3)^3)^3)^3$, $(2^3)^3)^3)^3)^3$....and so on: and innermost and all through by the irreducible common denominator electron module.

This is naturally bounded by the Cartesian segment (Fig. 1 b) and when approaching its width just symmetrically narrowing in its already constituted polyhedral mesh (Fig. 14). As seen from Fig.14, its frontal plane sheets are separated by the $2^{1/2}$ diagonal expanse of the tetra/octahedral singlet loop, which also imposes a $2^{1/2}$ lateral increase on each side of the forward axis. The vertical change per sheet is $2^{1/2}$ both upwards and downwards from the equatorial plane (Fig. 14). This means that the margin growth per segment is 2 over all Cartesian extensions. "Dense packing of the Platonic and Archimedean solids" (Torquato and Jiao [2009]) is an important research subject in its own right, and the present work shows that the tetrahedron, octahedron and truncated octahedron when wound together in this manner fills the space to 100% and thereby provides a credible vehicle for the global electron conveyance.

Furthermore, the modular nature of this "quantized electronic structure" is entirely in line with the most modern "standing-electron-wave states" observed at extremal junction layers, here the vicinity of the nucleon, where like "the discrete electronic structure of a thin metal film" the "thickness-dependent properties are…not a continuous variable but vary in steps given by the thickness of the atomic layer", and "the new type of electron wave extends through the film and into the substrate to a length determined by the doping level", and "heavily doped…rise to additional states that coherently span the

379

silver film" (Walldén [2006]). Exchange the film with the module and we have the identical situation as also "in quantum mechanics the electron is placed in a confining box", in which "at regular thickness intervals, new states become populated" which in parallel "the rules of quantum mechanics....specify" (Ib.). The individual Electron has an inherent spectral structure and together with its Proton moiety in the Nucleon constitutes the Hydrogen atom (Fig. 19). This may, also as H_2 and like Helium, continue its geodetical build-up of its Atom module (Fig. 22) smoothly and thus be gaseous, while, for instance, Lithium (Fig. 21) tips over the edge of the L shelf as it were, with correspondingly folded further motif outlining and hence in spite of its low atom number will be metallic in physical appearance.

In conclusion, by quoting a previous paper (Trell [2008]), the present report is but a prelude in need of further clarification, e.g. of crowding and overlapping properties in atoms rooted from central parts of larger cores. However, the findings are true and lasting, and open the way for a definite bottom-up reconstruction of rendered matter. It is high time to abandon the latter-day Western misconception of the regular solids as mere pastimes of ancient symposia. They are the genuine segments of an Ether built by its own and its deep philosophical and logical imperative, and it is the one and only substantial world that we are equal parts of.

And it is essential that its pieces are so distinct already at the ground because precise constructions need precise elements; still vastly larger than the 'not even wrong' superstrings currently in vogue. It may at last be a trivial observation but non-the-less one deserving and fascinating to contemplate full out, that all forms that we encounter and envisage and employ in outlines, combinations, assemblies and patterns, in Nature itself, lie in the span between Straight and Round, yes, bit by bit even more radically are steps of either of them. Whatever anatomy, layout, design, text, sculpture, frame, particle trail, contour – or string – you imagine, actually exists there and so, in maximally three real-world dimensions. Such is their ultimate font and types. This is not a sollipsism but on the contrary bears witness of the aptness of our senses – and common sense. Facts and findings can never be dismissed in objective empirical Science and the bottomline here is but firm reproducible results.

And as an extra spin-off in a basically three-dimensional world, the altogether plausible and paradox-free opportunity of temporo-spatial excursion therefore warrants constructive elaboration (Trell [1984, 2004e], US Patent nr. 4,851.688 [1989]). And due to the hyperbolic jungle's after all limited amount and permutations of workable Carbon lightweight alloys wherever else in endless Ether, no matter how rare and fragile the appropriate circumstances are destined to exist for the inventory and manufacture, not only of life's feasible structures and optimal organism design but in parallel of information and language (Johansen [2006 a,b, 2008]), it is much more than a fiction but a fair prediction that when one day we debark the time-adjusted spacecraft on that nearest Trafalmador planet we are greeted by the quite humanoid meeting officer with "Tellurians, I presume".

References

Adhikari, M.R., Adhikari, A. [2003]: *Groups, Rings and Modules With Applications,* 2nd Edition. Hyderabad: Universities Press.

Aizenberg, J., Muller, D.A., Grazul, J.L., Hamann, D.R. [2003]: 'Direct Fabrication of Micropatterned Single Crystals', *Science, 299,* pp 1205-8.

Ambjørn, J., Jurkiewics, J., Loll, R. [2008]: 'The self-organizing Quantum Universe', *Scientific American,* **July 2008**, pp 42-9.

Bahcall, N.A., Ostriker, J.P., Perlmutter, S., Steinhard, P.J. [1999]: 'The Cosmic Triangle: Revealing the State of the Universe', *Science, 284,* pp 1481-5.

Battener, E. and Florido, E. [1998]: 'The egg-carton Universe', *arXiv:astro-ph/9802009v1,* pp 1-7.

Cardano, G. [1545]: *Ars Magna.* Milan.

Carmeli, M. [1977]: *Group Theory and General Relativity.* New York, St. Louis, San Francisco, Toronto: McGraw-Hill International.

Cho, A. [2002]: 'Constructing Spacetime - No Strings Attached', *Science,* **298,** pp 1166-7.

Cho, A. [2004]: 'String Theory Gets Real – Sort Of', *Science,* **306,** pp 1460-2.

Cho, A. [2008a]: 'Insights Flow From Ultracold Atoms That Mimic Superconductors', *Science, 319,* pp 1180-1.

Cho, A. [2008b]: 'Does Fermilab Have a Future?', *Science, 320,* pp 1148-51.

Chodos, A., Jaffe R.L., Johnson, K., Thorn. C.B. [1974]: 'Baryon structure in the Bag theory', *Physical Review D, 10,* pp 2599-604.

Coey, J. M. D. [2004]: 'Georg Francis FitzGerald Millenium Discourse'. www.ted.ie/Physics/History/Fitzgerald/GFFG-JMDC/science.php

Davies, E.B. [2001]: 'Building Infinite Machines', *British Journal for the Philosophy of Science,* **52,** pp 671-8.

Devlin, K. [2002]: 'Kurt Gödel - Separating Truth from Proof in Mathematics', *Science,* **298,** pp 1899-1900.

De Weck O.L., Nadir, W.D., Wong, J.G., Bounova, G., et al. [2005]: 'Modular Structures for Manned Space Exploration: The Truncated Octahedron as a Building Block', *AIAA,* **2005-2764,** pp 1-26.

Douglas, T. [2003]: 'A Bright Bio-inspired Future', *Science,* **299,** pp 1192-3.

Duffy, M.C. [2004]: 'Geometrized space-time & the world-ether' *Moscow Satellite PIRT Meeting (30 June - 03 July 2003), Proceedings, Moscow: Bauman State University, pp 108-11.*

Earman, J. and Norton, F.D. [1996]: 'Infinite Pains: The Trouble with Supertasks' in: A. Morton and S.P. Stich (*eds*), 1996. *Benacerraf and his critics,* Cambridge MA: Blackwell, pp 231-61.

Fraser, J.T. [2001]: 'And take upon's the mystery of time', *Kronoscope Journal for the Study of Time,* **1,** pp. 7-13.

Fraser, J.T. [2003]: 'Mathematics and Time', *Kronoscope Journal for the Study of Time,* **3,** pp. 153-67.

Frey, S., Richter, R.P., Görlich, D. [2006]: 'FG-Rich Repeats of Nuclear Pore Proteins Form a Three-Dimensional Meshwork with Hydrogel-Like Properties', *Science,* **314**, pp 815-7.

Garasco, G.I. [2007]: 'On the World Function and the Relation Between Geometries'. In: (Ed.s) Pavlov, D.G., Atanasiu, Gh., Balan,V. *Space-Time Structure, Algebra and Geometry.* Russian Hypercomplex Society, Lilia Print, Moscow, pp 360-75.

Gilmore, R. [1974]: *Lie Groups, Lie Algebras and Some of their Applications.* NY, London, Toronto: John Wiley.

Gilson, J. G. [2004]: 'The fine structure constant, a 20th century mystery', *www.btinternet.com/ugah174/*

Goldfeld, D. [1996]: 'Beyond the Last Theorem', *The Sciences,* March/April, pp. 34-40.

Gregori, G.P. [2006]: 'The cognitive process in Physics'. *Physical Interpretations of Relativity Theory X,* 8-11 September 2006, Imperial College, London.

Guth, A.H. and Kaiser, D.I. [2005]: 'Inflatory Cosmology: Exploring the Universe from the Smallest to the Largest Scales', *Science,* **307**, pp 884-90.

Haselhurst, G. [2003]: 'On Absolute Space (Aether, Ether, Akasa) and its Properties as an Infinite Eternal Continuous Wave Medium'. *www.spaceandmotion.com/Physics- Space-Aether-Ether.htm*

Heath, T.L. [1964]: *Diophantus of Alexandria: A Study in the History of Greek Algebra.* New York: Dover Publications.

Hill, V. [2006]: Tetrahedral rendering of DNA structure. The joint NTNU/BCSCMsG Symposium 'Science and Philosophy Engaged' 31st March-2nd April, Sage Skysstasjon Trollheimer conference center.

Hill, V.J. and Rowlands, P. [2008]: 'Nature's Code'. *American Institute of Physics Conference Proceedings,* **1051**, pp. 127-26.

Hogben, L. [1937]: *Mathematics for the Million.* London: Georg Allen&Unwin.

Holden, C. [2004]: 'A Different String', *Science,* **51**, p 463.

Hossack, K. [2000]: 'Plurals and Complexes', *British Journal for the Philosophy of Science,* **51**, pp 411-43.

Huybers, P. [2007]: 'Nested Polyhedra, *IASS Newsletter,* **14**, pp 31-40.

http://classics.mit.edu/Aristotle/physics.1.i.html (a)

http://scholar.uwinnipeg.ca/courses/38/4500.6001/Cosmology/dimensionality.htm (b)

http://www.amazon.com/exec/obidos/tg/detail/-/037575766X/102-6497986-7237716 (c)

http://academic.udayton.edu/BradHume/hst340/aristotle.htm (d)

Ikkala, O. and ten Brinke, G. [2002]: 'Functional Materials Based on Self-Assembly of Polymeric Supramolecules', *Science,* **295**, pp 2407-9.

Jaffe, R.L. [1997]: 'Quark confinement', *Nature,* **268**, pp 201-8.

Johansen, S. [2006a]: 'Initiation of 'Hadronic Philosophy', the Philosophy Underlying Hadronic Mechanics and Chemistry', *Hadronic Journal,* **29**, pp 111-35.

Johansen, S. [2006b]: Outline of a Differential Epistemology. The joint NTNU/BCSCMsG Symposium 'Science and Philosophy Engaged' 31st March-2nd April, Sage Skysstasjon Trollheimer conference center.

Johansen, S. [2008]: *Grunnriss av en differentiell epistemologi.* Oslo: Abstract Forlag.

Kamien, R.D. [2003]: 'Topology from the Bottom Up', *Science,* **299**, pp 1671-3.

Kamionkowski, M. [2002]: 'A Hawking-eye View of the Universe', *Science,* **296**, p 267.

Kepler, J. [1619]: *Harmonice Mundi.* Linz.

Kleinert, H. [2008]: *Multivalued Fields in in Condensed Matter, Electrodynamics, and Gravitation.* Singapore: World Scientific.

Kukla, A. [1998]: *Studies in Scientific Realism.* Oxford: Oxford University Press.

Kurlansky, M. [2003]: *Salt. A World History.* London: Vintage, p 300.

Laubichler, M.D. [2003]: 'A Premodern Synthesis', *Science,* **299**, pp 516-7.

Lehto, A. [2006]: 'On the structure of space-time and matter as obtained from the Planck scale by period doubling in three and four dimensions'. *Physical Interpretations of Relativity Theory X,* 8-11 September 2006, Imperial College, London.

Lie, M.S. [1871]: *Ph.D. thesis: Over en Classe Geometriske Transformationer,* Kristiania (now Oslo): Kristiania University.

Mackenzie, D. [1997]: 'Number Theorists Embark on a New Treasure Hunt', *Science,* **279**, p 139.

Mackenzie, D. [2003a]: 'Shape Shifters Tread a Daunting Path Toward Reality', *Science,* **301**, pp 754-6.

Mackenzie, D. [2003b]: 'Topologists and Roboticists Explore an 'Inchoate World''', *Science,* **301**, p 756.

Mackenzie, D. [2004]: 'Taming the Hyperbolic Jungle by Pruning its Unruly Edges', *Science,* **306**, pp 2182-3.

Marcer, P. [2006]: Notes on 3-D quantum holography Universe. The joint NTNU/BCSCMsG Symposium 'Science and Philosophy Engaged' 31stMarch-2nd April, Sage Skysstasjon Trollheimer conference center.

Marcer, P. [2008]: 'A Mathematical Definition of Intelligence. Back to the Future: the Machines of Bletchley Park'. Manuscript.

Marjan, M., Kurik, M., Kikineshy, A., Watson, L.,M., Szász A. [1999]: 'Two-structure model of liquid water', *Modelling Simul. Mater. Sci. Eng.,* **7**, pp 321-31.

Martin, F., Fernández, J., Havermeier, T., Foucar, L. et al. [2007]: 'Single Photon-Induced Symmetry Breaking of H_2 Dissociation', *Science,* **315**, pp 629-33.

Martineau, J. [2001]: *A little book of coincidence.* Walkmill, Cascob, Presteigne, Powys, Wales: Wooden Books.

Mauldin, R. D. [1997---]: 'The Beal Conjecture and Prize', www.math.unt. edu/mauldin/ beal.html.

McGinnis, J. [2003]: 'For Every Time there is a Season: John Philoponus on Plato and Aristotle´s Conception of Time', *Kronoscope,* **3**, pp 83-111.

Murphy, C.J. [2002]: 'Nanocubes and Nanoboxes', *Science,* **298**, pp 2139-41.

Netz, R. [2002]: 'Proof, Amazement, and the Unexpected', *Science,* **298**, pp 967-8.

Noel, E. (Ed.) [1985]: *Le matin des mathématiciens – Entretiens sur l'histoire des mathématiques.* Paris: Belin – Radio France.

Parshall, K.H. [1988]: 'The Art of Algebra from Al-Khwarizmi to Viète: a Study in the Natural Selection of Ideas', *History of Science,* **26**, pp 129-64.

Pavlov, D.G. [2006]: 'Special world functions in polynumber Finsler spaces & precise solutions of Einstein's equations linked with them'. *Physical Interpretations of Relativity Theory X,* 8-11 September 2006, Imperial College, London.

Penrose, R. [1995]: *Shadows of the Mind.* Oxford: Oxford University Press.

Petti, R. [1995]: 'Why Math-Software Development Counts', *Computers in Physics,* **8**, pp 623-5.

Purcell, M. [2006]: Universal Klein bottle. The joint NTNU/BCSCMsG Symposium 'Science and Philosophy Engaged' 31/3 – 2/4, Sage Skysstasjon Trollheimer conference center.

Rees, M.J. [2000]: 'Piecing Together the Biggest Puzzle of All', *Science,* **290**, pp 1919-25.

Rowlands, P. [2003]: From Zero to the Dirac Equation. In: (Ed.s) Duffy MC, Gladyshev VO, Morozov AN. *Proceedings of International Scientific Meeting PIRT -2003,* Moscow 30/6–3/7 2003, Bauman StateUniversity, Moscow, Liverpool, Sunderland, pp.13-34.

Rowlands, P. and Diaz, B. [2003]: A Computational Path to the Nilpotent Dirac Equation, *Symposium 10, International Conference for Computing Anticipatory Systems,* HEC Liege, Belgium, August 11-16, 2003, *International Journal of Computing Anticipatory Systems,* editor Daniel Dubois, 203-218. Also at arXiv.cs.OH/0209026

Rowlands, P. [2006]: How close are we to a fundamental theory?. The joint NTNU/BCSCMsG Symposium 'Science and Philosophy Engaged' 31[st]March-2nd April, Sage Skysstasjon Trollheimer conference center.

Rowlands, P. [2006]: 'Nilpotent theory & vacuum'. *Physical Interpretations of Relativity Theory X,* 8-11 September 2006, Imperial College, London.

Rowlands, P. [2008]: *Zero to Infinity: The Foundation of Physics.* New Jersey, London, Singapore, Beijing, Shanghai, Hong Kong, Taipei, Chennai: World Scientific.

Royal Swedish Academy of Sciences. [2004]: 'Asymptotic Freedom and Quantum Chromodynamics: the Key to the Understanding of the Strong Nuclear Forces', *www.kva.se.*

Santilli, R.M. [2001]: *Foundations of hadronic chemistry with applications to new clean energies and fuels.* Boston, Dordrecht, London: Kluwer Academic Publishers.

Schempp, W. [1997]: *Magnetic Resonance Imaging: Mathematical Foundations and Applications.* New York: John Wiley and sons.

Schmidt, M. and Lipson, H. [2009]: 'Distilling Free-Form Natural Laws From Experimental Data', *Science,* **324**, pp 81-5.

Seife, C. [2002]: 'Eternal-Universe Idea Coming Full Circle', *Science,* **296**, p 639.

Seife, C. [2003]: 'Polyhedral Model Gives the Universe An Unexpected Twist', *Science,* **302**, p 209.

Seife, C. [2004]: 'Laurels to Three Who Tamed Equations of Quark Theory', *Science* **306**, p 400.

Service, R.F. [2005a]: 'Structural Genomics, Round 2', *Science,* **307**, pp 1554-7.

Service, R.F. [2005b]: 'A Dearth of New Folds', *Science,* **307**, p 1555.

Service, R.F. [2006]: 'New in Nanotech: Self-Folding Delivery Boxes', *Science*, **313**, pp 1032-3.

Sherrill, B.M., [2008]: 'Designer Atomic Nuclei', *Science*, **320**, pp 751-2.

Stocks, J.L. [1922]: 'De Caelo'. In: *(Smith. J.A., Ross, W.D. Ed.s), The works of Aristotle translated into English* [1908 – 1956], London: Oxford at the Clarendon Press.

Sutton, D. [1998]: *Platonic and Archimedean Solids. The Geometry of Space.* Presteigne, Powys, Wales: Wooden books.

Torquato, S. and Jiao, Y. [2009]: 'Dense packings of the Platonic and Archimedean solids', *Nature*, **460**, pp 876-9.

Trefil, J. [2004]: 'Untangling the Universe', *Science*, **304**, p 212.

Trell, E. [1982]: 'A calculation of the electron circular orbital radius, *Speculations in Science and Technology*, **5**, pp 533-5.

Trell, E. [1983]: 'Representation of particle masses in Hadronic SU(3) diagram, *Acta Physica Austriaca*, **55**, pp 97-110.

Trell, E. [1984]: 'Scheme for a time antenna in three-dimensional Hausdorff space', *Speculations in Science and Technology*, **7**, pp 269-77.

Trell, E. [1990]: 'Geometrical Reproduction of (u,d,s) Baryon, Meson, and Lepton Transformation Symmetries, Mass Relations, and Channels', *Hadronic Journal*, **13**, pp 277-97.

Trell, E. [1991]: 'On Rotational Symmetry and Real Geometrical Representations of the Elementary Particles With Special Reference to the N and Δ Series', *Physics Essays*, **4**, pp 272-83.

Trell, E. [1992]: 'Real Forms of the Elementary Particles with a Report of the Σ Resonances', *Physics Essays*, **5**, pp 362-73.

Trell, E. [1997]: 'An alternative solution to Fermat´s Last Theorem: Infinite ascent in isotopic geometry', *Hadronic Journal Supplement*, **12**, pp 217-240.

Trell, E. [1998a]: 'Isotopic proof and reproof of Fermat's Last Theorem verifying Beal's Conjecture', *Algebras Groups and Geometries*, **15**, pp 299-318.

Trell, E. and Santilli, R.M. [1998b]: 'Marius Sophus Lie's Doctoral Thesis Over en Classe Geometriske Transformationer', *Algebras Groups and Geometries*, **15**, pp 395-445.

Trell, E. [1998c]: 'The Eightfold Eightfold Way: Application of Lie's True Geometriske Transformationer to Elementary Particles', *Algebras Groups and Geometries*, **15**, pp 447-71.

Trell, E. [1999]: 'Real Charm of Form - Real Form of Charm. Duality in transition'. In: *(Gill, T., Liu, K., Trell, E., Ed.s), Fundamental Open Problems in Science at the End of the Millenium*, pp. 1-29, Palm Springs: Hadronic Press.

Trell, E. [2000]: 'The Eightfold Eightfold Way. A Lateral View on the Standard Model'. *Physical Interpretations of Relativity Theory (11-14 September 1998), Late Papers, London: British Society for the Philosophy of Science*, pp. 263-84.

Trell, E. [2003a]: 'Book Review: Foundations of Hadronic Chemistry with Applications to New Clean Energies and Fuels', *International Journal of Hydrogen Energy*, **28**, pp 251-3.

Trell, E. [2003b]: 'String and Loop Quantum Gravity Theories Unified in Platonic Ether. With Proof of Fermat´s Last Theorem and Beal´s Conjecture'. In: (Ed.s) Duffy MC, Gladyshev VO, Morozov AN. *Proceedings of International Scientific Meeting PIRT -2003,* Moscow 30 June – 03 July, 2003, Bauman State University, Moscow, Liverpool, Sunderland, pp 134-49.

Trell, E. [2003c]: 'Original Diophantine equations lodge BC without ABC' *International Symposium on Recent Advances in Mathematics and its Applications (ISRAMA 2003), Proceedings, Calcutta: Calcutta Mathematical Society.*

Trell, E. [2004a]: 'Original Diophantine equations lodge BC without ABC', *Rev. Bull. Cal. Math. Soc.,* **12**, pp. 29-54.

Trell, E. [2004b]: 'Tessellation of Diophantine Equation Block Universe'. *Physical Interpretations of Relativity Theory VIII (6-9 September 2002), Proceedings, London: British Society for the Philosophy of Science,* pp. 585-601.

Trell, E. [2004c]: 'Classical 3-d. Geometrical 'Vortex Sponge World-Ether Provides Natural Quantum Cavity Elementary Particle Standing Wave Incubation and Original Diophantine Equation Encapsulation'. *Physical Interpretations of Relativity Theory IX (3-6 September 2004), Proceedings, London: British Society for the Philosophy of Science,* pp. 503-30.

Trell, E. [2004d]: 'Cubit Isounits 'Tread a Daunting Path to Reality' While Proving Fermat's Last Theorem and Beal's Conjecture', *Hadronic Journal,* **26**, pp. 237-71.

Trell, E.[2004e]: 'Temporospatial transition – Back to go'. *Physical Interpretations of Relativity Theory (15-18 September 2000), Late Papers, London: British Society for the Philosophy of Science,* pp. 305-11.

Trell, E. [2005a]: 'Invariant Aristotelian Cosmology: Binary Phase Transition of the Universe from the smallest to the largest scales', *Hadronic Journal,* **28**, pp. 1-42.

Trell, E. [2005b]: 'An excursion in curvature I. Diophantine equations get real again in re-established flat Euclidean space', *Bull. Cal. Math. Soc.,* **97**, pp. 509-30.

Trell, E. [2005c]: 'An excursion in and betweem curvature II. From classical Lie algebra neighbourhood to QED and QCD of real elementary particles', *Bull. Cal. Math. Soc.,* **97**, pp. 509-30.

Trell, E. [2006a]: 'Regular Solid Universal Morphogenesis'. The joint NTNU/BCSCMsG Symposium 'Science and Philosophy Engaged' 31st March-2nd April, Sage Skysstasjon Trollheimer conference center.

Trell, E. [2006b]: 'Filling a Gap in Nilpotent Vacuum: How Close Are We to a Fundamental Reality?', *Physical Interpretations of Relativity Theory X (8-11 September 2006), Proceedings, London: British Society for the Philosophy of Science,* pp. 503-30.

Trell, E. [2006c]: 'Space-filling electron module is a truncated octaeder' *International Symposium on Recent Advances in Mathematics and its Applications (ISRAMA 2006), Proceedings, Calcutta: Calcutta Mathematical Society.*

Trell, E. [2008]: 'Elementary Particle Spectroscopy in Regular Solid Rewrite'. *American Institute of Physics Conference Proceedings,* **1051**, pp. 127-41.

Velikov, K. P., Christova, C. G., Dullens, R. A. and van Blaaderen, A [2002]: 'Layer-by-Layer Growth of Binary Colloid Crystals', *Science,* **296**, pp 106-9.

Walldén, L. [2006]: 'Beyond the particle in the box', *Science, 314*, pp 769-70.

Whitesides, G.M. and Grzybowski, B. [2002]: 'Self-Assembly at all Scales', *Science, 295*, pp 2418-21.

Whitney, C.K. [2006]: 'Algebraic chemistry based on a PIRT'. *Physical Interpretations of Relativity Theory X,* 8-11 September 2006, Imperial College, London.

Winterberg, F. [2000]: 'Elementary Particle Physics: Science or Dogma?', *Physical Interpretations of Relativity Theory (11-14 September, 1998), Supplementary Papers,* London: British Society for the Philosophy of Science, pp. 217-28.

wwwgroups.dcs.stand.ac.uk/~history/HistTopics/Fermat'slast_theorem.html [1996].

www.coe.uncc.edu/cas/flt.html [1997]: 'History of Fermat´s Last Theorem'.

www.simillium.com/Thelittlelibrary/References/Pythagsystem.html [2004]: The Pythagorean System.

Zerhusen, A. [1999]: 'Diophantine Equations'. www.ms.uky.edu/~carl/ ma330/projects/dioph

Torsion Fields, the Extended Photon, Quantum Jumps, The Eikonal Equation, The Twistor Geometry of Cognitive Space and the Laws of Thought

Diego L. Rapoport

DC&T,Univ. Nacional de Quilmes, Bernal, Buenos Aires, Argentina [1]

Keywords: Torsion fields, mind apeiron, singularities, eikonal, nilpotence, quantum jumps, quantum potential, spinors, twistors, cognitive space, multivalued logic, mind-matter problem, Klein bottle, time operator, self-reference, perception, Intelligence Code, matrix logic, neurocortex, Cartesian Cut, Kozyrev phenomenae, chronomes, wave genetics .

Abstract: A geometrical origin of quantum jumps in terms of torsion fields and the propagation of wave-front singularities given by the eikonal equation of geometrical optics, which lies at the basis of Fock's theory of gravitation, is introduced. A discussion on the connection between quantum jumps and a global time and space coordinates system is presented. The most general form of the solutions of the eikonal and wave equations in a quaternionic setting to obtain the representation of the photon as an extended singularity is formalized, as well as their twistor representations. Matrix logic and its connections to quantum field operators and hypernumbers are elaborated. The torsion geometry of matrix logic and the relations with quantum mechanical observables and quantum superposition, namely: the so called Schroedinger cat problem, the multivalued character of matrix logic and non-orientable surfaces - the Moebius band and the Klein bottle-, are presented. The plenum zero operator (defined by the matrix with all entries equal to 0), of matrix logic as a logical-quantum ground-state observable (which we shall call the *mind apeiron*) and its twistor eigenstates are introduced. The relation between the twistor representations of the quaternionic eikonal equation and those of the mind apeiron is discussed, establishing thus a relation between the extended structure of the photon and the eigenstates of the mind apeiron. This gives in principle a solution to the so-called mind-matter problem, surmounting Cartesian duality. We present a connection between the quaternionic structure in matrix logic and some metrics in cosmological models .

---

[1] Also Telesio-Galilei Academy of Sciences, London; Email: diego.rapoport@gmail.com.

1 Introduction.

In this article we shall deal with several issues which in principle might seem unconnected but are all related to apeiron. These issues are the space-time self-referential geometries with torsion [13] and the self-referential character of the photon -both terms to be explained below-. We shall treat them as the quaternionic solutions of the system given by the eikonal and wave equations, and the relation with the laws of thought. These laws will be addressed here, in terms of a multivalued logic given by matrix logic, an operator extension of the usual Aristotelian-Boolean connective two-valued logic. This matrix logic extends quantum, fuzzy, modal and Aristotelian-Boolean logics, showing that there is a close connection between quantum fields, logical operators and the torsion geometry of cognition. The latter appears as the coefficients in the commutator of the TRUE and FALSE operators, which turns out to be non-trivial due to their non-hermiticity and consequent non-duality. [2]

---

[2] We here introduce, for a sequential completeness of the presentation of this article, some elements we shall later retake. Logical operators are the representation of their truth values tables by 2×2 matrices acting on cognitive states given by Dirac-like bras or kets of the form $|q> = \begin{pmatrix} \bar{q} \\ q \end{pmatrix}$ with q an arbitrary real number (instead of being 0 or 1, as in Boolean logic) and $\bar{q} = 1 - q$ is the negation of q as in Boolean logic. Then, the definitions are: $\text{TRUE}|q> = |1>$ and $\text{FALSE}|q> = |0>$, with $|0>$ and $|1>$ the false and true *vectors*, respectively. The matrix representations -which we shall present further below- of these operators are non-self adjoint. In distinction with usual (hermitean) quantum observables, logical operators are generally non-hermitean although they have representations as quantum field operators and the reciprocal is also the case. A consequence of this non-hermiticity is that in contrast with the trivial duality of the true and false scalars of connective logic, 1 and 0 respectively (which is represented by the relations $\bar{1} = 0$ and $\bar{0} = 1$), by defining the *complement* $\bar{L}$ of a logical operator L, by $I - L$, where I is the identity operator, we obtain that $\overline{\text{TRUE}} \neq \text{FALSE}$ and $\overline{\text{FALSE}} \neq \text{TRUE}$. So, this notion of complementarity when restricted to scalar fields coincides with the dual operation of Boolean logic transforming conjuction into disjunction. This duality affirms the principle of non-contradiction of Aristotelian-Boolean logic: given any proposition, p, then p *and* not p is false, and thus the previous result proves the non-duality of TRUE and FALSE. Another important consequence of this non-hermiticity, is the appearance in matrix logic of a *logical* momentum operator and its relation to the Bohr-Sommerfeld quantization condition, as we shall present below. Finally, Lie symmetry groups (say of finite dimension n) have a canonical geometrical structure with non-null torsion which characterizes precisely the infinitesimal symmetries of their Lie algebras. Indeed, if we consider the structure coefficients C^a_{bc} of the Lie algebra defined by the commutator $[\xi_b, \xi_c] = C^a_{bc} \xi_a$ (summation convention of repeated indices whenever it applies), for any elements ξ_i (with $i = 1, \ldots, n$) of the Lie algebra, then the torsion of this canonical geometry of the Lie groups is given by $-C^a_{bc}$ [61]. In matrix logic FALSE and TRUE play the role of infinitesimal vectors (i.e. as vector fields, or linear operators) acting on the vector space of bras and kets under a superposition principle which we shall characterize below. We shall

The subject being a singularity (an irreducible form which is also a process) cognizes the world and simultaneously establishes himself as a self-aware observer through cognition and perception stemming from distinctions. These distinctions are primeval in being differences that make a difference in the sense introduced in [45]. These are distinctions which on being perceived, cognized, abstracted or interpreted, generate higher-order differences which amount to the universe of all manifestations, either virtual, processual, operational, algorithmic, formal, conceptual or real [47]; for further developments of an epistemology that departs from this notion of primeval distiction see [46]. Without distinctions in its manifold manifestations, the world would be homogeneous and imperceptible [13]. To introduce this conception we shall take the somewhat paradoxical approach of presenting it through a seemingly realistic approach, based on a geometrical theory for the characterization of quantum jumps. The latter will be characterized in terms of spacetime singularities produced by a torsion field. This field is the logarithmic differential of a wave function propagating on spacetime as a light-like singularity described by the eikonal equation of geometrical optics for light rays [11]. Yet, this realism follows from the peculiar embodiment of the fusion of object with subject that the absorbed photon is. Indeed, the absorbed photon is not an 'objective' structure, but rather a structure and a process constituted by its perception by the subject, and thus of second order. As stated simply [77], "...light is not seen; it is seeing", which in this article we shall prove to admit an extension: Light is seeing-thinking. Thus the photon is a difference, a quantum of action, which generates higher-order differences including its perception and the constitution of the self-observing subject , and thus a self-referential process, which is the embodiment of the fusion of object with subject, the seeing just mentioned [28, 13]. [3] At the level of visual perception at the neurocortex this is sustained by the self-referential topology of the Klein bottle and its paradoxical structure and is further related to Gabor wavelets and holography [48, 49, 50]. [4]

---

prove below that $[\text{FALSE}, \text{TRUE}] = 1.\text{FALSE} - 1.\text{TRUE}$, (so here $n = 2$); thus the structure coefficients reduce to a vector, which defines the torsion of cognitive space, namely the vector $(-1 \ 1)$, which we shall later associate with a normal vector to a Moebius band. For a proof of the legitimacy of this extension to cognitive space see below.

[3]Studies in mathematical psychology on visual perception have proved that there exists no such thing as a purely objective spacetime; see [13, 33, 34, 35]. Bohm has wisely thought the other way round, so to speak, to show the obvious and yet difficult to acknowledge fact that thought has an essential role in creating reality and perception [47], as the history and affairs of humankind shows reiteratively -would a better proof be left wanting-. Thus, we recover a central notion in mathematical psychology in which the construction of the geometry of visual representation depends on parameters proper to the subject [34].

[4]The Klein bottle as a 2-dimensional manifold (a Riemann surface) is not *defined* by its

Thus the conception which we shall present points to the demise of the Cartesian duality, which in logic is the Aristotelian-Boolean dual logic. Cartesian duality (also called the *Cartesian* or *epistemic cut*) appears in several guises: 1) In the formulation of the so-called mind-matter problem, separating the physical world from the *observer*, and more generally the world of *objects* from those of *subjects* (subjectivity). 2) In first-order cybernetics of observed systems, as the idealization of systems which are controlled by a detached subject vis-à-vis the integrative conception of second-order cybernetics. The latter is the cybernetics of observing systems. It is the basis of the mind-matter problem, though never addressed from the point of view of self-reference with some notable exceptions [13, 26]). 3) In the purported duality between form and function, or more generally between form and process. 4) In the duality between content and context, and an endless stream of fractures introduced by subjects, which Nature (which also evokes our nature) by no means abides to [47].

In the geometrical theory of this article, we shall show that quantum jumps are produced whenever the logarithm of the wave function -that acts as the source of the torsion singularity (a spacetime dislocation)- becomes singular on the node set of the wave function. These are the spacetime points where the wave *vanishes*. This establishes 0 as being essentially generative, and we shall see this all along the present article, in the generation of the IC and the mutual coding of light and cognitive states of the IC. We have already discussed that torsion is in distinction with the metric-based geometries of General Relativity (GR), a self-referential construction of spacetime and the subject [13]. If we remain inscribed in this realistic conception which is the daily bread of the working scientist, it is pertinent to remark that these geometrical structures with torsion fields include

---

embedding in 4-dimensional spacetime as is the case of geometries for the Cartesian conception of objects, though in the usual realist approach and for computations it can be taken this embedding. In this conception, objects occupy space rather than being singularities that generate it [28]. In contrast, the Klein bottle [72] is a manifold which is self-contained, so that there is no 'exterior' or 'interior' of it, but a transformation which stems from a singularity (the hole which allows in 3D the reentrance of the Klein bottle into itself) which is the subject as already indicated, which unfolds to the whole Klein bottle to return to the singularity in a form which is a process that incorporates Bohm's explicate and implicate orders [3] as we discussed in [13]. Thus, the Klein bottle is both content and context, form and process, subject and object [28] and thus in relation to it the principle of non-contradiction is invalid [13]. This is the paradoxical being of the Klein bottle which generates matrix logic as a mathematical representation of the laws of thought and cognition. We follow Stern in calling this representation as the *Intelligence Code* (IC) [26]. This paradoxical being will produce (topo)logical superpositions states transforming into the true and false states $|1>$ and $|0>$, and viceversa, and altogether these states allow the generation of all the operators of matrix logic and thus of the IC. This produces the multivalued logic character of the IC which we shall present below [13].

the Hertz potential that yields subluminal and superluminal solutions of the Maxwell equations, and its equivalence with the Dirac-Hestenes equation in the Clifford bundle setting. They furthermore yield a theory of unification of space-time geometries, non-relativistic and relativistic quantum mechanics [10], the weak interactions without a remaining Higgs field [43], fluid and magnetofluid-dynamics [9], non-equilibrium and equilibrium statistical thermodynamics [8], the strong interactions as characterized by Hadronic Mechanics [12], and most importantly Brownian motions[7]. [5]. (Torsion is also essential to the problem of spin precession, appearing in the formulation of the classical mechanics of spinning particles submitted to gravitational fields with torsion, which does nor rely on lagrangian nor hamiltoneans [78].) So torsion is closely related to the chaotic and ordered -and generally non-equilibrium- processes which coexist in apeiron. This contrasts with Einstein's conception in which he claimed by stating that 'God does not play with dice' the elimination of chaos as it could not be framed -at that time- as geometry [7]. This seemingly dual character of apeiron has been the source for the historical record of rejection that different cultures and conceptions had with relation to this untameable Being; for a philosophical discussion we refer to the work in [28]. In [10,12] we proposed these quantum fluctuations as a source for the space anisotropy, the strong nuclear interactions and the time fields (*chronomes* [74]) discovered in tens of thousands of experiments carried out in the last fifty years [44]. In giving a first indication on the nature of torsion as related to surmounting a Cartesian conception, we point out that torsion appears with a Janus face, as the primitive distinction in the calculus of distinctions in the protologic of Spencer-Brown. In this primitive setting, by further incorporating the reentrance of a form on itself and particularly the Klein bottle, the IC is generated [13]. In this code, quantum field operators and logical operators are represented by hypernumbers, establishing thus a connection between quantum field theory, nilpotents and matrix logic. Nilpotence, which in our conception we

---

[5]Indeed, Brownian motions are unified into the geometrical structure of torsion geometries: the metric conjugate vector field of the trace-torsion is the drift of the Brownian motions, and the noise density is a square root of the metric which can be trivially Minkowski or Euclidean. In this setting, Brownian motions determined the torsion geometry or alternatively are determined by it. They are further related to the linear and non-linear Schroedinger equations [10] and the isotopic lift of the former in the Hadronic Mechanics theory of the strong interactions [12]. In terms of them we proposed an explanation [10] of the extraordinary experiments by Kozyrev and Nasonov in Russia -repeated by others- [69], which lead to conceive time as a physical operator related to spin, and thus ultimately to torsion fields [56], yet they were not related to Brownian motions [69]. Kozyrev's conception is currently developed in geophysics [70], chronoastrobiology [74] and in consciousness and physics studies -the Kozyrev mirrors- carried out at ISRICA-Russian Academy of Sciences [71]. These experiments manifest a pervasive action of apeiron.

more accurately call *plenumpotence* -as much as the vacuum is to be called the *plenum*-, has a crucial role in the *nilpotent universal rewrite system* [15]. In this frame for logic, these plenumpotents act as polarizations (i. e. as factorizations) of the mind apeiron defined by the logical-quantum observable given by the identically null matrix. These polarizations turn to constitute the IC, similarly to the constitution of the manifested physical world from the Brownian fluctuations -which is the essential process-structure of apeiron-, and the generative role of the zero points of the torsion generating waves. In the course of this work we shall relate the mind apeiron with the twistor representations of the photon as an extended structure which characterizes the solution of the system given by the quaternionic wave and eikonal equations.

Returning to the torsion geometry, we remark that it introduces a quantization of the apeiron, since it signifies the non-conmutativity of the infinitesimal parallel transport of two vector fields. [6] This non-commutativity manifests as the impossibility of the closure of the infinitesimal parallelogram thus constructed; this closure is achieved by introducing the torsion (the antisymmetric coefficients of the Cartan connection; see Appendix and [61, 7]) with which the parallel transport was produced in the first place. This closure has an invariant meaning since the torsion is a tensor, and thus is invariant by invertible smooth coordinate transformations. [7] This non-commutativity of parallel transport with a connection with non-null torsion, is tantamount to the quantization of the geometry of spacetim. This naturally invites to present the relation between torsion and light, and in particular the photon. So, this article will start by *this* relation to *end* with the relation between light and the laws of thought when treating the relation between twistors (which were introduced by Penrose to construct a geometrical theory of physics starting with light) and the eigenstates of the mind apeiron. Surprisingly this relation will come out from the fact that when trying to localize the photon, we shall found that the singularities of the torsion which

---

[6]For a detailed presentation of this we direct the reader to the Appendix in this article.

[7]This invariance is essential to the joint constitution of the world as a process and geometrical structure, and the subject. It is a mathematical instru*ment* (i.e. an instruction by the mind) by which the subject establishes an 'objectivity' of the 'outer' world while keeps its own invariance which is further projected and retrieved from this 'outer' invariance. Thus, this invariance lies at the foundation of the generation of the spacetime manifold. It further establishes form and function, content and context, outside and inside, through the fusion of the 'outer' world with that of the subject which poses-discovers this joint constitution. For a geometry thus constructed we reiterate that the metric can be the Minkowski or Euclidean metrics, this is irrelevant to the invariant process of generation of an invariant spacetime and an invariant subject. An important example of torsion geometries is given by dislocated crystals, in which dislocations are represented by the torsion tensor [5].

will be supporting it, for scalar *complex* fields does *not* provide a pointlike photon, but rather an *extended* one. On extending these fields to *quaternion* (instead of real or complex) valued functions, this is still the case. In fact we shall give a representation for the photon which will allow us to give its most general structure and still to represent it by twistors which will finally appear as eigenstates of the apeiron mind (as a non-zero polarization of it). This will lead us to conclude with the establishment of a link between light and the IC, which we here recall that is related to quantum field operators through matrix logic.

We turn to discuss light in classical physics. In his theory of gravitation that stemmed from his criticism of General Relativity (GR), V. Fock showed that light rays described by the eikonal equations of geometrical optics, were at the basis for the possibility of introducing 'objective' [8] spacetime coordinates and furthermore for the construction of a theory of gravitation based on characteristic hypersurfaces of the Einstein equations of GR. These equations being hyperbolic partial differential equations have propagating wavefronts. They arise as *singularities* of spacetime which are identical to the wavefronts singular solutions of Maxwell's covariant equations of electromagnetism: they are all characterized by the solutions of the eikonal equation. These singular propagating fields stand for the inhomogenities of the otherwise uniform spacetime that the metric based geometry of GR leads to. As already discussed, this is also a common feature with a theory of spacetime conceived in terms of Cartan geometries with torsion (which is more fundamental than curvature of the latter geometries, as the Bianchi equations show [17]) rather than the curvature produced by a metric. Without inhomogenities it is impossible to give sense to a geometrical locus as argued by Fock, and we reiterate, both are essential features generated by torsion as we argued before [12, 13]. In fact, Fock further proved that the Lorentz transformations of special relativity arise together with the Moebius (conformal) transformations as the unique solutions of the problem of establishing a relativity principle for observers described by inertial fields. It is *not* the Lorentz invariance of the Maxwell's equation what makes this invariance so important in special relativity -paving the way to a diffeomorphism invariant theory of gravitation which Einstein insisted in relating to special relativity-. For Fock, it is rather the fact that the *singular* solutions of the Maxwell equations are invariant by the Lorentz transformations and still, by the full conformal group [4]. We must recall, that already in 1910, Bateman discovered the invariance of Maxwell's equations by

---

[8]Fock's takes an approach based in dialectical materialism. In the philosophical approach by the present author for surmounting the Cartesian cut, the photon is not an 'objective' particle, but the very signature of the fusion of object with subject, the latter being absent in the *geometry* of GR and unacknowledged in Fock due to his mantainance of the Cartesian cut .

this fifteen dimensional Lie group. [9] The equivalence class of reference systems transformable by Lorentz transformations preserve the singular solutions propagating at a finite constant invariant speed equal to c [13]. The velocity of light waves is no longer constant for observers transformable under conformal transformations, but can be infinite [2]. Thus, for all observers related by a Lorentz transformation, if any one would identify a propagating discontinuity with velocity c, all of them would likewise identify the phenomena. Thus, while the Maxwell equations are well defined with respect to *all* diffeomorphic observers, the singular solutions with speed c are well defined for all *Lorentz* group related observers. Most importantly, the singular sets $N(\phi) = \{x \in M : \phi(x) = 0\}$ were introduced by Fock in terms of scalar fields which are solutions ϕ of the eikonal equation

$$(\frac{\partial \phi}{\partial x})^2 + (\frac{\partial \phi}{\partial y})^2 + (\frac{\partial \phi}{\partial z})^2 - (\frac{\partial \phi}{\partial t})^2 = 0, \tag{1}$$

which in the more general case of a space-time manifold provided with an arbitrary Lorentzian metric, say g, can be written as $g(d\phi, d\phi) = 0$, from which in the case of g being the Minkowski metric lead to the light-cone differential equation $(dt)^2 - (dx^1)^2 - (dx^2)^2 - (dx^3)^2 = 0$. Notice that eq. (1) is a nilpotence (or as we stated above, plenumpotence) condition on the field $d\phi$ with respect to the Lorentzian metric g. But while the Maxwell equations are invariant by these two groups (Lorentz and Moebius-conformal) transformations, one could look for propagating waves that remain solutions of the propagation equation determined by the metric-Laplace-Beltrami operator, $\triangle_g$, which we shall describe below- under arbitrary perturbations: Instead of considering solutions of the wave equation $\triangle_g \phi = 0$, which form a linear space, we want to investigate the class of solutions which are further invariant under the action of *arbitrary* (with certain additional qualifications) perturbations f (real or complex valued) acting on by composition on the ϕ's, $f(\phi)$, that verify the same propagation equation: $\triangle_g f(\phi) = 0$. Notice

---

[9]We see here that the Lorentz group are fundamentally related to the invariance of singularities. Consider a primitive distinction as the semiotic (i.e. through a sign) codification of torsion [13] introduced by the fusion of subject -itself a singularity albeit unacknowledged- with spacetime- . If we attach values to the 'inside' and 'outside' of this distinction, then up to a scale factor, the numerical transformation of the distinction yields the Lorentz group; see page 462 [29] and [13]. Thus through 'radar coordinates' a 2D construction of spacetime results, yet with a twist, which amounts to a built-in spinor action on the space variable. Therefore the Lorentz group is related to a valued distinction introduced by this fusion, which is a more primitive and general introduction of this symmetry group that its usual introduction in Special Relativity. Bohm noted that the establishment or recognition of distinctions is a primeval act of thought and of its further projection in the creation of a real world [47].

that in these considerations we are concerned with singularities propagating on a spacetime which is *seemingly* torsionless; this will turn out not to be the case. We start by introducing the geometrical-analytical setting with torsion.

2 RCW Geometries, Laplacians and Torsion

In this section M denotes a smooth compact orientable n-dimensional manifold (without boundary) provided with a linear connection described by a covariant derivative operator $\tilde{\nabla}$ which we assume to be compatible with a given metric g on M, i.e. $\tilde{\nabla}g = 0$. Given a coordinate chart (x^α) $(\alpha = 1, \ldots, n)$ of M, a system of functions on M (the Christoffel symbols of $\tilde{\nabla}$) are defined by $\tilde{\nabla}_{\frac{\partial}{\partial x^\beta}}\frac{\partial}{\partial x^\gamma} = \Gamma(x)^\alpha_{\beta\gamma}\frac{\partial}{\partial x^\alpha}$. The Christoffel coefficients of $\tilde{\nabla}$ can be decomposed as [7,8,9,10]

$$\Gamma^\alpha_{\beta\gamma} = \left\{ \begin{matrix} \alpha \\ \beta\gamma \end{matrix} \right\} + \frac{1}{2}K^\alpha_{\beta\gamma}. \tag{2}$$

The first term in (2) stands for the metric Christoffel coefficients of the Levi-Civita connection ∇^g associated to g (which is the backbone of GR), i.e. $\left\{ \begin{smallmatrix} \alpha \\ \beta\gamma \end{smallmatrix} \right\} = \frac{1}{2}(\frac{\partial}{\partial x^\beta}g_{\nu\gamma} + \frac{\partial}{\partial x^\gamma}g_{\beta\nu} - \frac{\partial}{\partial x^\nu}g_{\beta\gamma})g^{\alpha\nu}$, and $K^\alpha_{\beta\gamma} = T^\alpha_{\beta\gamma} + S^\alpha_{\beta\gamma} + S^\alpha_{\gamma\beta}$, is the cotorsion tensor, with $S^\alpha_{\beta\gamma} = g^{\alpha\nu}g_{\beta\kappa}T^\kappa_{\nu\gamma}$, and $T^\alpha_{\beta\gamma} - (\Gamma^\alpha_{\beta\gamma} - \Gamma^\alpha_{\gamma\beta})$ the skew-symmetric torsion tensor. We are interested in (one-half) the Laplacian operator associated to $\tilde{\nabla}$, i.e. the operator acting on smooth functions, ϕ, defined on M by [10]

$$H(\tilde{\nabla})\phi := 1/2\tilde{\nabla}^2\phi = 1/2g^{\alpha\beta}\tilde{\nabla}_\alpha\tilde{\nabla}_\beta\phi. \tag{3}$$

A straightforward computation shows that $H(\tilde{\nabla})$ only depends in the trace of the torsion tensor and g, so that we shall write them as $H(g, Q)$, with

$$H(g, Q)\phi = \frac{1}{2}\triangle_g\phi + \hat{Q}(\phi) \equiv \frac{1}{2}\triangle_g + Q.\nabla\phi, \tag{4}$$

with $Q := Q_\beta dx^\beta = T^\nu_{\nu\beta}dx^\beta$ the trace-torsion one-form and where $\hat{Q}$ is the vector field associated to Q via g: $\hat{Q}(\phi) = g(Q, d\phi) = Q.\nabla\phi$, (the dot standing for the metric inner product) for any smooth function ϕ defined on M; in local coordinates, $\hat{Q}(\phi) = g^{\alpha\beta}Q_\alpha\frac{\partial\phi}{\partial x^\beta}$. Finally, $\triangle_g$ is the Laplace-Beltrami operator of g: $\triangle_g\phi = \mathrm{div}_g\nabla\phi$, $\phi \in C^\infty(M)$, with div_g and ∇ the Riemannian divergence and gradient operators $(\nabla\phi = g^{\alpha\beta}\partial_\alpha\phi\partial_\beta)$, respectively. Of course, on application on scalar fields, $\tilde{\nabla}, \nabla^g$ are identical: it is in taking the second derivative that the torsion term appears in the former case. Thus for any smooth function, we have

$\triangle_g \phi = (1/|det(g)|)^{\frac{1}{2}} g^{\alpha\beta} \frac{\partial}{\partial x^\beta}(|det(g)|^{\frac{1}{2}} \frac{\partial \phi}{\partial x^\alpha})$. Thus $H(g,0) = \frac{1}{2}\triangle_g$, is the Laplace-Beltrami operator, or still, $H(\nabla^g)$, the laplacian of Levi-Civita connection ∇^g given by the first term in eq. (2). The connections $\tilde{\nabla}$ defined by a metric g and a purely trace-torsion Q are called RCW (after Riemann-Cartan-Weyl) connections with Cartan-Weyl trace-torsion one-form, hereafter denoted by Q [7-10].

3   Quantum Jumps and Torsion

The following section follows our work [11]. In the following we shall take g to be a Lorentzian metric on a smooth time-oriented space-time four-dimensional manifold M which we assume compact and boundaryless; we have the associated volume n-form given by $vol_g = |det(g)|^{\frac{1}{2}} dx^1 \wedge \wedge dx^2 \wedge dx^3 \wedge dx^4$, where (x^1, x^2, x^3, x^4) is a local coordinate system. This is a more general case with regards to what the essential condition for the aether is, the Minkowski metric, which is thus incorporated into this setting, though we shall later consider compact submanifolds of Minkowski space, namely the Klein bottle. [10]

The solutions of the wave equation constitute a linear space. Furthermore, the germ of solutions of the wave equation in a neighborhood of a point form a linear space. Thus , the algebra generated by a single solution of the wave equation

$$\triangle_g \phi = 0, \tag{5}$$

consists of solutions of this equation if and only if ϕ satisfies in addition the eikonal equation of geometrical optics

$$(\nabla \phi)^2 := g(\nabla \phi, \nabla \phi) = 0. \tag{6}$$

Indeed, if f is twice continuously differentiable (we shall say a C^2 function) and ϕ is real-valued, or still, if f is analytic and ϕ is complex valued, then the following identity is valid

$$\triangle_g(f(\phi)) = f'\triangle_g\phi + f''(\nabla\phi)^2. \tag{7}$$

---

[10]In spite of the greater generality, it is this case what we have in mind: The Klein bottle as a self-contained manifold (in fact a Riemann surface) produced by a distinction in an homogeneous plane, which thus establishes a *limited* closed domain lifting to 3d, reenters itself by producing an essential singularity in its structure [72]; see footnote no. 3 above. So instead of a Cartesian aether (a conceptual impossibility [28, 13]), we have a bounded self-referential one. The boundedness is *quantized* by a Planck constant, which we recall that it may have different values, including cosmological scales [30, 31].

The solutions of the system of equations

$$\triangle_g \phi = 0 \tag{8}$$
$$(\nabla\phi)^2 = 0, \tag{9}$$

are called *monochromatic waves*. They represent pure light waves; we already discussed their relevance. A set of monochromatic waves having the structure of an algebra, will be called a *monochromatic algebra*. In Fock's approach, they are called *electromagnetic signals* [4]. Notice that the eikonal equation is a nilpotence condition for $d\phi$, the differential of ϕ, or equivalently its gradient, $\nabla\phi$, under the square multiplication defined by the metric. From the identity

$$e^{-i\phi}\triangle_g e^{i\phi} = i\triangle_g\phi - (\nabla\phi)^2, \tag{10}$$

we obtain , if $\triangle_g\phi = 0$,

$$(\nabla\phi)^2 = -e^{-i\phi}\triangle_g e^{i\phi}, \tag{11}$$

Let us consider the mapping $\phi \rightarrow e^{i\phi} = \psi$ which transforms the linear space of solutions of the wave equation into a multiplicative $U(1)$-group, in which the kinetic energy integrand in the lagrangian functional $(\nabla\phi)^2$ is transformed into $-\frac{\triangle_g\psi}{\psi}$, which has the familiar form of the quantum potential of Bohm, yet in a relativistic domain [3, 7]. If the ϕ are real valued, then the ψ are bounded and we can embed the above group in the Banach algebra under the supremum norm that it generates under pointwise operations and further completion [1]. To distinguish between them we call the original linear space the functional phase space **S** and the Banach algebra defined above as the algebra of wave states **A**, or simply the functional algebra of states. It is simple to see that the critical points of the functional

$$J(\psi) = \int \frac{\triangle_g\psi}{\psi}\mathrm{vol}_g, \tag{12}$$

are those ψ which satisfy

$$\triangle_g\ln\psi = 0, \tag{13}$$

i.e., those whose phase function satisfy the wave equation. Those intrinsic states will be called *elementary states*. The new representation has two advantages over the original one. It is richer in structure and in elements, as **S** is mapped into a subset of the set of invertible elements Ω of **A**. Thus, by taking the logarithm pointwise on the elements of Ω, we obtain an enlargement of **S** by possibly *multivalued* functions. The second advantage , that actually justifies the whole construction, is that the integrand of the lagrangian $-\frac{\triangle_g \psi}{\psi}$, when integrated, exhibits jumps across the boundary $\partial \Omega$ of Ω. These jumps correspond to kinetic energy changes. In the interpretation of the integrand as a quantum potential, they represent a change due to the *holographic* information of the system present in the whole Universe; see [3, 39]). Let $\mathbf{A}$ be a Banach algebra of continuous complex-valued functions defined on a four-dimensional Lorentzian manifold (M, g), containing the constant functions, closed under complex-conjugation, with the algebraic operations defined pointwise and the supremum norm and containing a dense subset $\mathbf{A_2}$ of C^2 functions which are mapped by the Laplace-Beltrami operator $\triangle_g$ into $\mathbf{A}$. Assume further $f \in \mathbf{A}$ is invertible with inverse $f^{-1} \in \mathbf{A}$ if and only if $\inf_M |f(x)| > 0$. The set of invertible elements is denoted by Ω. Furthermore, assume a positive linear functional, denoted by λ such that $\lambda : \mathbf{A_2} \cap \Omega \to \mathbf{C}$ (the complex numbers) defined by

$$\lambda(\phi) = \int \frac{\triangle_g \phi}{\phi} \text{vol}_g \tag{14}$$

The critical elements of λ are those u such that

$$\text{div}(\frac{\text{grad} u}{u}) = 0, \quad i.e. \quad \frac{\triangle_g u}{u} - (\frac{\text{grad} u}{u})^2 = 0. \tag{15}$$

If the linear functional is strictly positive, i.e. $\lambda(\phi) = 0$ if and only if $\phi \equiv 0$, these two identities are to hold in $\mathbf{A}$, otherwise in the sense of the inner product defined by λ on $\mathbf{A}$. By eq. (15) the set C of critical points of λ is clearly a subgroup of Ω. The monochromatic functions of $\mathbf{A}$ are as before, those $w \in \mathbf{A_2}$ satisfying the system of eqs. (8, 9) and their set is denoted by $\mathbf{M}$. From eq. (13) the composition function given by $f(w)$ belongs to $\mathbf{M}$ again if f is an analytic function on a neighborhood of the set of values taken by w on M. Since by eq. (15) $\mathbf{M} \cap \Omega \subset C$, we have that $u f(w) \in C$ if $w \in \mathbf{M}$ and $f(w) \in \Omega$. The spectrum $\sigma(v)$ for any $v \in \mathbf{A}$, is defined by $\sigma(v) = \{z \in C / |v - ze| \notin \Omega\}$ and therefore, by a previously assumed property, is the closure of the set of values $v(x)$ taken by v on M [1]. It is obviously a compact non-void subset of C. Ω has either one or else infinitely many maximal connected components, of which Ω_0 is the one containing the identity, e, defined by $e(x) \equiv 1$. Two elements f, h belong to the same component of Ω, if and only if $f h^{-1} \in \Omega_0$. Further, $f \in \Omega - \Omega_0$ if and only if its spectrum $\sigma(f)$ separates 0 and ∞. The logarithm function, as a mapping from $\mathbf{A}$ into $\mathbf{A}$ is defined only on Ω_0 [1]. With these preliminaries, we can now show that the quantum jumps arise as a generalized form of the standard argument principle of complex analysis. [11]

Theorem Let $u \in C, w \in \mathbf{M} \cap \Omega$, i.e, it is an invertible monochromatic function. Denote by $H_1, H_2, \ldots$, the maximal connected components of the complement of $\sigma(w)$. Then there exists fixed numbers $q_i, i = 1, \ldots$, depending on u and w only, such that for any function $f(z)$ analytic in a neighborhood of $\sigma(w)$ and with no zeros in $\sigma(w)$, we have

$$\lambda(uf(w)) = \lambda(u) + \sum_i (N_i - P_i)q_i, \tag{16}$$

where N_i, P_i are the number of zeros and poles, respectively, of f in H_i, $i = 1, 2, \ldots$. In particular choosing $\alpha_i \in H_i$, the q_i are given by

$$q_i = 2 \int g\left(\frac{\nabla u}{u}, \frac{\nabla w}{w - \alpha_i}\right)\mathrm{vol}_g, i = 1, 2, \ldots. \tag{17}$$

Proof.[12] Let $f = f(w) \in \mathbf{M}$ with $f(z)$ as in the hypothesis. A computation yields

$$\frac{\triangle_g(uf)}{uf} - \frac{\triangle_g u}{u} = 2g\left(\frac{\nabla u}{u}, \frac{\nabla f}{f}\right), \tag{18}$$

which we note that it is another way of writing

$$\frac{\triangle_g(uf)}{uf} = \frac{1}{u}H\left(g, \frac{df}{f}\right)(u), \tag{19}$$

where we have introduced in the r.h.s. of eq. (19) the laplacian defined in eq. (4) by a RCW connection defined by the metric g and the trace-torsion $Q = \frac{df}{f}$.

---

[11]The following result is a simpler geometrical version of a theorem proved by Nowosad in the more intricate setting of non-compact manifolds and functionals on generalized curves in L.C. Young's calculus of variations for curves with velocities having a probability distribution (Young measures)[6]. In our approach surmounting the Cartesian cut, we were interested in a particular Riemann surface, the Klein bottle. We may consider the embedding of this surface in a compact submanifold of Minkowski space and we are in the situation of the theorem below without the need of intricate variational problems nor the full Minkowski space. The latter in its unboundedness corresponds to a conception of spacetime which is associated to the Cartesian approach [28] and its epistemic cut that is surmounted by considering torsion as a self-referential construction of spacetime, logic and cognition [13].

[12]An example. Take a compact submanifold of Minkowski space and plane waves with adequate boundary periodicity conditions. Take $u = e^{ik \cdot x}, w = e^{ik_0 \cdot x}, k_0^2 = 0, k_0.k \neq 0$ and the spectrum $\sigma(w) = S^1$, where S^1 is the unit circle; then $\lambda(u) = -k^2$ (minus the mass squared) and eq. (16) becomes $-\lambda(e^{ik \cdot x} f(e^{ik_0 \cdot x})) = k^2 + 2(k_0.k)(N - P)$, where N and P are the number of zeros and poles of f inside of the unit circle.

Integrating eq. (18) yields,

$$\lambda(uf) - \lambda(u) = 2 \int g(\frac{\nabla u}{u}, \frac{\nabla f}{f}) \text{vol}_g. \tag{20}$$

In particular this shows that q_i in eq. (17) are well defined. From (20) one gets directly

$$\lambda(ufh) - \lambda(u) = [\lambda(uf) - \lambda(u)] + [\lambda(uh) - \lambda(u)] \tag{21}$$
$$\lambda(uf^{-1}) - \lambda(u) = -[\lambda(uf) - \lambda(u)], \tag{22}$$

where $h = h(w)$ as well as we recall $f = f(w)$, are the composition functions, from now onwards. Now if $f \in \Omega_0$, then $\ln f \in \mathbf{A}$ and $\nabla \ln f = \frac{\nabla f}{f}$, which substituted in (20) gives, upon integration,

$$\lambda(uf) - \lambda(u) = 2 \int g(\frac{\nabla u}{u}, \nabla \ln f) \text{vol}_g = -2 \int f \text{div}_g(\frac{\nabla u}{u}) \text{vol}_g = 0, \tag{23}$$

by eq. (15). Hence

$$\lambda(uf) = \lambda(u), \quad \text{if} \ \ f \in \Omega_0. \tag{24}$$

If now f, h belong to the same component of Ω we can write $uh = (uf)(hf^{-1})$, and since $hf^{-1} \in \Omega_0$, the previous result yields

$$\lambda(uh) = \lambda(uf). \tag{25}$$

This shows that $\lambda(uf(w))$ is locally constant in Ω as f varies in the set of analytic functions. Let now $f(z) = z - \nu$ with $\nu \in H_i$. Then $z - \nu$ can be changed analytically into $z - \alpha_i$ without ν leaving H_i, which means that $w - \nu e$ and $w - \alpha_i e$ are the same connected component of Ω, with $e \equiv 1$. Therefore from eqs. (25, 20) and eq. (17) follows that

$$\lambda(u(w - \nu e)) - \lambda(u) = q_i, \tag{26}$$

and by eq. (22)

$$\lambda(u(w - \nu e)^{-1})) - \lambda(u) = -q_i. \tag{27}$$

On the other hand, if ν belongs in the unbounded component of the complement of $\sigma(w)$, we may let $\nu \to \infty$ without crossing $\sigma(w)$ so that

$$\lambda(u(w - \nu e)) - \lambda(u) = 2 \int g(\frac{\nabla u}{u}, \frac{\nabla w}{w - \nu e}) \text{vol}_g$$
$$= \lim_{\nu \to \infty} 2 \int g(\frac{\nabla u}{u}, \frac{\nabla w}{w - \nu e}) \text{vol}_g = 0. \tag{28}$$

Therefore, if $f(z) = c_0 \Pi_{i=1}^N (z - a_i) . \Pi_{j=1}^p \frac{1}{z-b_j}, c_0 \neq 0, a_i, b_j \notin \sigma(w)$, then eq. (16) follows from eqs. (21, 27, 28) In the general case , if $f(z)$ is an holomorphic function in a neighbourhood of $\sigma(w)$,without zeros there, we can find a rational function $r(z)$ such that

$$|f(z) - r(z)| < \min_{\sigma(w)} |f(z)| \text{ in} \sigma(w), \tag{29}$$

by Runge's theorem in complex analysis. Then, $r(z)$ has no zeros in $\sigma(w)$ too, and $r(w)$ and $f(w)$ are in the same component of Ω, so that eq. (16) holds for $f(w)$ too. The proof is complete.

Observations. The quantization formula (16) tells us how the basic functional changes when we perturb the elementary state u into $uf(w)$ with f analytic near and on $\sigma(w)$. Changes occur only when zeros or poles of $f(z)$ reach and eventually cross the boundary of $\sigma(w)$, and these changes are integer multiples of fixed quanta q_i, each one attached to the hole H_i whose boundary is reached and crossed, while u, v remain fixed. Two more aspects are important. The first one being that the actual jump is measured modulo the product of the q_i by a classical difference (where by classical we stress we mean that it is the substraction, in distinction of the quantum difference given by the commutator of operators) of poles and zeros. At the level of second quantization quantum jumps appear in terms of the difference of the creation and annihilation operators. These in turn define the TIME operator in matrix logic in which the commutator of the FALSE and TRUE logical operators coincide with their classical difference, establishing thus a non-null torsion in cognitive space . The second aspect is the actual form of the q_i which are given by integrating the internal product of the trace-torsion one-form $Q = \frac{du}{u}$ defined by the critical state u, with another almost logarithmic differential of the form $dw/(w - \alpha_i)$. Finally, let C_u denote the linear operator $h \rightarrow uh, h \in \mathbf{A}, u \in \Omega$. The very simple analysis above hinges on the fact that $C_u^{-1} \circ \triangle_g \circ C_u - C_{\frac{\triangle_g u}{u}}$ is a derivation on the germ $\mathbf{F}(w)$ of functions of w (see eq. (14) and still eq. (19) to see how it is related to the torsion geometry), which are analytic in a neighbourhood of $\sigma(w)$, and it could have been performed abstractly without further mention to the special case under consideration. The general abstract theory of variational calculus extending the functional λ for quantum jumps when specialized to second order differential operators, say $\triangle_g$ or still $H(g, Q)$, shows that the condition $w \in \mathbf{M}$ in not only sufficient but also necessary in order to the quantum behaviour of λ occur [6]. Let us see next the relation with RCW geometries.

The set of linear mappings $C_{f^{-1}}$ of $\mathbf{A}$ defined by $h \rightarrow f^{-1}h, h \in \Omega$, f defined on M, is a group which maps each connected component of Ω onto another one. In terms of functions defined on M it changes locally the scale of the functions, i.e. the ratio of any function at two distinct points is changed in a given proportion, and it therefore a gauge transformation of the first kind. Under this transformation we have that

$$\triangle_g \to C_{f^{-1}} \triangle_g C_f = \triangle_g + 2\frac{\nabla f}{f} \cdot \nabla + \frac{\triangle_g f}{f} = 2H(g, \frac{df}{f}) + 2V_f, \qquad (30)$$

where $H(g, \frac{df}{f})$ is the RCW laplacian operator of eq. (4) with trace-torsion 1-form $Q = \frac{df}{f}$ and $V_f = \frac{\triangle_g f}{f}$ is the relativistic quantum potential defined by f^2 [10]. Now noting that for vectorfields $A = A^i \partial_i, B = B^i \partial_i$, with $A^i, B^i, i = 1, \ldots, 4$ complex valued functions on M, with the hermitean pairing defined by the metric g on M, i.e. $\int g(\bar{A}, B) \mathrm{vol}_g = \int g(B, \bar{A}) \mathrm{vol}_g$ so that $A^\dagger = \bar{A} = \bar{A}^i \partial_i$. Therefore for the gauge transformation $d \to d + \frac{df}{f}$, since $\triangle_g = -d^\dagger d$ (see [9,10,15]), we further have the transformation

$$-d^\dagger d \to -(d + \frac{df}{f})^\dagger (d + \frac{df}{f}) = -(d^\dagger + (\overline{\frac{df}{f}}).)(d + \frac{df}{f}). \qquad (31)$$

where $d^\dagger$ is the adjoint operator, the codifferential, of d with respect to this hermitean product so that $d^\dagger = -\mathrm{div}_g$ on vectorfields [15]. If we assume that $(\overline{\frac{df}{f}}) = -\frac{df}{f}$, so that $|f(x)| \equiv 1$ and thus f is a phase factor, $f(x) = e^{i\phi(x)}$, i.e. a section of the $U(1)$-bundle over M then the r.h.s. of eq. (31) can be written as

$$-(d^\dagger - \frac{\nabla f}{f}.)(d + \frac{df}{f}) = \triangle_g + 2\frac{\nabla f}{f}. + (\frac{df}{f})^2 + \frac{\triangle_g f}{f} - (\frac{df}{f})^2$$

$$= (\triangle_g + 2\frac{\nabla f}{f}.) + \frac{\triangle_g f}{f} = 2H(g, \frac{df}{f}) + \frac{\triangle_g f}{f} = C_{f^{-1}} \circ \triangle_g \circ C_f. \qquad (32)$$

Consequently, if f is a phase factor on M, then under the gauge transformation of the first kind $h \to f^{-1}h$, the change of $\triangle_g$ into $C_{f^{-1}} \circ \triangle_g \circ C_f$ can be completely determined by the transformation $d \to d + \frac{df}{f}$ which is nothing else than the gauge-transformation of second type, from the topological (metric and connection independent) operator d to the covariant derivative operator $d + \frac{df}{f}$, of a RCW connection whose trace-torsion is $\frac{df}{f}$, equivalent to the gauge transformation $d \to d + A$ in electromagnetism [15].

In summary, when f is a phase factor, the gauge transformations of the first and second type are equivalent, and gives rise to the exact Cartan-Weyl 1-form. If we further impose on f the condition similar to the one placed for the electromagnetic potential 1-form, A, to satisfy the Lorenz gauge $\delta A = 0$, i.e. $\delta(\frac{\mathrm{grad} f}{f}) = 0$,

404

we find that this is nothing else than the condition on f to be an elementary state i.e. a critical point of the the functional $\lambda(f)$ given by (14). Therefore, when f is a phase factor, both the first and second kind of gauge transformations are equivalent and they give rise to a Cartan-Weyl one-form $Q = \frac{df}{f}$. When $\frac{df}{f}$ cannot be written globally as $d\ln f$, f is said to be a non-integrable phase factor. When f belongs to the algebra $\mathbf{A}$, this is equivalent to saying that f does not have a logarithm in $\mathbf{A}$, which means that $f \in \Omega - \Omega_0$. In any case, the 2-form of intensity $F = d(\frac{df}{f})$ is always identically 0 because $\frac{df}{f}$ can be *locally* written as $d\mathrm{Log}f$, where Log is a pointwise locally defined logarithm determination.[13] Consider now all the connected components Ω_α of Ω. Any such component can be transformed into Ω_0 by a gauge-transformation of the first kind: it suffices to take $f \in \Omega_\alpha$ and consider $h \to f^{-1}h$, which is indeed a diffeomorphism of Ω. This choice of the component, is a choice of gauge, and of course, there is no preferred gauge. That is, the topological operator d of one observer becomes the covariant derivative operator $d + \frac{df}{f}$ of a RCW connection for the other observer. We can interpret the difference of gauges as being equivalent to the presence of the trace-torsion 1-form $\frac{df}{f}$ in the second's observer referential. However as the electromagnetic 2-form $F \equiv 0$, this is an instance of the Bohm-Aharonov phenomena: non-null effects associated with identically zero electromagnetic fields. That there are non-null effects is checked by our previous analysis of the functional $\lambda(uf(w))$, where u is any elementary state and f, besides being a phase factor, is also monochromatic. In this case λ, which is locally constant depends on which $\Omega_\alpha f$ belongs to, that is to say, on the choice of the gauge. Finally, according to the two ways of interpreting a linear operator (as a mapping on the vector space or as a change of referential frames) we have two possibilities. Indeed let $w \in \mathbf{M}$ and let $f_t(w), t \in [0,1]$ with $f_t(z)$ analytic in a neighbourhood of $\sigma(w)$, be a continuous curve on $\mathbf{A}$. For any $u \in C$ we consider the curve of elementary states $uf_t(w)$; we described in eq. (16) the behaviour of $\lambda(uf_t(w))$ along this curve. In particular we considered $uf_t(w)$ as a perturbation, or excitation, of u evolving in time (here time may not be the time coordinate of a Lorentzian manifold but the universal evolution parameter introduced first in quantum field theory with other important current formulations; see [16].) We can also regard $u \to C_{f_t(w)}u$ as a continuous curve of gauge transformations of first kind acting on a fixed elementary state u, which, when f_t crosses $\partial\Omega$, determines a change of gauge. When that happen f_t cannot be made a phase factor for all t obviously, so that no electromagnetic interpretation can be given all along the evolution in t. However

---

[13]The relation between Cartan torsion, singularities and dislocations in condensed matter physics is well known [5].

if, say, the initial states f_0 and f_1 are phase factors (i.e. $|f_i(x)| \equiv 1, i = 0, 1$), this change of gauge is equivalence to the appearance of a non-trivial trace-torsion one-form, which we can interpret as an electromagnetic potential, between the initial and final states. In any of these interpretations a non-null effect is detected by a jump in λ as given by eq. (16). This quantum transition is interpreted in the first case as an excitation of the state u. In the second case as a change of gauge of u. This materializes by the appearance of the corresponding Cartan-Weyl one-form as an electromagnetic Arahonov-Bohm potential with zero intensity and non-null effects. Thus, in this interpretation, *quantum jumps are the signature of a non-trivial geometrical structure, the appearance of torsion.*

Finally we examine the dimensions of singular sets $N(f)$ of monochromatic functions. Recall that a C^2 real or complex-valued function f defined on (M, g) is a monochromatic wave, $f \in \mathbf{M}$, if it satisfies the system given by eqs. (8, 9). In the real-valued case, all C^2 functions of f, and in the complex case, all analytic or anti-analytic functions of f belong to $\mathbf{M}$ again, by eq. (7) (we changed here our notation there, pointing precisely to $f = f(w)$ for $w \in \mathbf{M}$, as above) . If f is real, smooth and $df \neq 0$, then $N(f)$ is locally three-dimensional. If it is complex and $\mathrm{Re}(f)$ and $\mathrm{Im}(f)$ are functionally independent $N(f)$ is two-dimensional. Yet the Newtonian picture of a photon as an isolated point-like singularity moving with the speed of light in the vacuum, requires a one-dimensional singular set $N(f)$. Can we achieve this by going to hypercomplex, say quaternionic functions, or still Musès' hypernumbers which are rich in divisors of 0? The answer to the former question is negative; in the quaternionic framework, the photon is a propagating three-dimensional singularity with lower dimensional singularities, but still undivisely extended. We shall present this in the next section. For closing remarks we note that quantum jumps were obtained here in terms of the quantum potential which stands for an holographic in-formation of the whole Universe. In considering the semiclassical theory of gravitation, quantum jumps produce discontinuities in the energy-momentum tensor. These jumps produce a cosmological time associated with a quantum-jumps, in a *global* canonical decomposition of spacetime; see arXiv:gr-qc/0303046v1. [14]

---

[14]This work departs from the incompleteness of the Cauchy problem for the Einstein equations of GR: They provide only six equations for the ten components of the metric. For curved spacetime, it is proved that diffeomorphism invariance of the solutions of the Einstein equation is not valid; only in the case of Ricci flat spacetime this is assured. This underdetermination is resolved by the canonical complementary conditions. In the semiclassical approach they are provided by nonlocal quantum jumps; instead, in Fock's theory they are provided by four equations as eq. (5), the so-called harmonic coordinates. Thus, quantum jump nonlocality is essential for GR, it occurs in nonempty spacetime where the underdetermination problem arises

4 Monochromatic Hypercomplex Functions

This section will deal with the problem of the non-pointlike extended structure of the photon by expanding the field of **C**-valued to quaternion-valued propagating waves verifying the plenumpotence eikonal equation. Two results will appear: Firstly that the node set of these waves reduces to a *single* set, and furthermore, the *generic* form of these waves (Theorems 1 & 2 below), which will later play a crucial role in finding its spinor and twistor representations, which in this article will finally be associated with the Intelligence Code. Our presentation will be highly technical, following [6] and can be skipped - in a first reading if wished- to focus in the statements of these theorems.

A system S of hypercomplex numbers is a finite-dimensional vector space over **R** (or **C**) on which multiplication of any ordered pairs of elements is defined , taking into S it again, and being distributive with respect to vector addition. If $\{e_1, \ldots, e_n\}$ for a basis of this vector space we get

$$e_i e_j = \sum_{k=1}^{n} c_{ijk} e_k, (i, j = 1, \ldots, n), \tag{33}$$

with $c_{ijk} \in \mathbf{R}(or\mathbf{C})$. The constants c_{ijk} are called the constants of the multiplication table of S with respect to a given base, where we still have denoted the product by the juxtaposition. These constants are arbitrary and once fixed, define the multiplication according to the above rule. The product in S is associative if and only if $e_i(e_j e_k) = (e_i e_j)e_k$, for all $i, j, k = 1, \ldots, n$ and this imposes conditions on the c_{ijk}. Furthermore, S has a principal unit u, i.e. an element such that $ux = xu = x,, \forall x$, if and only if there are numbers $\alpha_1, \ldots, \alpha_n$ such that $\sum_i \alpha_i c_{ijk} = \delta_{jk(j,k=1,\ldots,n)}$. In general, S need not be commutative. However, the algebra generated by a single S-valued function f defined on M is always commutative provided S is associative. We shall assume next,that S is associative and has a principal unit. In this case S is isomorphic to a subalgebra of the algebra M_n of n times n matrices over **R** (or **C**) through the correspondence

$$a = a_1 e_1 + \ldots + a_n e_n \rightarrow C_a \in M_n \tag{34}$$

which associates with an element $a \in S$ the matrix C_a of the linear operation $x \rightarrow ax$ in S, with respect to the basis $\{e_1, \ldots, e_n\}$. Thus, C_a is given explicitly

and actually solves this problem. Furthermore, quantum jumps lead to a Universe with complete retrodiction in which only partial prediction is possible; see arXiv:gr-qc/0303046v1.

by

$$(C_a)_{jk} = \sum_{i=1}^{n} a_i c_{ijk} (j, k = 1, \ldots, n). \tag{35}$$

S may contain zero-divisors, i.e. non-invertible elements other than 0. An element a is non-invertible if and only if $\det C_a = 0$, which means that at least one of the eigenvalues λ_i of C_a is 0. Therefore, if f is an S-valued function on M, then $N(f)$ defined as the set of points $x \in M$ where $f(x)$ is not invertible, is the set of points where at least one of the (possibly complex) eigenvalues λ_i of C_a is zero. If the λ_is are locally smooth functions, $N(f)$ will be the finite union of the sets $N(\lambda_i)$, each of which will be at least two-dimensional (over the real numbers). Hence, so will $N(f)$ be at least two-dimensional (over the reals). Thus, we have proved that it is *impossible* to localize a photon to be a *one*-dimensional Newtonian singularity, by the provision of taking a hypercomplex field. Our task goes further to give a structure form of monochromatic quaternionic functions.

We shall say that an S-valued function f is locally smooth if its components f_i *and* eigenvalues λ_i can be chosen locally as smooth functions on M. Clearly the condition

$$\triangle_g f = e_1 \triangle_g f_1 + \ldots + e_n \triangle_g f_n = 0, \tag{36}$$

implies that

$$\triangle_g f_i = 0, \forall i = 1, \ldots, n. \tag{37}$$

As the entries of C_f are given by $\sum_{i=1}^{n} f_i c_{ijk}, j, k = 1, \ldots, n$ and the c's are constant, this implies that all the entries of C_f satisfies this equation again. The hypothesis that $f \in \mathbf{M}$ means that any entire function ϕ of f with real (or complex) coefficients, satisfies

$$\triangle_g \phi(f) = 0. \tag{38}$$

Combining this with the previous remark and with the fact that

$$\mathrm{tr} C_{\phi(f)} = \sum_i \phi(\lambda_i), \tag{39}$$

we get

$$\triangle_g \sum_i \phi(\lambda_i) = \sum_i [(\phi'(\lambda_i) \triangle_g \lambda_i + \phi''(\lambda_i)(\nabla \phi)^2)] = 0. \tag{40}$$

at the point p. Let n_p be the number of distinct eigenvalues of $C_{\phi(f)}$ at the point $p \in M$. By taking $\phi(\lambda) = \frac{\lambda^p}{p}, p = 1, \ldots, 2n_p$, in turn in eq. (40) we obtain $2n_p$ linear homogeneous equations at a point p in M. The unknowns are the sums $\sum_i \triangle_g \lambda_i^{(k)}, \sum_i (\nabla \lambda_i^{(k)})^2, k = 1, \ldots, n$ where $\lambda_i^{(k)}$ are the original eigenvalues grouped by the condition $\lambda_i^{(k)} = \lambda_k$ at $p, i = 1, \ldots, k = 1, \ldots, n_p$ (so called $\lambda(p)$-groups). Since the above system has non-zero determinant we get the $2n_p$ conditions holding at p,

$$\sum_i \triangle_g \lambda_i^{(k)} = 0, \sum_i (\nabla \lambda_i^{(k)})^2 = 0, k = 1, \ldots, n_p. \tag{41}$$

Simple eigenvalues therefore satisfy $\triangle_g \lambda = 0, (\nabla \lambda)^2 = 0$, i.e. $\lambda \in \mathbf{M}$. So do obviously the multiple eigenvalues of a group of eigenvalues that are coincident in an open set and remain distinct from the other in that set. More general situations arise as limiting combinations of both these cases. We therefore conclude that, generically speaking, the eigenvalues of an S-valued monochromatic function should be monochromatic itself. This therefore implies that

$$N(f) = \bigcup_{i=1}^n N(\lambda_i), \quad \text{with} \lambda_i \in M, i = 1, \ldots, n, \tag{42}$$

Consequently, the analysis of singular sets of monochromatic S-valued functions reduce to the analysis of those singular sets of real (complex)-valued functions defined on M.

We will now show that $N(f)$ reduces to a single set, $N(\lambda)$, λ real or complex, for all possible S-valued functions if and only if S is a division algebra over $\mathbf{R}$ or $\mathbf{C}$, i.e. S is either $\mathbf{R}$, $\mathbf{C}$ or $\mathbf{H}$, where $\mathbf{H}$ denotes the real quaternions (Hurwitz theorem) [18]. To show this we need the following facts. Any linear associative algebra has a uniquely determined maximal nilpotent ideal (its radical, R) and is isomorphic to the sum of R with the semisimple algebra S/R. Each semisimple algebra is the direct sum of simple algebras, and Cartan's fundamental theorem says that the simple algebras over $\mathbf{R}$ are just the matrix algebras $M_m(\mathbf{R})$, $M_m(\mathbf{C})$ and $M_m(\mathbf{H})$, and over $\mathbf{C}$ just $M_m(\mathbf{C})$, up to isomorphisms. In particular from this follows that the only real division algebras, i.e. real algebras with no zero divisors are $\mathbf{R}, \mathbf{C}$ and $\mathbf{H}$, and the only complex one is $\mathbf{C}$ itself [19].

To prove the above claim we make the following observations:

1. If $a \in S$ is invertible then so is $a + r$, for any $r \in R$, and viceversa, because $(1 - r')^{-1}$ exists if $r' \in R$ and is given by $1 + r' + \ldots + r'^n$ and so therefore so does

$$(a + r)^{-1} = (1 + a^{-1}r)^{-1}a^{-1}. \tag{43}$$

2. If $p(x)$ is a polynomial in the indeterminate x then,

$$p(a + r) = p(a) + r', r' \in R, \tag{44}$$

and so also for any analytic function f.

3. Take any basis of S formed by a basis $\{e_1, \ldots, e_p\}$ of R and a basis $\{e_{p+1}, \ldots, e_n\}$ of a linear space K complementary to R in S. Since R is an ideal, we have $e_i e_j \in R$ if not both i, j are bigger than p. This means in particular, if $i, j, k > p$ then $e_i e_j e_k$ has the same last $n - p$ coefficients that it would have if we had disregarded in the product $e_i e_j$ its coefficients with respect to $e_1, \ldots, e_p$ in the given basis (by induction, this extends to any number of factors). Therefore, if we define a new product in K given by the original multiplication table restricted to indices $i, j > p$ leaving the vector addition unmodified, K is then a concrete representation of S/R. Furthermore if p is a polynomial, $a \in K$ and $r \in R$ then

$$\pi p(a + r) = \tilde{p}(a), \tag{45}$$

where π is the projection on K along R and $\tilde{p}$ is the same polynomial p but computed on the element $a \in K$ with the restricted multiplication table defined above.

Therefore, let q be a smooth monochromatic S-valued function on (M, g). Decomposing it according to the subspaces K and R we get $q = a + r$, with smooth functions $a \in K, r \in R$. We claim that a is a monochromatic K-valued function under the restricted multiplication table. Indeed, by hypothesis $\triangle_g p(a + r) = 0$ for any polynomial p and this holds if and only if each of the coefficients of $p(a+r)$ with respect to the basis $e_1, \ldots, e_n$ satisfies the same equation. But this implies, in particular, $\triangle_g \pi p(a + r) = 0$ and by the last equation, then $\triangle_g \tilde{p}(a) = 0$, which proves the claim.

Combining remarks $1, 2, 3$

$$N(p(a + r)) = N(p(a)) = N(\pi p(a)) = N(\tilde{p}(a)). \tag{46}$$

for polynomials and so for analytic functions as well. This means that passing from S into K with the restricted multiplication, preserves the monochromatic functions and their singular sets. Since K with the new multiplication is semisimple, the claim above now follows from the fact that it is then a direct sum of the matrix algebras given by the Cartan theorem. In the case of division algebras, as $\mathbf{R}, \mathbf{C} \subset \mathbf{H}$, it suffices that we obtain the general form for the quaternion-valued monochromatic function, because the real and complex ones are then obtained by restriction and complexification. Further the knowledge of $N(\lambda)$ for real and complex monochromatic λ gives $N(f)$ for general hypercomplex functions f, according to the decomposition formula (42) above.

4.1 Monochromatic Quaternion-Valued Functions

Let us introduce the quaternionic units $\vec{i_1}, \vec{i_2}, \vec{i_3}$ given by the multiplication rules

$$\begin{aligned}
\vec{i_1}\vec{i_2} &= \vec{i_3}, \vec{i_2}\vec{i_3} = \vec{i_1}, \vec{i_3}\vec{i_1} = \vec{i_2} \\
\vec{i_j}\vec{i_k} &= -\vec{i_k}\vec{i_j}, k \neq j, \vec{i_k}^2 = -1, j, k = 1, 2, 3.
\end{aligned} \tag{47}$$

Notice here that we could chose here the logical quaternions introduced in matrix logic [13], and thus the structures we shall produce below, can be conceived as spacetime structures which are both 'inner' and ' outer' representations of the self-referential character of photons (though the neutrino is also considered below). We shall introduce the notation $(\phi, \psi) \in \mathbf{M}$ to mean that $\phi, \psi \in \mathbf{M}$ (i.e. they satisfy eqs. (7,8)) *and* furthermore

$$g(\nabla\phi, \nabla\psi) = 0, \tag{48}$$

which is the requirement that any algebraic combination of ϕ, ψ belong in $\mathbf{M}$ as well. It will also be assumed that ϕ and ψ are functionally independent, to rule out the trivial cases. We then have the following theorem.

Theorem 1. Any monochromatic quaternion valued function F defined on (M, g) is determined by a triple of real valued functions (ϕ, f, ρ) such that

$$(\phi, f + i\rho) \in \mathbf{M}, \quad \text{i.e.} \quad g(\nabla\phi, \nabla f + i\nabla\rho) = 0, \tag{49}$$

and each of ϕ, f, ρ satisfy the system

$$\triangle_g \kappa = 0, \quad (\nabla\kappa)^2 = 0, \tag{50}$$

has the form

$$F = f + \rho[\vec{i_1}G(\phi) + \vec{i_2}H(\phi) + \vec{i_3}P(\phi)] \tag{51}$$

where G, H, P are real valued functions satisfying

$$P^2 + H^2 + G^2 = 1. \tag{52}$$

411

Thus, F is a section of a $R \times R \times S^2$-bundle over (M, g), where S^2 denotes the two-dimensional sphere. [15]

Proof: Let $F = f + \vec{i_1}k + \vec{i_2}h + \vec{i_3}p \in \mathbf{M}$ be a smooth quaternion-valued function, with $k^2 + h^2 + p^2 \neq 0$. The condition $\triangle_g F = 0$ requires equivalently

$$\triangle_g f = \triangle_g k = \triangle_g h = \triangle_g p = 0. \tag{53}$$

Since by assumption

$$\begin{aligned} (\nabla F)^2 &= (\nabla f)^2 - (\nabla k)^2 - (\nabla h)^2 - (\nabla h)^2 \\ &+ 2\vec{i_1}g(\nabla f, \nabla k) + 2\vec{i_2}g(\nabla f, \nabla h) + 2\vec{i_3}g(\nabla f, \nabla p) = 0 \end{aligned} \tag{54}$$

then it follows that

$$(\nabla f)^2 = (\nabla k)^2 + (\nabla h)^2 + (\nabla p)^2. \tag{55}$$

and

$$g(\nabla f, \nabla k) = g(\nabla f, \nabla h) = g(\nabla f, \nabla p) = 0. \tag{56}$$

The eigenvalues of a quaternion, namely the real or complex numbers λ such that $f + \vec{i_1}k + \vec{i_2}h + \vec{i_3}\rho - \lambda$ is not invertible, or equivalently such that

$$(f - \lambda)^2 + k^2 + h^2 + p^2 = 0 \tag{57}$$

are obviously of the form

$$\lambda_\pm = f \pm i\rho, \tag{58}$$

where

$$\rho = (k^2 + h^2 + p^2)^{\frac{1}{2}} > 0, \tag{59}$$

where we remark that i is the commutative square root of minus 1, of complex numbers. It is easy to check that from our previous analysis it follows that

$$\lambda_\pm \in \mathbf{M}, \tag{60}$$

---

[15] We have constructed the quaternions in terms of logical operators in matrix logic [13]. So we can represent this result as an 'objective' space representation of the objective-subjective photon (the seeing process), or -inclusively (surmounting dualism) - as a 'subjective' representation of it in terms of a quaternionic structure which stems from the laws of thought.

which implies that in to addition to eq. (102) we have

$$\triangle_g \rho = 0, \tag{61}$$
$$(\nabla f)^2 = (\nabla \rho)^2 \tag{62}$$
$$g(\nabla f, \nabla \rho) = 0, \tag{63}$$

as one obtains from specializing eqs. (53, 55, 56) to the complex case. We now consider the algebra over the reals generated by F (we have to restrict ourselves to real coefficients because the real quaternions $\mathbf{H}$ is a division algebra over $\mathbf{R}$, but over $\mathbf{C}$ it is not).

The analytic functions in the complex plane generated by polynomial with real coefficients are those whose domain is symmetric about the real axis and which satisfy $\Phi(z) = \bar{\Phi}(\bar{z})$ (called intrinsic functions on $\mathbf{C}$ [20]). If $\Phi(x+iy) = u(x,y) + iv(x,y)$ is an intrinsic entire function, and u and v are its real and complex part, respectively, then if x_0, x_1, x_2, x_3 are real, then we have the decomposition of the form

$$\Phi(x_0 + \vec{i_1}x_2 + \vec{i_2}x_2 + \vec{i_3}x_3) = u(x_0, q) + Jv(x_0, q), \tag{64}$$

where

$$q = (x_1^2 + x_2^2 + x_3^2)^{\frac{1}{2}}, \tag{65}$$

and

$$J = \vec{i_1}\frac{x_1}{q} + \vec{i_2}\frac{x_2}{q} + \vec{i_3}\frac{x_3}{q}. \tag{66}$$

This follows directly from the observation that in $x_0 + Jq$, the powers of J behave alike those of i, i.e. $J^2 = -1, J^3 = -J, \ldots$, plus the fact that the coefficients of Φ are real. This is also shown in Theorem. 7.1, page 14 of [20], as a consequence of the fact that the intrinsic functions over $\mathbf{H}$ may be characterized as those that are invariant under the automorphisms or antiautomorphisms ([73]) of $\mathbf{H}$. Using eq. (64) we get

$$\Phi(f + \vec{i_1}k + \vec{i_2}h + \vec{i_3}\rho) = u(f, \rho) + v(f, \rho)(\vec{i_1}K + \vec{i_2}H + \vec{i_3}P) \tag{67}$$

where we have set

$$K = \frac{k}{\rho}, H = \frac{h}{\rho}, P = \frac{p}{\rho}. \tag{68}$$

413

We now show that

$$\triangle_g K = \triangle_g H = \triangle_g P = 0, \tag{69}$$

and

$$g(\nabla K, \nabla K) + g(\nabla H, \nabla H) + g(\nabla P, \nabla P) = 0. \tag{70}$$

We start by noticing that from eqs. (56, 63) follows that

$$g(\nabla f, \nabla K) = g(\nabla f, \nabla H) = g(\nabla f, \nabla P) = 0. \tag{71}$$

Now

$$\triangle_g \Phi(F) = 0, \tag{72}$$

implies that

$$\triangle_g (vK) = \triangle_g (vH) = \triangle_g (vP) = 0. \tag{73}$$

Since $\nabla v = (\partial_x v)\nabla f + (\partial_y v)\nabla\rho$ and so

$$
\begin{aligned}
\triangle_g v &= \operatorname{div}\nabla v = (\partial_x v)\triangle_g f + (\partial_y v)\triangle_g\rho \\
&+ (\partial_{xx}v)(\nabla f)^2 + (\partial_{yy}v)(\nabla\rho)^2 + 2(\partial_{xy}v)g(\nabla f, \nabla\rho) \\
&= (\partial_{xx}v + \partial_{yy}v)(\nabla f)^2 = 0,
\end{aligned}
\tag{74}
$$

where we used eqs.(53, 61, 62, 63) and the fact that v is harmonic in (x, y) . Therefore,

$$
\begin{aligned}
0 &= \triangle_g(vK) + 2g(\nabla v, \nabla K) + K\triangle_g v \\
&= v\triangle_g K + 2\partial_x v g(\nabla f, \nabla K) + 2\partial_y v g(\nabla\rho, \nabla K) \\
&= v\triangle_g K + 2\partial_y v g(\nabla\rho, \nabla K)
\end{aligned}
\tag{75}
$$

by eq. (71). Since we have v at our disposal, this equation implies that

$$\triangle_g K = 0, \quad g(\nabla K, \nabla\rho) = 0. \tag{76}$$

Indeed, take the intrinsic function

$$\Phi(x + iy) = e^{\lambda(x+iy)} = e^{\lambda x} + ie^{\lambda x}\sin(\lambda y), \lambda \in R. \tag{77}$$

Then, $v = e^{\lambda x}\sin(\lambda y)$ and eq. (74) becomes

$$(\triangle_g K)\sin(\lambda\rho) = -2\lambda g(\nabla\rho, \nabla K)\cos(\lambda\rho). \tag{78}$$

Since $\rho \neq 0$ we can choose $\lambda \neq 0$ so that $\cos(\lambda\rho) = 1$, which gives $g(\nabla K, \nabla \rho) = 0$, and then $\sin(\lambda\rho) = 1$ which gives $\triangle_g K = 0$. Similarly for H and P, so that eq. (69) is proved, together with

$$g(\nabla\rho, \nabla K) = g(\nabla\rho, \nabla H) = g(\nabla\rho, \nabla P) = 0. \tag{79}$$

To prove eq. (68) we first apply eq.(55) to the function in eq.(67),obtaining

$$(\nabla u)^2 = (\nabla(vk))^2 + (\nabla(vH))^2 + (\nabla(vP))^2. \tag{80}$$

Now in view of eqs. $(62, 63)$

$$(\nabla u)^2 = (\partial_x u)\nabla f + (\partial_y u)\nabla\rho)^2 = ((\partial_x u)^2 + (\partial_y u)^2)(\nabla f)^2. \tag{81}$$

Similarly

$$\begin{aligned}
(\nabla(vK))^2 &= v^2(\nabla K)^2 + 2vKg(\nabla v, \nabla K) + K^2(\nabla v)^2 \\
&= v^2(\nabla K)^2 + K^2((\partial_x v)^2 + (\partial_y v)^2)(\nabla f)^2,
\end{aligned} \tag{82}$$

because $g(\nabla v, \nabla K) = 0$ in view of eqs.$(71, 79)$. Analogous expressions for H and P hold, since $(\partial_x u)^2 + (\partial_y u)^2 = (\partial_x v)^2 + (\partial_y v)^2$ by the Cauchy-Riemann equations, and then eq. (80) becomes

$$\begin{aligned}
((\partial_x u)^2 + (\partial_y u)^2)(\nabla f)^2 &= v^2[(\nabla)^2 K + (\nabla H)^2 + (\nabla P)^2] \\
&+ ((\partial_x u)^2 + (\partial_y u)^2))(\nabla f)^2,
\end{aligned} \tag{83}$$

because $K^2 + H^2 + P^2 = 1$. This gives eq. (70). Therefore we have proved that K, H and P all belong to $\mathbf{M}$.

We now show that there is a real valued function ϕ such that K, H and P are functions of ϕ. For this purpose we will show necessarily that

$$\begin{aligned}
(\nabla K)^2 &= (\nabla H)^2 = (\nabla P)^2 = 0, \tag{84} \\
g(\nabla K, \nabla H) &= g(\nabla K, \nabla P) = g(\nabla H, \nabla P) = 0 \tag{85}
\end{aligned}$$

everywhere. First we note that the group of automorphisms and antiautomorphisms of $\mathbf{H}$, which are precisely the rotations that leave the real unit $1 = 1 + 0\vec{i_1} + 0\vec{i_2} + 0\vec{i_3}$, possibly combined with reflections preserve eqs. $(53, 55, 56, 59)$ as well as the condition $K^2 + H^2 + P^2 = 1$. Furthermore, the intrinsic functions on the quaternions are invariant under this group. Therefore we may always apply a constant rotation on the space of $\vec{i_1}, \vec{i_2}, \vec{i_3}$ to make K, H, P not zero at a particular point $p \in M$. Then it is clear that the new H, K, P will satisfy eqs.

(84, 85) if and only if the original ones they are satisfied by the original functions. Suppose that this is the case at p. Since $P = (1 - H^2 - K^2)^{\frac{1}{2}} > 0$ at p, and so they satisfy it in a neighbourhood of p, and $\triangle_g P = 0$, we get by differentiation

$$2(1 - K^2 - H^2)(-\triangle_g K^2 - \triangle_g H^2) = (\nabla(K^2 + H^2))^2, \tag{86}$$

i.e.

$$\begin{aligned} 2 \quad & (K^2 + H^2 - 1)[2K\triangle_g K + 2(\nabla K)^2 + 2H\triangle_g H + 2(\nabla H)^2] \\ = \quad & 4K^2(\nabla K)^2 + 4H^2(\nabla H)^2 + 8HKg(\nabla H, \nabla K). \end{aligned} \tag{87}$$

Using eq. (53) and simplifying

$$(1 - K^2)(\nabla H)^2 + (1 - H^2)(\nabla K)^2 = -2HKg(\nabla H, \nabla K). \tag{88}$$

Now if ∇H and ∇K are space-like [16], Schwarz's inequality

$$|g(\nabla K, \nabla H))| \leq |\nabla K|^2 |\nabla H|^2 \tag{89}$$

applies. Therefore taking absolute values in eq. (88) we get, since $(\nabla H)^2$ and $(\nabla K)^2$ have the same sign and $1 - K^2 > 0$, $1 - H^2 > 0$,

$$(1 - K^2)|\nabla H|^2 + (1 - H^2)|\nabla K|^2 \leq 2|HK||\nabla K||\nabla H|, \tag{90}$$

i.e.

$$(1 - K^2)|\nabla H|^2 - 2|HK||\nabla H||\nabla K| + (1 - H^2)|\nabla K|^2 \leq 0. \tag{91}$$

The determinant of the matrix of this quadratic form in $(|\nabla K|, |\nabla H|)$ is $(1 - K^2)(1 - H^2) - K^2 H^2 = 1\text{-}K^2 H^2 - P^2 > 0$ and its trace is

$$2 - K^2 - H^2 - P^2 > 1. \tag{92}$$

so its eigenvalues are positive. This implies in eq. (91) that $|\nabla H| = |\nabla K| = 0$, i.e.

$$(\nabla K)^2 = (\nabla H)^2 = 0, \tag{93}$$

and so also

$$(\nabla P)^2 = 0, \tag{94}$$

---

[16] For this condition the case of g being positive-definite is automatically satisfied, while in the Lorentzian case it has to be assumed.

by eq.(70) and

$$g(\nabla H, \nabla K) = 0 \tag{95}$$

by eq. (90). Interchanging the roles of H, K, P in eq. (90) we get now the remaining equations in eq. (85).

Therefore, eqs. (84, 85) hold when any two of the vectors $\nabla K, \nabla H, \nabla P$ are space-like (as we said, for g Riemannian this is always the case), since this property is preserved under the small rotation that may be needed to make $H, K, P \neq 0$ at a given point $p \in M$.

In the Lorentzian case, we are left to consider the case when just one of them is space-like, say ∇H, one is time-like, say ∇P, and the third one ∇K is time-like or isotropic. Now the small rotation that may be necessary to make $H, K, P \neq 0$ at a given $p \in M$, may change the character of ∇K if it is isotropic. If it becomes space-like, we are back into the previous case, so we need consider only the remaining case whenever M is Lorentzian (for the Riemannian case, this case is empty).

Clearly then the subspace determined by ∇K and ∇H cuts the light-cone and so we may rotate ∇H and ∇K by the above procedure till ∇H cuts the light-cone at the point p, while leaving P and ∇P unchanged. Since ∇H becomes isotropic, ∇K must then get space-like so as to compensate $(\nabla P)^2$ in eq. (85). Therefore, by continuity, just before ∇H touches the light-cone at the point p, both ∇K and ∇H will be space-like and since $P \neq 0$, this then reduces the problem to the previous case. This proves that eqs. (84, 85) hold everywhere.

To complete the proof we only need observe that any two real isotropic vectors which are orthogonal in the Lorentzian manifold (M, g), are necessarily parallel. Hence if $\nabla H \neq 0$, necessarily $\nabla P = \mu \nabla H, \nabla K = \lambda \nabla P$ with μ, λ real functions, and therefore $P = P(H), K = K(H)$ locally, as claimed. In view of eqs.(71, 79, 84), we conclude that

$$(H, f + i\rho) \in \mathbf{M} \tag{96}$$

concluding thus with the proof of Theorem 1.

4.2 Maximal Monochromatic Algebras

A monochromatic algebra is called maximal monochromatic if it is not a proper subalgebra of a monochromatic algebra. The importance of maximal monochromatic algebras in our context is obvious, in particular with respect to the question of singular sets. The main result in this respect is

Theorem 2. The maximal C^2 algebras in (M, g) are precisely those generated by a single pair (see (49,50))

$$(\phi, f + i\rho) \in \mathbf{M}, \tag{97}$$

with ϕ, f, ρ real, and are C^2 functions of the form

$$\xi(f, \phi, \rho) + \eta(f, \phi, \rho)[\vec{i_1}K(\phi) + \vec{i_2}H(\phi) + \vec{i_3}P(\phi)] \tag{98}$$

in the quaternionic case, and

$$\xi(f, \rho, \phi) + i\eta(f, \rho, \phi), \tag{99}$$

in the complex case, where for each fixed ϕ, $\xi + i\eta$ is an intrinsic analytic (or antianalytic [17]) function of $f + i\rho$, of class C^2 on ϕ is arbitrary and $K^2 + H^2 + P^2 = 1$, with K, H, P of class C^2, but otherwise arbitrary. Thus, in the quaternionic case, it is given by a C^2-section of a $R \times R \times S^2$-bundle over M. In the complex case, non intrinsic functions are allowed.

Proof. We first prove that the most general quaternionic valued monochromatic of class C^2 function of a pair $(\phi, f + i\rho) \in \mathbf{M}$ has the form (89). Indeed, let $F(f, \rho, \phi) \in \mathbf{M}$ be a continuously differentiable up to order two quaternionic function. By Theorem 1, it has the expression

$$F = \xi + \eta[\vec{i_1}\Gamma_1(\Phi) + \vec{i_2}\Gamma_2(\Phi) + \vec{i_3}\Gamma_3(\Phi)] \tag{100}$$

with $(\Phi, \xi + i\eta)) \in M, \Gamma_1^2 + \Gamma_2^2 + \Gamma_3^2 = 1, \Gamma_i$ real valued.

By assumption ξ, η and Φ are functions of f, ρ, ϕ . Since $\xi + i\eta \in \mathbf{M}$ and $\Phi \in \mathbf{M}$, it suffices therefore that we analyze the problem for these particular functions, and this reduces the problem to the case when F is a real or complex-valued function of the pair $(\phi, f + i\rho) \in \mathbf{M}$. Now, using the properties of this pair we get from $\nabla F = F_\phi \nabla \phi + F_f \nabla f + F_\rho \nabla \rho$, that

$$(\nabla F)^2 = (F_f^2 + F_\rho^2)(\nabla f)^2. \tag{101}$$

As f is independent of ϕ, then $(\nabla f)^2 \neq 0$ as remarked earlier. Therefore, necessarily

$$F_f^2 + F_\rho^2 = (F_f + iF_\rho)(F_f - iF_\rho) = 0, \tag{102}$$

---

[17]A function defined on an open set in the complex plane is called antianalytic (or antiholomorphic) if its derivative with respect to $\bar{z}$ exists at all points in that set, where $\bar{z}$ is the complex conjugate.

i.e. F must be analytic or anti-analytic function of $f + i\rho$. No additional restriction is placed on F as a function of ϕ. The condition $\nabla_g F = 0$ is automatically satisfied since

$$
\begin{aligned}
\triangle_g F &= F_\phi \triangle_g \phi + F_\rho \triangle_g \rho + F_f \triangle_g f + 2F_{\phi\rho} g(\nabla\phi, \nabla\rho) \\
&+ 2F_{\rho f} g(\nabla f, \nabla\rho) + 2F_{f\phi} g(\nabla f, \nabla\phi) + F_{\phi\phi}(\nabla\phi)^2 \\
&+ F_{\rho\rho}(\nabla\rho)^2 + F_{ff}(\nabla f)^2 \\
&= (F_{ff} + F_{\rho\rho})(\nabla f)^2 = 0,
\end{aligned}
\tag{103}
$$

as F is harmonic in (f, ρ).

Therefore $\xi(f, \rho, \phi)$ and $\eta(f, \rho, \phi)$ satisfy the stated conditions, and so do the Γ_is. Since Γ_i are real-valued they are therefore constant on $f + i\rho$, i.e. they depend on ϕ only.

The complex case is obtained by specializing $H \equiv P \equiv 0$, $K \equiv 1$, and by complexification, non-intrinsic functions are obtained.

It is easy to check that eq. (98) belongs to the real algebra generated by ϕ and $f + i\rho$ (or, on ϕ and $f - i\rho$ if it anti-analytic in $f + i\rho$). Similarly, eq. (99) belongs to the complex algebra generated by ϕ and $f - i\rho$, as before). The same applies trivially to functions of ϕ only.

We finally prove the maximality condition. In any of the two cases above let a monochromatic algebra contain $(\phi, f + i\rho) \in \mathbf{M}$ and a third function F. By Theorem 1 this function is given in terms of a pair $(\tilde{\phi}, \tilde{f} + i\tilde{\rho}) \in \mathbf{M}$. If the function is trivial, it is expressible as a function of $(\phi, f + i\rho)$ too. If not, it depends non-trivially on at least one of $\tilde{\phi}, \tilde{f} + i\tilde{\rho}$. In that case, since the functions belong to a monochromatic algebra, the corresponding $\nabla\tilde{\phi}$ and/or $\nabla(\tilde{f} + i\tilde{\rho})$ must be orthogonal to both $\nabla\phi$ and $\nabla(f + i\rho)$. (Notice that the latter commute with ∇F , in the scalar product).

If $g(\nabla\tilde{\phi}, \nabla\phi) = 0$ locally, then necessarily $\tilde{\phi} = \tilde{\phi}(\phi)$ as both are real monochromatic and so $\tilde{\phi}$ belongs to the algebra of ϕ.

If $g(\nabla(\tilde{f} + i\tilde{\rho}), \nabla(f + i\rho)) = 0$ and $g(\nabla(\tilde{f} + i\tilde{\rho}), \nabla\phi) = 0$, then we have

$$
\nabla(\tilde{f} + i\tilde{\rho}) = \alpha(x)\nabla\phi + \beta(x)\nabla(f + i\rho),
\tag{104}
$$

with α, β complex-valued functions defined on M by the Lemma below. This implies that $\tilde{f} + i\tilde{\rho} \in \mathbf{M}$ is (locally) a function of (f, ρ, ϕ) and so by Theorem 1, belongs to the algebra generated by $(\phi, f + i\rho)$.

Therefore, Theorem 2 is proved once we prove the following Lemma, valid only for g a Lorentzian metric (i.e. only the degenerate metric case).

419

Lemma. If two isotropic vectors [18] v_1, v_2 are orthogonal to a real isotropic vector v_3 in Minkowski space, then either v_1, v_2, v_3 or $v_1, \tilde{v}_2, v_3$ are linearly dependent. If v_2 is orthogonal to v_1 then the first case holds.

Proof. We may assume that the real vector is $(1, 1, 0, 0)$, the signature being $\mathrm{diag}(-1, 1, 1, 1)$. Since linear combinations of v_1, v_2 with v_3 preserve their stated properties we can make the first components of v_1, v_2 into zero by adding a convenient multiple of v_3. But then since they are orthogonal to the real vector also their second components are zero. So they are of the form $(0, 0, a, b), (0, 0, c, d)$ and by isotropy $a^2 + b^2 = c^2 + d^2 = 0$, i.e. $b = \pm ia, d = \pm ic$. Hence they are multiples of $(0, 0, \pm i, 1)$ and $(0, 0, 1, \pm i)$. For any choice of sign, these vectors are equal or one is equal to the complex conjugate of the other. They can be orthogonal only in the first case. The result follows.

Clearly the result holds pointwise for an arbitrary Lorentzian manifold (M, g) since we can always make $g_{\alpha\beta} = \mathrm{diag}(-1, 1, 1, 1)$ at a fixed point p.

Remarks.1. In view of Theorem 2, all functions of $f + i\rho$ in the same algebra must be simultaneously analytic or anti-analytic in the same connected regions. For simplicity we refer to them as analytic bearing in mind these two possibilities. 2. It is clear that $(\phi, f + i\rho) \in \mathbf{M}$ implies that $(\phi, f - i\rho) \in \mathbf{M}$. However since the analytic functions in $f - i\rho$ are precisely the antianalytic functions in $f + i\rho$ and viceversa, we will not consider these two pairs as distinct, because they generate the same maximal monochromatic algebras, according to Theorem 2.

5 General Form of Singular Sets, and Their Physical Interpretations

We are now in conditions for completing the objective of the previous Section, namely, the characterization of the node set of complex and quaternionic monochromatic functions. According to the above results the most general form for singular sets N of monochromatic complex or quaternionic functions is given by the conditions

$$
\begin{aligned}
\xi(f, \rho, \phi) &= 0, \\
\eta(f, \rho, \phi) &= 0, \quad (\phi, f + i\rho) \in \mathbf{M}
\end{aligned} \tag{105}
$$

Although N is locally at least two-dimensional we have now the possibility of locating a higher-order zero on a bicharacteristic line.

---

[18]Later we shall name them as *null vectors*, i.e. their length is 0.

For instance, the singular set of $\phi.(f + i\rho)$ is the union of the 2-dimensional set defined by $f = \rho = 0$, and the 3-dimensional set $\phi = 0$, and since $(\nabla\phi)^2 = 0$, their intersection $f = \rho = \phi = 0$ is a bicharacteristic line carrying an isolated zero of higher order. The corresponding phase function has a higher order singularity located at a single point in three-space, moving with the speed of light along the singular line $f = \rho = 0$, accompanied by the wave-front singularity $\phi = 0$.

Observations. This result is remarkable in many ways. Firstly, in the present analytic approach, it is apparent that the photon cannot exist per-se as a point-like singularity, since in fact the most general maximal monochromatic algebra is three-dimensional when we go to the quaternionic case and then we can describe it as built in the larger singularity. Thus, we have in a four-dimensional Lorentzian manifold, three real dimensions to describe the eikonal wave discontinuities. This precisely gives a representation for a recent conjecture by Kiehn that the photon is to be regarded as a three-dimensional singularity [23]. Furthermore, in Kiehn's point of view, the photon is associated with a spinor, which are the natural representations for propagating singularities; we shall present another approach to this below. Furthermore, spinors are related to conjugate minimal surfaces [19], which for the photon are described by Falaco solitons [23]. Now, regarding higher dimensional singularities, they always give rise to the de Rham's topological rules. For three dimensional singularities, we have the topological torsion first Poincaré invariant, while for four dimensional singularities, we have the spin-torsion second Poincaré invariant , the Euler number of the manifold. It is apparent from the present theorem, that for the construction of the three dimensional singularities, monochromatic algebras are enough for their generation. These represent thermodynamically irreversible processes which exchange energy but no matter with the environment (the photon), while the four-dimensional highest order case, represent the case in which there is exchange of matter [23]. We shall retake in detail these issues elsewhere.

5.1 On the Intelligence Code and Some Metrics in Cosmology

We recall or initial discusion on Fock's critique to GR [4] as a theory of uniform space, and the need of singularities to establish an 'objective' spacetime. We have discussed already that the primeval distinction that encodes the torsion field is such a singularity [13]. Thus, in analyzing the hyperbolic nature of the Einstein's partial differential equations of GR and the Maxwell equations, Fock was lead to propose as a starting point the eikonal and propagation wave equations,

---

[19]This is most remarkable since it points out to the existence of a Platonian world, with generic geometrical surfaces associated to the subjective-objective photon.

whose wavefronts correspond precisely to propagating singularities, that we have further associated with torsion. Furthermore, using the functions defining the maximal monochromatic algebra it is possible to establish a coordinate system for spacetime without recourse to an ad-hoc non-geometric energy-momentum tensor as Einstein's inception of it in GR (the "right hand side made of mud", in Einstein's words). This is done using the energy-momentum tensor of the electromagnetic field by solving the Einstein's equations of GR with light as source for the gravitational field described by the curvature derived from the metric in the Levi-Civita connection. This provides a self-referential construction of the metric which is absent in the conception of GR. Further below we shall relate this maximal monochromatic algebra to cognitive states which would thus generate a spacetime metric inverting our common undestanding of the 'exterior' world as being passively represented by the mind but disregarding the inverse direction of constitution of reality as *jointly* operational in a Kliein bottle sense. This understanding is further supported by the fact that the quaternion field **H** can be constructed in matrix logic [13]. Now, the natural metric in the Lie group of the invertible quaternions can be parametrized as the closed Friedmann-Lemaitre-Robertson-Walker metrics [75] which constitute one of the most important classes of solutions of Einsteins equations and furthermore, as the Carmeli metric of rotational relativity. We recall that the latter was introduced to explain spiral galaxies rotation curves and 'dark matter' [76]. We stress that these derivations do *not* require solving the Einstein's equations of GR but are intrinsic to **H**. So the Intelligence Code has some remarkable built-in metrics that are purported to describe cosmological phenomenae.

5.1.1   Photon, Nodal Lines, Monopoles

Until know we have described the singular sets of quaternionic and complex solutions of the eikonal and wave propagation equations. A typical case is to establish a Cartesian coordinate system $(x, y, z, t) \in \mathbf{R}^{1,3}$ (in Minkowski space) given by taking

$$f + i\rho = y + iz, \tag{106}$$

and

$$\phi = f(r) - t, \text{with} \quad r = (x^2 + y^2 + z^2)^{\frac{1}{2}}, \tag{107}$$

and f is a monotonic function of the radius r. In this case, for the function

$$(y + iz)(f(r) - t) \tag{108}$$

the singular set consists of a spherical wave front in 3-space moving with the speed of light and cutting the singular x-axis $y = z = 0$ at a single point in the positive semi-axis $0 \leq x$, where therefore lies a higher-order singularity. This higher-order singular point, piloting a lower order singular spherical wave along a lower order singular line, is now liable to represent the photon. Here the photon is conceived as a moving point singularity carrying energy, in agreement with the experimentally observed corpuscular behaviour of the photon at a metallic plate, and obtain pictures of its trajectories in cloud chambers. On the other hand the weaker singularity carried by the spherical wave front $f(r) - t = 0$ is responsible for the diffraction patterns in the typical slit experiments, according to Huygens's law of propagation of singularities (eikonal equation), and so accounts for the experimentally observed wave nature of the photon. In this way the purely analytical characterization of the maximal monochromatic algebras leads us unequivocally, to the correct conclusions as regards the physical nature of the photon and express its *purportedly* dual wave-corpuscular nature as a simple mathematical fact. It is essentially a wave, the particle being a factor of it, but not dual in any intrinsic sense. Remarkably, this stands in contrast with the de Broglie-Vigier double solution theory [81], in which the wavelike pattern is associated to a linear propagation (alike eq. (8)) while the particle was treated as a propagating singularity ascribed to a non-linear equation, which in the present theory is eq. (9). Thus the present theory fleshes out in a completely geometric setting, the double solution theory, which appears in the torsion geometry of the linear and nonlinear Dirac-Hestenes equation [10], (2005), yet it relinquishes duality.

The line $y = z = 0$ in 3-space carries a singularity too, but this is a standing one, independent on time, and therefore, its presence is detected through different effects. Actually this line is so-called a nodal line of the wave function $\phi \equiv y + iz$ ([24]) or a dislocation line of the planes of constant phase of ψ. Around this line occur vortices of the flux of the trace torsion one-form $d\ln\phi$ of the phase function (when the circulation of this flux around a nodal line is non-zero), described in detail in [32]. Alternatively Dirac found these nodal lines when considering singularities of wave functions, upon imposing the only requirement that the complex-valued functions ψ (in our example equal to $y + iz$) be single-valued and smooth, but not necessarily with single-valued argument, and then quantized them in terms of the winding number of the vector-field $(\nabla \ln Re(\psi), \nabla \ln Im(\psi))$ along a closed curve around the line. He then found that one could remove the non-zero circulation by means of a gauge transformation of the second kind, and that the electromagnetic vector potential associated with this transformation was precisely the same electromagnetic potential produced by a magnetic monopole at the initial point. He then equated the effect of the circulation around the nodal line in the original gauge to the effect of a monopole in the new gauge. His quantization by the winding number is actually just a *special case of the general quantization theorem* above, and his gauge interpretation is thus a concrete exemplification of the meaning of the analysis given there.

The variety of types of singular sets defined by the representation given in eqs. (105) is very great, as exemplified in the pioneering work [24]. Besides the singular sets that we previously identified with the photon (spherical wave front plus a nodal line) there is also a remarkable singular set of the monochromatic wave constructed out of $\phi = f(r) - t$ and $f + i\rho = y + iz \in \mathbf{R}^{1,3}$, by the following sum

$$\epsilon e^{i\omega[f(r)-t]} - (y + iz), \epsilon > 0. \tag{109}$$

Its singular set is given by

$$y = \epsilon \cos\omega[f(r) - t], z = \epsilon \operatorname{sen}\omega[f(r) - t]. \tag{110}$$

This represents an helicoidal line lying on the cylinder $y^2 + z^2 = \epsilon^2$, and moving with (variable) speed of light along its tangent direction at each of its points. (For simplicity, we can assume that $f(r) = t$ in order to get a better visualization: the speed is then constant and the helicoid has then a constant step.) Taking $y - iz$ instead, we get a screw motion with opposite handedness. The singular set is thus a moving screw in 3-space that can be right or left handed, and may carry the energy associated with a quantum jump, as shown above. It seems therefore that a monochromatic wave line like this can represent appropiately a right or left handed neutrino, concretely identified with its singular set. It has then quite distinct properties from those associated with a photon. For it is given by an infinitely long moving right of left handed helicoidal line in 3-space (which by the way, it is a minimal surface; more on dislocations and minimal surfaces and turbulence, shall be presented elsewhere) while the photon is given by a point piloting a spherical wave. In particular if the singular screw line of above is associated with an elementary state u and carries energy E in the manner described in [61] it also carries the angular momentum $\epsilon^2 E\omega/c^2$ directed along a x-axis, in the given referential. Hence the neutrino carries angular momentum while the photon does not. On the other hand, according to this description, the neutrino should not have (primary) diffraction patterns as the photon does, which should explain why it is so difficult of detect. The infinitely long screw line seems to agree, in principle, with the experimentally estimated fact that the neutrino has an extremely long absorption path.

5.1.2 Distinction of Maximal Monochromatic Algebras

Lemma 2. The maximal monochromatic algebras $\mathbf{M_1}$ and $\mathbf{M_2}$ with generators $(f, \phi + i\psi)$ and $(\tilde{f}, \tilde{\phi} + i\tilde{\psi})$ respectively, are distinct if and only if

$$g(\nabla f, \nabla \tilde{f}) = 0. \tag{111}$$

Proof. If $g(\nabla f, \nabla \tilde{f}) = 0$, then necessarily f is a function of $\tilde{f}$ and therefore

$$g(\nabla f, \nabla \tilde{\phi} + i \nabla \tilde{\psi}) = 0, \tag{112}$$

besides

$$g(\nabla f, \nabla \phi + i \nabla \psi) = 0. \tag{113}$$

But then from Lemma 1 follows that either ∇f, $\nabla \phi + i \nabla \psi$, $\nabla \tilde{\phi} + i \nabla \tilde{\psi}$ or ∇f, $\nabla \phi + i \nabla \psi$ and $\nabla \tilde{\phi} - i \nabla \tilde{\psi}$ are linearly dependent. In any case, $\tilde{\phi} + i\tilde{\psi}$ is a function of $(f, \phi + i\psi)$ and so is $\tilde{f}$, the two algebras are the same (here again we are considering the possibility of having to consider analytic functions in one and anti-analytic in the other).

Conversely, assume the algebras are not distinct. Then by Theorem 2, $\tilde{f}$ is a function of $(f, \phi + i\psi)$, analytic or anti-analytic in $\phi + i\psi$, and then $g(\nabla f, \nabla \tilde{f}) = 0$. c.q.d.

5.2 Spinor and Twistor Description of Maximal Chromatic Algebras

If (x^0, x^2, x^2, x^3) is a vector in Minkowski space $\mathbf{R}^{1,3}$ we may associáte with it the 2×2 hermitean matrix

$$x = (x^0, x^1, x^2, x^3)X = \frac{1}{\sqrt{2}} \begin{pmatrix} x^0 + x^1 & x^2 + ix^3 \\ x^2 - ix^3 & x^0 - x^1 \end{pmatrix} \tag{114}$$

This is obviously a linear isomorphism of $\mathbf{R}^{1,3}$ with the real space of 2 times 2 hermitean matrices. Direct computation shows then that when the vector is acted upon by a proper Lorentz transformation L, the associated hermitean matrix undergoes multiplication by a 2 times 2 complex unimodular matrix on the left and by its transpose conjugate on the right. The unimodular matrix is uniquely determined by L, except for sign of course. This correspondence gives an isomorphism between the group $Sl(2, \mathbf{C})$ of 2 by 2 unimodular complex matrices and the two-fold universal covering group of the connected subgroup of the Lorentz group $O(1,3)$. A spin vector on (M, g) is defined locally by taking a smooth moving orthonormal frame, i.e., such that on each point the metric has the standard form $(dx^0)^2 - (dx^1)^2 - (dx^2)^2 - (dx^3)^2$, and assigning, in a smooth way, a spin vector at each point of the corresponding tangent manifold. Since in the above correspondence we have

$$(x^0)^2 - (x^1)^2 - (x^2)^2 - (x^3)^2 = \frac{1}{2}\det(X), \tag{115}$$

then any (real) isotropic vector (i.e. a non-zero vector whose Minkowski length is 0, also called a null vector, or still, a Cartan spinor) corresponds to a hermitean matrix given by the tensor product of a spin vector ω^A with its complex conjugate $\bar{\omega}^{A'}$ (the vector is then called future-pointing) or with $-\bar{\omega}^{A'}$ (the vector is called past pointing), $(A, A' = 1, 2)$.

The spin vector ω^A is determined by the vector, up to a factor $e^{i\theta}, \theta$ real, obviously. This extra degree of freedom relates to a possible polarization of the objects involved. Clearly complex isotropic vectors are given by the tensor product $\omega^A \pi^{A'}$ of spin vectors. Finally we remark that two isotropic vectors are orthogonal if and only if either the first or their second associated spin vectors are parallel. Indeed, let in matrix form

$$u = \begin{pmatrix} a \\ b \end{pmatrix} \begin{pmatrix} c & d \end{pmatrix}, v = \begin{pmatrix} e \\ f \end{pmatrix} \begin{pmatrix} g & h \end{pmatrix} \tag{116}$$

Using polarization of bilinear forms and eq. (115) we get

$$\begin{aligned} u^i v_i &= \frac{1}{4}[(u_i + iv_i)(u^i + v^i) - (u^i - v^i)(u_i - v_i)] \\ &= \frac{1}{4}\det[\begin{pmatrix} a \\ b \end{pmatrix} \begin{pmatrix} c & d \end{pmatrix} + \begin{pmatrix} e \\ f \end{pmatrix} \begin{pmatrix} g & h \end{pmatrix}] \\ &\quad - \frac{1}{4}\det[\begin{pmatrix} a \\ b \end{pmatrix} \begin{pmatrix} c & d \end{pmatrix} - \begin{pmatrix} e \\ f \end{pmatrix} \begin{pmatrix} g & h \end{pmatrix}] \\ &= \frac{1}{2}(af - be)(ch - dg). \end{aligned} \tag{117}$$

Hence, $u^i v_i = 0$ if and only if $af = be$ or $ch = dg$, as claimed.

5.2.1 Twistors and Maximal Monochromatic Algebras

Consider the generators $(\phi, f + i\rho) \in \mathbf{M}$ of a maximal monochromatic algebras. The vectorfields $\nabla\phi$ and $\nabla(f + i\rho)$ are, respectively, real and complex isotropic fields, mutually orthogonal on (M, g); here M is a generic spin-manifold provided with a Lorentzian metric g. By the previous analysis $\nabla \phi$ is given by a spin vectorfield ω^A in the spinor form

$$\nabla \phi = \omega^A \bar{\omega}^{A'} \quad (\text{or} - \omega^A \bar{\omega}^{A'}). \tag{118}$$

and since $\nabla(f + i\rho)$ is isotropic and orthogonal to $\nabla \phi$ then we have

$$\nabla(f + i\rho) = \omega^A \bar{\pi}^{A'}, \tag{119}$$

where π^A is another spin vectorfield. Consequently the pair $(\nabla \phi, \nabla(f + i\rho))$ of vectorfields is completely determined by the ordered pair of spin vectorfields

$$(\omega^A, \pi^{A'}), \tag{120}$$

but we have a fourfold map here since we have already altogether four different ways of building the vectorfields according to eq. $(118, 119)$ out of the ordered pair (120). The correspondence (118), extended to complex vectors x shows that the second choice in (147) reverses $\nabla \phi$ from, say , a future-pointing isotropic vector to a past-pointing isotropic vector while in (118) it chooses the complex-conjugate $\nabla(f - i\rho)$ instead of $\nabla(f + i\rho)$, reversing the roles of analytic and anti-analytic functions, which means inversion of handedness. Choosing locally a given time orientation and a given handedness, corresponds to a particular choice of the assignements in eqs. $(118, 119)$. The ordered pair (120) of spin vectorfields at a point in (M, g) is called a local twistor and the corresponding field a local twistor field. From the twistor field we determine the real and complex vectorfields by eqs. $(118, 119, 114)$, which, upon integration yield an equivalent pair $(\phi, f + i\rho)$. This means that we can characterize completely a maximal monochromatic algebra (and consequently the light quanta it represents) in terms of a twistor field with divergence free associated vectorfields.[20]. The new representation is even richer as it has built in an extra degree of freedom, namely, polarization, due to the

---

[20]This is remarkable in relation to the fact that the Intelligence Code which we shall present below is a *nilpotent universal rewrite system* in the sense of [15] albeit generated by the Klein bottle, yet due to the representation by Musès hypernumbers of logical operators, non-trivial square roots of $+1$ in addition of those of -1 appear; the latter is the case in [15]. We recall that photons appear as propagating singularities for both the eikonal and the Maxwell equations [4]. The latter equations, under certain conditions -that amount to reduce 4 to 2 degrees of freedom (in the Dirac algebra)-, similarly as in the (Gupta-Bleuler) quantization of the electromagnetic field, are equivalent to the Dirac-Hestenes linear and non-linear equation of relativistic quantum mechanics [10] (2005). In spite that we enlarged the original complex fields to quaternion-valued ones to derive the characterization of the extended photons by a maximal monochromatic algebra and its spinor and twistor descriptions, we showed above that it only requires the canonical

factor $e^{i\phi}$ mentioned before. This result, showing that we have identified as light quanta are indeed given by twistor fields, substantiates the belief of Penrose that twistors are the appropiate tool to describe zero rest-mass particles and to effect the connection between gravitation and quantum mechanics, and particularly, Kiehn's conjecture [23]. Further below we shall see that this connection extends to the laws of thought.

5.3   Classical Interpretation, Helicity and Spin

We follow the original definition of Penrose of twistors in Minkowski space, which starts with the fact that if a zero rest-mass particle has linear momentum p^a $(a = 0, 1, 2, 3)$ and angular momentum M^{ab} with $M^{ab} = -M^{ba}$ $(a, b = 0, 1, 2, 3,)$ with respect to some point taken as an origin, then we can write them in spinor form as

$$p_{AA} = \bar{\pi}_A \pi_{A'}, M^{AA'BB'} = i\omega^{(A}\bar{\pi}^{B)}\epsilon^{A'B'} - i\epsilon^{AB}\bar{\omega}^{(A'}\bar{\pi}^{B')}, \tag{121}$$

where brackets stand for symmetrization, $Z^\alpha = (\omega^A, \pi_{A'})$ is a twistor, and

$$\epsilon^{AB} = \begin{pmatrix} 0 & 1 \\ -1 & 0 \end{pmatrix} \tag{122}$$

is the spinor index raising operator; further below in eq. (144) we shall re-encounter is as a TIME operator in logic. We recall that the twistor is only defined up to a real phase change $e^{i\theta}$, with $\theta \in \mathbf{R}$. The vector p_a is an eigenvector of M^{ab}, i.e.

$$\frac{1}{2}e_{abcd}p^b M^{cd} = sp_a, \tag{123}$$

where $e_{0123} = 1$ and e_{abcd} is antisymmetric in all indices. The eigenvalue s is the helicity of the particle and $|s|$ its spin. It is also given by

$$2s = Z^\alpha \bar{Z}_\alpha = \omega^A \bar{\pi}_A + \pi_{A'}\bar{\omega}^{A'}, \tag{124}$$

---

commutative square root of -1 of complex numbers (whose matrix representation will appear to be the TIME operator in matrix logic). This is also the case of the *infinite*-dimensional representations of spinors and the Dirac equation for them [62]. This explicit independence of the *non*-commutative square roots of -1 will carry out to the codification of the mind apeiron in terms of twistors, and thus at the primeval zero observable of thought (which we called the mind apeiron) they play no role. As well known, the non-commutative square roots of -1 naturally only show up for the *finite* dimensional Dirac algebra which is the one used in [15].

where the second term stands for the scalar product of the twistor Z^α with its complex conjugate $\bar{Z}_\alpha = (\bar{\pi}_A, \bar{\omega}^{A'})$. Let the twistor $Z^\alpha \neq 0$ be null, i.e. $Z^\alpha \bar{Z}_\alpha = 0$, then there is a single line Z of points with respect to which $M^{ab} = 0$, and it is parallel to p^a, and therefore isotropic. If $X^\alpha \neq 0$ is another null twistor, with isotropic line X, then X and Z meet if and only if $X^\alpha \bar{Z}_\alpha = 0$ [22]. The isotropic line Z describing the twistor Z^α up to a factor, is thus completely characterized by the congruence given by the isotropic line that meet Z, i.e. by the set

$$\{X | X^\alpha \bar{X}_\alpha = 0, X^\alpha \bar{Z}_\alpha = 0\} \tag{125}$$

If $Z^\alpha \bar{Z}_\alpha \neq 0$ we can again describe Z^α by the congruence of isotropic lines that satisfy the previous equation but now there is no isotropic line assoicated with Z^α and the lines associated with X^α twist about one another (right-handedly when $s > 0$, left-handedly when $s < 0$) and never intersect.

6 Self-reference, the Klein Bottle, Torsion, the Laws of Thought and the Twistor Structure of the Cognitive Plenum

6.1 Introduction

Up to know we have elaborated a theory of torsion and photons, which we departed presenting it as a theory of an 'objective' realm that has its standing in the Cartesian cut mindset. This mindest corresponds to a world in which subjectivity does not participate, or altogether does not exist in the universe of discourse. Yet, we could not keep this cut, having shown that both torsion and the photon are very closely related to self-reference, and thus to consciousness [13]. Furthermore, the semiotic codification of torsion as a distinction sign produces in incorporating paradox, a multivalued logic which is associated with the Klein bottle and time waves [13]. From this logic, it was proved that the most general matrix-tensor logic that has as particular cases quantum, fuzzy, modal and Boolean logics [26] stem from these time waves. In this theory which stems from abandoning the scalar logic theory of Aristotle and Boole, promoting it to logical operators, we find that the Klein bottle plays a fundamental role as an in-formation operator, which coincides with the Hadamard gate of quantum computation. The role of this gate is to transform the vector Boolean states to superposed states, the latter being associated with the torsion of cognitive space and the non-orientability of this space due to its constitution in terms of the self-referential non-orientable Moebius and Klein bottle surfaces. Furthermore, the logical cognitive operator which leads to quantization of cognition, is generated by the torsion produced from the commutator of the FALSE and TRUE logical operators which self-referentially involutes to give the difference between these two operators, as we shall see below. The picture that stems is that matrix logic can be seen as the self-referential logical code which stands at the foundation of quantum physics to which is indisolubly related. We have elaborated the relations between matrix logic, self-reference, non-orientability and the Klein bottle, nilpotent hypernumber representations of quantum fields that represent some logical operators [13]. Thus, in this theory, matter quantum field theories are logical operators, and viceversa, and a transformation between quantum and cognitive logical observables has been established. This theory has produced a new fundamental approach to the so-called mind-matter problem , establishing its non-separateness, and the primacy of consciousness which thus *cannot* be claimed to be an epiphenomenon of physical or other complex fields [13]. By promoting the 'truth tables' of usual Boolean logic to matrix representations, the founder of matrix logic, A. Stern, was able to produce an operator logic theory in which logical operators may admit inverses, and the operations of commutation and anticommutation are natural [26]. Furthermore, logical operators can interact by multiplication or addition and, in some cases,being invertible, they yield thus to a more complex representation of the laws of thought that the one provided by the usual Boolean theory of logical connectives. This representation is the *Intelligence Code*. In this conception the meaning of *intelligence* is essentially related to self-reference, i.e. related to recognition. [21] The Intelligence Code is related to quantum mechanics for two-state sytems and to quantum fields. Matrix logic is naturally quantized, since its eigenvalues take discrete values which are $\pm 1, 0, 2, \pm \Phi$, with Φ the Golden number [26]. In this setting, the null quantum-cognitive observable is the 2×2 matrix, $\mathbf{0}$, with identical entries given by 0, the *mind apeiron*. The relation with quantum field operators and this observable which represents the apeiron observable, is their role in polarizing this cognitive-quantum apeiron

---

[21]There is no cognition unvinculated to a subject, in contrast with the basic tenants of the Theory of Information. In other words, cognition is embodied, instead of received. The latter conception is, of course, another example of the Cartesian Cut, which proposes a receiver, a desingularized unstructured subject, instead of a cocreator of meaning. In that alienated conception, there is no actual physics, and the bottomline is the replacement of measurements and recognition by registrations. These are carried out by a subject turned into an object, operating as a physics independent machine. This is also the (mis)conception operating in the current standard dogma of genetics, which stands in sharp contrast with wave genetics [37].

through non-null square roots which can be represented by plenumpotents, i.e. Musès hypernumbers whose square is **0**. In distinction with the other cognitive-quantum observables, is that the eigenstates of **0** are no longer quantized, but rather give an orthogonal complex two-dimensional nullvector space. In this way the plenum is no longer represented by a single point, **0**, but rather becomes an extended object or zero-brane. This will allow to map the twistor representations of the extended photon presented in (120) with its representation in a cognitive state and viceversa!

6.2 Torsion of Cognitive Space, Schroedinger Entanglement, Non-Orientable Manifolds, the Klein Bottle, Quantum Field Theory, Logic and Hypernumbers

We consider a space of all possible cognitive states (which in this context replace the Boolean logical variables) represented in this plenum as the set of all Dirac bras $< q| = (\bar{q} \quad q)$, and kets $|q >$, with $\bar{q} + q = 1$, $q \in \mathbf{R}$ [22], is a continuous cognitive logical value not restricted to the false and true scalar values, represented by the numbers 0 and 1 respectively. Still, the standard logical connectives admit a 2×2 matrix representation of the their 'truth tables' and now we have that for such an operator, L, we have the action of L on a ket $|q >= \begin{pmatrix} \bar{q} \\ q \end{pmatrix}$ is denoted by $L|q >$ alike the formalism in quantum mechanics, and still we have a scalar truth value given by $< p|L|q >$, where $< p|$ denotes another logical vector. We can further extend the usual logical calculus by considering the Truth and False operators, defined by the eigenvalue equations $\text{TRUE}|q >= |1 >$

---

[22]Notice that a difference with the definition of qubits in quantum computation, is that for them we have the normalization condition for complex numbers of quantum mechanics. In this case, the values of q are arbitrary real numbers, which leads to the concept of non-convex probabilities. While this may sound absurd in the usual frequentist interpretation, when observing probabilities in non-orientable surfaces, say, Moebius surfaces, it turns out to be very natural. If we start by associating to both sides of an orientable surface -from which we construct the Moebius surface by the usual procedure of twisting and gluing with both sides identified-the notion of say Schroedinger's cat being dead or alive in each side, then for each surface the probability of being in either state equals to 1 and on passing to the non-orientable case, the sum of these probabilities is 2. While this is meaningless in an orientable topology, in the non-orientable case which actually exist in the macroscopic world, this value is a consequence of the topology. In this case, the superposed state ' being alive and being dead' or 'true plus false' which is excluded in Aristotelian dualism by the principle of non-contradiction, is here the case very naturally supported by the fact that we have a non-trivial topology and non-orientability. As for the case of negative probabilities, we see in the previous example that -1 is the probability value complement of the value 2.

and FALSE$|q \ge \ = |0 >$, where $|1 >= \begin{pmatrix} 0 \\ 1 \end{pmatrix}$ and $|0 >= \begin{pmatrix} 1 \\ 0 \end{pmatrix}$ are the true and false vectors. It is easy to verify that the eigenvalues of these operators are the scalar truth values of Boolean logic. We can represent these operators by the matrices

$$\text{TRUE} = \begin{pmatrix} 0 & 0 \\ 1 & 1 \end{pmatrix}, \text{FALSE} \simeq \begin{pmatrix} 1 & 1 \\ 0 & 0 \end{pmatrix} \tag{126}$$

We note that the spaces of bras and kets do not satisfy the additivity property of vector spaces -while keeping the property that one is the dual of the other- due to the fact that normalization is not preserved under addition. A superposition principle is necessary. If $|p <$ and $|r >$ are two normalized states, then the superposition defined as follows

$$|q >= c|p > +\bar{c}|r >, \quad \text{where} \quad \bar{c} + c = 1, \tag{127}$$

also defines a normalized logical state. We can interpret these coefficients as components of a logical state $|c >$ or still a probability vector, termed *denktor*, a German-English hybrid for a *thinking vector*. The normalization condition is found as follows: Multiply the states $|p >$ and $|r >$ by $\bar{c}$ and c, respectively. By definition, the normalization condition on the sum $|q >$ with coefficients $\bar{c}, c$ leads to

$$\begin{pmatrix} \bar{q} \\ q \end{pmatrix} = c \begin{pmatrix} \bar{p} \\ p \end{pmatrix} + \bar{c} \begin{pmatrix} \bar{r} \\ r \end{pmatrix} = \begin{pmatrix} c\bar{p} + \bar{c}\bar{r} \\ cp + \bar{c}r \end{pmatrix}, \tag{128}$$

yet, since $\bar{q}+q = c\bar{p}+\bar{c}\bar{r}+cp+\bar{c}r = c(\bar{p}+p)+\bar{c}(\bar{r}+r) = c.1+\bar{c}.1$ and thus $c+\bar{c} = 1$ since $|q >$ is a normalized state by assumption. So through this superposition principle is that we can give a vector space structure to normalized cognitive states. We now can identify under these prescriptions, the tangent space to the space of bras (alternatively, kets) with the space itself. [23]

Returning to the vector space structure provided by the superposition principle, and thus the identification of its tangent space with the vector space itself, it follows that a vector field as a section of the tangent space can be seen as a transforming a bra (ket) vector into a bra (ket) vector through a 2×2 matrix, so we can identify the tangent space which with the space of logical operators. We have as usual the commutator of any such matrices $[A, B] = AB - BA$ and

---

[23]Here it is simple to see that if $|q >, |q' >$ are two superpositions, then for any operator L, $L|(q + q') >= L|q > +L|q' >$.

432

the anticommutator $\{A, B\} = AB + BA$. In particular we take the case of $A = $ FALSE, $B = $ TRUE and we compute to obtain

$$[\text{FALSE}, \text{TRUE}] = \text{FALSE} - \text{TRUE}, \tag{129}$$

$$\{\text{FALSE}, \text{TRUE}\} = \text{FALSE} + \text{TRUE}. \tag{130}$$

Thus in the subspace spanned by TRUE and FALSE we find that the commutator that here coincides with the Lie-bracket of vectorfields defines a torsion vector given by the vector $(1 \; -1)$, and that this subspace is integrable in the sense of Frobenius: Indeed, $[\text{FALSE}, \text{TRUE}], \text{TRUE}] = [\text{FALSE}, \text{TRUE}]$ and $[[\text{FALSE}, \text{TRUE}], \text{FALSE}] = [\text{TRUE}, \text{FALSE}]$. Furthermore, on account that $\text{TRUE}^2 = \text{TRUE}$ and $\text{FALSE}^2 = \text{FALSE}$, i.e. both operators are idempotent, then the anticommutators also leaves this subspace invariant.

The remarkable aspect here is that the quantum distinction produced by the commutator, exactly coincides with the classical distinction produced by the difference (eq. (25)), while the same is valid for the anticommutator with a classical distinction which is represented by addition (eq, (26)). We notice that in distinction of quantum observables, these logical operators are not hermitean and furthermore they are noninvertible. Furthermore, we shall see below how this torsion is linked with the creation of cognitive superposed states, very much like the coherent superposed states that appear in quantum mechanics. Now, if we denote by M the commutator $[\text{FALSE}, \text{TRUE}]$ so that from eqs. $(22, 25)$ we get

$$M = \begin{pmatrix} 1 & 1 \\ -1 & -1 \end{pmatrix}, \tag{131}$$

we note that it is nilpotent, (in fact a nilpotent hypernumber, since $M = \epsilon_2 + i_1 = \sigma_z + i_1$)

$$M^2 = \begin{pmatrix} 0 & 0 \\ 0 & 0 \end{pmatrix} \equiv \mathbf{0}, \tag{132}$$

thus yielding the identically zero matrix. $\mathbf{0}$ represents the universe of all possible cognitive states created by a non-null divisor of zero; we have already called it the mind apeiron. M creates a polarization of the mind apeiron through the fact that the torsion is a superposed state which cannot be fit into the scheme of Boolean logic but can be obtained independently by the loss of orientability of a surface which thus allows for paradox. Since M coincides with the classical difference between FALSE and TRUE, which are not hermitean, then we can think of this non-invertible operator as a cognitive operator related to the variation of truth value of the cognitive state, as we shall prove further below that $M = -\frac{d}{dq}$.

We would like to note that this polarization of the plenum $\mathbf{0}$ is not unique, there are many divisors of $\mathbf{0}$, the mind apeiron, for instance the operator

$$\text{ON} = \begin{pmatrix} 0 & 0 \\ 1 & 0 \end{pmatrix} := a^\dagger, \tag{133}$$

and

$$\text{OFF} = \begin{pmatrix} 0 & 1 \\ 0 & 0 \end{pmatrix} := a \tag{134}$$

satisfy

$$a^2 = \mathbf{0}, (a^\dagger)^2 = \mathbf{0}, \tag{135}$$

and furthermore, $\{a, a^\dagger\} = I$, so they can be considered to be matrix representations of creation and annihilation operators, $a^\dagger$ and a as in quantum field theory. In fact, if we consider the wave operators given by the exponentials of $a, a^\dagger$ we have

$$e^a = I + a = \begin{pmatrix} 1 & 1 \\ 0 & 1 \end{pmatrix} = \text{IMPLY}, e^{a^\dagger} = I + a^\dagger = \begin{pmatrix} 1 & 0 \\ 1 & 1 \end{pmatrix} = \text{IF}, \tag{136}$$

where IMPLY $=\rightarrow$ is the implication, and IF $=\leftarrow$ is the converse implication: $x \leftarrow y = \bar{x} \rightarrow \bar{y}$. Thus the implication and the converse implication logical operators are both wave-like logical operators given by the exponentials of divisors of $\mathbf{0}$, and in fact they are derived from quantum field operators of creation and annhilation in second-quantization theory, $a^\dagger$ and a, respectively, which in fact can be represented by nilpotent hypernumbers. Indeed, $a = \frac{1}{2}(\epsilon_3 - i_1) = \frac{1}{2}(\sigma_x - i_1)$ and $a^\dagger = \frac{1}{2}(\epsilon_3 + i_1) = \frac{1}{2}(\sigma_x + i_1)$; see [13].

6.3 The Quantization of Matrix Logic

Now we wish to prove that the interpretation of M as the logical momentum operator is natural since $M = -\frac{d}{dq}$. Indeed,

$$-\frac{d}{dq}|q> = -\frac{d}{dq}\begin{pmatrix} 1 - q \\ q \end{pmatrix} = \begin{pmatrix} 1 \\ -1 \end{pmatrix} = \begin{pmatrix} 1 & 1 \\ -1 & -1 \end{pmatrix}\begin{pmatrix} \bar{q} \\ q \end{pmatrix} = M|q> \tag{137}$$

so that for any normalized cognitive state $|q>$ we have the identity

$$M = -\frac{d}{dq}, \tag{138}$$

434

which allows to interpret the cognitive operator as a kind of logical momentum. Thus, in this setting which is more general but less primitive than the calculus of distinctions from which it can be derived [13], it is the non-duality of TRUE and FALSE what produces cognition as variation of the continuous cognitive state; cf. footnote no. 1. We certainly are here with a situation that is far from the one contemplated by Aristotle with his conception of a trivial duality of (scalar) true and false, and which lead the elimination (and consequent trivialization) of time and of subjectivity, as argued in [13].

Now consider a surface given by a closed oriented band projecting on the xy plane. Thus to each side of the surface we can associate its normal unit vectors, (1 0) and (0 1). Suppose that we now cut this surface and introduce a twist on the band and we glue it to get thus a Moebius surface. Now the surface has lost its orientability and we can identify one side with the other so that we can generate the superpositions

$$< 0|+ < 1| =< (1 \ \ 1)| =< S_+|, \quad < 0|- < 1| =< (1 \ \ -1)| =< S_-|. \tag{139}$$

which we note that the latter corresponds to the torsion produced by the commutator of TRUE and FALSE operators. Theses states are related by a change of phase by rotation of 90 degrees. What the twisting and loss of orientability produced, can be equivalently produced by the fact that TRUE and FALSE are no longer dual as in Boolean logic and the Aristotelian frame. What is relevant is their difference (and we return to the Introduction's motto of a difference that produces differences), which in the case of scalar truth values does not exist. The other state also can be interpreted as a state that represents the fact that the states as represented by vectors, have components standing for truth and falsity values which are independent, so that the Aristotelian link that makes one the trivial reflexive value of the other one is no longer present: they each have a value of their own. In that case then (0 0) is another state, 'false and true' (which is the case of the Liar paradox as well as Schroedinger's cat), which together with (1 1) , 'nor false nor true' state together with (0 1), true, and (1 0) false states we have a 4-state logic in which the logical connectives have been promoted to operators.

Now consider for an arbitrary normalized cognitive state q the expression

$$
\begin{aligned}
[q, M]|q> &= [q, -\frac{d}{dq}]|q> = -q\frac{d}{dq}|q> +\frac{d}{dq}q|q> = -q\frac{d}{dq}\begin{pmatrix} 1-q \\ q \end{pmatrix} \\
&+ \frac{d}{dq}\begin{pmatrix} q-q^2 \\ q^2 \end{pmatrix} = \begin{pmatrix} q \\ -q \end{pmatrix} + \begin{pmatrix} 1-2q \\ 2q \end{pmatrix} = |q>, \tag{140}
\end{aligned}
$$

for any normalized cognitive state q so that we have the quantization rule

$$[q, M] = I, \tag{141}$$

where $I = \begin{pmatrix} 1 & 0 \\ 0 & 1 \end{pmatrix}$, the identity operator. Instead of the commutation relations of quantum mechanics $[q, p] = i\hbar$ for $p = -i\frac{\partial}{\partial q}$ and those of diffusion processes associated to the Schroedinger equation, $[q, p] = \sigma$ where $p = \sigma\frac{\partial}{\partial x}$ with σ the diffusion tensor given by the square-root of the metric g on the manifold with coordinates x on which the diffusion takes place so that $\sigma \times \sigma^\dagger = g$ [7], we have that the commutation of a normalized cognitive state with the cognitive (momentum) operator is always the identity yielding thus a fixed point. Indeed, consider the function $F_M(q) = [q, M]$, then $F_M(F_M(F_M(FM(\ldots))))(|q>) = |q>$, for any normalized cognitive state $|q>$. Thus, $F_M(q)$ defines what is called in system's theory an eigenform, albeit one which does not require infinite recursion but achieves a fix point already in the first step of the process, by the formation of the commutator $[q, M]$ [67]. This is the structure of the Self, which whatever operation may suffer by the action of logical operators, it retains its invariance by the quantization of logic as expressed above by eq. (141).

Now we want to return to the superposed states, S_+ and S_-, the latter being the torsion produced by the commutator of the TRUE and FALSE operators, to see how they actually construct the cognitive operator. First a slight detour to introduce the usual tensor products of two cognitive states, $|p><q|$ which as the tensor product of a vector space and its dual is isomorphic to the space of linear transformations between them, we can think as an operator L acting by left multiplication on kets and by right multiplication on bras. So that if $L = |p><q|$ then $<y|L|x> = <y, p><q|x>$, for any $<y| = \bar{y}<0| + y<1|$ and $|x> = \bar{x}|0> + x|1>$, where $<x|y> = \delta_{xy}$ equal to 1 for $x = y$, and equal to 0 for $x \neq y$ and $\sum_i |x_i><x_i| = I$. Then,

$$M = |S_+><S_-|, \tag{142}$$

which shows that the cognitive operator that arises from the quantum-classical difference between the TRUE and FALSE operators can be expressed in terms of the tensor products of the superposition states, being the sum of the true and false states and the torsion produced in the quantum commutator of the TRUE and FALSE operators.

Starting with the logical momentum M, that satisfies $[q, M] = I$ for any cognitive variable q, we can link the quantization rule in cognitive space to the quantization rule of Bohr-Sommerfeld. The logical potential carrying the logical energy could be linked to the Bohr energy of atomic structures in the following way: $\infty(k) = \oint M dq = 2\pi(n + 1/2) = k\pi$, where q is a logical variable (if it is zero than the contour integral runs a full great circle on the Riemann sphere of zeros), n is the winding number specifying the numbers of times the closed curve runs round in an anticlockwise sense; n runs the bosonic numbers $0, 1, 2 \ldots$ and $(n + 1/2)$ the fermionic numbers, $\frac{1}{2}, \frac{3}{2}, \frac{5}{2}, \ldots$. The topological potential is an odd multiple $(2n + 1)\pi$ of the elemental (topo)logical phase π and is $\hbar^{-1}$ times the Bohr energy of the quantum oscillator: $\oint p dx = 2\pi\hbar(n + 1/2)$, where the position and momentum operator satisfy the standard quantum commutation relation: $[x, p] = i\hbar$. As we see, the topological potential, multiplied by the factor $\hbar$, gives the Bohr quantum energy opening up the possibility to treat atomic structure as a dynamical logic in a fundamental sense, where quantization stems from the closed topology or self-observation feature at this fundamental level of reality. Another interesting conjecture which follows is, since matter, as energy, ($E = mc^2$) is a topologically transformed logical energy, the mass of an object is basically the information contained in the holomatrix which projects it out from the ground state.

6.4 The Time and Spin Operators, Quantum Mechanics and Cognition

Let us now introduce the operator defined by

$$\triangle = a - a^\dagger \tag{143}$$

so that is follows that its matrix representation is

$$\triangle = \begin{pmatrix} 0 & 1 \\ -1 & 0 \end{pmatrix}. \tag{144}$$

and furthermore

$$\triangle = \rightarrow - \leftarrow . \tag{145}$$

We shall call $\triangle$ the TIME operator. [24] We notice that it is unitary and antisymmetric:

$$\mathrm{TIME}^\dagger = \mathrm{TIME}^{-1} = -\mathrm{TIME}. \tag{146}$$

---

[24]Remarkably, $-2i\mathrm{TIME}$ is the hamiltonian operator of the damped quantum oscillator in the quantum theory of open systems; see N. Gisin and I. Percival, arXiv:quant-ph/9701024v1. In this theory based on the stochastic Schroedinger equation the role of torsion is central [10] (2007)].

437

As an hypernumber TIME $= -i_1$, minus the unique 2×2 matrix representing a 90 degrees rotation, the old commutative square root of -1 from which complex numbers appeared. The reason for considering this operator given by the difference of nilpotents is because it plays the role of a comparison operator. Indeed, we have

$$< p|\text{TIME}|q > \quad = \quad \bar{p}q - \bar{q}p = (1 - p)q - (1 - q)p = q - p = \bar{p} - \bar{q}. \quad (147)$$

TIME appears to be unchanged for unaltered states of consciousness:

$$< q|\text{TIME}|q >= 0, \quad (148)$$

and if we have different cognitive states p, q, then $< p|\text{TIME}|q > \neq 0$. So this operator does represent the appearance of a primitive difference on cognitive states (another example of the motto in the Introduction). It is antisymmetric and unitary. It is furthermore linked with a difference between annihilation and creation operators and thus stands for what we argued already as a most basic difference that leads to cognition and perception: the *appearance of quantum jumps*. Without them, no inhomogeneities nor events are accesible to consciousness. The very nature of self-reference as consciousness of consciousness requires such an operator for the joint constitution of the subject and the world. Thus its name, a TIME operator operator. It stands clearly in the subject side of the construction of a conception that overcomes the Cartesian cut, yet a subject that has superposed paradoxical states. Yet, we have seen that it plays a major role in the representation of the extended photon.

Let us consider next the eigenvalues of TIME, i.e. the numbers λ such that TIME$|q >= \lambda|q >$; they are obtained by solving the characteristic equation $\det|\text{TIME} - \lambda I| = \lambda^2 + 1 = 0$, so that they are $\lambda = \pm i$ with complex eigenstates

$$\begin{pmatrix} 1 \\ i \end{pmatrix}, \begin{pmatrix} i \\ 1 \end{pmatrix}. \quad (149)$$

They are not orthogonal, but self-orthogonal; thus, they are spinors, and the complex space generated by them generates a two-dimensional null space. We diagonalize TIME by taking

$$\begin{pmatrix} 1 & i \\ i & 1 \end{pmatrix} \text{TIME} \begin{pmatrix} 1 & i \\ i & 1 \end{pmatrix}^{-1} = \begin{pmatrix} i & 0 \\ 0 & -i \end{pmatrix} \quad (150)$$

so that

$$\text{TIME}_{\text{diag}} = \begin{pmatrix} i & 0 \\ 0 & -i \end{pmatrix} \quad (151)$$

which as an hypernumbers we have that $\text{TIME}_{\text{diag}} = i_2$, so that $\text{TIME}_{\text{diag}}^2 = -I$. We want finally to comment that TIME is not a traditional clock, yet it allows to distinguish between after and before ($\rightarrow - \leftarrow$), forward and backwards. There is no absolute logical time, nor a priviliged direction of it. To have a particular direction it must be asymetrically balanced towards creation or annihilation. This can be computed as the complement of the operator phase[25]

$$\overline{cos(2a^\dagger) + sen(2a^\dagger)} = a^\dagger - a, \tag{152}$$

from which it follows that $\text{TIME} = \overline{\leftarrow^2} = \rightarrow - \leftarrow$, as we stated before.

Let us now retake M and decompose it as

$$M = \text{TIME} + \sigma, \text{ or still} \tag{153}$$

$$\begin{pmatrix} 1 & 1 \\ -1 & 1 \end{pmatrix} = \begin{pmatrix} 0 & 1 \\ -1 & 0 \end{pmatrix} + \begin{pmatrix} 1 & 0 \\ 0 & -1 \end{pmatrix}. \tag{154}$$

Then we have that

$$< q|M|q >=< q|\sigma|q > . \tag{155}$$

Indeed, since $< q|\text{TIME}|q >= 0$, so that the proof of eq. (155) follows. Furthermore we note that

$$< q|\sigma|q >= \bar{q}^2 - q^2 = (\bar{q} - q)(\bar{q} + q) = \bar{q} - q. \tag{156}$$

from the normalization condition. Note here that the identity given by eq. (156) is a kind of quadratic metric in cognitive space which due to the normalization condition looses its quadratic character to become the difference in the cognitive values: $\bar{q} - q = 1 - 2q$ which becomes trivial in the undecided state in which $\bar{q} = q = \frac{1}{2}$.

The role of σ is that of a SPIN operator, as we shall name it henceforth, which coincides with the hypernumber ϵ_2 (or as a Pauli matrix is σ_z), so that $\sigma^2 = I$ the non-trivial square root of hypernumber $I = \epsilon_0$, which is the usual Pauli matrix σ_z in the decomposition of a Pauli spinor in the form $\sigma_x e_x + \sigma_y e_y + \sigma_z e_z$, for e_x, e_y, e_z the standard unit vectors in $\mathbf{R}^3$ and we write their representations as hypernumbers

$$\sigma_x = \begin{pmatrix} 0 & 1 \\ 1 & 0 \end{pmatrix} = \epsilon_3, \sigma_y = \begin{pmatrix} 0 & -i \\ i & 0 \end{pmatrix} = \epsilon_1, \sigma_z = \begin{pmatrix} 1 & 0 \\ 0 & -1 \end{pmatrix} = \epsilon_2. \tag{157}$$

---

[25]The complement of a logical operator L, is defined by $\bar{L} = I - L$.

439

We can rewrite this average equation $< q|M|q >=< q|\sigma|q >$ as an average equation which the l.h.s. takes place in cognitive space of normalized states $|q >$ and the r.h.s. in a Hilbert space of a two-state quantum system, say, spin-up $\psi(\uparrow)$, spin-down $\psi(\downarrow)$, so that the generic element is of the form

$$\psi = \psi(\uparrow)|0 > +\psi(\downarrow)|1 > . \tag{158}$$

Indeed, if we write

$$|q >= \overline{\psi(\uparrow)}\psi(\uparrow)|0 > +\overline{\psi(\downarrow)}\psi(\downarrow)|1 >, \tag{159}$$

then the r.h.s. of eq. (156) is $\bar{q}^2 - q^2$, with $\bar{q} = \overline{\psi(\uparrow)}\psi(\downarrow)$, and $q = \overline{\psi(\downarrow)}\psi(\uparrow)$, so that eq. (156) can be written as

$$< q|M|q >=< \psi|\sigma|\psi > \tag{160}$$

where the average of M is taken in cognitive states while that of the SPIN operator is taken in the two-state Hilbert space.

We review the previous derivation for which the clue is the relation between cognitive states $|q >$ and elements of two-state of Hilbert state $|\psi >$: The former are derived from the latter by taking the complex square root of the latter. Hence, probability$(|0 >) = \bar{q} = \overline{\psi(\uparrow)}\psi(\uparrow)$ and probability$(|1 >) = q = \overline{\psi(\downarrow)}\psi(\downarrow)$, so that $< \psi|\sigma|\psi >= \bar{q} - q = (\bar{q} - q)(\bar{q} + q) = \bar{q}^2 - q^2$. Therefore, by using the transformation between real cognitive states q defined by the complex square root of ψ, i.e. $q = \bar{\psi}\psi$, we have a transformation of the average of the cognitive operator M on cognitive states on the average of SPIN on two-states quantum elements in Hilbert state, i.e. eq. (159). This is a very important relation, established by an average of the cognition operator (which transforms an orientable plane into a non-orientable Moebius surface due to the torsion introduced by M, as represented by eq.(139), and SPIN on the Hilbert space of two-state quantum mechanics. It is an identity between the action of the cognizing self-referential mind and the quantum action of spin. Thus the cognitive logical processes of the subject become related with the physical field of spin on the quantum states. This is in sharp contrast with the Cartesian cut, and we remark again that this is due to the self-referential classical-quantum character of M as evidenced by eq. (139) which produces a torsion on the orientable cognitive plane of coordinates (true, false) to one to yield a superposed state, S_-. The relation given by eq. (155, 159) establishes a link between the operations of cognition and the quantum mechanical spin. This link is an interface between the in-formational and quantum realms, in which topology, torsion, logic and the quantum world operate jointly. Yet, due to fact that for the Klein bottle there is no inside nor outside, the exchange can go in both ways, i.e. the quantum realm can be incorporated into the classical cognitive dynamics, while the logical elements can take part in the quantum evolution. Indeed, if we have a matrix-logical string which contains the momentum product, say, $\ldots < x|A|y >< q|M|q >< z|B|s > \ldots = \ldots < x|A|y >< \psi|\sigma|\psi >< z|B|s > \ldots$. Thus, the factor $< \psi|\sigma|\psi >$ entangles with the rest of the classical logical string creating a Schrödinger cat superposed state, since we have a string of valid propositions where one may be the negation of the other [26]

6.5 The Klein Bottle, Quantum Computation and the Intelligence Code

There is still another very remarkable role of these superposed states in producing a topological representation of a higher order form of self-reference, produced from oppositely twisted Moebius surfaces. So we shall consider the Cartesian modulo 2 sum of the superposed states

$$\mathcal{H} := |S_+ > \oplus |S_- >= \begin{pmatrix} 1 & 1 \\ 1 & -1 \end{pmatrix}, \tag{161}$$

which we call the topological in-formation operator which is an hypernumber; indeed, $\mathcal{H} = \sigma_x + \sigma_z = \epsilon_3 + \epsilon_2$. We could have chosen the opposite direct sum or still place the minus sign on the first row in any of the columns and obtain a similar theory, but for non-hermitean operators unless the minus sign is on the first matrix element. Notice that it is a hermitean operator, which essentially represents the topological (or still, logo-topological) in-formation of a Klein bottle formed by two oppositely twisted Moebius surfaces [72]. [27] The in-formation matrix satisfies $\mathcal{H}\mathcal{H}^\dagger = \mathcal{H}\mathcal{H}^{-1} = 2I$. We recognize in taking $1\sqrt{2}\mathcal{H}$ the Hadamard gate in quantum computation [36], which due to the introduction of the $\frac{1}{\sqrt{2}}$ factor is hermitean and unitary. Now we have two orthogonal basis given

---

[26]This primordial role of spin as as protopsychic as well as protophysical is found also in [27], though not mathematically based. In this work it is claimed that spin is "the linchpin between mind and brain", though in a certain Cartesian way, associating spinor fields to processes in the brain and not to the processes of the mind. They further link it with self-referential processes alike the Klein bottle [27]; see also [82].

[27]Alternatively we can introduce instead of $\mathcal{H}$ another in-formation matrix for the Klein bottle, namely

$$.\mathcal{H} := |S_+ > \oplus |S_- >= \begin{pmatrix} 1 & 1 \\ -1 & 1 \end{pmatrix}, \tag{162}$$

which is non-hermitean.

by the sets $\{|0>, |1>\}$ and $\{|S_->, |S_+>\}$ of classical and superposed states respectively, the latter un-normalized for which a factor $\frac{1}{2}$ has to be introduced but still does not give a unitary system as in quantum theory. An important role of the Klein bottle is precisely to transform these orthogonal basis, from classical states to superposed states which are nor classical nor quantum, but become quantized by appropiate normalization with the $\frac{1}{\sqrt{2}}$ factor. Indeed,

$$\mathcal{H}|0>= |S_+>, \mathcal{H}|1>= |S_->, \tag{163}$$

and

$$\frac{1}{2}\mathcal{H}|S_+>= |0>, \frac{1}{2}\mathcal{H}|S_->= |1>. \tag{164}$$

In the logical space coordinates $(true, false)$ we have rotated the state $|0>$ clockwise by 45 degrees through the action of $\mathcal{H}$ and multiplied it its norm by 2, and for the state $|1>$ we have rotated it likewise after being flipped. In reverse, the superposed states are transformed into the classical states by halving the in-formation matrix of the Klein bottle, producing 45 degrees counterclockwise rotations, one with a flip. Now classical and quantum states are functionally complete sets of eigenstates spanning each other. The classical states $|0>$ and $|1>$ can be easily determined to be the eigenstates of AND, and and the superposed states $|S_->, |S_+>$ are the eigenstates of NOT. It is known that the logical basis of operators $\{\text{AND}, \text{OR}\}$ is functionally complete, generating all operators. Hence our system of classical and superposed (or still, quantum by appropiate normalization by $\frac{1}{\sqrt{2}}$) eigenstates constitute together a functionally complete system: all operators of matrix logic can be obtained from them. This system is self-referential. Furthermore, there are operators which produce the rotation of one orthogonal system on the other orthogonal system. The logical differentiation operator M defined by the commutator [FALSE, TRUE] or still eq. (129) transforms classical states $|x>= \bar{x}|0> +x|1>$ into $|S_-$ and still the anticommutator $\{\text{FALSE}, \text{TRUE}\}$ which coincides with the matrix $\mathbf{1} = \begin{pmatrix} 1 & 1 \\ 1 & 1 \end{pmatrix}$ transforms $|x>$ into $|S_+>$, i.e.

$$M|x>= |S_->, \mathbf{1}|x>= |S_+>. \tag{165}$$

which can be rephrased by saying that M evidences on its action on a classical state the torsion in the quantum commutator of FALSE and TRUE while the ONE operator $\mathbf{1}$ transforms $|1>$ into $\begin{pmatrix} 1 \\ 1 \end{pmatrix} = |S_->$. Since both M and $\mathbf{1}$ are non-invertible, we shall use instead the fact that $\mathcal{H}^{-1} = \frac{1}{2}\mathcal{H}$, so that in addition of the transformation by the Klein bottle of the classical basis in eq. (163), the reversed transformation from the superposed to the classical states is achieved by

$$\frac{1}{2}\mathcal{H}|S_+> = |0>, \frac{1}{2}\mathcal{H}|S_-> = |1>. \tag{166}$$

Yet, we stress again that these transformations are not unitary which is easily resolved by the $\frac{1}{\sqrt{2}}$ factor and then we have a transformation of classical into quantum states and viceversa. In the latter case, the renormalized Klein bottle acts like a quantum operator producing coherent quantum states, a topological Schroedinger "cat" state which does not decohere. Therefore to resume, from these four states, or alternatively, from the matrix representation of the Klein bottle, it is possible to generate the Intelligence Code [26, 13]. We note that it is essentially self-referential.

7   The Eigenstates of the Cognitive Plenum, Twistors and the Extended Photon, and the So-Called Mind-Matter Problem

We shall now discuss the eigenstates of the mind apeiron, namely the 2×2 identicaly zero matrix which we denoted as $\mathbf{0}$; this was first partially and roughly sketched in [25]. In distinction with the other logical operators the eigenstates of $\mathbf{0}$, as a linear transformation from $\mathbf{C}$ on $\mathbf{C}$, which thus becomes a point of $\mathbf{C}^2$, its origin, are no longer quantized, but rather give an orthogonal complex two-dimensional nullvector space. [28] In this way the plenum is no longer represented by a single point, $\mathbf{0}$, but rather becomes an extended object or zero-brane. This phenomenon of the blowup of a point or more generally a manifold (here $\mathbf{C}^2$) is well known in complex Clifford bundles and is the most fundamental operation in algebraic geometry (which was the origin of twistors by R. Penrose [22]). Namely, it consists in replacing each point of the manifold by the projectivized tangent space at that point [68]. In the case of the mind plenum represented by the origin in $\mathbf{C}^2$ it amounts to replace it by the projectivized tangent space at the origin. Let Z be the origin in n-dimensional complex space, $\mathbf{C}^n$. That is, Z is the point where the n coordinate functions $(x_1, \ldots, x_n) \in \mathbf{C}^n$ simultaneously vanish. Let

---

[28]We have already seen this an identical situation in the eigenstates of TIME. Thus the eigenstates of the mind apeiron are given by a plenumpotence condition alike the eikonal equation; we shall see that this similarity is in fact an identity.

443

$\mathbf{CP}(n-1)$ be the (n-1)-dimensional complex projective space with homogeneous coordinates $(y_1, \ldots, y_n)$. The blowup of Z is the subset of $\mathbf{C}^n \times \mathbf{CP}(n-1)$ that satisfies the equation $x_i y_j = x_j y_i$, for all $i, j = 1, \ldots, n$. In the case $n = 2$ where (y_1, y_2) are complex numbers not both zero, homogeneous coordinates of $\mathbf{CP}(1)$, which can thus be also can described by the single coordinate $\xi = \frac{y_1}{y_2}$. Since $\mathbf{CP}(1)$ is the familiar Riemann (-Argand-Euler) sphere of complex analysis, S, then the blowup of the origin in $\mathbf{C}^2$ is its replacement by the Riemann sphere, or still by the complex 2-sphere, S^2, on which we represent the spinor eigenstates of the mind apeiron.[29] Indeed, a cross-section of the blowup of the origin in $\mathbf{C}^2$ represents the spinor vectors in S or in its isomorphic two-sphere, S^2, giving a 2- complex-dimensional vector space, which can be mapped to the 2-dimensional logic space of matrix logic by stereographic projection. [30]. We apply this to the twistor representation of the extended photon through the maximal monochromatic algebra as described by $(118, 119)$ which has an equivalent representation as a pair of divergenceless orthogonal spinor vectors $(\omega^A, \pi^{A'})$, $A, A' = 1, 2$ by (120). By stereographic projection of the spinors ω^A, π^A which form the twistor representation of the extended photon, we obtain a representation of it in cognitive space in the basis $|0>$ and $|1>$. Viceversa, by taking the inverse of the stereographic

---

[29]There is a certain ambiguity on regards of **0** being also interpretable as the origin in $\mathbf{C}^4$ rather $\mathbf{C}^2$, after all it has four entries! In this case, the blowup of the origin in $\mathbf{C}^4$ has no longer for crossection $\mathbf{CP}(1)$ but instead $\mathbf{CP}(3)$, which is the three dimensional complex projective space of twistors [22] (1979). In this case, the eigenstates of **0** are (projective) twistors, elements of a nullspace. The difference in this interpretation is that for the effect of the association between the maximal monochromatic algebra of the extended photon is characterized by eqs. (120) representing the *pair* of spinors characterized by eqs. (118, 119) as cross-sections of the blowup of **0** as the origin in $\mathbf{C}^2$.

[30]The blowing up of the origin, transforming its point-like structure to yield a manifold has profound consequences. For example, the blowing up of the origin in $\mathbf{R}^2$ is the Moebius surface, which as we already saw is basic to the generation of Intelligence Code: its normal vectors defines logical momentum and also the Klein bottle. We recall that two oppositely twisted Moebius bands generate the Klein bottle [72], the high order (in relation to the Moebius band [28]) surface of paradox, whose matrix representation, up to a normalizing constant, is the Hadamard gate of quantum computation, which together with the phase conjugator, allows to generate all quantum gates [36]. Thus, embedded in the blowup of the mind apeiron as the origin of $\mathbf{C}^2$ lies the generation of the Intelligence Code from the blowup of the origin in $\mathbf{R}^2$. DNA performs quantum computations [37] which is further related to holography [38, 39]. We recall that holography is already performed by the Klein bottle visual processing of the neurocortex. This evidences the importance of the Hadamard gate in quantum computation and the Klein bottle multivalued logic we presented. It can also be derived from the semiotic codification of torsion as a distinction and the time-waves related to paradox; see [13]. For technological implementations of the Moebius band, the Klein bottle and the generation of Kozyrev fields see [85]; for its relation to anthropology [79]. An important contribution to the geometrical studies of consciousness, though in a different setting is [86].

projection we reconstruct the maximal monochromatic algebra. In any case via the normalized Klein bottle Hadamard in-formation matrix, all the operators of matrix logic are generated (the completitude we mentioned before). In this we see how the extended photon which we claimed to be a subjective-objective fused structure-process is represented as a basis for cognitive space, and conversely, from cognitive space we are able to codify the maximal monochromatic algebra representation of the extended photon. This establishes the full self-referential construction of a world which is perceived through quantum jumps, i.e. distinctions, or still, in terms of cognitive states that belong to states of cognition of the mind. Yet, we have seen above that the role of the Planck constant $\hbar$ is precisely to connect the transformation of the quantum world into the world of the mind, bridging thus the material and mind domains. Since $\hbar$ can be associated to a cosmological scale [30, 31], we can speak about *cosmological consciousness*. Thus, we can modify the quotation in the Introduction, "...light is seeing", to light is seeing-thinking, as these two actions become inseparable at the mind apeiron level. More complex levels operate through convolution, and perhaps through other processes. The Riemann sphere is not only instrumental to this joint constitution by codifying the extended photon as a cognitive state. It is also the manifold in which the logarithmic function takes multivalued complex values to quantize the quantum jumps in terms of the different branches of the logarithm, allowing thus to codify the 'outer' and 'inner' worlds. For further ellaborations in relation to the transactional interpretation of Quantum Mechanics, the important notion of anticipatory systems [80], cosmological Kozyrev torsion fields and entanglement, and brain synchronization in binocular vision, see [43]. [31]

For closing remarks is enough to summarize by saying that the plenumpotence of 0 and **0** have been shown to be related to quantum jumps through the *extended* singularities of the torsion field related to propagating *extended* eikonal singularities and to the *extended* eigenstates of the mind apeiron, enclosing and generating a self-referential world which amounts to the *extended* character of 0 and that of the subject. This extension is the fusion of the res cogitans and res extensa of Descartes: apeiron.

---

[31]For a different conception in terms of Endophysics which does not incorporate explicitly self-reference we refer to [40]. The theory of fractal time is relevant [41]. Remarkably, inasmuch matrix logic has a projective structure as well as the eigenstates of the mind apeiron, a theory of altered mind states in terms of Cremona transformations - which arise as well as blowups - has been developed [42].

8   Appendix: Torsion, Non-Commutativity of Space-time and New Energies

We want to introduce torsion in terms of the self-referential definition of the manifold structure in terms of the concept of difference or distinction derived from the operation of comparison to establish a difference that makes a difference, as discussed in the Introduction. We shall assume that there are two observers on a manifold (of dimension n), say observer 1 and observer 2, which may not be moving inertially. To compare measurements and to establish thus a sense of objectivity (identity of their results), they need to compare their measurements which take place in the tangent space at different points of the n-dimensional manifold M in which they are placed, so they have to establish the difference between their reference frames, i.e. the difference between the set of orthogonal (or pseudo-orthogonal) vectors at their locations, the so-called n-beins . Let $e_i(\mathcal{P}_0) = e_i^\alpha(\mathcal{P}_0)\partial_\alpha, i = 1, \ldots, n$ be the basis for observer 1 at point $\mathcal{P}_0$, and similarly $e_i(\mathcal{P}_1) = e_i^\alpha(\mathcal{P}_1)\partial_\alpha$ the reference frame for observer 2 at $\mathcal{P}_1$; let us denote the reference frame at the tangent space to the point $\mathcal{P}_1$ when parallely transported (without changing its length and angle) from $\mathcal{P}_0$ to $\mathcal{P}_1$ by $e_i(\mathcal{P}_0 \to \mathcal{P}_1)$ along a curve joining $\mathcal{P}_0$ to $\mathcal{P}_1$ with an affine connection, whose covariant derivative operator we denote as $\tilde{\nabla}$ as in Section 2. Then, $\tilde{\nabla}e_1$ is the difference between $e_i(\mathcal{P}_0 \to \mathcal{P}_1)$ and $e_i(\mathcal{P}_1)$. This gap defect originates either from: 1) the deformation of $e_i(\mathcal{P}_0)$ along its path to $\mathcal{P}_1$, which cannot be transformed away by a change of coordinates, or, 2) by a change of coordinates from $\mathcal{P}_0$ to $\mathcal{P}_1$, which is not intrinsic and thus can be transformed away, or finally, 3) by a combination of both. Let us move observer's one frame over two different paths. Parallel displacing an incremental vector $dx^b e_b$ from the point $\mathcal{P}_0$ along the basis vector e_a over an infinitesimal distance dx^a to the point $\mathcal{P}_1 = \mathcal{P}_0 + dx^a$ gives the vector

$$e_b dx^b(\mathcal{P}_0 \to \mathcal{P}_1) = dx^b e_b(\mathcal{P}_0) + \Gamma_{ba}^c dx^b \wedge dx^b e_c. \tag{167}$$

Similarly, the parallel transport of the incremental vector $dx^a e_a$ from the point $\mathcal{P}_0$ to $\mathcal{P}_2$ along the frame e_b over an infinitesimal distance dx^b to the point $\mathcal{P}_2 = \mathcal{P}_0 + dx^b$ gives the vector $e_a(\mathcal{P}_1 \to \mathcal{P}_2) = dx^a e_a(\mathcal{P}_1) + \Gamma_{ab}^c dx^a \wedge dx^b e_c$. The gap defect between $e_a(\mathcal{P}_0 \to \mathcal{P}_1)$ and the value of $dx^b e_b(\mathcal{P}_1)$ is

$$dx^b \tilde{\nabla} e_b(\mathcal{P}_1) = dx^b(\frac{\partial e_b}{\partial x^a}) \wedge dx^a - \Gamma_{ba}^c dx^a \wedge dx^b e_c, \tag{168}$$

and the gap defect between the vector $e_b(\mathcal{P}_1 \to \mathcal{P}_2)$ and $e_b dx^b(\mathcal{P}_2)$ is

$$dx^a{}_a(\mathcal{P}_2) = dx^a(\frac{\partial e_a}{\partial x^b}) \wedge dx^a - \Gamma_{ab}^c dx^a \wedge dx^b e_c. \tag{169}$$

Therefore, the total gap defect between the two vectors is (the comparison already mentioned)

$$dx^b \tilde{\nabla} e_b(\mathcal{P}_1) - dx^a D e_a(\mathcal{P}_2) = (\frac{\partial e_b}{\partial x^a} - \frac{\partial e_a}{\partial x^b}) dx^a \wedge dx^b$$

$$+ [\Gamma^c_{ab} - \Gamma^c_{ba}] dx^a \wedge dx^b e_c, \qquad (170)$$

where we recognize in the first term the Lie-bracket

$$[e_a, e_b] = (\frac{\partial e_b}{\partial x^a} - \frac{\partial e_a}{\partial x^b}) dx^a \wedge dx^b, \qquad (171)$$

which we can write still as

$$[e_a, e_b] = C^c_{ab} e_c, \qquad (172)$$

where C^c_{ab} are the coefficients of the anholonomity tensor, and then finally we can write the difference in eq.(170) as

$$dx^b e_b(\mathcal{P}_1) - dx^a \tilde{\nabla} e_a(\mathcal{P}_2) = (C^c_{ab} + [\Gamma^c_{ab} - \Gamma^c_{ba}]) dx^a \wedge dx^b e_c. \qquad (173)$$

If we further introduce the vector-valued torsion two form

$$T = \frac{1}{2} T^c_{ab} dx^a \wedge dx^b e_c := \tilde{\nabla} e_b(e_a) - \tilde{\nabla} e_a(e_b) - [e_a, e_b]^c e_c \qquad (174)$$

we find that the components T^c_{ab} are given by the so-called torsion tensor

$$T^c_{ab} = C^c_{ab} + [\Gamma^c_{ab} - \Gamma^c_{ba}] \qquad (175)$$

Thus, we have two possibilities for the non-closure of infinitesimal parallelograms. Either by anholonomity, or due to the non-symmetricity of the Christoffel coefficients. These are radically different. The former can in some instances be set to be equal to zero, while the other term cannot. Say we have a coordinate transformation continuously differentiable $(x^1, \ldots, x^n) \rightarrow (y^1, \ldots, y^n)$ so we have that an holonomous transformation, i.e. we have that each dy^i is exact of the form

$$dy^i = \frac{\partial y^i}{\partial x^j} dx^j. \qquad (176)$$

Then, if we take an holonomous basis $e_j = (\frac{\partial y^i}{\partial x^j}) \frac{\partial}{\partial y_i}$, then the anholonomity vanishes, $[e_i, e_j] = 0$ identically on M, and we are left for the expression for the torsion tensor

$$T^c_{ab} = \Gamma^c_{ab} - \Gamma^c_{ba}. \qquad (177)$$

447

Observations. Anholonomity is related to the Sagnac effect and to the Thomas precession [51]. Nowadays, relativistic rotation has become an issue of great interest, and the interest lays in rotating anholonomous frames, in distinction with non-rotating holonomous frames. The torsion tensor evidences how the manifold is folded or dislocated, and the latter situation can be produced by tearing the manifold of by the addition of matter or fields to it. These are the well known Volterra operations of condensed matter physics, initially, introduced in metalurgy [52]. This was the first technological implementation of torsion. The second example was elaborated in the pioneering work by Gabriel Kron in the geometrical representation of electric networks; it lead to the concept of negative resistance [53]. Contemporarily, negative resistance has become an important issue, after the discovery of its existence in some materials, with an accompanying apparent phenomenon of superconductivity [54]. In [10] it was proved that Brownian motions -which are associated to torsion geometries- produce rotational fields. This encompasses the Brownian motions produced by the wave function of arbitrary quantum systems, and the case of viscous fluids, magnetized or not [7, 10, 11]. These examples are independent of any scale, from the galactic to the quantum scales. In the galactic scales, vortices can explain the red-shift without introducing any big-bang hypothesis [66]. Thus, we have a modified form of Le Sage's kinetic theory [65] producing universal fluctuations which have additionally rotational fields associated to them. Due to the universality of quantum wave functions, either obeying the rules of linear or non-linear Quantum Mechanics, Hadronic Mechanics (HM) and Hadronic Chemistry (HQ) [63, 64], these vortices are rather common. Then it is no surprise that vortices and superconductivity appear as universal coherent structures. Superconductivity is usually related to a non-linear Schroedinger equation with a Landau-Ginzburg potential, which is also an example of the Brownian motions related to torsion fields with further noise related to the metric [7, 10]. Furthermore, atoms and molecules have spin-spin interactions which will produce a contribution to the torsion field; we have seen already that the torsion geometry exists in the realm of HM and HQ. [32] This is the case of the compressed hydrogen atom model of the neutron in the Rutherford-Santilli model in HM, in which their is a spin alignment with opposite direction and magnetic moments for the electron and the proton (topologically,

---

[32] A different approach relates spin-torsion fields [56] to the teleparallel geometries in Minkowski space also explored by the author in [62]. In that work the torsion polarization of the vacuum which also shows up in the phantom DNA effect [37], is related to an hypothetical particle known as the *phyton*. Further experiments related to torsion fields have been carried out [59] and the phenomenae revealed by the Kozyrev experiments have been interpreted in terms of torsion [69, 10].

this yields the Klein bottle), which produces a stable state which leads to fusion [63]. This is *not* a cold fusion process since it appears to occur at temperatures of the order of $5,000$ degrees Celsius. Yet in electrochemical reactions, there are sources of torsion which are given by the the wave functions of the components involved, but furthermore the production of vortex structures. Gas bubbles appear after switching off the electrochemical potentials, and sonoluminescence have been observed at the Oak Ridge National Laboratory, USA [57]. There is a surprising phenomenae of remnant heat that persists after death which could be produced by the vortex dynamics of the tip effect [60]. These experimental findings have been claimed to be observed in different laboratories across the world [55, 60], and explained in terms of torsion fields [58, 59]. Superconductors of class II present also some surprising phenomenae such as low-frequency noise, history-dependent dynamic response, and memory direction, amplitude direction and frequency of the previously applied current [58]. Would these findings be reproduced systematically, we would have a new class of sources of energy, which stem from apeiron. Other important source may occur as a resonant coupling of the torsion generated apeiron Brownian motions with especially designed circuits [83], and in the so-called *cavity structural effect* discovered by research in entomology, which are being developed with widespread applications in Russia and Ukraine [84].

Acknowledgements: I dedicate this work to my son Rodrigo Tsafrir Rapoport, for a life time of philosophical discussions which here appear somehow reflected, to my wife, Sonia, and my daughter Tania. Without their support and love this work would not have been possible. My gratitude to Prof. Stein Johansen for his warm support and encouragement; through his work I came to grasp in March 2008 the potential importance of the Klein bottle. My gratitude to Prof. Steven M. Rosen for sending to me an important reference and his works, to the editors of this book, Prof. Michael Duffy and Prof. Joseph Levy, for their kind invitation to contribute to it, and to the referees for requesting more clarifications to the original manuscript.

References

[1] http://en.wikipedia.org/wiki/Banach-algebra, and references therein.

[2] H. Bateman, *The Theory of Electric and Optical Wave Motions from Maxwell's Equations,*(Cambridge University Press, 1915).

[3] D. Bohm & B. Hiley, *Wholeness and the Implicate Order*, (Routledge-Kegan, London, 1980); Paavo T. I. Pylkkanen, *Mind, Matter and the Implicate Order*, (The Frontiers Collection, Springer, Berlin 2006).

[4] V. Fock, *The Theory of Gravitation*, (Pergamon Press, London, 1962).

[5] H. Kleinert, *Gauge Fields in Condensed Matter*, (World Sc., Singapore, 1990); ibid. *Multivalued Fields: In Condensed Matter, Electromagnetism, and Gravitation*, (World Scientific, Singapore, 2008).

[6] P.Nowosad, MRC-Wisconsin Report, 1982; ibid. *Comm.Pure Appl. Math.***21** p. 401-65, 1968. L. C. Young & P. Nowosad, J.Opt. Th. App.**41**, p.261, 1983.

[7] D.L. Rapoport, "On the Unification of Geometric and Random Structures through Torsion Fields: Brownian Motions, Viscous and Magneto-fluid-dynamics", *Found. Phys.* **35**, no.7, pags. 1205-1244, 2005 ; ibid. D Rapoport, "On the derivation of the stochastic processes associated to Lie isotopic gauge theory", pags. 359-374, in *Hadronic Mechanics and Nonpotential Interactions V, vol. II, Proceedings of the Fifth International Conference, Univ. of Iowa, August 1990*, Hyo Myung (ed.), 359-374, (Nova Science Publs., New York/Budapest, 1992); ibid. "Stochastic processes in conformal Riemann-Cartan-Weyl gravitation", *Int. J. Theor. Phys.* **30**, no. 11, (1991), 1497-1515; ibid. " The Lie-Isotopic theory of Stochastic Mechanics associated to a torsion potential", *Algebras, Groups and Geometries*, **8**, no. 1, pags. 1-60 (1991); ibid. " The Riemann-Cartan-Weyl Quantum geometries II: The Cartan stochastic copying method, Fokker-Planck operator and Maxwell-de Rham equations", *Int. J. Theor. Phys.* **36**, No. 10, 1997, 2215-2152; ibid. "The Cartan Structure of Classical and Quantum Gravity", in *Gravitation, the Space-time structure, Proceedings VIII th. SILARG*, 220-229, P.Letelier and W. Rodrigues (eds.), World Scientific , Singapore, 1994.

[8] D. L. Rapoport,"Covariant Thermodynamics and the Ergodic Theory of Stochastic and Quantum Flows" in *Instabilities and Nonequilibrium Structures VI*,pags. 359-370, E. Tirapegui and W. Zeller (edts.), (Kluwer, Dordrecht/Boston, 2000); ibid. in *Frontiers of Algebras, Groups and Geometries*, G. Tsagas (ed.), Proceedings of the Monteroduni Conference (Monteroduni, Italy, August 1997), Hadronic Press, Palm Harbor, 1998.

[9] D. Rapoport,"Random Diffeomorphisms and integration of the classical Navier-Stokes equations" *Rep. Math. Phys.* **49**, no. 1, p. 1-27, 2002; **50**, no.2, 211-250, 2002; ibid. *Rand. Oper. Stoch. Eqs.*, **11**, no.2, 109-150, no.4, 351-382, 2003.

[10] D. L. Rapoport, " Torsion Fields, CartanWeyl SpaceTime and State-Space Quantum Geometries, their Brownian Motions", and the Time Variables, *Found. Phys.* **37**, nos. 4-5, 813-854, 2007; ibid. "CartanWeyl Dirac and Laplacian Operators, Brownian Motions: The Quantum Potential and Scalar Curvature, Maxwells and Dirac-Hestenes Equations, and Supersymmetric Systems", *Found. Phys.***35**, no. 8,

pags. 1383-1431, 2005; ibid., in *Foundations of Probability and Physics IV*, A. Khrennikov et al (eds.), (AIP Proceedings, Springer,Berlin, 2007).

[11] D. L. Rapoport," Torsion Fields, Propagating Singularities, Nilpotence, Quantum Jumps and the Eikonal Equations", in *Proceedings of CASYS09 (Conference on Anticipative Systems, Univ. of Liège, August 2009)*, D. Dubois et al (ed.), AIP-Conferences Springer Series, Berlin, 2009.

[12] D.L. Rapoport, in *Hadronic Mathematics, Mechanics and Chemistry III*, R.M. Santilli, International Academic Press, in press, 2009.

[13] D. L.Rapoport, "Surmounting the Cartesian Cut Through Philosophy, Physics, Logic, Cybernetics and Geometry: Self-reference, Torsion, the Klein Bottle, the Time Operator, Multivalued Logics and Quantum Mechanics", Found. Phys. **40**, 2010.

[14] C. Musès, "Applied Hypernumbers: Computational Concepts",*Applied Mathematics and Computation* **3**, pp.211-226 (1976); ibid. Hypernumbers-II, **4**, pp. 45-66 (1978).

[15] P.Rowlands, *From Zero to Infinity*, (World Scientific, Singapore, 2008); V. Hill and P. Rowlands, "Nature's Fundamental Symmetry Breaking", ibid. "The Numbers of Nature's Codes", in *Proceedings of CASYS 09*, AIP, D. Dubois (ed), 2009 .

[16] E. C. Stueckelberg, *Helv. Physica Acta* **14**(1941), 322,588; L.P. Horwitz and C. Piron, *Helv. Physics Acta*, **46** (1973), 316 and **66** (1993), 694; L.P. Horwitz and N. Shnerb, Found. of Phys. **28**(1998), 1509 and references therein.

[17] T. Frankel, *The Geometry of Physics*, (Cambridge Univ.Press, Cambridge, 2001).

[18] I. Porteous, *Topological Geometry*, (Springer, Berlin, 2001)

[19] S. Helgason ,*Differential geometry, Lie groups, and symmetric spaces*, (AMS, Providence, 2001).

[20] R. F. Rinehart, "Elements of a Theory of Intrinsic Functions on Algebras", *Duke Math.J.* vol.**27**, 1-19, (1960).

[21] P. M. Dirac, "Quantized Singularities in the Electromagnetic Field",*Proc. Royal Soc. London* **133**, 6-72 (1931); J.O. Hirschfelder, A.C. Christoph and W.E. Pale, "Quantum Mechanical Streamlines", *J. Chem. Phys.***65**, 470-486 (1976) and references therein. J. F. Nye and M. Berry, "Dislocations in Wave Trains", *Proc. R. Soc. Lond. A* January 22, 1974 336:165-190.

[22] R. Penrose and W. Rindler, *Spinor and Twistor Methods in Spinors and Space-Time, I* (Cambridge Univ. Press, Cambridge, 2008); R.O.G. Wells Jr. "Complex Manifolds and Mathematical Physics", *Bull. A.M.S.* **1**, no.2, March 1979.

[23] R. M. Kiehn, "The Photon Spin and other Topological Features of Classical Electromagnetism" in *Gravitation and Cosmology: From the Hubble Radius to the Planck Scale* , R.M. Amoroso et al, (Springer, Berlin , 2003); R. M. Kiehn, *Topological Torsion and Macroscopic Spinors Vol. 5*, (Lulu Publs., 2008).

[24] J. Nye and M. Berry, "Dislocations in Wave Trains", *Proc. Royal Soc. London A* **336**, (1974), pp. 165-190.

[25] I. Dienes, "Consciousness - Holomatrix Quantized Dimensional Mechanics", in R.L. Amoroso K.H. Pribram (eds.) *The Complementarity of Mind and Body: Realizing the Dream of Descartes, Einstein and Eccles*, p. 463-471, Cambridge, MIT Univ. Press, to appear, 2010; http://www.slideshare.net/Dienes/consciousness-as-holographic-quantised-dimension-mechanics-presentation.

[26] A. Stern, *Matrix Logic and Mind*, (Elsevier , Amsterdam, 1992); ibid. *Quantum Theoretic Machines*, (Elsevier, Amsterdam, 2000).

[27] H. Hu and M. Wu, "Spin as Primordial Self-Referential Process Driving Quanum Mechanics, Spacetime Dynamics and Consciousness," *J. of Neuroquantology*, **2**, 41 (2004).

[28] S. Rosen, *Dimensions of Apeiron: A Topological Phenomenology of Space, Time and Individuation*, (Rodopi, Amsterdam-New York , 2004); ibid. *The Self-Evolving Cosmos: A Phenomenological Approach to Nature's Unity-in-Diversity* (Series on Knots and Everything), (World Scientific, Singapore, 2008); ibid. *Topologies of the Flesh: A Multidimensional Exploration of the Lifeworld* (Series In Continental Thought), (Ohio University Press, 2006).

[29] L. H. Kauffman, *Knots and Physics*, second edition, (World Sc., Singapore, 1993).

[30] L. Nottale, *Scale Relativity and Fractal Space-Time*, (World Sc., Singapore, 1993).

[31] M. Pitkanen, *Topological Geometrodynamics*, Luniver, Beckington, U.K., 2006.

[32] J. Hirschfelder, A. Christophe and W. Palke, "Quantum Mechanical Streamlines", *J.Chem.Phys.* **61**, (1974), pp. 5435-5345.

[33] R. Luneburg , *Mathematical Analysis of Binocular Vision*, (Edward Brothers, 1948); ibid. *Astron. Astrophys. Lett.***315**, L9 (1996); ibid. *Chaos, Solitons & Fractals* **7**, 877 (1996).

[34] T. Indow, *The Global Structure Of Visual Space*, (World Scientific, 2004).

[35] P. Heelan, *Space Perception and the Philosophy of Science*, (Univ. of California Press, Berkeley, 1989).

[36] J. Audretsch, *Entangled Systems*, (Wiley-VCH, Bonn, 2006).

[37] A.A. Berezin and P.P. Gariaev, "Is it Possible to Create Laser Based on Information Biomacromolecules?", *Laser Physics*, vol. **6**, no. 6, 1996, pp. 1211-1213; Gariaev, P., Tertishny, G. and Leonova, K. (2001), "The Wave: Probabilistic and Linguistic Representation of Cancer and HIV", *J. of Non-Locality and Remote Mental Interactions (JNLRMI)*, **I**, no. 2, May 2002, at http://www.emergentmind.org; Maslov, M.U. and Gariaev, P.P. (1994), "Fractal Presentation of Natural Language Texts and Genetic Code", *2nd Intl. Conference on Quantitative Linguistics*, 20-24 September 1994; Gariaev P.P., Vasiliev A.A., Berezin A.A., 1994, "Holographic associative memory and information transmission by solitary waves in biological systems", SPIE - The International Society for Optical Engineering. CIS Selected Papers. Coherent Measuring and Data Processing Methods and Devices. v.1978, pp.249-259. 6. S.A. Reshetnyak, V.A. Shcheglov, V.I. Blagodatskikh, P.P. Gariaev, and M.Yu.Maslov, 1996, "Mechanism of interaction of electromagnetic radiation with a biosystem", *Laser Physics* **6**, 2,1996, pp.621-653.

[38] P. Marcer and W. Schempp, "A Mathematically Specified Template for DNA and the Genetic Code, in Terms of the Physically Realizable Processes of Quantum Holography", in *Proceedings of the Greenwich Symposium on Living Computers* (A. Fedorec and P. Marcer eds.), (1996), pp. 45-62; ibid. "The Model of the Prokaryote Cell as an Anticipatory System Working by Quantum Holography", in *Proceedings of CASYS '97, 1115 August 1997, HEC, Liège, Belgium, Inter. J. of Comp. Ant. Sys.*, **2**, (1997) pp. 307-315. W. Schempp, *Magnetic Resonance Imaging: Mathematical Foundations and Applications*, (John Wiley, NY, 1997) .

[39] I. Miller and R. A. Miller, "From Helix to Hologram, an Ode on the Human Genome", *Nexus*, August-September, 2003; www.nexusmagazine.com .

[40] O. Rossler, *Endophysics: The World As an Interface*, (World Sc., Singapore, 1998).

[41] S. Vrobel and O. Rossler, *Simultaneity: Temporal Structures and Observer Perspectives* , (World Sc., Singapore, 2008).

[42] R. Bucccheri, V. di Gesù and M. Saniga,*Studies on the Structure of Time: From Physics to Psycho(patho)logy* , (Springer, Berlin, 2000); R. Buccheri, M. Saniga, and W.M. Stuckey, *The Nature of Time: Geometry, Physics and Perception*, (Springer, Berlin , 2003).

[43] D.L. Rapoport, "Self-reference, the Moebius and Klein bottle surfaces, multivalued logics and cognition", in *Proceedings of CASYS09, Conference on Anticipative Systems, Liège, August 03-09, 2009*, D. Dubois et al (eds.).

[44] S. E. Shnoll, K. I. Zenchenko, I. I. Berulis, N. V. Udaltsova and I. A. Rubinstein; arxiv: physics/0412007 (December 2004); V. A. Panchelyuga, V. A. Kolombet, M. S. Panchelyuga and S. E. Shnoll; physics/0612055 (December 2006);A. V. Kaminsky and S. Shnoll, physics/0605056 (May 2006); S. Snoll, physics/0602017; ibid.

Phys. Letts. A**359**, 4, 249-251 (2004); ibid. Uspekhi **43**, no. 2, 205-209 (2000); ibid. *Progress in Physics* , January 2007.

[45] G. Bateson, *Steps to an Ecology of Mind*, Paladin Books (1973); ibid. *Mind and Nature:A Necessary Unity*, (Bantam, New York ,1988).

[46] S. Johansen, *Outline of a Differential Epistemology*, English translation of the original Norwegian published by University of Trondheim (1991) to appear; ibid. "Initiation of 'Hadronic Philosophy', the philosophy underlying Hadronic Mechanics and Chemistry", *Hadronic Journal* **29**, 2, 2006, 111-135.

[47] D. Bohm, *Thought as a System*, (Routledge, London, 1993).

[48] N. V. Swindale,"Visual Cortex: Looking into a Klein bottle", *Current Biology* **6**, No 7, 1996, 776779.

[49] S. Tanaka, "Topological Analysis of Point Singularities in Stimulus Preference Maps of the Primary Visual Cortex", *Proc. Royal Soc. London (Biol)* **261**, 1995, 8188.

[50] S. Marcelja S, "Mathematical Description of the Responses of Simple Cortical Cells", *J. Opt. Soc. Am.* **70**, 1980, pp. 12971300; M. Liebling, T. Blu and M. Unser, "Fresnelets: New Multiresolution Wavelet Bases for Digital Holography", *Image Processing, IEEE Transactions*, **12**, Issue 1, Jan 2003, ps. 29 - 43.

[51] J.F. Corum, "Relativistic Rotation and the Anholonomic Object", *J.Math. Phys* **18**, no.4 , 77 (1976); ibid. "Relativistic Covariance and Electrodynamics", *J. Math.Phys.* **21**, no. 9, 83 (1980).

[52] K. Kondo, *Memoirs of the Unifying Study of the Basic Problems in Engineering Sciences by Means of Geometry*, Tokyo, Gakujutsu Bunken Fukyu-Kai (1955).

[53] K. Kondo and Y. Ishizuka, "Recapitulation of the Geometrical Aspects of Gabriel Kron's Non- Riemannian Electrodynamics", *Memoirs of the Unifying Study of the Basic Problems in Engineering Sciences by Means of Geometry (RAAG Memoirs)*, vol.1, Division B (1955); G.Kron, "Non-Riemannian Dynamics of Stationary Electric Networks", *RAAG Memoirs* vol. 4, 156 (1958); ibid. "Electric Ciruit Models of the Schrdinger Equation",*Phys. Rev.* **67**, 39-43 (1945).

[54] S. Wand and D.D.L. Chung, *Composites Part B* (1999), 579-590.

[55] Xing-liu Jiang, Jin-zhi Lei and Li-juan Han, "Torsion Field Effects and Zero-Point Energy in Electrical Discharge Systems", , *Journal of Theoretics***4-5**, Dec. 2003/Jan 2004; T. Matsuda, K. Harada, H. Kasai, O. Kamimura and A. Tomomura, "Observation of Dynamic Interaction of Vortices with Pinning by Lorentz Microscopy", *Science* **271**, 1393 (1996)

[56] A. E. Akimov and G. Shipov, "Torsion Fields and their Experimental Manifestations", *Journal of New Energy* **2**, no. 2, 67 (1999).

[57] C. Eberlein, "Theory of Quantum Radiation Observed as Sonoluminisence",*Phys. Rev. Letts.* **53**, 2772 (1996).

[58] Y. Paltiel, E. Zeldov et al, "Dynamic Instabilities and Memory Effects in Vortex Matter", *Nature* **403**, 398 (2000).

[59] V. F. Panov, V. I. Kichigin, G. V. Khaldeev, A. V. Klyuev, B. V. Testov, T. A. Yushkova, V. V. Yushkov, "Torsion Fields and Experiments", *Journal of New Energy* **2**, nos. 2,3 (1997).

[60] T. Mizuno, T. Akimoto, T. Ohmori," Neutron and Heat Generation Induced by Electric Discharge", *Journal of New Energy* **3**, Number 1, Spring 1998.

[61] J. Schouten, *Ricci Calculus*, (Springer Verlag, Berlin,1951).

[62] D. Rapoport and M. Tilli, "Scale Fields as a Simplicity Principle",in Proceedings of the Third Workshop on Hadronic Mechanics, vol. II (Patras, 1986), *Hadronic J. Suppl.* **6**,no.4, 682-778 (1986). I. M. Gel' fand and Shilov, *Generalized Functions IV*, (Academic Press, NY, 1967).

[63] R.M. Santilli, *Elements of Hadronic Mechanics III*, (Hadronic Press-Naukova, Palm Harbor-Kiev, 1993); ibid.*Isodual Theory of Antimatter, with Applications to Antigravity, Spacetime Machine and Grand Unification*, (Springer, New York ,2006); ibid. "The Physics of New Clean Energies and Fuels According to Hadronic Mechanics", *Journal of New Energy (Special Issue: Twenty Years of Research Extending Quantum Mechanics for New Industrial and Energy Products* **4**, no. 1 (1999); ibid. *Hadronic Mathematics, Mechanics and Chemistry, vol. I: Iso-, Geno-, Hyper-Formulations for Matter, and Their Isoduals for Antimatter*, International Academic Press, in press.

[64] R.M. Santilli, *Foundations of Hadronic Chemistry*, (Kluwer Series in the Fundamental Theories of Physics, Dordrecht-Boston, 2001); A.O. E. Animalu, " A New Theory on the Structure of the Rutherford -Santilli Neutron", *Hadronic Journal***17**, 349 (2004).

[65] M. R. Edwards, *Pushing Gravity: New perspectives on Le Sages theory of gravitation*, Apeiron,Quebec (2001).

[66] H. Arp, *Seeing Red: Redshifts, Cosmology and Academic Science*,(Apeiron, Quebec, 1998).

[67] L. Kauffman, "Formal Systems, Eigenforms", *Kybernetes* **34**, No. 1/2, pp. 129-150 (2004).

[68] P. Griffiths and J. Harris, *Principles of Algebraic Geometry* (John Wiley & Sons, NY, 1978).

[69] N. A. Kozyrev, "On the possibility of experimental investigation of the properties of time", in *Time in Science and Philosophy*, pp. 111132, Prague, 1971, www.chronos.msu.ru ; A.P. Levich, "A Substantial Interpretation of N.A. Kozyrev's Conception of Time", in *On the Way to Understanding the Time Phenomenon: the Constructions of Time in Natural Science. Part 1. Interdisciplinary Time Studies* (World Scientific, 1995); in www.chronos.msu.ru; I.A. Eganova, "The World of Events Reality: Instantaneous Action as a Connection of Events Through Time", in *Relativity, Gravitation, Cosmology*, V.Dvoeglazov (ed), pag. 149-163, (Nova Sc. Publ., New York, 2004). A. Dadaev," Astrophysics and Causal Mechanics", Galilean Electrodynamics, **11**, Special Issues 1, p. 4, 2000. B. Stanislavovich, "On Torsion Fields", Galilean Electrodynamics, **12**, Special Issue 1, p. 5-9, (2000).

[70] S.M. Korotaev, V.O. Serdyuk and M.O.Sorokin, " Experimental Verification of Kozyrev's Interaction of Natural Processes",*Galilean Electrodynamics* **11** Special Issues 2, p. 23, 2000. "Experimental estimation of macroscopic nonlocality effect in solar and geomagnetic activity", *Phys. Wave Phenomena*, **11** (1), 46-54 (2003); S.Korotaev, in Casys, Int.J. Comp.Anticip. Systems **22**, 2008.

[71] V.P. Kaznatcheyev, " Consciousness and Physics" ,*Physics of Consciousness and Life, Cosmology and Astrophysics*, **2**, 2002; V.P. Kaznatcheyev and A.V. Trofimov, " The experimental works on the problem intellect as the cosmoplanet phenomenon", ibid. **4**(5), 2005.

[72] http://www.geom.uiuc.edu/zoo/toptype/klein/standard/; http://plus.maths.org/issue26/features/mathart/Build.html.

[73] http://en.wikipedia.org/wiki/Antiautomorphism.

[74] F. Halberg, http://www.msi.umn.edu/ halberg/bib.html. E. Páles, *Seven Archangels: Rythms of Inspiration in the History of Culture and Nature*, (Sophia Foundation, Bratislava, 2009).

[75] V. Trifonov, "Natural geometry of nonzero quaternions",*Int. J.Theor. Phys.* **46** (2) (2007) 251-257; ibid. *Europhys. Lett.* **32**(8) (1995) 621-626.

[76] V. Christianto and F. Smarandache, "Kaluza-Klein-Carmeli metric from quaternion-Clifford space, Lorenz force and some observables", *Progress in Physics* **2**, April 2002, 144-150.

[77] A. Young, *The Reflexive Universe*, (Robert Briggs Associates, 1976).

[78] D. Rapoport and S. Sternberg "On the interactions of spin with torsion", *Annals of Physics* **158**, no.11, (1984), p. 447-475; ibid. "Classical Mechanics without lagrangians nor hamiltoneans", *Nuovo Cimento* **80A** (1984), 371-383.

[79] M.C. Purcell, http://www.towardsanewera.net/ .

[80] D. Dubois, in *Proceedings of CASYS (Conferences on Anticipatory Sytems), AIP, Liège, 1997, 1999, 2000,2002,2005,2007,2009.*

[81] L. de Broglie, *Non-linear Wave Mechanics: A Causal Interpretation.* (Elsevier, Amsterdam, 1960); J.P. Vigier, *Physics Letters A,* **135**, Issue 2, 13 February 1989, 99-105.

[82] E. Conte, A. Khrennikov, O. Tadarello, A. Federici and J. Zbilut, J. Neuroquantology **7**, n.2, 204-212 (2009) and references therein.

[83] http://et3m.net/.

[84] V. Grebennikov, *Sibirskii Vestnik Selskokhoziastvennoi Nauki* no. 3, 1984; ibid. *The Mysteries of the World of Insects,* Novosibirsk, 1990; V. Krasnoholovetz, private communication.

[85] I. M. Shakhparonov, " Kozyrev-Dirac Emanation: Interaction with Matter and Methods of Detecting". http://www.rexresearch.com/kozyrev2/2-1.pdf

[86] R. Amoroso and E. Rauscher, *The Holographic Anthropic Multiverse: Formalizing the Complex Geometry of Reality* (Series on Knots and Everything), (World Sc., Singapore, 2009).

Contents of Volume one

459

Eddington Ether and Number
Raul A. Simon, LAMB, Santiago Chile

-257-
The dynamical Space-time as a Field Configuration in a Background Space-time
A.N. Petrov, Department of Physics and Astronomy, University of Missouri,-Columbia, Columbia MO 65211, USA and Sternberg Astronomical Institute, Universitetskii pr., 13 Moskow 119992 Russia E-mail: anpetrov@rol.ru

-305-
Locality and Electromagnetic Momentum in Critical Tests of Special Relativity
Gianfranco Spavieri, Jesus Quintero, Arturo Sanchez, José Ayazo, Centro de Fisica Fundamental, Universidad de los Andes, Merida 5101-Venezuela E-mail : Spavieri@ula.ve
And Georges T. Gillies Department of Mechanical and Aerospace Engineering, University of Virginia, P.O. Box 400746, Charlottesville, Virginia 22904, USA E-mail: gtg@Virginia.edu

-357-
Correlations Leading to Space-time Structure in an Ether
J.E. Carroll,
Engineering Department, University of Cambridge, CB2 1PZ, United Kingdom, E.mail: jec1000@cam.ac.uk

-407-
Reasons for Gravitational Mass and the Problem of Quantum Gravity
V. Krasnoholovets
Institute for basic research, 90 East Winds Court, Palm Harbor, Fl 34683, USA

Contents of Volume 2

461

Made in the USA
Monee, IL
10 June 2026

53029990R00260